PROGRAMMING THE FINITE ELEMENT METHOD

2ND EDITION

PROGRAMMING THE FINITE ELEMENT METHOD

SECOND EDITION

I. M. Smith
and
D. V. Griffiths
University of Manchester, UK

JOHN WILEY & SONS

Chichester . New York . Brisbane . Toronto . Singapore

Library of Congress Cataloging-in-Publication Data:

Smith, I. M.
Programming the finite element method.
Includes bibliographies and indexes.
1. Finite element method—Computer programs.
2. Engineering—Data processing. 3. Soil mechanics—Data processing.
I. Griffiths, D. V. II. Title.
TA347.F5S64 1988 620′.001′515353 87-8123

ISBN 0 471 91552 1 (cloth)
ISBN 0 471 91553 X (paper)

British Library Cataloguing in Publication Data:

Smith, I. M. (Ian Moffat)
Programming the finite element method.
—2nd ed.
1. Finite element method—Data processing
I. Title II. Griffiths, D. V.
515.3′53 TA347.F5

ISBN 0 471 91552 1 (cloth)
ISBN 0 471 91553 X (paper)

Typeset at Thomson Press (India) Limited, New Delhi.
Printed and bound in Great Britain by Anchor Brendon Ltd, Tiptree, Essex

Contents

Preface to Second Edition

Following the success of the First Edition, considerable modifications and improvements have been made. In retrospect it was probably a mistake to combine in one book a programming philosophy and a survey of other workers' applications to geomechanics, so the latter has been dropped. However, the geomechanics flavour remains, particularly in Chapters 6, 7, 8, 9 and 12.

Although the programming philosophy remains intact, Chapter 1 reflects the rapid developments in the hardware and (to a lesser extent) software fields. Emphasis is given to what are generally known today as 'micro' computers while the programming language has been updated from FORTRAN IV (FORTRAN 66) to FORTRAN V (FORTRAN 77).

Chapter 2 has been extended to give a fuller treatment of structural applications and the equations of fluid flow including the Navier–Stokes equations.

In the original Chapter 3, choice of storage strategy was rather limited, so more versatility has been introduced, in the form of a 'profile' ('skyline') option.

Chapter 4 has been extended to embrace elastic, elasto-plastic and elastic stability analyses of all commonly encountered skeletal structures, while in Chapter 5 a much greater range of plane and solid finite elements is provided.

In Chapter 6 more realistic examples of analyses of geotechnical problems—involving walls, foundations and slopes—have been incorporated.

The original Chapter 7 dealt both with steady state and with transient flow problems. In the present edition, transient problems are removed to Chapter 8 while the steady state confined and unconfined flow problems analysed in Chapter 7 are more typical of those encountered in geotechnical engineering practice.

Chapter 8 also includes relatively recent developments of element-by-element algorithms for first order transient problems.

The original Chapter 8, now Chapter 9, has been extended to deal with two typical 'coupled' problems—those arising from the (steady state) Navier–Stokes equations and those arising from Biot's equations for transient analysis of porous elastic solids.

The original Chapter 9 has been subdivided, so that eigenvalue analysis (new Chapter 10) is divorced from the solution of second order transient problems (new Chapter 11). In the new Chapter 10 a solver based on the Lanczos method for large eigenvalue problems is demonstrated. In the new Chapter 11 second order differential equations are integrated by a greater range of methods, for example mixed integration operators.

The final chapter (new Chapter 12) has been expanded to embrace the analysis of pile groups as well as of single piles.

COMPUTER PROGRAMS

All software (libraries and programs) described in this book are available from:

Numerical Algorithms Group Ltd,
NAG Central Office,
Mayfield House,
256 Banbury Road,
Oxford OX2 7DE,
UK

Specify the medium (tape, disc) and format when applying. A small handling charge is made.

CHAPTER 1
Preliminaries: Computer Strategies

1.0 INTRODUCTION

Many textbooks exist which describe the principles of the finite element method of analysis and the wide scope of its applications to the solution of practical engineering problems. Usually, little attention is devoted to the construction of the computer programs by which the numerical results are actually produced. It is presumed that readers have access to pre-written programs (perhaps to rather complicated 'packages') or can write their own. However, the gulf between understanding in principle what to do, and actually doing it, can still be large for those without years of experience in this field.

The present book attempts to bridge this gulf. Its intention is to help readers assemble their own computer programs to solve particular engineering problems by using a 'building block' strategy specifically designed for computations via the finite element technique. At the heart of what will be described is not a 'program' or a set of programs but rather a collection (library) of procedures or subroutines which perform certain functions analogous to the standard functions (SIN, SQRT, ABS, etc.) provided in permanent library form in all useful scientific computer languages. Because of the matrix structure of finite element formulations, most of the building block routines are concerned with manipulation of matrices.

The aim of the present book is to teach the reader to write intelligible programs and to use them. Super-efficiency has not been paramount in program construction. However, all building block routines (numbering over 100) and all test programs (numbering over 50) have been verified on a wide range of computers. Their efficiency is also believed to be reasonable, given a tolerably well written compiler. In order to justify the choice of high level language used (FORTRAN 77) and the way in which it is asserted programs should be written, it is necessary briefly to review the development of computer 'architecture' and its relationship to program writing.

1.1 COMPUTER HARDWARE

In principle, any computing machine capable of compiling and running FORTRAN programs can execute the finite element analyses described in this

Table 1.1

Designation	Example	Wordlength (bits)	Core storage (words)	Virtual memory?	Approximate cost 1985 (£UK)
Micro	IBM PC	16	128K	No	2K
Super micro	ICL PERQ	32	64K	Yes	20K
Mini	VAX 11/780	32	64K	Yes	100K
Mainframe	CDC 7600	64	100K	No	1000K
Super mainframe	CDC 10GF	64	46,000K	Yes	10,000K

book. In practice, hardware will range from 'micro' computers for small analyses and teaching purposes to 'super' computers for very large (especially non-linear) analyses. It is a powerful feature of the programming strategy proposed that the *same* software will run on all machine ranges. The special features of vector and array processors are described in a separate section (1.3).

Table 1.1 shows that, in very general terms, five categories of useful hardware can be distinguished at the time of writing, graded using an 'order of magnitude' rule with respect to cost. While the trend in costs, certainly for 'minis' and below, is downwards some such cost-based tabulation will always apply.

At the lower end of the cost range, it is of course possible to code finite element analysis algorithms for very cheap machines, by employing machine code or an interpreter such as BASIC. The complete lack of portability of the former and the lack of a true subroutine structure in the latter leads the authors to exclude such hardware from consideration.

The user's choice of hardware is a matter of accessibility and of cost. For example, in very rough terms, a 'mini' computer (Table 1.1) has a processing speed about 30 times slower than that of a 'mainframe'. Thus a job taking two minutes on the mainframe takes one hour on the mini. Which hardware is 'better' clearly depends on individual circumstances. The main advice that can be tendered is against using hardware that is too weak for the task; that is the user is advised not to operate at the extremes of the hardware's capability.

1.2 STORE MANAGEMENT

In the programs in this book it will be assumed that sufficient main random access core store is available for the storage of data and the execution of programs. However, the arrays processed in very large finite element calculations can be of size, say, 10,000 by 200. Thus a computer would have to have a main store of 2×10^6 words to hold this information, and while some such computers exist, they are still comparatively rare. A much more typical core size is still of the order of 1×10^5 words.

Two distinct strategies are used by computer manufacturers to get round the problem of limited random access core. In the first (simplest and earliest) technique different 'levels' of store are provided with transfer of data to core from various secondary storage devices such as discs and tape being organized by the programmer. The CDC 7600 is a computer that used this strategy. Programs are executed in a small fast store while data can be stored in a medium-fast 'large core memory'. Further back-up storage is available on a big disc but of course when the programmer decides to transfer information to and from disc the central processor would be idle. A job calling for disc transfers is therefore automatically swapped out of fast memory, and is replaced by the next job waiting in the queue.

The second strategy removes store management from the user's control and vests it in the system hardware and software. The programmer sees only a single level of store of very large capacity and his information is moved from backing store to core and out again by the supervisor or executive program which schedules the flow of work through the machine. This concept, namely of a very large 'virtual' core, was first introduced on the ICL ATLAS in 1961. It can be seen from Table 1.1 that this strategy of store management is now very common. The ATLAS and most subsequent machines of this type also permit multi-programming in that there can be several partly completed programs in the core at any one time. The aim in any such system is to keep the hardware (CPU, peripherals, etc.) fully utilised.

Clearly it is necessary for the system to be able to translate the virtual address of variables into a real address in core. This translation usually involves a complicated bit-pattern matching called 'paging'. The virtual store is split into segments or pages of fixed or variable size referenced by page tables, and the supervisor program tries to 'learn' from the way in which the user accesses data in order to manage the store in a predictive way. However, store management can *never* be totally removed from the user's control. It must always be assumed that the programmer is acting in a reasonably logical manner, accessing array elements in sequence (by rows or columns as organised by the compiler). If the user accesses a virtual store of 10^6 words in a random fashion the paging requests will readily ensure that very little execution of the program can take place.

1.3 VECTOR AND ARRAY PROCESSORS

In later chapters it will be shown that finite element computations consist largely of handling matrices and vectors—multiplying them together, adding, transposing, inverting and so on. It has seemed appropriate to computer manufacturers, therefore, to provide special hardware capable of doing vector or matrix operations at high speed. This hardware can be in the form of central processing vector registers such as on the Cray and CDC Cyber 205 or in peripheral form such as the ICL 2900 DAP concept. Other array processors can also be connected as peripherals, one such being the Floating Point Systems FPS264, and this is a very likely future development.

The drawback associated with specialised hardware of this kind is that specialised programs usually have to be written to exploit it. This detracts from such important software considerations as program portability and machine independence. However, at the very least, software can be written to pre-process standard high level language code to see if it can be re-organised to make best use of a particular machine's vector capabilities.

Future trends in this direction are difficult to anticipate. It may be that, ultimately, array processing will become so fast and cheap, even for very large arrays, that much of the special purpose software which has been written to date, for example to process arrays with various sparsity patterns, will become redundant. However, such a trend would be entirely consistent with the philosophy of the present book, and would merely necessitate a change of emphasis. At present, though, only standard, completely portable high level languages will be considered in the following sections on software.

1.4 SOFTWARE

Since all computers have different hardware (instruction formats, vector capability, etc.) and different store management strategies, programs that would make the most effective use of these varying facilities would of course differ in structure from machine to machine. However, for excellent reasons of program portability and programmer training, engineering computations on all machines are usually programmed in 'high level' languages which are intended to be machine-independent. The high level language (FORTRAN, ALGOL, etc.) is translated into the machine order code by a program called a 'compiler'.

It is convenient to group the major high level languages into those which are FORTRAN-like (for example PASCAL) and those which are ALGOL-like (for example ADA). The major difference between these groups is that the latter allow dynamic management of storage at run-time whereas the former do not. This difference has immediate implications for store management strategy because it is clear that unless a computer permits genuine multi-programming and a one-level store in which programs are dynamically expanding and contracting there is not much point in having a language that allows these facilities. Thus, on CDC 7600 machines, FORTRAN was an obvious choice of language whereas on other computers there is greater room for manoeuvre. Dynamic store management seems to be a uniform trend for the future, so one would expect ALGOL-like languages to become more and more attractive.

There are many texts describing and comparing computer languages, often from the point of view of computer scientists, e.g. Barron (1977). What matters in the context of the present book is not syntax, which is really remarkably constant among the various languages, but program structure. The following example illustrates the two crucial differences between ALGOL-like and FORTRAN-like languages as far as the engineer-programmer is concerned.

Suppose we want to write a program to multiply two matrices together, a task that occurs hundreds of times during execution of a typical finite element program. In ALGOL 60 the program might read as follows:

```
begin   integer l, m, n;
        statements which read or compute l, m, n;
        begin   array a[1:l,1:m], b[1:m,1:n], c[1:l,1:n];
        integer i, j, k; real x;
        statements which initialise a, b;
        for i:=1 step 1 until l do
        for j:=1 step 1 until n do
          begin x:=.0;
                for k:=1 step 1 until m do
                x:=x+a[i,k]*b[k,j];
          end;
                c[i,j]:=x
        statements which do something with c
        end
end
```

while in FORTRAN 77 it might be written:

```
      INTEGER I, J, K, L, M, N
      REAL X
      REAL A(20,20), B(20,20), C(20,20)
      Statements which read or compute L, M, N
      Statements which initialise A, B
      DO 1 I=1, L
      DO 1 J=1, N
      X=.0
      DO 2 K=1, M
    2 X=X+A(I,K)*B(K,J)
      C(I,J)=X
    1 CONTINUE
      Statements which do something with C
      END
```

Note that the declarations of the simple integers and the real in lines 1 and 2 of this program are unnecessary.

Comparison of the two codes will easily show that the syntaxes of the two languages are very similar, e.g. compare the ALGOL

```
for i:=1 step 1 until l do
```

with the corresponding FORTRAN

```
DO 1 I=1, L
```

or the assignment statements

```
x:=x+a[i,k]*b[k,j];
```

and

$$X = X + A(I, K) * B(K, J).$$

Therefore, anyone familiar with one syntax can learn the other in a very short time.

However, consider the program structures. The main difference is that in ALGOL, the array dimensions *l*, *m*, *n* do not need to be known at compile time whereas in FORTRAN they do. Indeed, our FORTRAN program is useless if we want to multiply together arrays with dimension greater than 20. It is this lack of 'dynamic' storage which is FORTRAN's major limitation. It can partially be alleviated by various devices such as allocating all storage to a large vector and keeping track of the starting addresses of the various arrays held in the vector, but one can never get away without a dimensioning statement of some kind *within* a FORTRAN program.

This has various bad effects, among which is a tendency for FORTRAN users to declare their arrays to have the maximum size that will ever be encountered, subject to the limitation of the core size of the machine. Since most large machines allow multi-programming this can be inefficient.

1.5 PORTABILITY OF 'SUBROUTINES' OR 'PROCEDURES'

The second major deficiency of FORTRAN becomes apparent when we wish to construct a program to multiply together matrices of *varying size* within the *same* program. This sort of activity occurs very frequently in finite element programs. To avoid the repetition of sections of code within a program, most languages allow these sections to be hived off into self-contained sub-programs which are then 'called' by the main program. These are termed 'procedures' in ALGOL and 'subroutines' in FORTRAN and their use is central to efficient and readable programming. For example, it would be very tedious to include the code required to compute the sine of an angle every time it is required, so ALGOL and FORTRAN provide a special, completely portable routine (actually a function) called SIN. This is held in a permanent library and is accessible to any user program. Why not do the same thing with matrix multiplication?

In ALGOL the technique is straightforward and encouraged by the structure of the language. For example, an ALGOL 60 matrix multiplication procedure might be as follows:

```
procedure matmult (a, b, c, l, m, n);
value l, m, n; array a, b, c; integer l, m, n;

begin integer i, j, k; real x;
      for i:= 1 step 1 until l do
      for j:= 1 step 1 until n do
      begin x:= .0;
            for k:= 1 step 1 until m do
            x:= x + a[i,k] * b[k,j];
```

```
        end
                c[i, j] := x
end
```

Suppose we now wish to multiply the $(l \times m)$ array a by the $(m \times n)$ array b to give the $(l \times n)$ array c as before, and then to multiply c by the $(n \times l)$ array d to yield the $(l \times l)$ array e. The ALGOL 60 program would be as follows:

```
begin integer l, m, n;
        procedure matmult (a, b, c, l, m, n);
        statements which read or compute l, m, n;
        begin array a[1:l, 1:m], b[1:m, 1:n],
                c[1:l, 1:n], d[1:n, 1:l], e[1:l, 1:l];
                statements which initialise a, b;
                matmult (a, b, c, l, m, n);
                matmult (c, d, e, l, n, l);
                statements which do something with e
        end
end
```

In the above, the procedure *matmult* is assumed to be available in a user library in exactly the same way as LOG, COS, etc., are available in the permanent library. The program structure can be seen to be simple and logical, reflecting exactly what the user wants to do and requiring a minimum of information about array dimensions to do it.

Let us now try to do the same thing in FORTRAN 77. Reasoning from our previous program we write:

```
  SUBROUTINE MATMUL (A, B, C, L, M, N)
  REAL A(20, 20), B(20, 20), C(20, 20), X
  INTEGER I, J, K, L, M, N,
  DO 1 I = 1, L
  DO 1 J = 1, N
  X = .0
  DO 2 K = 1, M
2 X = X + A(I, K)*B(K, J)
  C(I, J) = X
1 CONTINUE
  RETURN
  END
```

Now the dimensioning statement in line 2 of this subroutine is merely a device for passing the addresses in memory of the arrays locally called A, B, C back and forth between the subroutine and the main program. Since the FORTRAN subroutine only needs to know a starting address and a column size, we could as well have written

```
REAL A(20, *), B(20, *), C(20, *), X
```

More importantly, since in FORTRAN the local A, B, C are mapped onto the arrays of the main program, it can be seen that our subroutine is useless, unless arrays processed by it have a first dimension of 20. This is an impossible restriction akin to saying we can only work out the sines of angles less than 20° using SIN.

A way round this deficiency in FORTRAN is to make the subroutine completely portable at the expense of including extra parameters in the call, associated with each array. These parameters, which are constants, contain on entry to the subroutine the column sizes of the particular arrays to be multiplied at that time as declared in the main program. Thus our portable FORTRAN subroutine becomes:

```
      SUBROUTINE MATMUL (A, IA, B, IB, C, IC, L, M, N)
      REAL A(IA, *), B(IB, *), C(IC, *), X
      INTEGER IA, IB, IC, L, M, N, I, J, K
      DO 1 I = 1, L
      DO 1 J = 1, N
      X = .0
      DO 2 K = 1, M
    2 X = X + A(I, K)*B(K, J)
      C(I, J) = X
    1 CONTINUE
      RETURN
      END
```

We could now construct our general matrix multiplication program as follows, assuming MATMUL to be available in a user library:

```
INTEGER L, M, N, IA, IB, IC, ID, IE
REAL A(20, 5), B(5, 15), C(20, 15), D(15, 20), E(20, 20)
DATA IA/20/, IB/5/, IC/20/, ID/15/, IE/20/
Statements which read or compute L, M, N
Statements which initialise A, B
CALL MATMUL (A, IA, B, IB, C, IC, L, M, N)
CALL MATMUL (C, IC, D, ID, E, IE, L, N, L)
Statements which do something with E
END
```

We have now succeeded in making FORTRAN look reasonably like ALGOL. The differences lie only in the need to dimension arrays in the main program and to pass the extra column size parameters in the subroutine calls. In fact we will only be concerned in our finite element programs with adjusting the sizes of a few large arrays at run-time, and so the inconvenience can really be restricted to changing a line at the beginning of the main program.

This is most easily done by the PARAMETER statement in FORTRAN 77. Using this facility the second and third lines of the preceding program can be replaced by:

```
      PARAMETER (IA = 20, JA = 5, IB = 5, JB = 15, IC = 20, JC = 15,
     * ID = 15, JD = 20, IE = 20, JE = 20)
      REAL A(IA,JA), B(IB,JB), C(IC,JC), D(ID,JD), E(IE,JE)
```

so that only the PARAMETER line need ever be changed. This is the method used in this book.

1.6 LIBRARIES

The concept that will be described in subsequent chapters is one of having a user library containing completely portable subroutines such as MATMUL, which was described in the previous section. The idea is not new, of course, and many such libraries already exist, notably the NAG mathematical subroutine library (Ford and Sayers, 1976). The present library is completely compatible with the NAG mathematical subroutine library and uses the same parameter passing convention, but it differs from the NAG mathematical subroutine library in two ways. First, it contains only routines used in finite element analysis, and second it uses mnemonic names such as MATMUL to describe its operations (an equivalent routine in the NAG mathematical subroutine library is called FO1CKF). This greatly enhances program readability.

The library routines which are the 'building blocks' from which the finite element programs are constructed are described in Chapter 3. Some of them may be considered as 'black box' routines, that is the user need not understand or see them (how many users know how SIN(X) is computed?). Some of these routines are not even listed in the book but the source code is available on tape or floppy disc in all the usual formats. Other routines are, however, specifically related to the descriptions of finite element operations, and these routines are listed for reference purposes in Appendix 5, and of course on tape also.

The test programs which are built up from the library routines are described in Chapters 4 to 12 and also listed there. Interested readers can substitute their own routines at any point, because of the modular nature of the whole system. Such routines could be in machine code for greater efficiency or could be linked to the use of an array processor. For example, the Floating Point Systems Array Processor can look to the user like a set of library subroutine calls which could be substituted for the FORTRAN subroutines if required.

The programs described in this book are all assumed to be processed 'in core'. Naturally, one could construct out of core versions of the library routines, which process large arrays for use on non-virtual machines.

1.7 STRUCTURED PROGRAMMING

The finite element programs which will be described are strongly 'structured' in the sense of Dijkstra (1976). The main feature exhibited by our programs will be seen to be a *nested* structure and we will use representations called 'structure charts' (Lindsey, 1977) rather than flow charts to describe their actions.

The main features of these charts are:

(i) The Block

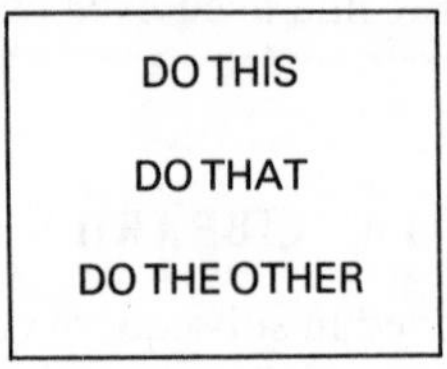

This will be used for the outermost level of each structure chart. Within a block, the indicated actions are to be performed sequentially.

(ii) The Choice

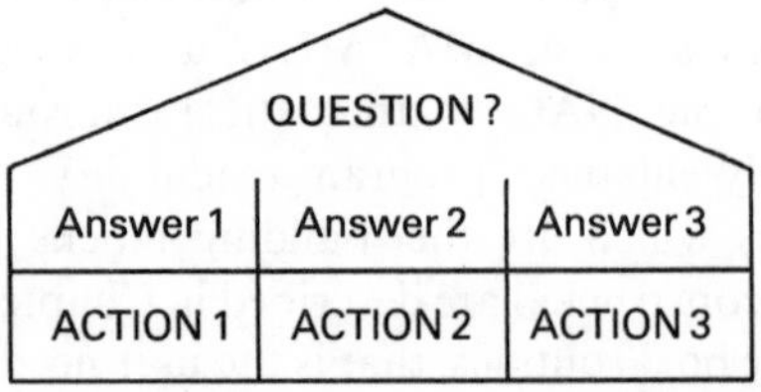

This corresponds to the **if** ... **then** ... **else if** ... **then** ... **end if** kind of construct.

(iii) The Loop

This comes in various forms, but we shall usually be concerned with the 'DO' loop

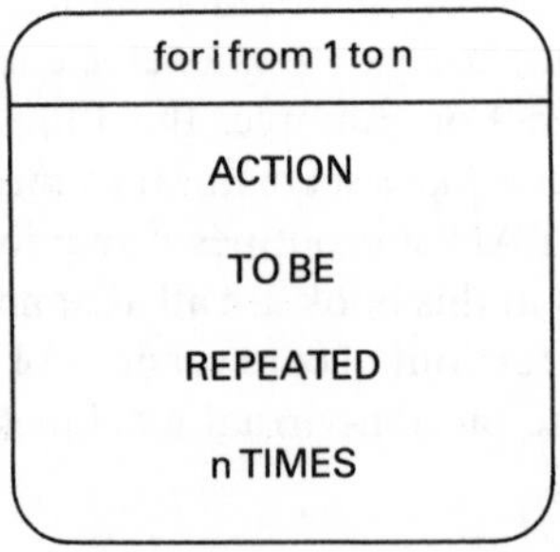

In particular, the structure chart notation encourages the use only of those GOTO statements that are safe in a structured program.

Using the notation, our matrix multiplication program would be represented as follows:

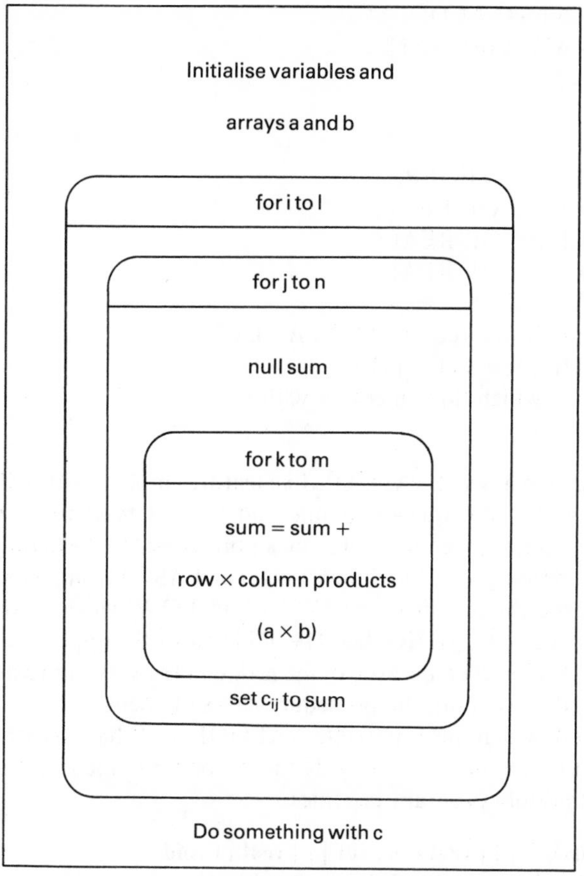

The nested nature of a typical program can be seen quite clearly.

1.8 OTHER LANGUAGES

New programming languages appear from time to time, usually at the instigation of computer scientists rather than engineers. For example, PASCAL is popular as a teaching language and has a simple structure enabling compact compilers to be written. In PASCAL 3 our matrix multiply algorithm becomes:

```
TYPE ARR = ARRAY [INTEGER, INTEGER] OF REAL;
PROCEDURE MATMUL (VAR A, B, C: DYNAMIC ARR; L, M, N:
INTEGER)
VAR I, J, K: INTEGER; X:REAL;
BEGIN
  FOR I:= 1 TO L DO
  FOR J:= 1 TO N DO
  BEGIN X:= 0.0;
```

```
        FOR K:= 1 TO M DO
        X:= X + A[I, K] * B[K, J];
        C[I, J]:= X;
        END;
    END;
    VAR L, M, N: INTEGER;
        A:[1..10, 1..10] OF REAL;
        B:[1..10, 1..10] OF REAL;
        C:[1..10, 1..10] OF REAL;
    BEGIN
        Statements which read L, M, N, A, B, C
        MATMUL (A, B, C, L, M, N);
        Statements which do something with C
    END.
```

Thus the language has some ALGOL-like features in that, within the procedure, arrays A, B and C are quasi-dynamic and do not need to be dimensioned. However, there is no real concept of a stack pointer and in the main program A, B and C need to be assigned fixed dimensions. The PASCAL language does possess the built-in procedures (functions) HIGH and LOW which enable bounds of arrays to be inspected; e.g. HIGH(A, 1) would deliver the upper bound of the first dimension of A. However, because of the lack of a truly dynamic structure, these cannot be used to simplify the procedure call statements.

A language in which this is possible is ALGOL 68. It has an extremely flexible structure and points the way towards the probable shape of future languages. Our matrix multiply program becomes:

```
(proc matmul = ([ , ] real a, b, ref [ , ] real c) void;
    (int l = 1 upb a, m = 1 upb b, n = 2 upb; b
        loc int i, j, k;
        for i to l do for j to n do
        loc real x := .0;
            for k to m do
            x + := a[i, k] * b[k, j] od;
            c[i, j] := x od od
    );
    loc int l, m, n;
    read ((l, m, n));
    loc [1:l, 1:m] real a, loc [1:m, 1:n] real b, loc [1:l, 1:n] real c;
    read ((a, b));
    matmul (a, b, c);       print (c)
)
```

The procedure calling parameters have been reduced to three (compared with FORTRAN's nine) and the language has many more useful features. For example, operations between matrices can be defined so that procedure calls can be dispensed with and operation $c = a \times b$ would implicitly mean matrix multiply. One can therefore anticipate languages specially adapted for matrix manipulation.

1.9 CONCLUSIONS

Computers on which finite element computations can be done vary widely in their size and architecture. Because of its entrenched position FORTRAN is the language in which computer programs for engineering applications had best be written in order to assure maximum readership and portability. (There are inevitably minor differences between machines, for example in their handling of free format input which is used exclusively in the test programs, but these are not overwhelming.) However, FORTRAN possesses fundamental limitations which can only partially be overcome by writing a structure of FORTRAN which is as close to ALGOL as can be achieved. This is particularly vital in subroutine calling. Using this structure of FORTRAN, a library of subroutines can be created which is then held in compiled form on backing store and accessed by programs in just the way that a manufacturer's permanent library is.

Using this philosophy, a library of over 100 subroutines has been assembled, together with some 50 example programs which access the library. These programs are written in a reasonably 'structured' style, and tapes or discs of library and programs will be supplied to readers on request. Versions are at present available for all the common machine ranges.

The structure of the remainder of the book is as follows. Chapter 2 shows how the differential equations governing the behaviour of solids and fluids are programs are written in a reasonably 'structured' style, and tapes or discs of library and programs will be supplied to readers on request. Versions are at present available for all the common machine ranges.

Chapter 3 describes the sub-program library and the basic techniques by which main programs are constructed to solve the equations listed in Chapter 2.

The remaining Chapters 4 to 12 are concerned with applications, partly in the authors' field of geomechanics. However, the methods and programs described are equally applicable in many other fields such as structural mechanics, water resources, electromagnetics and so on. Chapter 4 leads off with static analysis of skeletal structures and plate problems. Chapter 5 deals with static analysis of linear solids, whilst Chapter 6 discusses extensions to deal with material non-linearity. Chapter 7 is concerned with problems of fluid flow in the steady state while transient states with inclusion of transport phenomena (diffusion with advection) are treated in Chapter 8. In Chapter 9, coupling between solid and fluid phases is treated, with applications to consolidation processes. A second type of 'coupling' involves the Navier–Stokes equations. Chapter 10 contains programs for the solution of steady state vibration problems, involving the determination of natural modes, by various methods. Integration of the equations of motion in time by direct integration, by modal superposition and by 'complex response' is described in Chapter 11. Finally, Chapter 12 deals with problems of pile capacity and driveability, for single piles and including group effects.

In every applications chapter, test programs are listed and described, together with specimen input and output.

1.10 REFERENCES

Barron, D. W. (1977) *An Introduction to the Study of Programming Languages.* Cambridge University Press.

Dijkstra, E. W. (1976) *A Discipline of Programming.* Prentice-Hall, Englewood Cliffs, New Jersey.

Ford, B., and Sayers, D. K. (1976) Developing a single numerical algorithms library for different machine ranges. *ACM Trans. Math. Software*, **2**, 115.

Lindsey, C. H. (1977) Structure charts: a structured alternative to flow charts. *SIGPLAN Notices*, **12**, No. 11, 36.

CHAPTER 2
Spatial Discretisation by Finite Elements

2.0 INTRODUCTION

The finite element method is a technique for solving partial differential equations by first discretising these equations in their space dimensions. The discretisation is carried out locally over small regions of simple but arbitrary shape (the finite elements). This results in matrix equations relating the input at specified points in the elements (the nodes) to the output at these same points. In order to solve equations over large regions, the matrix equations for the smaller sub-regions are usually summed node by node, resulting in global matrix equations. The method is already described in many texts, for example Zienkiewicz (1977), Strang and Fix (1973), Cook (1981), Connor and Brebbia (1976) and Rao (1982), but the principles will briefly be described in this chapter in order to establish a notation and to set the scene for the later descriptions of programming techniques.

2.1 ROD ELEMENT

Figure 2.1(a) shows the simplest solid element, namely an elastic rod, with end nodes 1 and 2. The element has length L while u denotes the longitudinal displacements of points on the rod which is subjected to axial loading only.

If P is the axial force in the rod at a particular section and F is an applied body force with units of force per unit length then

$$P = \sigma A = EA\varepsilon = EA\frac{\partial u}{\partial x} \tag{2.1}$$

and for equilibrium from Figure 2.1(b),

$$\frac{\partial P}{\partial x} + F = 0 \tag{2.2}$$

Hence the differential equation to be solved is

$$EA\frac{\partial^2 u}{\partial x^2} + F = 0 \tag{2.3}$$

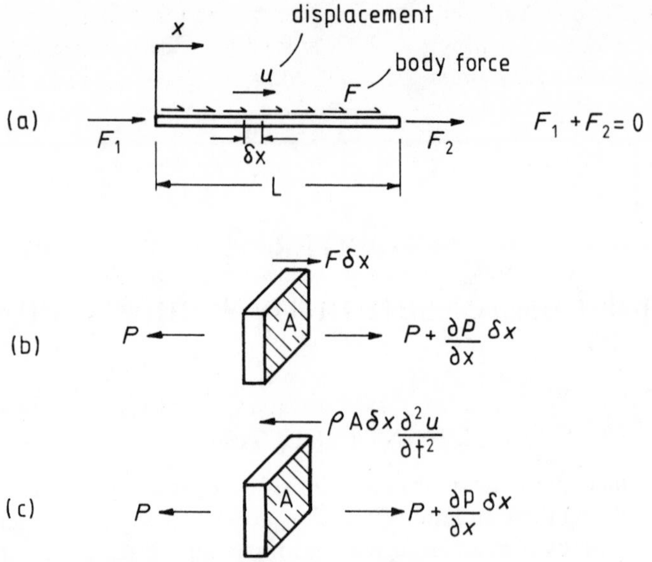

Figure 2.1 Equilibrium of a rod element

(Although this is an ordinary differential equation, future equations, for example (2.13), preserve the same notation.)

In the finite element technique, the continuous variable u is approximated in terms of its nodal values, u_1 and u_2, through simple functions of the space variable called 'shape functions'. That is

$$u \simeq N_1 u_1 + N_2 u_2$$

or

$$u \simeq [N_1 \quad N_2] \begin{Bmatrix} u_1 \\ u_2 \end{Bmatrix} = \mathbf{Nu} \tag{2.4}$$

In simple examples the approximate equality in (2.4) could be made exact by setting

$$N_1 = \left(1 - \frac{x}{L}\right), \qquad N_2 = \frac{x}{L} \tag{2.5}$$

if the true variation of u is linear. In general N_1 and N_2 will be of higher order or more linear element subdivisions will be necessary.

When (2.4) is substituted in (2.3) we have

$$EA \frac{\partial^2}{\partial x^2} [N_1 \quad N_2] \begin{Bmatrix} u_1 \\ u_2 \end{Bmatrix} + F = 0 \tag{2.6}$$

so that the partial differential equation has been replaced by an equation in the discretised space variables u_1 and u_2. The problem now reduces to one of finding 'good' values for u_1 and u_2.

Many methods could be used to achieve this, and Crandall (1956) discusses collocation, subdomain, Galerkin and least squares techniques. Of these, Galerkin's method, e.g. Finlayson (1972), is the most widely used in finite element work. The method consists in multiplying equation (2.6) by the shape functions in turn and integrating the resulting residual over the element. Thus

$$\int_0^L \begin{Bmatrix} N_1 \\ N_2 \end{Bmatrix} EA \frac{\partial^2}{\partial x^2} [N_1 \quad N_2] \begin{Bmatrix} u_1 \\ u_2 \end{Bmatrix} dx + \int_0^L \begin{Bmatrix} N_1 \\ N_2 \end{Bmatrix} F\, dx = 0 \tag{2.7}$$

Note that in the present example, double differentiation of the shape functions would cause them to vanish and yet we know the correct shape may not be of higher order than linear. This difficulty is resolved by applying Green's theorem (integration by parts) to yield typically

$$\int N_i \frac{\partial^2 N_j}{\partial x^2} dx = -\int \frac{\partial N_i}{\partial x} \frac{\partial N_j}{\partial x} dx + \text{boundary terms which we ignore} \tag{2.8}$$

Hence (2.7) becomes

$$EA \int_0^L \begin{bmatrix} \dfrac{\partial N_1}{\partial x} \dfrac{\partial N_1}{\partial x} & \dfrac{\partial N_1}{\partial x} \dfrac{\partial N_2}{\partial x} \\ \dfrac{\partial N_2}{\partial x} \dfrac{\partial N_1}{\partial x} & \dfrac{\partial N_2}{\partial x} \dfrac{\partial N_2}{\partial x} \end{bmatrix} dx \begin{Bmatrix} u_1 \\ u_2 \end{Bmatrix} - F \int_0^L \begin{Bmatrix} N_1 \\ N_2 \end{Bmatrix} dx = 0 \tag{2.9}$$

On evaluation of the integrals,

$$EA \begin{bmatrix} \dfrac{1}{L} & -\dfrac{1}{L} \\ -\dfrac{1}{L} & \dfrac{1}{L} \end{bmatrix} \begin{Bmatrix} u_1 \\ u_2 \end{Bmatrix} - F \begin{Bmatrix} \dfrac{L}{2} \\ \dfrac{L}{2} \end{Bmatrix} = \begin{Bmatrix} 0 \\ 0 \end{Bmatrix} \tag{2.10}$$

The above case is for a uniformly distributed force F acting along the rod. For the case in Figure 2.1(a) where the loading is applied only at the nodes we have

$$EA \begin{bmatrix} \dfrac{1}{L} & -\dfrac{1}{L} \\ -\dfrac{1}{L} & \dfrac{1}{L} \end{bmatrix} \begin{Bmatrix} u_1 \\ u_2 \end{Bmatrix} = \begin{Bmatrix} F_1 \\ F_2 \end{Bmatrix} \tag{2.11}$$

which are the familiar 'stiffness' equations for a uniform prismatic rod. In matrix notation we may write

$$\mathbf{KMu} = \mathbf{F} \tag{2.12}$$

where **KM** is called the 'element stiffness matrix'.

2.2 ROD INERTIA MATRIX

Consider now the case of an unrestrained rod in longitudinal motion. Figure 2.1(c) shows the equilibrium of a segment in which the body force is now given by Newton's law as mass times acceleration. If the mass per unit volume is ρ, the differential equation becomes

$$EA\frac{\partial^2 u}{\partial x^2} - \rho A\frac{\partial^2 u}{\partial t^2} = 0 \tag{2.13}$$

On discretising u by finite elements as before, the first term in (2.13) clearly leads again to **KM**. The second term takes the form

$$-\int_0^L \begin{Bmatrix} N_1 \\ N_2 \end{Bmatrix} \rho A \frac{\partial^2}{\partial t^2} [N_1 \quad N_2] \begin{Bmatrix} u_1 \\ u_2 \end{Bmatrix} \mathrm{d}x \tag{2.14}$$

or

$$-\rho A \int_0^L \begin{bmatrix} N_1 N_1 & N_1 N_2 \\ N_2 N_1 & N_2 N_2 \end{bmatrix} \mathrm{d}x \frac{\partial^2}{\partial t^2} \begin{Bmatrix} u_1 \\ u_2 \end{Bmatrix} \tag{2.15}$$

Evaluation of integrals yields

$$-\rho AL \begin{bmatrix} \frac{1}{3} & \frac{1}{6} \\ \frac{1}{6} & \frac{1}{3} \end{bmatrix} \frac{\partial^2}{\partial t^2} \begin{Bmatrix} u_1 \\ u_2 \end{Bmatrix} \tag{2.16}$$

or in matrix notation

$$-\mathbf{MM}\frac{\partial^2 \mathbf{u}}{\partial t^2}$$

where **MM** is the 'element mass matrix' or 'element inertia matrix'. Thus the full matrix statement of equation (2.13) is

$$\mathbf{KMu} + \mathbf{MM}\frac{\partial^2 \mathbf{u}}{\partial t^2} = 0 \tag{2.17}$$

Note that **MM** as formed in this manner is the 'consistent' mass matrix and differs from the 'lumped' equivalent which would lead to $\frac{1}{2}\rho AL$ terms on the diagonal with zeros off-diagonal.

2.3 THE EIGENVALUE EQUATION

Equation (2.17) is sometimes integrated directly (Chapter 11) but is also the starting point for derivation of the eigenvalues of an element or mesh of elements.

Suppose the elastic rod element is undergoing free harmonic motion. Then all nodal displacements will be harmonic, of the form

$$\mathbf{u} = \mathbf{a}\sin(\omega t + \psi) \tag{2.18}$$

where $\mathbf{a}$ are amplitudes of the motion, ω its frequencies and ψ its phase shifts. When (2.18) is substituted in (2.17) the equation

$$\mathbf{KMa} - \omega^2\mathbf{MMa} = 0 \tag{2.19}$$

is obtained, which can easily be rearranged as a standard eigenvalue equation. Chapter 10 describes solution of equations of this type.

2.4 BEAM ELEMENT

As a second one-dimensional solid element, consider the slender beam in Figure 2.2. The end nodes 1 and 2 are subjected to shear forces and moments which result in translations and rotations. Each node, therefore, has two degrees of freedom.

The element shown in Figure 2.2 has length L, flexural rigidity EI and carries a uniform transverse load of q per unit length. The well known equilibrium equation for this system is given by

$$EI\frac{\partial^4 w}{\partial x^4} = q \tag{2.20}$$

Again the continuous variable, w in this case, is approximated in terms of discrete nodal values, but we introduce the idea that not only w itself but also its derivatives can be used in the approximation. Thus we choose to write

$$w \simeq [N_1 \quad N_2 \quad N_3 \quad N_4]\begin{Bmatrix} w_1 \\ \theta_1 \\ w_2 \\ \theta_2 \end{Bmatrix} \tag{2.21}$$

where $\theta_1 = \partial w/\partial x$ at node 1 and so on. In this case, equation (2.21) can often be made exact by choosing the cubic shape functions:

$$\begin{aligned} N_1 &= \frac{1}{L^3}(L^3 - 3Lx^2 + 2x^3) \\ N_2 &= \frac{1}{L^2}(L^2 x - 2Lx^2 + x^3) \\ N_3 &= \frac{1}{L^3}(3Lx^2 - 2x^3) \\ N_4 &= \frac{1}{L^2}(x^3 - Lx^2) \end{aligned} \tag{2.22}$$

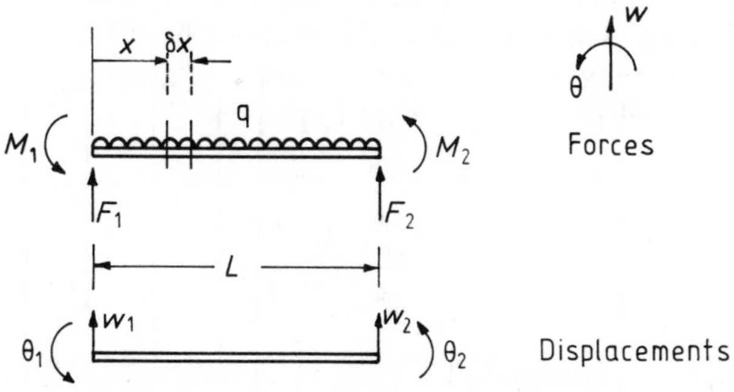

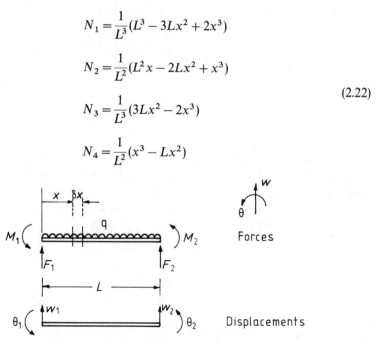

Figure 2.2 Slender beam element

Note that the shape functions have the property that they or their derivatives in this case equal one at a specific node and zero at all others.

Substitution in (2.20) and application of Galerkin's method leads to the four element equations:

$$\int_0^L \begin{Bmatrix} N_1 \\ N_2 \\ N_3 \\ N_4 \end{Bmatrix} EI \frac{\partial^4}{\partial x^4} [N_1 \quad N_2 \quad N_3 \quad N_4] \begin{Bmatrix} w_1 \\ \theta_1 \\ w_2 \\ \theta_2 \end{Bmatrix} \mathrm{d}x = \int_0^L \begin{Bmatrix} N_1 \\ N_2 \\ N_3 \\ N_4 \end{Bmatrix} q \, \mathrm{d}x \qquad (2.23)$$

Again Green's theorem is used to avoid differentiating four times; for example

$$\int N_i \frac{\partial^4 N_j}{\partial x^4} \mathrm{d}x = -\int \frac{\partial N_i}{\partial x} \frac{\partial^3 N_j}{\partial x^3} \mathrm{d}x = \int \frac{\partial^2 N_i}{\partial x^2} \frac{\partial^2 N_j}{\partial x^2} \mathrm{d}x + \text{neglected terms} \qquad (2.24)$$

Hence assuming that EI and q are not functions of x, (2.23) become

$$EI \int_0^L \frac{\partial^2 N_i}{\partial x^2} \frac{\partial^2 N_j}{\partial x^2} \mathrm{d}x \; i, j = 1, 2, 3, 4 \begin{Bmatrix} w_1 \\ \theta_1 \\ w_2 \\ \theta_2 \end{Bmatrix} = q \int_0^L \begin{Bmatrix} N_1 \\ N_2 \\ N_3 \\ N_4 \end{Bmatrix} \mathrm{d}x \qquad (2.25)$$

Evaluation of the integrals gives

$$EI \begin{bmatrix} \frac{12}{L^3} & \frac{6}{L^2} & -\frac{12}{L^3} & \frac{6}{L^2} \\ & \frac{4}{L} & -\frac{6}{L^2} & \frac{2}{L} \\ \text{Symmetrical} & & \frac{12}{L^3} & -\frac{6}{L^2} \\ & & & \frac{4}{L} \end{bmatrix} \begin{Bmatrix} w_1 \\ \theta_1 \\ w_2 \\ \theta_2 \end{Bmatrix} = q \begin{Bmatrix} \frac{L}{2} \\ \frac{L^2}{12} \\ \frac{L}{2} \\ -\frac{L^2}{12} \end{Bmatrix} \qquad (2.26a)$$

which recovers the standard slope-deflection equation for beam elements.

The above case is for a uniformly distributed load applied to the beam. For the case where loading is applied only at the nodes we have

$$EI \begin{bmatrix} \frac{12}{L^3} & \frac{6}{L^2} & -\frac{12}{L^3} & \frac{6}{L^2} \\ & \frac{4}{L} & -\frac{6}{L^2} & \frac{2}{L} \\ & & \frac{12}{L^3} & -\frac{6}{L^2} \\ \text{Symmetrical} & & & \frac{4}{L} \end{bmatrix} \begin{Bmatrix} w_1 \\ \theta_1 \\ w_2 \\ \theta_2 \end{Bmatrix} = \begin{Bmatrix} F_1 \\ M_1 \\ F_2 \\ M_2 \end{Bmatrix} \qquad (2.26b)$$

which represents the element stiffness relationship.

Hence, in matrix notation we have

$$\mathbf{KMw} = \mathbf{F} \tag{2.27}$$

Beam-column elements, in which axial and bending effects are superposed from (2.26b) and (2.11), are described further in Chapter 4.

2.5 BEAM INERTIA MATRIX

If the element in Figure 2.2 were vibrating transversely it would be subjected to an additional restoring force $-\rho A(\partial^2 w/\partial t^2)$. The inertia or mass matrix, by analogy with (2.15), is just

$$-\rho A \int_0^L \begin{bmatrix} N_1 N_1 & N_1 N_2 & N_1 N_3 & N_1 N_4 \\ N_2 N_1 & N_2 N_2 & N_2 N_3 & N_2 N_4 \\ N_3 N_1 & N_3 N_2 & N_3 N_3 & N_3 N_4 \\ N_4 N_1 & N_4 N_2 & N_4 N_3 & N_4 N_4 \end{bmatrix} \mathrm{d}x \, \frac{\partial^2}{\partial t^2} \begin{Bmatrix} w_1 \\ \theta_1 \\ w_2 \\ \theta_2 \end{Bmatrix} \tag{2.28}$$

Evaluation of the integrals yields

$$-\frac{\rho A L}{420} \begin{bmatrix} 156 & 22L & 54 & -13L \\ & 4L^2 & 13L & -3L^2 \\ & & 156 & -22L \\ & \text{Symmetrical} & & 4L^2 \end{bmatrix} \frac{\partial^2}{\partial t^2} \begin{Bmatrix} w_1 \\ \theta_1 \\ w_2 \\ \theta_2 \end{Bmatrix} \tag{2.29}$$

In this instance, the neglect of the consistent mass terms leads to large errors in the prediction of beam frequencies as shown by Leckie and Lindburg (1963). Superposition of (2.29) and (2.16) is required in the dynamic analysis of framed structures and this is described further in Chapter 10.

2.6 BEAM WITH AN AXIAL FORCE

If the beam element in Figure 2.2 is subjected to an additional axial force P (Figure 2.3), a simple modification to (2.20) results in the differential equation

$$EI\frac{\partial^4 w}{\partial x^2} \pm P\frac{\partial^2 w}{\partial x^2} = q \tag{2.30}$$

where the positive sign corresponds to a compressive axial load and vice versa.

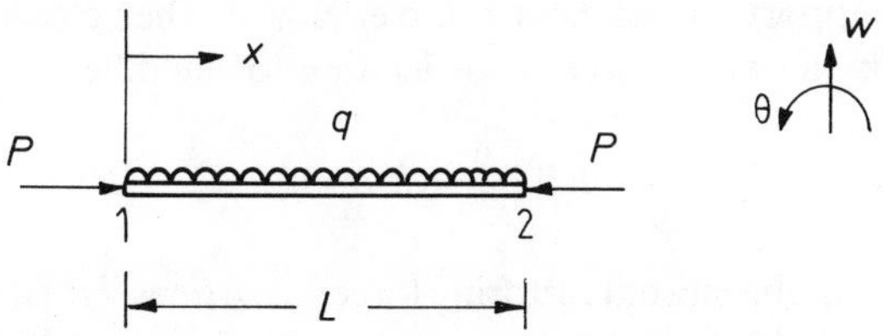

Figure 2.3 Beam with axial force

Finite element discretisation and application of Galerkin's method leads to an additional matrix associated with the axial force contribution

$$\mp P \int_0^L \frac{\partial N_i}{\partial x}\frac{\partial N_j}{\partial x}\mathrm{d}x \; i, j = 1, 2, 3, 4 \begin{Bmatrix} w_1 \\ \theta_1 \\ w_2 \\ \theta_2 \end{Bmatrix} \tag{2.31}$$

Evaluation of these integrals yields for compressive P

$$+\frac{P}{30}\begin{bmatrix} \frac{36}{L} & 3 & -\frac{36}{L} & 3 \\ & 4L & -3 & -L \\ & & \frac{36}{L} & -3 \\ \text{Symmetrical} & & & 4L \end{bmatrix}\begin{Bmatrix} w_1 \\ \theta_1 \\ w_2 \\ \theta_2 \end{Bmatrix} \tag{2.32}$$

If this matrix is designated by **KP** the equilibrium equation becomes

$$(\mathbf{KM} - \mathbf{KP})\mathbf{w} = \mathbf{F} \tag{2.33}$$

Buckling of a member can be investigated by solving the eigenvalue problem when $\mathbf{F} = \mathbf{0}$, by increasing the compressive force on the element (corresponding to **KP** in 2.33) until large deformations result or in simple cases by determinant search. Program 12.1 uses the **KP** matrix to assess the effect of axial loading on the response of laterally loaded piles. Equations (2.32) and (2.33) represent an approximation of the approach to modifying the element stiffness involving stability functions (Lundquist and Kroll, 1944; Horne and Merchant, 1965). The accuracy of the approximation depends on the value of P/P_E for each member, where P_E is the Euler load. Over the range $-1 \leqslant P/P_E \leqslant 1$ the approximation introduces errors no greater than 7% (Livesley, 1975). For larger positive values of P/P_E, however, equation (2.33) can become highly inaccurate unless more element subdivisions are used. Program 4.8 therefore uses the stability function approach to modify the element stiffness matrices in analysing the stability of plane frames.

2.7 BEAM ON ELASTIC FOUNDATION

In Figure 2.4 a continuous elastic support has been placed beneath the basic element. If this support has stiffness k (force/length^2) then clearly the transverse load is resisted by an extra force $+kw$ leading to the differential equation

$$EI\frac{\partial^4 w}{\partial x^4} + kw = q \tag{2.34}$$

By comparison with the inertial restoring force $-\rho A(\partial^2 w/\partial t^2)$ it will be apparent that application of the Galerkin process to (2.34) will result in a foundation

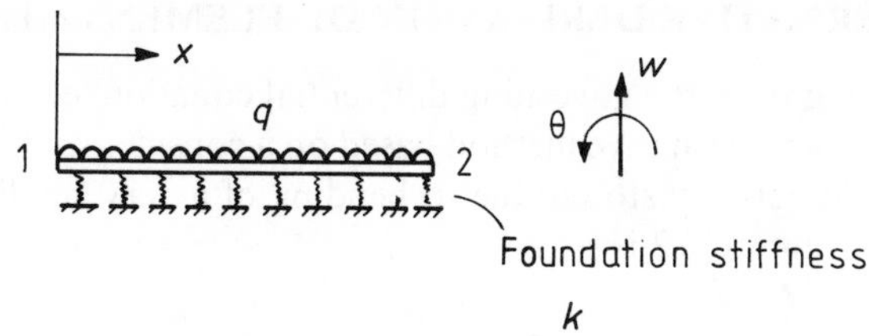

Figure 2.4 Beam on continuous elastic foundation

stiffness matrix that is identical to the consistent mass matrix apart from the multiple $+kL$ in place of $-\rho AL$. An example of a consistent finite element solution to (2.34) is given by Program 4.2 in Chapter 4. A 'lumped mass' approach to this problem is also possible by simply adding the appropriate spring stiffness to the diagonal terms of the beam stiffness matrix and Chapter 12 describes programs of this type.

2.8 GENERAL REMARKS ON THE DISCRETISATION PROCESS

Enough examples have now been described for a general pattern to emerge of how terms in a differential equation appear in matrix form after discretisation. Table 2.1 gives a summary, N being the shape functions.

In fact, first order terms such as $\partial u/\partial x$ have not yet arisen. They are unique in Table 2.1 in leading to matrix equations which are not symmetrical, as indeed would be the case for any odd order of derivative. We shall return to terms of this type in due course, in relation to advection in fluid flow.

Table 2.1 Semi-discretisation of partial differential equations

Term appearing in differential equation	Typical term in finite element matrix equation
u	$\int N_i N_j \, dx$
$\dfrac{\partial u}{\partial x}$	$\int N_i \dfrac{\partial N_j}{\partial x} dx$
$\dfrac{\partial^2 u}{\partial x^2}$	$-\int \dfrac{\partial N_i}{\partial x}\dfrac{\partial N_j}{\partial x} dx$
$\dfrac{\partial^4 u}{\partial x^4}$	$\int \dfrac{\partial^2 N_i}{\partial x^2}\dfrac{\partial^2 N_j}{\partial x^2} dx$

2.9 ALTERNATIVE DERIVATION OF ELEMENT STIFFNESS

Instead of working from the governing differential equation, element properties can be derived by an alternative method based on a consideration of energy. For example, the strain energy stored due to bending of a very small length of the beam element in Figure 2.2 is

$$\delta U = \frac{1}{2}\frac{M^2}{EI}\delta x \tag{2.35}$$

where M is the 'bending moment' and by conservation of energy this must be equal to the work done by the external loads q; thus

$$\delta W = \tfrac{1}{2} q w \delta x \tag{2.36}$$

To proceed, we discretise the displacements in terms of the element shape functions (2.21 and 2.22) to give

$$w = N_1 w_1 + N_2 \theta_1 + N_3 w_2 + N_4 \theta_2 = \mathbf{N}\mathbf{w} \tag{2.37}$$

The bending moment M is related to w through the 'moment-curvature' expression

$$M = -EI\frac{\partial^2 w}{\partial x^2}$$

or

$$M = \mathbf{D}\mathbf{A}w \tag{2.38}$$

where $\mathbf{D}$ is the material property EI and $\mathbf{A}$ is the operator $-\partial^2/\partial x^2$. Writing (2.35) in the form

$$\delta U = \frac{1}{2}\left(-\frac{\partial^2 w}{\partial x^2}\right) M\,\delta x \tag{2.39}$$

we have

$$\delta U = \tfrac{1}{2}(\mathbf{A}w)^{\mathrm{T}} M\,\delta x \tag{2.40}$$

Using (2.37) and (2.38) this becomes

$$\begin{aligned} \delta U &= \tfrac{1}{2}(\mathbf{ANw})^{\mathrm{T}}\mathbf{DANw}\,\delta x \\ &= \tfrac{1}{2}\mathbf{w}^{\mathrm{T}}(\mathbf{AN})^{\mathrm{T}}\mathbf{DANw}\,\delta x \end{aligned} \tag{2.41}$$

The total strain energy of the element is thus

$$U = \tfrac{1}{2}\int_0^L \mathbf{w}^{\mathrm{T}}(\mathbf{AN})^{\mathrm{T}}\mathbf{DANw}\,\mathrm{d}x \tag{2.42}$$

The matrix $\mathbf{AN}$ is usually written as $\mathbf{B}$, and since $\mathbf{w}$ are nodal values and therefore constants

$$U = \tfrac{1}{2}\mathbf{w}^{\mathrm{T}}\int_0^L \mathbf{B}^{\mathrm{T}}\mathbf{D}\mathbf{B}\,\mathrm{d}x\,\mathbf{w} \tag{2.43}$$

Similar operations on (2.36) lead to the total external work done and hence the stored potential energy of the beam is given by

$$\begin{aligned}\Pi &= U - W\\ &= \tfrac{1}{2}\mathbf{w}^{\mathrm{T}}\int_0^L \mathbf{B}^{\mathrm{T}}\mathbf{D}\mathbf{B}\,\mathrm{d}x\,\mathbf{w} - \tfrac{1}{2}\mathbf{w}^{\mathrm{T}}q\int_0^L \mathbf{N}^{\mathrm{T}}\,\mathrm{d}x\end{aligned} \tag{2.44}$$

A state of stable equilibrium is achieved when π is a minimum with respect to all $\mathbf{w}$. That is

$$\frac{\partial\Pi}{\partial w^{\mathrm{T}}} = \int_0^L \mathbf{B}^{\mathrm{T}}\mathbf{D}\mathbf{B}\,\mathrm{d}x\,\mathbf{w} - q\int_0^L \mathbf{N}^{\mathrm{T}}\,\mathrm{d}x = 0 \tag{2.45}$$

or

$$\int_0^L \mathbf{B}^{\mathrm{T}}\mathbf{D}\mathbf{B}\,\mathrm{d}x\,\mathbf{w} = q\int_0^L \mathbf{N}^{\mathrm{T}}\,\mathrm{d}x \tag{2.46}$$

which is simply another way of writing (2.25).

Thus we see from (2.27) that the element stiffness matrix **KM** can be written in the form

$$\mathbf{KM} = \int_0^L \mathbf{B}^{\mathrm{T}}\mathbf{D}\mathbf{B}\,\mathrm{d}x \tag{2.47}$$

which will prove to be a useful general matrix form for expressing stiffnesses of all solid elements. The computer programs for analysis of solids developed in the next chapter use this notation and method of stiffness formation.

2.10 TWO-DIMENSIONAL ELEMENTS: PLANE STRESS AND STRAIN

The elements so far described have not been true finite elements because they have been used to solve differential equations in one space variable only. Thus the real problem involving two or three space variables has been replaced by a hypothetical, equivalent one-dimensional problem before solution. The elements we have considered can be joined together at points (the nodes) and complete continuity (compatibility) and equilibrium achieved. In this way we can often obtain exact solutions to our hypothetical problems and these solutions will be unaffected by the number of elements chosen to represent uniform line segments.

This situation changes radically when problems in two or three space dimensions are analysed. For example, consider the plane shear wall with openings shown in Figure 2.5(a). The wall has been subdivided into rectangular elements of side $a \times b$ of which Figure 2.5(b) is typical. These elements have four corner nodes so that when the idealised wall is assembled, the elements will only be attached at these points.

If the wall can be considered to be of unit thickness and in a state of plane stress, see Timoshenko and Goodier (1951), the equations to be solved are the following:

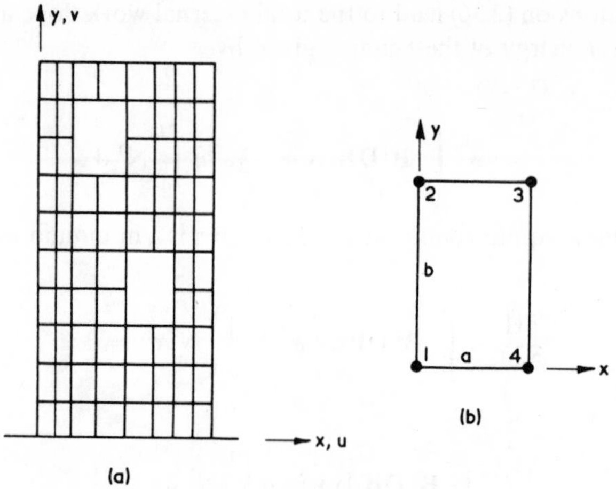

Figure 2.5 (a) Shear wall with openings. (b) Typical rectangular element with four corner nodes

(i) Equilibrium

$$\begin{aligned} \frac{\partial \sigma_x}{\partial x} + \frac{\partial \tau_{xy}}{\partial y} + F_x = 0 \\ \frac{\partial \tau_{xy}}{\partial x} + \frac{\partial \sigma_y}{\partial y} + F_y = 0 \end{aligned} \tag{2.48}$$

where σ_x, σ_y and τ_{xy} are the only non-zero stress components and F_x, F_y are body forces, per unit volume.

(ii) Constitutive (plane stress)

$$\text{(Generalised Hooke)} \quad \begin{Bmatrix} \sigma_x \\ \sigma_y \\ \tau_{xy} \end{Bmatrix} = \frac{E}{1 - \nu^2} \begin{bmatrix} 1 & \nu & 0 \\ \nu & 1 & 0 \\ 0 & 0 & \dfrac{1-\nu}{2} \end{bmatrix} \begin{Bmatrix} \varepsilon_x \\ \varepsilon_y \\ \gamma_{xy} \end{Bmatrix} \tag{2.49}$$

where E is Young's modulus, ν is Poisson's ratio and $\varepsilon_x, \varepsilon_y$ and γ_{xy} are the independent small strain components.

(iii) Strain-displacement

$$\begin{Bmatrix} \varepsilon_x \\ \varepsilon_y \\ \gamma_{xy} \end{Bmatrix} = \begin{bmatrix} \dfrac{\partial}{\partial x} & 0 \\ 0 & \dfrac{\partial}{\partial y} \\ \dfrac{\partial}{\partial y} & \dfrac{\partial}{\partial x} \end{bmatrix} \begin{Bmatrix} u \\ v \end{Bmatrix} \tag{2.50}$$

where u, v are the components of displacement in the x, y directions.

Using the notation of the previous section with **A** as the strain-displacement operator and **D** as the stress–strain matrix, these three equations become

$$\begin{aligned} \mathbf{A}^{\mathrm{T}}\boldsymbol{\sigma} &= -\mathbf{F} \\ \boldsymbol{\sigma} &= \mathbf{D}\boldsymbol{\varepsilon} \\ \boldsymbol{\varepsilon} &= \mathbf{A}\mathbf{e} \end{aligned} \tag{2.51}$$

where

$$\mathbf{e}=\begin{Bmatrix} u \\ v \end{Bmatrix}, \quad \mathbf{A}=\begin{Bmatrix} \dfrac{\partial}{\partial x} & 0 \\ 0 & \dfrac{\partial}{\partial y} \\ \dfrac{\partial}{\partial y} & \dfrac{\partial}{\partial x} \end{Bmatrix}, \quad \mathbf{D}=\frac{E}{1-\nu^2}\begin{bmatrix} 1 & \nu & 0 \\ \nu & 1 & 0 \\ 0 & 0 & \dfrac{1-\nu}{2} \end{bmatrix} \tag{2.52}$$

We shall only be concerned in this book with 'displacement' formulations in which $\boldsymbol{\sigma}$ and $\boldsymbol{\varepsilon}$ are eliminated from (2.51) as follows:

$$\begin{aligned} \mathbf{A}^{\mathrm{T}}\boldsymbol{\sigma} &= -\mathbf{F} \\ \mathbf{A}^{\mathrm{T}}\mathbf{D}\boldsymbol{\varepsilon} &= -\mathbf{F} \\ \mathbf{A}^{\mathrm{T}}\mathbf{D}\mathbf{A}\mathbf{e} &= -\mathbf{F} \end{aligned} \tag{2.53}$$

Writing out (2.53) in full we have

$$\frac{E}{1-\nu^2}\begin{Bmatrix} \dfrac{\partial^2 u}{\partial x^2}+\dfrac{1-\nu}{2}\dfrac{\partial^2 u}{\partial y^2}+\dfrac{1+\nu}{2}\dfrac{\partial^2 v}{\partial x \partial y} \\ \dfrac{1+\nu}{2}\dfrac{\partial^2 u}{\partial x \partial y}+\dfrac{1-\nu}{2}\dfrac{\partial^2 v}{\partial x^2}+\dfrac{\partial^2 v}{\partial y^2} \end{Bmatrix}=\begin{Bmatrix} -F_x \\ -F_y \end{Bmatrix} \tag{2.54}$$

which is a pair of simultaneous partial differential equations in the continuous space variables u and v.

As usual these can be solved by discretising over each element using shape functions

$$u=[N_1 \quad N_2 \quad N_3 \quad N_4]\begin{Bmatrix} u_1 \\ u_2 \\ u_3 \\ u_4 \end{Bmatrix}=\mathbf{N}\mathbf{u} \tag{2.55}$$

and

$$v=[N_1 \quad N_2 \quad N_3 \quad N_4]\begin{Bmatrix} v_1 \\ v_2 \\ v_3 \\ v_4 \end{Bmatrix}=\mathbf{N}\mathbf{v} \tag{2.56}$$

where in the case of the rectangular element shown in Figure 2.5(b) the N_i

functions were first derived by Taig (1961) to be

$$
\begin{aligned}
N_1 &= \left(1-\frac{x}{a}\right)\left(1-\frac{y}{b}\right)\\
N_2 &= \left(1-\frac{x}{a}\right)\frac{y}{b}\\
N_3 &= \frac{x}{a}\frac{y}{b}\\
N_4 &= \frac{x}{a}\left(1-\frac{y}{b}\right)
\end{aligned}
\tag{2.57}
$$

These result in linear variations in strain across the element which is sometimes called the 'linear strain rectangle'.

Discretisation and application of Galerkin's method (Szabo and Lee, 1969), using Table 2.1, leads to the stiffness equations for a typical element:

$$
\frac{E}{1-v^2}\int_0^a\int_0^b \begin{bmatrix} \left(\frac{\partial N_i}{\partial x}\frac{\partial N_j}{\partial x}+\frac{1-v}{2}\frac{\partial N_i}{\partial y}\frac{N_j}{\partial y}\right) & \left(v\frac{\partial N_i}{\partial x}\frac{\partial N_j}{\partial y}+\frac{1-v}{2}\frac{\partial N_i}{\partial y}\frac{\partial N_j}{\partial x}\right)\\ \left(v\frac{\partial N_i}{\partial y}\frac{\partial N_j}{\partial x}+\frac{1-v}{2}\frac{\partial N_i}{\partial x}\frac{\partial N_j}{\partial y}\right) & \left(\frac{\partial N_i}{\partial y}\frac{N_j}{\partial y}+\frac{1-v}{2}\frac{\partial N_i}{\partial x}\frac{\partial N_j}{\partial x}\right)\end{bmatrix}
$$

$$
\cdot \mathrm{d}y\,\mathrm{d}x,\quad i, j = 1, 2, 3, 4 \begin{Bmatrix}\mathbf{u}\\ \mathbf{v}\end{Bmatrix} = \begin{Bmatrix}\mathbf{F}_x\\ \mathbf{F}_y\end{Bmatrix} \tag{2.58}
$$

or

$$
\mathbf{KMr} = \mathbf{F}
$$

Evaluation of the first term in the stiffness matrix yields

$$
\mathrm{KM}_{1,1} = \frac{E}{1-v^2}\left(\frac{b}{3a}+\frac{1-v}{2}\frac{a}{3b}\right) \tag{2.59}
$$

and so on.

In the course of this evaluation, integration by parts now involves integrals of the type

$$
\iint N_i\frac{\partial^2 N_j}{\partial x^2}\mathrm{d}x\,\mathrm{d}y = -\iint \frac{\partial N_i}{\partial x}\frac{\partial N_j}{\partial x}\mathrm{d}x\,\mathrm{d}y + \oint N_i\frac{\partial N_j}{\partial x} l\,\mathrm{d}S \tag{2.60}
$$

where l is the direction cosine of the normal to boundary S and we assume that the contour integral in (2.60) is zero between elements. This assumption is generally reasonable but extra care is needed at mesh boundaries. Only if the elements become vanishingly small can our solution be the correct one (an infinite number of elements).

Physically, in a displacement method, it is usual to satisfy compatibility everywhere in a mesh but to satisfy equilibrium only at the nodes. It is also possible to violate compatibility, but none of the elements described in this book does.

2.11 ENERGY APPROACH

As was done in the case of the beam element, the principle of minimum potential energy can be used to provide an alternative derivation of (2.58). The element strain energy per unit thickness is

$$
\begin{aligned}
U &= \iint \tfrac{1}{2}\boldsymbol{\sigma}^{\mathrm{T}}\boldsymbol{\varepsilon}\,\mathrm{d}x\,\mathrm{d}y \\
&= \tfrac{1}{2}\mathbf{r}^{\mathrm{T}} \iint (\mathbf{AN})^{\mathrm{T}}\mathbf{D}(\mathbf{AN})\,\mathrm{d}x\,\mathrm{d}y\,\mathbf{r} \\
&= \tfrac{1}{2}\mathbf{r}^{\mathrm{T}} \iint \mathbf{B}^{\mathrm{T}}\mathbf{D}\mathbf{B}\,\mathrm{d}x\,\mathrm{d}y\,\mathbf{r}
\end{aligned}
\tag{2.61}
$$

where **A** and **D** are defined in (2.52) and

$$
\mathbf{N} = \begin{bmatrix} N_1 & N_2 & N_3 & N_4 & 0 & 0 & 0 & 0 \\ 0 & 0 & 0 & 0 & N_1 & N_2 & N_3 & N_4 \end{bmatrix} \tag{2.62}
$$

Thus we have again for this element

$$
\mathbf{KM} = \iint \mathbf{B}^{\mathrm{T}}\mathbf{D}\mathbf{B}\,\mathrm{d}x\,\mathrm{d}y \tag{2.63}
$$

which is the form in which it will be computed in Chapter 3.

Exactly the same expression holds in the case of plane strain, but the **D** matrix becomes (Timoshenko and Goodier, 1951), for unit thickness,

$$
\mathbf{D} = \frac{E(1-\nu)}{(1+\nu)(1-2\nu)} \begin{bmatrix} 1 & \dfrac{\nu}{1-\nu} & 0 \\ \dfrac{\nu}{1-\nu} & 1 & 0 \\ 0 & 0 & \dfrac{1-2\nu}{2(1-\nu)} \end{bmatrix} \tag{2.64}
$$

2.12 PLANE ELEMENT INERTIA MATRIX

When inertia is significant (2.54) are supplemented by forces $-\rho(\partial^2 u/\partial t^2)$ and $-\rho(\partial^2 v/\partial t^2)$ respectively, where ρ is the mass of the element per unit volume. For an element of unit thickness this leads, in exactly the same way as in (2.14), to the element mass matrix which has terms given by

$$
MM_{ij} = \rho \iint N_i N_j\,\mathrm{d}x\,\mathrm{d}y \tag{2.65}
$$

and hence to an eigenvalue equation the same as (2.19).

2.13 AXISYMMETRIC STRESS AND STRAIN

Solids of revolution subjected to axisymmetric loading possess only two independent components of displacement and can be analysed as if they were two-dimensional. For example, Figure 2.6(a) shows a thick tube subjected to radial pressure p and axial pressure q. Only a typical radial cross-section need be analysed and is subdivided into rectangular elements in the figure. The cylindrical coordinate system, Figure 2.6(b), is the most convenient and when it is used the element stiffness equation equivalent to (2.58) is

$$\mathbf{KM} = \iiint \mathbf{B}^{\mathrm{T}}\mathbf{DB}r\,\mathrm{d}r\,\mathrm{d}\theta\,\mathrm{d}z \tag{2.66}$$

which, when integrated over one radian, becomes

$$\mathbf{KM} = \iint \mathbf{B}^{\mathrm{T}}\mathbf{DB}r\,\mathrm{d}r\,\mathrm{d}z$$

where the strain-displacement relations are now (Timoshenko and Goodier, 1951)

$$\begin{Bmatrix} \varepsilon_r \\ \varepsilon_z \\ \gamma_{rz} \\ \varepsilon_\theta \end{Bmatrix} = \begin{bmatrix} \dfrac{\partial}{\partial r} & 0 \\ 0 & \dfrac{\partial}{\partial z} \\ \dfrac{\partial}{\partial z} & \dfrac{\partial}{\partial r} \\ \dfrac{1}{r} & 0 \end{bmatrix} \begin{Bmatrix} u \\ v \end{Bmatrix} \tag{2.67}$$

or

$$\boldsymbol{\varepsilon} = \mathbf{Ae}$$

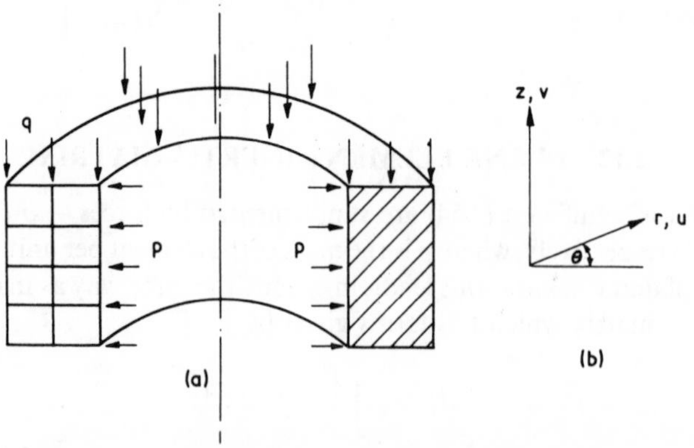

Figure 2.6 Cylinder under axial and radial pressure

and as usual $\mathbf{B} = \mathbf{AN}$, where for elements of rectangular cross-section $\mathbf{N}$ could again be defined by (2.57) and (2.62). The stress–strain matrix must be redefined as

$$\mathbf{D} = \frac{E(1-\nu)}{(1+\nu)(1-2\nu)} \begin{bmatrix} 1 & \frac{\nu}{1-\nu} & 0 & \frac{\nu}{1-\nu} \\ \frac{\nu}{1-\nu} & 1 & 0 & \frac{\nu}{1-\nu} \\ 0 & 0 & \frac{1-2\nu}{2(1-\nu)} & 0 \\ \frac{\nu}{1-\nu} & \frac{\nu}{1-\nu} & 0 & 1 \end{bmatrix} \tag{2.68}$$

2.14 THREE-DIMENSIONAL STRESS AND STRAIN

When (2.48), (2.49) and (2.50) are extended to the three space-displacement variables u, v, w, three simultaneous partial differential equations equivalent to (2.54) result. Discretisation proceeds as usual, and again the familiar element stiffness properties are derived as

$$\mathbf{KM} = \iiint \mathbf{B}^{\mathrm{T}} \mathbf{DB} \, \mathrm{d}x \, \mathrm{d}y \, \mathrm{d}z \tag{2.69}$$

where the full strain-displacement relations are (Timoshenko and Goodier, 1951)

$$\begin{Bmatrix} \varepsilon_x \\ \varepsilon_y \\ \varepsilon_z \\ \gamma_{xy} \\ \gamma_{yz} \\ \gamma_{zx} \end{Bmatrix} = \begin{bmatrix} \frac{\partial}{\partial x} & 0 & 0 \\ 0 & \frac{\partial}{\partial y} & 0 \\ 0 & 0 & \frac{\partial}{\partial z} \\ \frac{\partial}{\partial y} & \frac{\partial}{\partial x} & 0 \\ 0 & \frac{\partial}{\partial z} & \frac{\partial}{\partial y} \\ \frac{\partial}{\partial z} & 0 & \frac{\partial}{\partial x} \end{bmatrix} \begin{Bmatrix} u \\ v \\ w \end{Bmatrix} \tag{2.70}$$

or

$$\boldsymbol{\varepsilon} = \mathbf{Ae}$$

and as before $\mathbf{B} = \mathbf{AN}$. For example, the rectangular brick-shaped element shown in Figure 2.7, which has eight corner nodes, would have shape functions of the

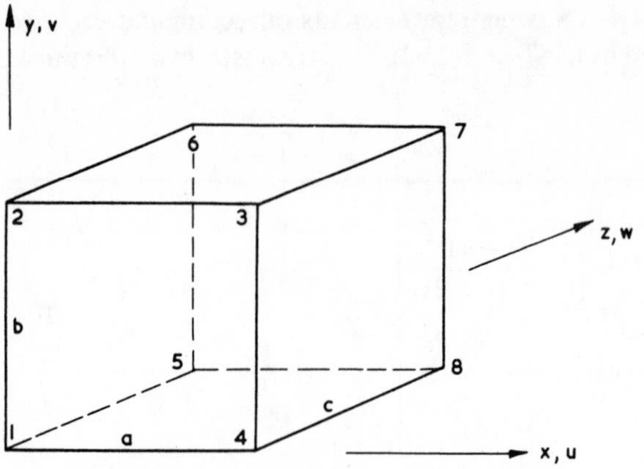

Figure 2.7 Rectangular brick element

type

$$N_1 = \left(1 - \frac{x}{a}\right)\left(1 - \frac{y}{b}\right)\left(1 - \frac{z}{c}\right) \tag{2.71}$$

and so on. The full **N** matrix would be

$$\mathbf{N} = \begin{bmatrix} \mathbf{N}_u & \mathbf{0} & \mathbf{0} \\ \mathbf{0} & \mathbf{N}_v & \mathbf{0} \\ \mathbf{0} & \mathbf{0} & \mathbf{N}_w \end{bmatrix} \tag{2.72}$$

where

$$\mathbf{N}_u = \mathbf{N}_v = \mathbf{N}_w = [N_1 \quad N_2 \quad N_3 \quad N_4 \quad N_5 \quad N_6 \quad N_7 \quad N_8] \tag{2.73}$$

leading to the assembly of **B**. The stress–strain matrix is in this case

$$\mathbf{D} = \frac{E(1-v)}{(1+v)(1-2v)} \begin{bmatrix} 1 & \frac{v}{1-v} & \frac{v}{1-v} & 0 & 0 & 0 \\ \frac{v}{1-v} & 1 & \frac{v}{1-v} & 0 & 0 & 0 \\ \frac{v}{1-v} & \frac{v}{1-v} & 1 & 0 & 0 & 0 \\ 0 & 0 & 0 & \frac{1-2v}{2(1-v)} & 0 & 0 \\ 0 & 0 & 0 & 0 & \frac{1-2v}{2(1-v)} & 0 \\ 0 & 0 & 0 & 0 & 0 & \frac{1-2v}{2(1-v)} \end{bmatrix} \tag{2.74}$$

2.15 PLATE BENDING ELEMENT

The bending of a thin plate is governed by the differential equation

$$D\nabla^4 w = q \tag{2.75}$$

where D is the flexural rigidity of the plate, w its deflection in the transverse (z) direction and q the applied transverse load. The flexural rigidity is given by

$$D = \frac{Eh^3}{12(1-\nu^2)} \tag{2.76}$$

where h is the plate thickness.

Solution of equation (2.75) directly, for example by Galerkin minimisation, appears to imply that for a fixed D, a thin plate's deflection is unaffected by the value of Poisson's ratio. This is in fact true only for certain boundary conditions and in general the integration by parts in the Galerkin process will supply extra terms which are dependent on ν.

This is a case in which the energy approach provides a simpler formulation. The strain energy stored in a piece of bent plate is (Timoshenko and Woinowsky-Krieger, 1959)

$$U = \tfrac{1}{2}D\iint\left\{\left(\frac{\partial^2 w}{\partial x^2}+\frac{\partial^2 w}{\partial y^2}\right)^2 - 2(1-\nu)\left[\frac{\partial^2 w}{\partial x^2}\frac{\partial^2 w}{\partial y^2}-\left(\frac{\partial^2 w}{\partial x\partial y}\right)^2\right]\right\}\mathrm{d}x\,\mathrm{d}y \tag{2.77}$$

or

$$U = \tfrac{1}{2}D\iint\left\{\left(\frac{\partial^2 w}{\partial x^2}\right)^2+\left(\frac{\partial^2 w}{\partial y^2}\right)^2 + 2\nu\frac{\partial^2 w}{\partial x^2}\frac{\partial^2 w}{\partial y^2} + 2(1-\nu)\left(\frac{\partial^2 w}{\partial x\partial y}\right)^2\right\}\mathrm{d}x\,\mathrm{d}y \tag{2.78}$$

Consider, for example, the rectangular element shown in Figure 2.8. If there are assumed to be four degrees of freedom per node, namely

$$w,\quad \frac{\partial w}{\partial x}=\theta_x,\quad \frac{\partial w}{\partial y}=\theta_y \quad\text{and}\quad \frac{\partial^2 w}{\partial x\,\partial y}=\theta_{xy}$$

then the appropriate element shape functions can be shown to be products of the

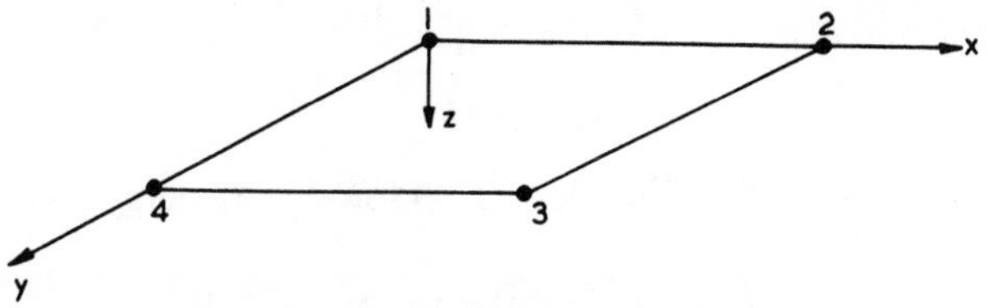

Figure 2.8 Rectangular plate bending element

beam shape functions already described. That is, as usual

$$w = \mathbf{N}\mathbf{w} \tag{2.79}$$

where, if the freedoms are numbered $(w, \theta_x, \theta_y, \theta_{xy})_{\text{node}=1,2,3,4}$, the first term in $\mathbf{N}$ would be

$$N_1 = \frac{1}{a^3}(a^3 - 3ax^2 + 2x^3)\frac{1}{b^3}(b^3 - 3by^2 + 2y^3) \tag{2.80}$$

Defining

$$\begin{aligned} P_1 &= \frac{1}{a^3}(a^3 - 3ax^2 + 2x^3), & Q_1 &= \frac{1}{b^3}(b^3 - 3by^2 + 2y^3) \\ P_2 &= \frac{1}{a^2}(a^2x - 2ax^2 + x^3), & Q_2 &= \frac{1}{b^2}(b^2y - 2by^2 + y^3) \\ P_3 &= \frac{1}{a^3}(3ax^2 - 2x^3), & Q_3 &= \frac{1}{b^3}(3by^2 - 2y^3) \\ P_4 &= \frac{1}{a^2}(x^3 - ax^2), & Q_4 &= \frac{1}{b^2}(y^3 - by^2) \end{aligned} \tag{2.81}$$

the full list of shape functions becomes

$$\begin{aligned} N_1 &= P_1Q_1, & N_9 &= P_3Q_3 \\ N_2 &= P_2Q_1, & N_{10} &= P_4Q_3 \\ N_3 &= P_1Q_2, & N_{11} &= P_3Q_4 \\ N_4 &= P_2Q_2, & N_{12} &= P_4Q_4 \\ N_5 &= P_1Q_3, & N_{13} &= P_3Q_1 \\ N_6 &= P_2Q_3, & N_{14} &= P_4Q_1 \\ N_7 &= P_1Q_4, & N_{15} &= P_3Q_2 \\ N_8 &= P_2Q_4, & N_{16} &= P_4Q_2 \end{aligned} \tag{2.82}$$

Using the same energy formulation terminology as before, define

$$\begin{Bmatrix} M_x \\ M_y \\ M_{xy} \end{Bmatrix} = D \begin{bmatrix} 1 & \nu & 0 \\ \nu & 1 & 0 \\ 0 & 0 & \dfrac{1-\nu}{2} \end{bmatrix} \begin{Bmatrix} \dfrac{\partial^2}{\partial x^2} \\ \dfrac{\partial^2}{\partial y^2} \\ 2\dfrac{\partial^2}{\partial x \partial y} \end{Bmatrix} w \tag{2.83}$$

or

$$\mathbf{M} = \mathbf{D}\mathbf{A}w$$

It can readily be verified that equation (2.78) for strain energy can be written

$$U = \tfrac{1}{2}\iint (\mathbf{A}w)^{\mathrm{T}}\mathbf{D}(\mathbf{A}w)\,\mathrm{d}x\,\mathrm{d}y \tag{2.84}$$

and that the stiffness matrix becomes

$$\mathbf{KM} = \iint \mathbf{B}^{\mathrm{T}}\mathbf{D}\mathbf{B}\,\mathrm{d}x\,\mathrm{d}y \tag{2.85}$$

which is again the familiar equivalent of equation (2.47), with $\mathbf{B} = \mathbf{AN}$.

A typical value in the element stiffness matrix is given by

$$KM_{ij} = D\iint \left\{ \frac{\partial^2 N_i}{\partial x^2}\frac{\partial^2 N_j}{\partial x^2} + v\frac{\partial^2 N_i}{\partial x^2}\frac{\partial^2 N_j}{\partial y^2} + v\frac{\partial^2 N_j}{\partial x^2}\frac{\partial^2 N_i}{\partial y^2} + \frac{\partial^2 N_i}{\partial y^2}\frac{\partial^2 N_j}{\partial y^2} + 2(1-v)\frac{\partial^2 N_i}{\partial x \partial y}\frac{\partial^2 N_j}{\partial x \partial y} \right\} \mathrm{d}x\,\mathrm{d}y \tag{2.86}$$

and these integrals are performed using Gaussian quadrature in two dimensions in Chapter 4. For some boundary conditions, the terms involving Poisson's ratio will cancel out.

As usual, typical values in the mass matrix are given by

$$MM_{ij} = \rho h \iint N_i N_j \,\mathrm{d}x\,\mathrm{d}y \tag{2.87}$$

2.16 SUMMARY OF ELEMENT EQUATIONS FOR SOLIDS

The preceding sections have demonstrated the essential similarity of all problems in solid mechanics when formulated in terms of finite elements. The statement of element properties is to be found in two expressions, namely the element stiffness matrix

$$\mathbf{KM} = \int \mathbf{B}^{\mathrm{T}}\mathbf{D}\mathbf{B}\, d\text{ (element)} \tag{2.88}$$

and the element mass matrix

$$\mathbf{MM} = \rho \int \mathbf{N}^{\mathrm{T}}\mathbf{N} d\text{ (element)} \tag{2.89}$$

These expressions then appear in the three main classes of problem which concern us in engineering practice, namely

(i) Static equilibrium problems $\mathbf{KMr} = \mathbf{F}$ (2.90)

(ii) Eigenvalue problems $(\mathbf{KM} - \omega^2\mathbf{MM})\mathbf{r} = \mathbf{0}$ (2.91)

(iii) Propagation problems $\mathbf{KMr} + \mathbf{MM}\dfrac{\partial^2 \mathbf{r}}{\partial t^2} = \mathbf{F}(t)$ (2.92)

Equations (2.90) are simultaneous equations which can be solved for known forces $\mathbf{F}$ to give equilibrium displacements $\mathbf{r}$. Equations (2.91) may be solved by various techniques (iteration, QR algorithm, etc.; see Bathe and Wilson (1976) or Jennings (1977) to yield mode shapes $\mathbf{r}$ and natural frequencies ω of elastic systems, while equations (2.92) can be solved by advancing step by step in time from a known initial condition or by transformation to the frequency domain.

Later chapters in the book describe programs that enable the user to solve practical engineering problems which are governed by these three basic equations. Additional features, such as treatment of non-linearity, damping, etc., will be dealt with in these chapters as they arise.

2.17 FLOW OF FLUIDS: NAVIER–STOKES EQUATIONS

We shall be concerned only with the equations governing the motion of viscous, incompressible fluids. These equations are widely developed elsewhere, e.g. Schlichting (1960). Preserving an analogy with previous sections on two-dimensional solids, u and v now become velocities in the x and y directions respectively and ρ is the mass density as before. Also as before, F_x and F_y are body forces in the appropriate directions.

Conservation of mass leads to

$$\frac{\partial \rho}{\partial t}+\frac{\partial}{\partial x}(\rho u)+\frac{\partial}{\partial y}(\rho v)=0 \tag{2.93}$$

but due to incompressibility this may be reduced to

$$\frac{\partial u}{\partial x}+\frac{\partial v}{\partial y}=0 \tag{2.94}$$

Conservation of momentum leads to

$$\begin{aligned}\rho\left(\frac{\partial u}{\partial t}+u\frac{\partial u}{\partial x}+v\frac{\partial u}{\partial y}\right)&=F_x+\left(\frac{\partial \sigma_x}{\partial x}+\frac{\partial \tau_{xy}}{\partial y}\right)\\ \rho\left(\frac{\partial v}{\partial t}+u\frac{\partial v}{\partial x}+v\frac{\partial v}{\partial y}\right)&=F_y+\left(\frac{\partial \sigma_y}{\partial y}+\frac{\partial \tau_{xy}}{\partial x}\right)\end{aligned} \tag{2.95}$$

where σ_x, σ_y and τ_{xy} are stress components as previously defined for solids.

Introducing the simplest constitutive parameters μ (the molecular viscosity), λ (taken to be $-\frac{2}{3}\mu$) and p (the fluid pressure) the following form of the stress equations is reached:

$$\begin{aligned}\sigma_x&=-p+\lambda\left(\frac{\partial u}{\partial x}+\frac{\partial v}{\partial y}\right)+2\mu\frac{\partial u}{\partial x}\\ \sigma_y&=-p+\lambda\left(\frac{\partial u}{\partial x}+\frac{\partial v}{\partial y}\right)+2\mu\frac{\partial v}{\partial y}\\ \tau_{xy}=\tau_{yx}&=\mu\left(\frac{\partial u}{\partial y}+\frac{\partial v}{\partial x}\right)\end{aligned} \tag{2.96}$$

Combining (2.94) to (2.96) a form of the 'Navier–Stokes' equations can be written

$$\frac{\partial u}{\partial t}+u\frac{\partial u}{\partial x}+v\frac{\partial u}{\partial y}=\frac{1}{\rho}F_x-\frac{1}{\rho}\frac{\partial p}{\partial x}+\frac{1}{3}\frac{\mu}{\rho}\frac{\partial}{\partial x}\left(\frac{\partial u}{\partial x}+\frac{\partial v}{\partial y}\right)+\frac{\mu}{\rho}\left(\frac{\partial^2 u}{\partial x^2}+\frac{\partial^2 u}{\partial y^2}\right)$$

$$\frac{\partial v}{\partial t}+u\frac{\partial v}{\partial x}+v\frac{\partial v}{\partial y}=\frac{1}{\rho}F_y-\frac{1}{\rho}\frac{\partial p}{\partial y}+\frac{1}{3}\frac{\mu}{\rho}\frac{\partial}{\partial y}\left(\frac{\partial u}{\partial x}+\frac{\partial v}{\partial y}\right)+\frac{\mu}{\rho}\left(\frac{\partial^2 v}{\partial x^2}+\frac{\partial^2 v}{\partial y^2}\right) \quad (2.97)$$

On introduction of the incompressibility condition, these can be further simplified to

$$\frac{\partial u}{\partial t}+u\frac{\partial u}{\partial x}+v\frac{\partial u}{\partial y}=\frac{1}{\rho}F_x-\frac{1}{\rho}\frac{\partial p}{\partial x}+\frac{\mu}{\rho}\left(\frac{\partial^2 u}{\partial x^2}+\frac{\partial^2 u}{\partial y^2}\right)$$

$$\frac{\partial v}{\partial t}+u\frac{\partial v}{\partial x}+v\frac{\partial v}{\partial y}=\frac{1}{\rho}F_y-\frac{1}{\rho}\frac{\partial p}{\partial y}+\frac{\mu}{\rho}\left(\frac{\partial^2 v}{\partial x^2}+\frac{\partial^2 v}{\partial y^2}\right) \quad (2.98)$$

For steady state conditions the terms $\partial u/\partial t$ and $\partial v/\partial t$ can be dropped resulting in 'coupled' equations in the 'primitive' variables u, v and p. The equations are also, in contrast to those of solid elasticity, non-linear due to the presence of products like $u(\partial u/\partial x)$.

Ignoring body forces for the present, the steady state equations to be solved are

$$u\frac{\partial u}{\partial x}+v\frac{\partial u}{\partial y}+\frac{1}{\rho}\frac{\partial p}{\partial x}-\frac{\mu}{\rho}\left(\frac{\partial^2 u}{\partial x^2}+\frac{\partial^2 u}{\partial y^2}\right)=0$$

$$u\frac{\partial v}{\partial x}+v\frac{\partial v}{\partial y}+\frac{1}{\rho}\frac{\partial p}{\partial y}-\frac{\mu}{\rho}\left(\frac{\partial^2 v}{\partial x^2}+\frac{\partial^2 v}{\partial y^2}\right)=0 \quad (2.99)$$

Proceeding as before, and for the moment assuming the same shape functions are applied to all variables, $u = \mathbf{N}\mathbf{u}$, $v = \mathbf{N}\mathbf{v}$, $p = \mathbf{N}\mathbf{p}$ we have for a single element, treating the terms u and v as constants $\bar{u} = \mathbf{N}\mathbf{u}_0$ and $\bar{v} = \mathbf{N}\mathbf{v}_0$ for the purposes of integration:

$$\bar{u}\frac{\partial \mathbf{N}}{\partial x}\mathbf{u}+\bar{v}\frac{\partial \mathbf{N}}{\partial y}\mathbf{u}+\frac{1}{\rho}\frac{\partial \mathbf{N}}{\partial x}\mathbf{p}-\frac{\mu}{\rho}\frac{\partial^2 \mathbf{N}}{\partial x^2}\mathbf{u}-\frac{\mu}{\rho}\frac{\partial^2 \mathbf{N}}{\partial y^2}\mathbf{u}=\mathbf{0} \quad (2.100)$$

$$\bar{u}\frac{\partial \mathbf{N}}{\partial x}\mathbf{v}+\bar{v}\frac{\partial \mathbf{N}}{\partial y}\mathbf{v}+\frac{1}{\rho}\frac{\partial \mathbf{N}}{\partial y}\mathbf{p}-\frac{\mu}{\rho}\frac{\partial^2 \mathbf{N}}{\partial x^2}\mathbf{v}-\frac{\mu}{\rho}\frac{\partial^2 \mathbf{N}}{\partial y^2}\mathbf{v}=\mathbf{0}$$

Multiplying by the weighting functions and integrating as usual yields

$$\iint \mathbf{N}\bar{u}\frac{\partial \mathbf{N}}{\partial x}\mathbf{u}\,\mathrm{d}x\,\mathrm{d}y+\iint \mathbf{N}\bar{v}\frac{\partial \mathbf{N}}{\partial y}\mathbf{u}\,\mathrm{d}x\,\mathrm{d}y+\frac{1}{\rho}\iint \mathbf{N}\frac{\partial \mathbf{N}}{\partial x}\mathbf{p}\,\mathrm{d}x\,\mathrm{d}y$$

$$-\frac{\mu}{\rho}\iint \mathbf{N}\frac{\partial^2 \mathbf{N}}{\partial x^2}\mathbf{u}\,\mathrm{d}x\,\mathrm{d}y-\frac{\mu}{\rho}\iint \mathbf{N}\frac{\partial^2 \mathbf{N}}{\partial y^2}\mathbf{u}\,\mathrm{d}x\,\mathrm{d}y=\mathbf{0}$$

and (2.101)

$$\iint \mathbf{N}\bar{u}\frac{\partial \mathbf{N}}{\partial x}\mathbf{v}\,\mathrm{d}x\,\mathrm{d}y + \iint \mathbf{N}\bar{v}\frac{\partial \mathbf{N}}{\partial y}\mathbf{v}\,\mathrm{d}x\,\mathrm{d}y + \frac{1}{\rho}\iint \mathbf{N}\frac{\partial \mathbf{N}}{\partial y}\mathbf{p}\,\mathrm{d}x\,\mathrm{d}y$$
$$-\frac{\mu}{\rho}\iint \mathbf{N}\frac{\partial^2 \mathbf{N}}{\partial x^2}\mathbf{v}\,\mathrm{d}x\,\mathrm{d}y - \frac{\mu}{\rho}\iint \mathbf{N}\frac{\partial^2 \mathbf{N}}{\partial y^2}\mathbf{v}\,\mathrm{d}x\,\mathrm{d}y = \mathbf{0}$$

Integrating products by parts where necessary and neglecting resulting contour integrals gives

$$\iint \mathbf{N}\bar{u}\frac{\partial \mathbf{N}}{\partial x}\mathrm{d}x\,\mathrm{d}y\,\mathbf{u} + \iint \mathbf{N}\bar{v}\frac{\partial \mathbf{N}}{\partial y}\mathrm{d}x\,\mathrm{d}y\,\mathbf{u} + \frac{1}{\rho}\iint \mathbf{N}\frac{\partial \mathbf{N}}{\partial x}\mathrm{d}x\,\mathrm{d}y\,\mathbf{p}$$
$$+\frac{\mu}{\rho}\iint \frac{\partial \mathbf{N}}{\partial x}\frac{\partial \mathbf{N}}{\partial x}\mathrm{d}x\,\mathrm{d}y\,\mathbf{u} + \frac{\mu}{\rho}\iint \frac{\partial \mathbf{N}}{\partial y}\frac{\partial \mathbf{N}}{\partial y}\mathrm{d}x\,\mathrm{d}y\,\mathbf{u} = \mathbf{0}$$
$$\iint \mathbf{N}\bar{u}\frac{\partial \mathbf{N}}{\partial x}\mathrm{d}x\,\mathrm{d}y\,\mathbf{v} + \iint \mathbf{N}\bar{v}\frac{\partial \mathbf{N}}{\partial y}\mathrm{d}x\,\mathrm{d}y\,\mathbf{v} + \frac{1}{\rho}\iint \mathbf{N}\frac{\partial \mathbf{N}}{\partial y}\mathrm{d}x\,\mathrm{d}y\,\mathbf{p} \qquad (2.102)$$
$$+\frac{\mu}{\rho}\iint \frac{\partial \mathbf{N}}{\partial x}\frac{\partial \mathbf{N}}{\partial x}\mathrm{d}x\,\mathrm{d}y\,\mathbf{v} + \frac{\mu}{\rho}\iint \frac{\partial \mathbf{N}}{\partial y}\frac{\partial \mathbf{N}}{\partial y}\mathrm{d}x\,\mathrm{d}y\,\mathbf{v} = \mathbf{0}$$

The set of equations is completed by the continuity condition

$$\iint \mathbf{N}\left(\frac{\partial \mathbf{N}}{\partial x}\mathbf{u} + \frac{\partial \mathbf{N}}{\partial y}\mathbf{v}\right)\mathrm{d}x\,\mathrm{d}y = \mathbf{0} \qquad (2.103)$$

Collecting terms in **u**, **p** and **v** respectively leads to an equilibrium equation (Taylor and Hughes, 1981):

$$\begin{bmatrix} \mathbf{C}_{11} & \mathbf{C}_{12} & \mathbf{C}_{13} \\ \mathbf{C}_{21} & \mathbf{C}_{22} & \mathbf{C}_{23} \\ \mathbf{C}_{31} & \mathbf{C}_{32} & \mathbf{C}_{33} \end{bmatrix} \begin{Bmatrix} \mathbf{u} \\ \mathbf{p} \\ \mathbf{v} \end{Bmatrix} = \begin{Bmatrix} \mathbf{0} \\ \mathbf{0} \\ \mathbf{0} \end{Bmatrix} \qquad (2.104a)$$

where

$$\mathbf{C}_{11} = \iint \left(\mathbf{N}\bar{u}\frac{\partial \mathbf{N}}{\partial x} + \mathbf{N}\bar{v}\frac{\partial \mathbf{N}}{\partial y} + \frac{\mu}{\rho}\frac{\partial \mathbf{N}}{\partial x}\frac{\partial \mathbf{N}}{\partial x} + \frac{\mu}{\rho}\frac{\partial \mathbf{N}}{\partial y}\frac{\partial \mathbf{N}}{\partial y}\right)\mathrm{d}x\,\mathrm{d}y$$
$$\mathbf{C}_{12} = \iint \frac{1}{\rho}\mathbf{N}\frac{\partial \mathbf{N}}{\partial x}\mathrm{d}x\,\mathrm{d}y$$
$$\mathbf{C}_{13} = \mathbf{0}$$
$$\mathbf{C}_{21} = \iint \mathbf{N}\frac{\partial \mathbf{N}}{\partial x}\mathrm{d}x\,\mathrm{d}y \qquad (2.104b)$$
$$\mathbf{C}_{22} = \mathbf{0}$$
$$\mathbf{C}_{23} = \iint \mathbf{N}\frac{\partial \mathbf{N}}{\partial y}\mathrm{d}x\,\mathrm{d}y$$

$$\mathbf{C}_{31} = \mathbf{0}$$

$$\mathbf{C}_{32} = \iint \frac{1}{\rho} \mathbf{N} \frac{\partial \mathbf{N}}{\partial y} \mathrm{d}x \, \mathrm{d}y$$

$$\mathbf{C}_{33} = \mathbf{C}_{11}$$

Referring to Table 2.1 we now have many terms of the type $N_i(\partial N_j/\partial x)$ which imply unsymmetrical structures for C_{ij}. Thus special solution algorithms will be necessary. Computational details are left until Chapter 9.

2.18 SIMPLIFIED FLOW EQUATIONS

In many practical instances it may not be necessary to solve the complete coupled system described in the previous section. The pressure p can be eliminated from (2.98) and if vorticity ω is defined as

$$\omega = \frac{\partial u}{\partial y} - \frac{\partial v}{\partial x} \tag{2.105}$$

this results in a single equation:

$$\frac{\partial \omega}{\partial t} + u\frac{\partial \omega}{\partial x} + v\frac{\partial \omega}{\partial y} = \frac{\mu}{\rho}\left(\frac{\partial^2 \omega}{\partial x^2} + \frac{\partial^2 \omega}{\partial y^2}\right) \tag{2.106}$$

Defining a stream function ψ such that

$$\begin{aligned} u &= \frac{\partial \psi}{\partial y} \\ v &= -\frac{\partial \psi}{\partial x} \end{aligned} \tag{2.107}$$

an alternative coupled system involving ψ and ω can be devised, given here for steady state conditions:

$$\begin{aligned} \frac{\partial^2 \psi}{\partial x^2} + \frac{\partial^2 \psi}{\partial y^2} &= \omega \\ \frac{\mu}{\rho}\left(\frac{\partial^2 \omega}{\partial x^2} + \frac{\partial^2 \omega}{\partial y^2}\right) &= \frac{\partial \psi}{\partial y}\frac{\partial \omega}{\partial x} - \frac{\partial \psi}{\partial x}\frac{\partial \omega}{\partial y} \end{aligned} \tag{2.108}$$

This clearly has the advantage that only two unknowns are involved rather than the previous three. However, the solution of (2.108) is still a relatively complicated process and flow problems are sometimes solved via equation (2.106) alone, assuming that u and v can be approximated by some independent means. In this form, equation (2.106) is an example of the 'diffusion-convection' equation, the second order space derivatives corresponding to a 'diffusion' process and the first order ones to a 'convection' process. The equation arises in various areas of engineering, for example sediment transport and pollutant disposal (Smith, 1976, 1979).

If there is no convection, the resulting equation is of the type

$$\frac{\partial \omega}{\partial t} = \frac{\mu}{\rho}\left(\frac{\partial^2 \omega}{\partial x^2} + \frac{\partial^2 \omega}{\partial y^2}\right) \tag{2.109}$$

which is the 'heat conduction' or 'diffusion' equation well known in many areas of engineering.

A final simplification is a reduction to steady state conditions, in which case

$$\frac{\partial^2 \omega}{\partial x^2} + \frac{\partial^2 \omega}{\partial y^2} = 0 \tag{2.110}$$

leaving the familiar 'Laplace' equation. In the following sections, finite element formulations of these simplified flow equations are described, in order of increasing complexity.

2.19 SIMPLIFIED FLUID FLOW: STEADY STATE

The form of Laplace's equation (2.110) which arises in geomechanics, for example concerning groundwater flow in an aquifer (Muskat, 1937), is

$$k_x \frac{\partial^2 \phi}{\partial x^2} + k_y \frac{\partial^2 \phi}{\partial y^2} = 0 \tag{2.111}$$

where ϕ is the fluid potential and k_x, k_y are permeabilities in the x and y directions. The finite element discretisation process reduces the differential equation to a set of equilibrium type simultaneous equations of the form

$$\mathbf{KP}\boldsymbol{\phi} = \mathbf{Q} \tag{2.112}$$

where **KP** is symmetrical and **Q** is a vector of net nodal inflows/outflows.

Reference to Table 2.1 shows that typical terms in the matrix **KP** are of the form

$$\iint \left(k_x \frac{\partial N_i}{\partial x}\frac{\partial N_j}{\partial x} + k_y \frac{\partial N_i}{\partial y}\frac{\partial N_j}{\partial y} \right) \mathrm{d}x\,\mathrm{d}y \tag{2.113}$$

With the usual finite element discretisation

$$\phi = \mathbf{N}\boldsymbol{\phi} \tag{2.114}$$

a convenient way of expressing the matrix **KP** in (2.112) is

$$\mathbf{KP} = \iint \mathbf{T}^{\mathrm{T}}\mathbf{K}\mathbf{T}\,\mathrm{d}x\,\mathrm{d}y \tag{2.115}$$

where the property matrix **K** is analogous to the stress strain matrix **D** in solid mechanics; thus

$$\mathbf{K} = \begin{bmatrix} k_x & 0 \\ 0 & k_y \end{bmatrix} \tag{2.116}$$

(assuming that the axes of the permeability tensor coincide with x and y). The **T** matrix is similar to the **B** matrix of solid mechanics and is given by

$$\mathbf{T}=\begin{bmatrix}\frac{\partial N_1}{\partial x} & \frac{\partial N_2}{\partial x} & \frac{\partial N_3}{\partial x} & \frac{\partial N_4}{\partial x}\\ \frac{\partial N_1}{\partial y} & \frac{\partial N_2}{\partial y} & \frac{\partial N_3}{\partial y} & \frac{\partial N_4}{\partial y}\end{bmatrix} \tag{2.117}$$

The similarity between (2.115) for a fluid and (2.63) for a solid enables the corresponding programs to look similar in spite of the governing differential equations being quite different. This unity of treatment is utilised in describing the programming techniques in Chapter 3.

Finally, it is worth noting that (2.115) can also be arrived at from energy considerations. The equivalent energy statement is that the integral

$$\iint\left[\tfrac{1}{2}k_x\left(\frac{\partial\phi}{\partial x}\right)^2+\tfrac{1}{2}k_y\left(\frac{\partial\phi}{\partial y}\right)^2\right]\mathrm{d}x\,\mathrm{d}y \tag{2.118}$$

shall be a minimum for all possible $\phi(x, y)$.

Example solutions to steady state problems described by (2.111) are given in Chapter 7.

2.20 SIMPLIFIED FLUID FLOW: TRANSIENT STATE

Transient conditions must be analysed in many physical situations, for example in the case of Terzaghi 'consolidation' in soil mechanics or heat conduction. The governing consolidation diffusion equation for excess pore pressure u_w takes the form

$$c_x\frac{\partial^2 u_w}{\partial x^2}+c_y\frac{\partial^2 u_w}{\partial y^2}=\frac{\partial u_w}{\partial t} \tag{2.119}$$

where c_x, c_y are the coefficients of consolidation. Discretisation of the left hand side of (2.119) clearly follows that of (2.111) while the time derivative will be associated with a matrix of the 'mass matrix' type without the multiple ρ. Hence the discretised system is

$$\mathbf{KPu_w}+\mathbf{PM}\frac{\mathrm{d}\mathbf{u_w}}{\mathrm{d}t}=\mathbf{0} \tag{2.120}$$

This set of first order, ordinary differential equations can be solved by many methods, the simplest of which discretise the time derivative by finite differences. The algorithms are described in Chapter 3 with example solutions in Chapter 8.

2.21 SIMPLIFIED FLUID FLOW WITH ADVECTION

If pollutants, sediments, tracers, etc., are transported by a laminar flow system they are at the same time translated or 'advected' by the flow and diffused within it. The governing differential equation for the two-dimensional case is (Smith *et al.*, 1973)

$$c_x \frac{\partial^2 \phi}{\partial x^2} + c_y \frac{\partial^2 \phi}{\partial y^2} - u \frac{\partial \phi}{\partial x} - v \frac{\partial \phi}{\partial y} = \frac{\partial \phi}{\partial t} \tag{2.121}$$

where u and v are the fluid velocity components in the x and y directions (compare equation 2.106).

The extra advection terms $-u(\partial \phi/\partial x)$ and $-v(\partial \phi/\partial y)$ compared with (2.119) lead, as shown in Table 2.1, to unsymmetric components of the 'stiffness' matrix of the type

$$\iint \left(-uN_i \frac{\partial N_j}{\partial x} - vN_i \frac{\partial N_j}{\partial y} \right) \mathrm{d}x \, \mathrm{d}y \tag{2.122}$$

which must be added to the symmetric, diffusion components given in (2.113). When this has been done, equilibrium equations like (2.112) or transient equations like (2.120) are regained.

Mathematically, equation (2.121) is a differential equation which is not self-adjoint (Berg, 1962), due to the presence of the first order spatial derivatives. From a finite element point of view, equations which are not self-adjoint will always lead to unsymmetrical stiffness matrices.

A second consequence of non-self-adjoint equations is that there is no energy formulation equivalent to (2.118). It is clearly a benefit of the Galerkin approach that it can be used for all types of equation and is not restricted to self-adjoint systems.

Equation (2.121) can be rendered self-adjoint by using the transformation

$$\phi = h \exp\left(\frac{ux}{2c_x}\right) \exp\left(\frac{vy}{2c_y}\right) \tag{2.123}$$

but this is not recommended unless u and v are small compared with c_x and c_y, as shown by Smith *et al.* (1973).

Equation (2.121) and the use of (2.123) are described in Chapter 8.

2.22 FURTHER COUPLED EQUATIONS: BIOT CONSOLIDATION

Thus far in this chapter, analyses of solids and fluids have been considered separately. However, Biot formulated the theory of coupled solid–fluid interaction which finds applications in soil mechanics (Smith and Hobbs, 1976). The soil skeleton is treated as a porous elastic solid and the laminar porefluid is coupled to the solid by the conditions of compressibility and of continuity. Thus Biot's governing equation is given by

$$\frac{K'}{\gamma_w} \left[k_x \frac{\partial^2 u_w}{\partial x^2} + k_y \frac{\partial^2 u_w}{\partial y^2} + k_z \frac{\partial^2 u_w}{\partial z^2} \right] = \frac{\partial u_w}{\partial t} - \frac{\partial p}{\partial t} \tag{2.124}$$

where K' is the soil bulk modulus and p is the mean total stress.

For two-dimensional equilibrium in the absence of body forces, the gradient of effective stress from (2.48) must be augmented by the gradients of the fluid pressure u_w as follows:

$$\frac{\partial \sigma'_x}{\partial x}+\frac{\partial \tau_{xy}}{\partial y}+\frac{\partial u_w}{\partial x}=0$$
$$\frac{\partial \tau_{xy}}{\partial x}+\frac{\partial \sigma'_y}{\partial y}+\frac{\partial u_w}{\partial y}=0 \qquad (2.125)$$

The constitutive laws are those previously defined for the solid and fluid respectively; hence in plane strain

$$\begin{Bmatrix}\sigma'_x\\ \sigma'_y\\ \tau_{xy}\end{Bmatrix}=\frac{E'(1-\nu')}{(1+\nu')(1-2\nu')}\begin{bmatrix}1 & \dfrac{\nu'}{1-\nu'} & 0\\ \dfrac{\nu'}{1-\nu'} & 1 & 0\\ 0 & 0 & \dfrac{1-2\nu'}{2(1-\nu')}\end{bmatrix}\begin{Bmatrix}\varepsilon_x\\ \varepsilon_y\\ \gamma_{xy}\end{Bmatrix} \qquad (2.126)$$

and

$$\begin{Bmatrix}q_x\\ q_y\end{Bmatrix}=\frac{1}{\gamma_w}\begin{bmatrix}k_x & 0\\ 0 & k_y\end{bmatrix}\begin{Bmatrix}\partial u_w/\partial x\\ \partial u_w/\partial y\end{Bmatrix} \qquad (2.127)$$

where q_x amd q_y are the volumetric flow rates per unit area into and out of the element and γ_w is the unit weight of water. The solid strain-displacement relations are still given by (2.50), and the final condition is that for full saturation and, in this case incompressibility, outflow from an element of soil equals the reduction in volume of the element. Hence

$$\frac{\partial q_x}{\partial x}+\frac{\partial q_y}{\partial y}=-\frac{\mathrm{d}}{\mathrm{d}t}\left(\frac{\partial u}{\partial x}+\frac{\partial v}{\partial y}\right) \qquad (2.128)$$

and from (2.127) the third differential equation is given by

$$\frac{k_x}{\gamma_w}\frac{\partial^2 u_w}{\partial x^2}+\frac{k_y}{\gamma_w}\frac{\partial^2 u_w}{\partial y^2}+\frac{\mathrm{d}}{\mathrm{d}t}\left(\frac{\partial u}{\partial x}+\frac{\partial v}{\partial y}\right)=0 \qquad (2.129)$$

As usual in a displacement method, $\boldsymbol{\sigma}$ and $\boldsymbol{\varepsilon}$ are eliminated in terms of u and v so that the final coupled variables are u, v and u_w. These are now discretised in the normal way:

$$u=\mathbf{Nu}$$
$$v=\mathbf{Nv} \qquad (2.130)$$
$$u_w=\mathbf{Nu}_w$$

In practice, it may be preferable to use a higher order of discretisation for u and

v compared with u_w but, for the present, the same shape functions are used to describe all three variables.

When discretisation and the Galerkin process are completed, (2.125) and (2.129) lead to the pair of equilibrium and continuity equations:

$$\begin{aligned} \mathbf{KMr} + \mathbf{Cu_w} &= \mathbf{F} \\ \mathbf{C}^{\mathrm{T}}\frac{\mathrm{d}\mathbf{r}}{\mathrm{d}t} - \mathbf{KPu_w} &= \mathbf{0} \end{aligned} \tag{2.131}$$

where, for a four-noded element,

$$\mathbf{r} = \begin{Bmatrix} u_1 \\ v_1 \\ u_2 \\ v_2 \\ u_3 \\ v_3 \\ u_4 \\ v_4 \end{Bmatrix} \quad \text{and} \quad \mathbf{u_w} = \begin{Bmatrix} u_{w1} \\ u_{w2} \\ u_{w3} \\ u_{w4} \end{Bmatrix} \tag{2.132}$$

KM and **KP** are the elastic and fluid 'stiffness' matrices and **C** is a rectangular coupling matrix consisting of terms of the form

$$\iint N_i \frac{\partial N_j}{\partial x} \mathrm{d}x\,\mathrm{d}y \tag{2.133}$$

F is the external loading vector. Equations (2.131) must be integrated in time by some method such as finite differences and this is described further in Chapter 3. Examples of such solutions in practice are given in Chapter 9.

2.23 CONCLUSIONS

When viewed from a finite element standpoint, all equilibrium problems, whether involving solids or fluids, take the same form, namely

$$\mathbf{KMr} = \mathbf{F}$$

or (2.134)

$$\mathbf{KP}\boldsymbol{\phi} = \mathbf{Q}$$

For simple uncoupled problems the solid **KM** and fluid **KP** matrices have similar symmetrical structures and computer programs to construct them will be similar. However, for coupled problems such as are described by the Navier–Stokes equations, **KM** (or **KP**) is unsymmetrical and appropriate alternative software will be necessary.

In the same way, eigenvalue, propagation and transient problems all involve the mass matrix **MM** (or a simple multiple of it, **PM**). Therefore, coding of these different types of solution can be expected to contain sections common to all three problems.

So far, single elements have been considered in the discretisation process, and only the simplest line and rectangular elements have been described. The next chapter is mainly devoted to a description of programming strategy, but before this, the finite element concept is extended to embrace meshes of interlinked elements and elements of general shape.

2.24 REFERENCES

Bathe, K. J., and Wilson, E. L. (1976) *Numerical Methods in Finite Element Analysis.* Prentice-Hall, Englewood Cliffs, New Jersey.

Berg, P. N. (1962) Calculus of variations. In *Handbook of Engineering Mechanics*, Chapter 16, ed. W. Flugge, McGraw-Hill, New York.

Connor, J. J., and Brebbia, C. A. (1976) *Finite Element Techniques for Fluid Flow.* Newnes–Butterworth, London.

Cook, R. D. (1981) *Concepts and Applications of Finite Element Analysis*, 2nd edition. Wiley, New York.

Crandall, S. H. (1956) *Engineering Analysis.* McGraw-Hill, New York.

Finlayson, B. A. (1972) *The Method of Weighted Residuals and Variational Principles.* Academic Press, New York.

Horne, M. R., and Merchant, W. (1965) *The Stability of Frames.* Pergamon Press, Oxford.

Jennings, A. (1977) *Matrix Computation for Engineers and Scientists.* John Wiley, London.

Leckie, F. A., and Lindberg, G. M. (1963) The effect of lumped parameters on beam frequencies. *The Aeronautical Quarterly*, **14**, 234.

Livesley, R. K. (1975) Matrix methods of structural analysis. Pergamon Press.

Lundquist, E. E., and Kroll, W. D. (1944) NACA Report ARR, No. 4824.

Muskat, M. (1937) *The Flow of Homogeneous Fluids through Porous Media.* McGraw-Hill, New York.

Rao, S. S. (1982) *The Finite Element Method in Engineering*, Pergamon Press, Oxford.

Schlichting, H. (1960) *Boundary Layer Theory.* McGraw-Hill, New York.

Smith, I. M. (1976) Integration in time of diffusion and diffusion-convection equations. In *Finite Elements in Water Resources* ed. W. G. Gray, G. F. Pinder and C. A. Brebbia, pp. 1.3–1.20. Pentech Press.

Smith, I. M. (1979) The diffusion-convection equation. In *A Survey of Numerical Methods for Partial Differential Equations*, ed. I. Gladwell and R. Wait, pp. 195–211. Oxford University Press.

Smith, I. M., Farraday, R. V., and O'Connor, B. A. (1973) Rayleigh–Ritz and Galerkin finite elements for diffusion-convection problems. *Water Resources Research*, **9**, No. 3, 593.

Smith, I. M., and Hobbs, R. (1976) Biot analysis of consolidation beneath embankments. *Géotechnique*, **26**, No. 1, 149.

Strang, G., and Fix, G. J. (1973) *An Analysis of the Finite Element Method.* Prentice-Hall, Englewood Cliffs, New Jersey.

Szabo, B. A., and Lee, G. C. (1969) Derivation of stiffness matrices for problems in plane elasticity by the Galerkin method. *Int. J. Num. Meth. Eng.*, **1**, 301.

Taig, I. C. (1961) Structural analysis by the matrix displacement method. English Electric Aviation Report No. SO17.

Taylor, C., and Hughes, T. G. (1981) *Finite Element Programming of the Navier–Stokes Equation.* Pineridge Press Ltd, Swansea.

Timoshenko, S. P., and Goodier, J. N. (1951) *Theory of Elasticity*. McGraw-Hill, New York.

Timoshenko, S. P., and Woinowsky-Krieger, S. (1959) *Theory of Plates and Shells*. McGraw-Hill, New York.

Zienkiewicz, O. C. (1977) *The Finite Element Method in Engineering Science*, 3rd edition. McGraw-Hill, London.

CHAPTER 3
Programming Finite Element Computations

3.0 INTRODUCTION

In Chapter 2, the finite element discretisation process was described, whereby partial differential equations can be replaced by matrix equations which take the form of linear and nonlinear algebraic equations, eigenvalue equations or ordinary differential equations in the time variable. The present chapter describes how programs can be constructed in order to formulate and solve these kinds of equations.

Before this, two additional features must be introduced. First, we have so far dealt only with the simplest shapes of elements, namely lines and rectangles. Obviously if differential equations are to be solved over regions of general shape, elements must be allowed to assume general shapes as well. This is accomplished by introducing general triangular and quadrilateral elements together with the concept of a coordinate system local to the element.

Second, we have so far considered only a single element, whereas useful solutions will normally be obtained by many elements, usually hundreds or thousands in practice, joined together at the nodes. In addition, various types of boundary conditions may be prescribed which constrain the solution in some way.

Local coordinate systems, element assembly and incorporation of boundary conditions are all explained in the sections that follow.

3.1 LOCAL COORDINATES FOR QUADRILATERAL ELEMENTS

Figure 3.1 shows two types of four-noded quadrilateral elements. The shape functions for the rectangle were shown to be given by equations (2.57), namely $N_1 = (1 - x/a)(1 - y/b)$ and so on. If it is attempted to construct similar shape functions in the 'global' coordinates (x, y) for the general quadrilateral, very complex algebraic expressions will result, which are at the very least tedious to check (Irons and Ahmad, 1980).

Instead it is better to work in a local coordinate system as shown in Figure 3.2, originally proposed by Taig (1961). The general point $P(\xi, \eta)$ within the

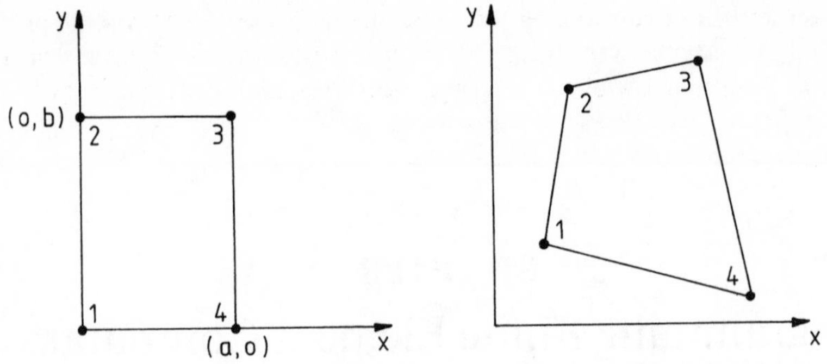

Figure 3.1 (a) Plane rectangular element. (b) Plane general quadrilateral element

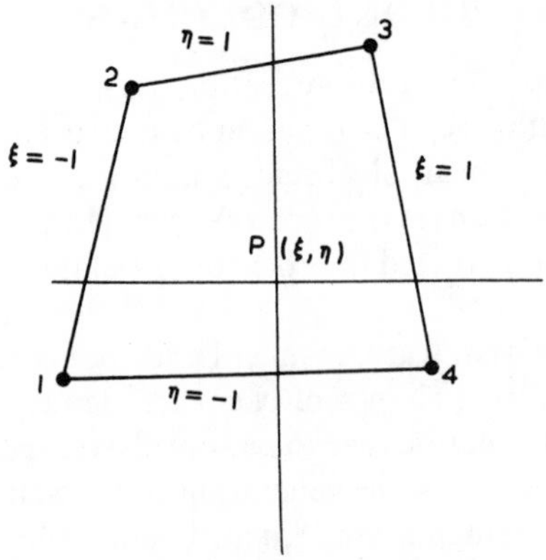

Figure 3.2 Local coordinate system for quadrilateral elements

quadrilateral is located at the intersection of two lines which cut opposite sides of the quadrilateral in equal proportions. For reasons associated with subsequent numerical integrations it proves to be convenient to 'normalise' the coordinates so that side 1 2 has $\xi = -1$, side 3 4 has $\xi = +1$, side 1 4 has $\eta = -1$ and side 2 3 has $\eta = +1$. In this system the intersection of the bisectors of opposite sides of the quadrilateral is the point (0, 0), while the corners 1, 2, 3, 4 are $(-1, -1)$, $(-1, 1)$, $(1, 1)$, $(1, -1)$ respectively.

When this choice is adopted, the shape functions for a four-noded quadrilateral with corner nodes take the simple form:

$$
\begin{aligned}
N_1 &= \tfrac{1}{4}(1-\xi)(1-\eta) \\
N_2 &= \tfrac{1}{4}(1-\xi)(1+\eta) \\
N_3 &= \tfrac{1}{4}(1+\xi)(1+\eta) \\
N_4 &= \tfrac{1}{4}(1+\xi)(1-\eta)
\end{aligned}
\tag{3.1}
$$

and these can be used to describe the variation of unknowns such as displacement or fluid potential in an element as before.

Under special circumstances the same shape functions can also be used to specify the relation between the global (x, y) and local (ξ, η) coordinate systems. If this is so the element is of a type called 'isoparametric' (Ergatoudis *et al.*, 1968; Zienkiewicz *et al.*, 1969); the four-node quadrilateral is an example. The coordinate transformation is therefore

$$\begin{aligned} x &= N_1 x_1 + N_2 x_2 + N_3 x_3 + N_4 x_4 \\ &= \mathbf{N}\mathbf{x} \\ y &= N_1 y_1 + N_2 y_2 + N_3 y_3 + N_4 y_4 \\ &= \mathbf{N}\mathbf{y} \end{aligned} \tag{3.2}$$

where the $\mathbf{N}$ are given by (3.1).

In the previous chapter it was shown that element properties involve not only $\mathbf{N}$ but also their derivatives with respect to the global coordinates (x, y) which appear in matrices such as $\mathbf{B}$ and $\mathbf{T}$. Further, products of these quantities need to be integrated over the element area or volume.

Derivatives are easily converted from one coordinate system to the other by means of the chain rule of partial differentiation, best expressed in matrix form by

$$\begin{Bmatrix} \dfrac{\partial}{\partial \xi} \\ \dfrac{\partial}{\partial \eta} \end{Bmatrix} = \begin{bmatrix} \dfrac{\partial x}{\partial \xi} & \dfrac{\partial y}{\partial \xi} \\ \dfrac{\partial x}{\partial \eta} & \dfrac{\partial y}{\partial \eta} \end{bmatrix} \begin{Bmatrix} \dfrac{\partial}{\partial x} \\ \dfrac{\partial}{\partial y} \end{Bmatrix} = \mathbf{J} \begin{Bmatrix} \dfrac{\partial}{\partial x} \\ \dfrac{\partial}{\partial y} \end{Bmatrix} \tag{3.3}$$

or

$$\begin{Bmatrix} \dfrac{\partial}{\partial x} \\ \dfrac{\partial}{\partial y} \end{Bmatrix} = \mathbf{J}^{-1} \begin{Bmatrix} \dfrac{\partial}{\partial \xi} \\ \dfrac{\partial}{\partial \eta} \end{Bmatrix} \tag{3.4}$$

where $\mathbf{J}$ is the Jacobian matrix. The determinant of this matrix, $\det|\mathbf{J}|$, must also be evaluated because it is used in the transformed integrals as follows:

$$\iint \mathrm{d}x\,\mathrm{d}y = \int_{-1}^{1}\int_{-1}^{1} \det|\mathbf{J}|\,\mathrm{d}\xi\,\mathrm{d}\eta \tag{3.5}$$

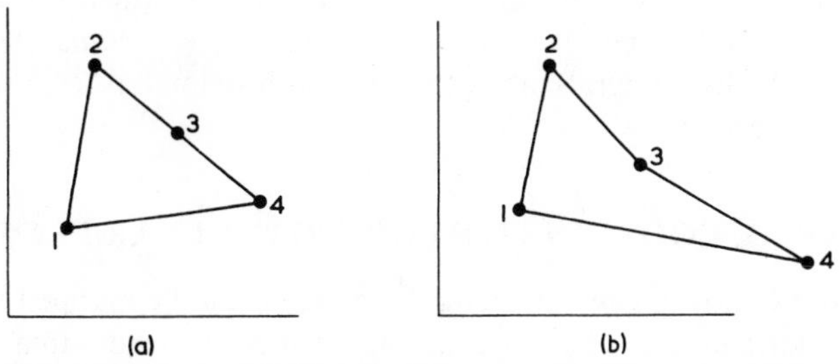

Figure 3.3 (a) Degenerate quadrilateral. (b) Unacceptable quadrilateral

Table 3.1 Coordinates and weights in Gaussian quadrature formulae

n	$\xi_i \eta_j$	$w_i w_j$
1	0	2
2	$\pm\frac{1}{\sqrt{3}}$	1
3	$\pm\frac{\sqrt{3}}{\sqrt{5}}$	$\frac{5}{9}$
	0	$\frac{8}{9}$

Under certain circumstances, for example that shown in Figure 3.3(b), the Jacobian becomes indeterminate. When using quadrilateral elements, reflex interior angles should be avoided.

3.2 NUMERICAL INTEGRATION FOR QUADRILATERALS

Although some integrals of this type could be evaluated analytically, it is impractical for complicated functions, particularly in the general case when (ξ, η) become curvilinear. In practice (3.5) are evaluated numerically, using Gaussian quadrature over quadrilateral regions (Irons, 1966a, 1966b). The quadrature rules in two dimensions are all of the form

$$\int_{-1}^{1}\int_{-1}^{1} f(\xi,\eta)\,\mathrm{d}\xi\,\mathrm{d}\eta \simeq \sum_{i=1}^{n}\sum_{j=1}^{n} w_i w_j f(\xi_i,\eta_j) \tag{3.6}$$

where w_i and w_j are weighting coefficients and ξ_i, η_j are coordinate positions within the element. These values for n equal to 1, 2 and 3 are shown in Table 3.1 and complete tables are available in other sources, e.g. Kopal (1961). The table assumes that the range of integration is ± 1; hence the reason for normalising the local coordinate system in this way.

The approximate equality in (3.6) is exact for cubic functions when $n = 2$ and for quintics when $n = 3$. Usually one attempts to perform integrations over finite elements exactly, but in special circumstances (Zienkiewicz *et al.*, 1971) 'reduced' integration whereby integrals are evaluated approximately can improve the quality of solutions.

3.3 LOCAL COORDINATES FOR TRIANGULAR ELEMENTS

Figure 3.4 shows how a general triangular element can be mapped into a right-angled isosceles triangle. This approach is identical to using area coordinates (Zienkiewicz *et al.*, 1971) in which any point within the triangle can be referenced using three local coordinates (L_1, L_2, L_3). Clearly for a plane region, only two

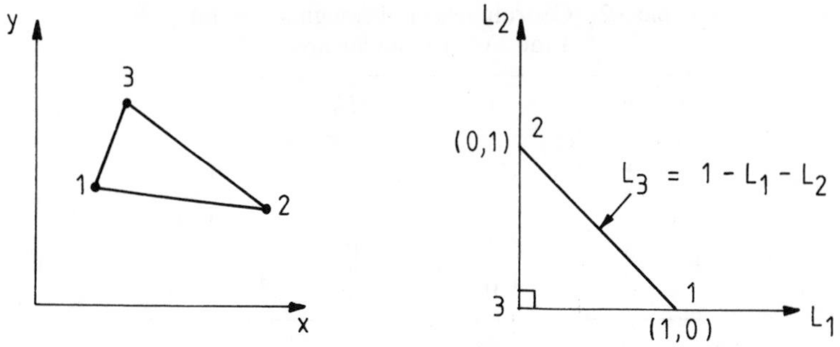

Figure 3.4 (a) General triangular element. (b) Element mapped in local coordinates

independent coordinates are necessary; hence the third coordinate is a function of the other two:

$$L_3 = 1 - L_1 - L_2 \tag{3.7}$$

However, it often leads to more elegant formulations algebraically if all three coordinates are retained. For example, the shape functions for a three-noded triangular element ('constant strain' triangle) take the form

$$\begin{aligned} N_1 &= L_1 \\ N_2 &= L_2 \\ N_3 &= L_3 \end{aligned} \tag{3.8}$$

and the isoparametric property is that

$$\begin{aligned} x &= N_1 x_1 + N_2 x_2 + N_3 x_3 \\ y &= N_1 y_1 + N_2 y_2 + N_3 y_3 \end{aligned} \tag{3.9}$$

Equations (3.3) and (3.4) from the previous paragraph still apply regarding the Jacobian matrix but equation (3.5) must be modified for triangles to give

$$\iint \mathrm{d}x\,\mathrm{d}y = \int_0^1 \int_0^{1-L_1} \det|\mathbf{J}|\,\mathrm{d}L_2\,\mathrm{d}L_1 \tag{3.10}$$

3.4 NUMERICAL INTEGRATION FOR TRIANGLES

Numerical integration over triangular regions is similar to that for quadrilaterals except that the sampling points do not occur in such 'convenient' positions.

The quadrature rules take the general form

$$\int_0^1 \int_0^{1-L_1} f(L_1, L_2)\,\mathrm{d}L_2\,\mathrm{d}L_1 \simeq \tfrac{1}{2} \sum_{i=1}^{n} w_i f(L_1^i, L_2^i) \tag{3.11}$$

where w_i is the weighting coefficient corresponding to the sampling point (L_1^i, L_2^i) and n represents the number of sampling points. Typical values of the weights and sampling points are given in Table 3.2.

Table 3.2 Coordinates and weights for integration over triangular areas

n	L_1^i	L_2^i	w_i
1	$\frac{1}{3}$	$\frac{1}{3}$	1
3	$\frac{1}{2}$	$\frac{1}{2}$	$\frac{1}{3}$
	$\frac{1}{2}$	0	$\frac{1}{3}$
	0	$\frac{1}{2}$	$\frac{1}{3}$

As with quadrilaterals, numerical integration can be exact for certain polynomials. For example, in Table 3.2, the one-point rule is exact for integration of first order polynomials and the three-point rule is exact for polynomials of second order. Reduced integration can again be beneficial in some instances.

Computer formulations involving local coordinates, transformation of coordinates and numerical integration are described in subsequent paragraphs.

3.5 ELEMENT ASSEMBLY

Properties of elements in isolation have been shown to be given by matrix equations, for example the equilibrium equation

$$\mathbf{KP}\boldsymbol{\phi} = \mathbf{Q} \tag{3.12}$$

describing steady laminar fluid flow. In Figure 3.5 is shown a small mesh containing three quadrilateral elements, all of which have properties defined by (3.12). The next problem is to assemble the elements and so derive the properties of the three-element system. Each element possesses node numbers, shown in parentheses, which follow the scheme in Figures 3.1 to 3.3, namely numbering clockwise from the lower left hand corner. Since there is only one unknown at

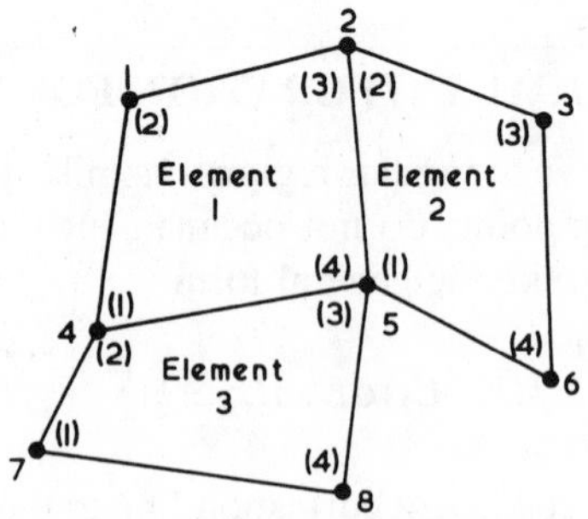

Figure 3.5 Mesh of quadrilateral elements

every node, the fluid potential, each individual element equation can be written

$$\begin{bmatrix} KP_{1,1} & KP_{1,2} & KP_{1,3} & KP_{1,4} \\ KP_{2,1} & KP_{2,2} & KP_{2,3} & KP_{2,4} \\ KP_{3,1} & KP_{3,2} & KP_{3,3} & KP_{3,4} \\ KP_{4,1} & KP_{4,2} & KP_{4,3} & KP_{4,4} \end{bmatrix} \begin{Bmatrix} \phi_1 \\ \phi_2 \\ \phi_3 \\ \phi_4 \end{Bmatrix} = \begin{Bmatrix} Q_1 \\ Q_2 \\ Q_3 \\ Q_4 \end{Bmatrix} \tag{3.13}$$

However, in the mesh numbering system, not in parentheses, mesh node 4 corresponds to element node (1) in element 1 and to element node (2) in element 3. The total number of equations for the mesh is 8 and, within this system, term $KP_{1,1}$ from element 1 and term $KP_{2,2}$ from element 3 would be added together and would appear in location 4,4 and so on. The total system matrix for Figure 3.5 is given in Table 3.3, where the superscripts refer to element numbers.

This total system matrix is symmetrical provided its constituent matrices are symmetrical. The matrix also possesses the useful property of 'bandedness' which means that the terms are concentrated around the 'leading diagonal' which stretches from the upper left to the lower right of the table. In fact no term in any row can be more than four locations removed from the leading diagonal so the system is said to have a 'semi-bandwidth' IW of 4. This can be obtained by inspection from Figure 3.5 by subtracting the lowest from the highest freedom number in each element. Complicated meshes have variable bandwidths and useful computer programs make use of banding when storing the system matrices.

The importance of efficient mesh numbering is illustrated for a mesh of line elements in Figure 3.6 where the scheme in parentheses has $IW = 13$ compared with the other scheme with $IW = 2$.

If system symmetry exists it should also be taken into account. For example, if the system in Table 3.3 is symmetrical, there are only 30 unique components (the leading diagonal terms plus a maximum of four terms to the left or right of the diagonal in each row). Often, with a slight decrease of efficiency, the symmetrical half of a band matrix is stored as a rectangular array with a size equal to the

Table 3.3 System stiffness matrix for mesh in Figure 3.5. Superscripts indicate element numbers

$KP^1_{2,2}$	$KP^1_{2,3}$	0	$KP^1_{2,1}$	$KP^1_{2,4}$	0	0	0
$KP^1_{3,2}$	$KP^1_{3,3}+KP^2_{2,2}$	$KP^2_{2,3}$	$KP^1_{3,1}$	$KP^1_{3,4}+KP^2_{2,1}$	$KP^2_{2,4}$	0	0
0	$KP^2_{3,2}$	$KP^3_{3,3}$	0	$KP^2_{3,1}$	$KP^2_{3,4}$	0	0
$KP^1_{1,2}$	$KP^1_{1,3}$	0	$KP^1_{1,1}+KP^3_{2,2}$	$KP^1_{1,4}+KP^3_{2,3}$	0	$KP^3_{2,1}$	$KP^3_{2,4}$
$KP^1_{4,2}$	$KP^1_{4,3}+KP^2_{1,2}$	$KP^2_{1,3}$	$KP^1_{4,1}+KP^3_{3,2}$	$KP^1_{4,4}+KP^2_{1,1}$ $+KP^3_{3,3}$	$KP^2_{1,4}$	$KP^3_{3,1}$	$KP^3_{3,4}$
0	$KP^2_{4,2}$	$KP^2_{4,3}$	0	$KP^2_{4,1}$	$KP^2_{4,4}$	0	0
0	0	0	$KP^3_{1,2}$	$KP^3_{1,3}$	0	$KP^3_{1,1}$	$KP^3_{1,4}$
0	0	0	$KP^3_{4,2}$	$KP^3_{4,3}$	0	$KP^3_{4,1}$	$KP^3_{4,4}$

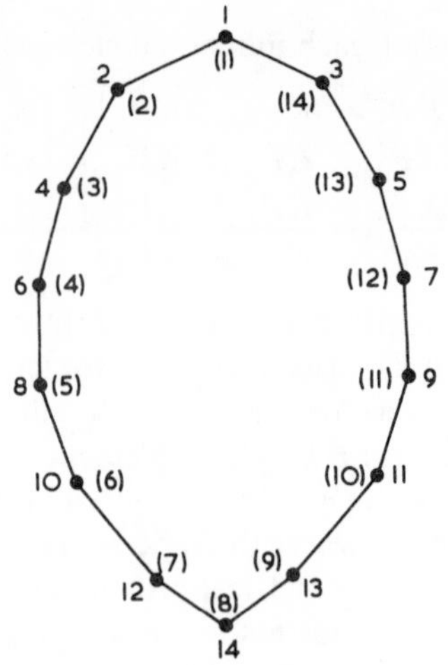

Figure 3.6 Alternative mesh numbering schemes

number of system equations times the semi-bandwidth plus 1. In this case zeros are filled into the extra locations in the first few or last few rows depending on whether the lower or upper 'half' of the matrix is stored. Special storage schemes, for example 'skyline' techniques (Bathe and Wilson, 1976), are also considered later in the chapter (see Figure 3.17) and are important in cases where bandwidths vary greatly.

Later in this chapter, procedures are described whereby system matrices like that in Table 3.3 can be automatically assembled in band form, with or without the 'skyline' option, from the constituent element matrices.

3.6 INCORPORATION OF BOUNDARY CONDITIONS

Eigenvalues of freely floating elements or meshes are sometimes required but normally in eigenvalue problems and always in equilibrium and propagation problems additional boundary information has to be supplied before solutions can be obtained. For example, the system matrix defined in Table 3.3 is singular and this set of equations has no solution.

The simplest type of boundary condition occurs when the dependent variable in the solution is known to be zero at various points in the region (and hence nodes in the finite element mesh). When this occurs, the equation components associated with these nodes are not required in the solution and information is given to the assembly routine which prevents these components from ever being assembled into the final system. Thus only the non-zero nodal values are solved for.

A variation of this condition occurs when the dependent variable has known, but non-zero, values at various locations. Although an elimination procedure could be devised, the way this condition is handled in practice is by adding a 'large' number, say 10^{12}, to the leading diagonal of the 'stiffness' matrix in the row in which the prescribed value is required. The term in the same row of the right hand side vector is then set to the prescribed value multiplied by the augmented 'stiffness' coefficient. For example, suppose the value of the fluid head at node 5 in Figure 3.5 is known to be 57.0 units. The unconstrained set of equations (Table 3.3) would be assembled and term (5,5) augmented by adding 10^{12}. In the subsequent solution there would be an equation

$$(K_{5,5} + 10^{12})\phi_5 + \text{small terms} = 57.0 \times (K_{5,5} + 10^{12}) \tag{3.14}$$

which would have the effect of making ϕ_5 equal to 57.0. Clearly this procedure is only successful if indeed 'small terms' are small relative to 10^{12}.

Boundary conditions can also involve gradients of the unknown in the forms

$$\frac{\partial \phi}{\partial n} = 0 \tag{3.15}$$

$$\frac{\partial \phi}{\partial n} = C_1 \phi \tag{3.16}$$

$$\frac{\partial \phi}{\partial n} = C_2 \tag{3.17}$$

where n is the normal to the boundary and C_1, C_2 are constants.

To be specific, consider a solution of the diffusion-advection equation (2.121) subject to boundary conditions (3.15), (3.16) and (3.17) respectively. When the second order terms $c_x(\partial^2\phi/\partial x^2)$ and $c_y(\partial^2\phi/\partial y^2)$ are integrated by parts, boundary integrals of the type

$$\int_s c_n \mathbf{N}^{\mathrm{T}} \frac{\partial \phi}{\partial n} l_n \, \mathrm{d}s \tag{3.18}$$

arise, where s is a length of boundary and l_n the direction cosine of the normal. Clearly the case $\partial\phi/\partial n = 0$ presents no difficulty since the contour integral (3.18) vanishes.

However, (3.16) gives rise to an extra integral, which for the boundary element shown in Figure 3.7 is

$$\int_j^k c_y \mathbf{N}^{\mathrm{T}} C_1 \phi l_y \, \mathrm{d}s \tag{3.19}$$

When ϕ is expanded as $\mathbf{N}\boldsymbol{\phi}$ we get an additional matrix

$$\frac{-C_1 c_y (x_k - x_j)}{6} \begin{bmatrix} 0 & 0 & 0 & 0 \\ 0 & 2 & 1 & 0 \\ 0 & 1 & 2 & 0 \\ 0 & 0 & 0 & 0 \end{bmatrix} \tag{3.20}$$

which must be added to the left hand side of the element equations.

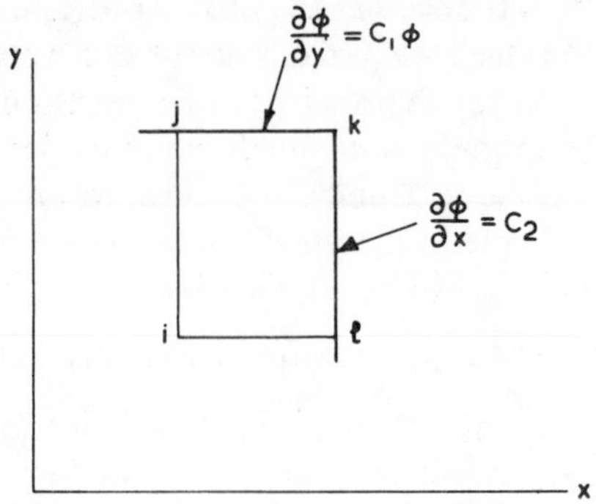

Figure 3.7 Boundary conditions involving non-zero gradients of the unknown

For boundary condition (3.17) the additional term is

$$\int_k^l c_x \mathbf{N}^{\mathrm{T}} C_2 l_x \, \mathrm{d}s \tag{3.21}$$

which is just a vector

$$\frac{C_2 c_x (y_k - y_1)}{2} \begin{Bmatrix} 0 \\ 0 \\ 1 \\ 1 \end{Bmatrix} \tag{3.22}$$

which would be added to the right hand side of the element equations. For a further discussion of boundary conditions see Smith (1979).

In summary, boundary conditions of the type $\phi = 0$ or $\partial\phi/\partial n = 0$ are the most common and are easily handled in finite element analyses. The cases $\phi =$ constant or $\partial\phi/\partial n =$ constant $\times\ \phi$ are somewhat more complicated but can be appropriately treated. Examples of the use of these types of boundary specification are included in the examples chapters.

3.7 PROGRAMMING USING BUILDING BLOCKS

The programs in subsequent chapters are constituted from 100 or so building blocks in the form of FORTRAN subroutines which perform the tasks of computing and integrating the element matrices, assembling these into system matrices and carrying out the appropriate equilibrium, eigenvalue or propagation calculations.

It is anticipated that most users will elect to pre-compile all of the building blocks and to hold these permanently in a library on backing store (usually disc). The library should then be automatically accessible to the calling programs by means of a simple attach in the job description.

A summary of the building blocks is listed in Appendix 4 where their actions and input/output parameters are described. A separation has been made into ‘black box’ routines (concerned with some matrix operations), whose mode of action the reader need not necessarily know in detail, and special purpose routines which are the basis of finite element computations. These special purpose routines are listed in Appendix 5. The black box routines should be thought of as an addition to the permanent library functions such as SIN or ABS, and could well be replaced from a mathematical subroutine library such as NAG.

3.8 BLACK BOX ROUTINES

The first group of subroutines is concerned with standard matrix handling, and the action of these should be self-explanatory. The routines, usually assuming two-dimensional matrices, are

MATCOP (copies)	MATRAN (transposes)	MATSUB (subtracts)
MATMUL (multiplies)	NULL (nulls)	TWOBY2 (inverts 2 × 2)
MVMULT (multiplies matrix by vector)	MATADD (adds)	TREEX3 (inverts 3 × 3)
MSMULT (multiplies matrix by scalar)		MATINV (inverts N × N)
		NULL3 (3-d array)

Similar routines associated with vectors are

VECCOP
VVMULT (cross-product)
VECADD
NULVEC

The second batch of subroutines is concerned with the solution of linear algebraic equations (required in equilibrium and propagation problems). The subroutines have been split into reduction and forward/backward re-substitution phases.

	BANRED BACK1 BACK2	CHOLIN CHOBK1 CHOBK2	GAUSBA SOLVBA
Method	Gauss	Choleski	Gauss
Equation coefficients	Symmetrical	Symmetrical	Unsymmetrical

COMRED COMBK1 COMBK2	SPARIN SPARB1 SPARB2
Gauss	Choleski
Symmetrical complex	Symmetrical skyline

In the majority of programs, the forward and backward substitution is enshrined in the single subroutines BACSUB, CHOBAC, COMBAC and SPABAC respectively.

Several subroutines are associated with eigenvalue and eigenvector determination, for example

Tridiagonalisation	TRIDIA	BANDRD
Finds eigenvalues	EVECTS	BISECT
Method	Q-R factorisation	Jacobi
Equation coefficients	Symmetrical full N × N stored	Symmetrical, banded, upper triangle stored

The first of each pair tridiagonalises the matrix and the second extracts all of the eigenvalues. The eigenvectors are found only by EVECTS. It should be noted that these routines, although very robust and accurate, can be inefficient both in storage requirements and in run-time and should not be used for solving very large problems, in which in any case it is unlikely that the full range of eigenmodes would be required. The various vector iteration methods (Bathe and Wilson, 1976) should be resorted to in those cases.

One of the most effective of these is the Lanczos method, and routines

LANCZ1
LANCZ2

are used to calculate the Lanczos vectors and eigenvectors of a matrix.

Although simple matrix-by-vector multiplications can be accomplished by MVMULT, advantage is usually taken of banding of the matrix coefficients whenever possible. To allow for this, four special matrix-by-vector multiplication routines are provided:

	LINMUL	BANMUL	BANTML	LINMLS
Matrix coefficients	Symmetrical	Symmetrical	Unsymmetrical	Symmetrical
Storage of matrix	Vector	Lower triangle	Full band	Skyline

Further information on when these routines should be used is given in Section 3.10.

During eigenvalue computation a couple of special matrix operations are catered for by

LBKBAN
LBBT

whose action is described in Appendix 4.

In a teaching text such as this, elaborate input and output procedures are avoided. It is expected that users may pre-process their input data using independent programs and plot the output graphically. At present these operations are very machine-dependent. Therefore simple free format input to 'TAPE 5' is adopted in all programs. The results of calculations are usually vectors or arrays and two subroutines enable the output of these in a standard E format to 'TAPE 6'. They are called

PRINTV
PRINTA

respectively.

In order to describe the action of the remaining special purpose subroutines, which are listed in full in Appendix 5, it is necessary first to consider the properties of individual finite elements and then the representation of continua from assemblages of these elements. Static linear problems (including eigenproblems) are considered first. Thereafter modifications to programs to incorporate time dependence are added.

3.9 SPECIAL PURPOSE ROUTINES

The job of these routines is to compute the element matrix coefficients, for example the 'stiffness', to integrate these over the element area or volume and finally to assemble the element submatrices into the global system matrix or matrices. The black box routines for equation solution, eigenvalue determination and so on then take over to produce the final results.

3.9.1 Element matrix calculation

In the remainder of this chapter the notation adopted is that used in the subroutine listings in Appendix 5. Wherever possible, mnemonics are used so that local coordinate ξ becomes XI in the subroutines and so on.

3.9.1.1 *Plane strain (stress) analysis of elastic solids using quadrilateral elements*

As an example of element matrix calculation, consider the computation of the element stiffness matrix for plane elasticity given by (2.63):

$$\mathbf{KM} = \iint \mathbf{B}^{\mathrm{T}}\mathbf{D}\mathbf{B}\,\mathrm{d}x\,\mathrm{d}y \tag{3.23}$$

In program terminology this becomes

$$\mathrm{KM} = \iint \mathrm{BEE}^{\mathrm{T}} * \mathrm{DEE} * \mathrm{BEE}\,\mathrm{d}x\,\mathrm{d}y \tag{3.24}$$

and its formation is described by the inner loop of the structure chart in Figure 3.8.

It is assumed for the moment that the element nodal coordinates (x, y) have been calculated and stored in the array COORD. For example, we would have for a four-node quadrilateral

$$\mathrm{COORD} = \begin{bmatrix} x_1 & y_1 \\ x_2 & y_2 \\ x_3 & y_3 \\ x_4 & y_4 \end{bmatrix} \tag{3.25}$$

The shape functions **N** are held in array FUN, in terms of local coordinates, as

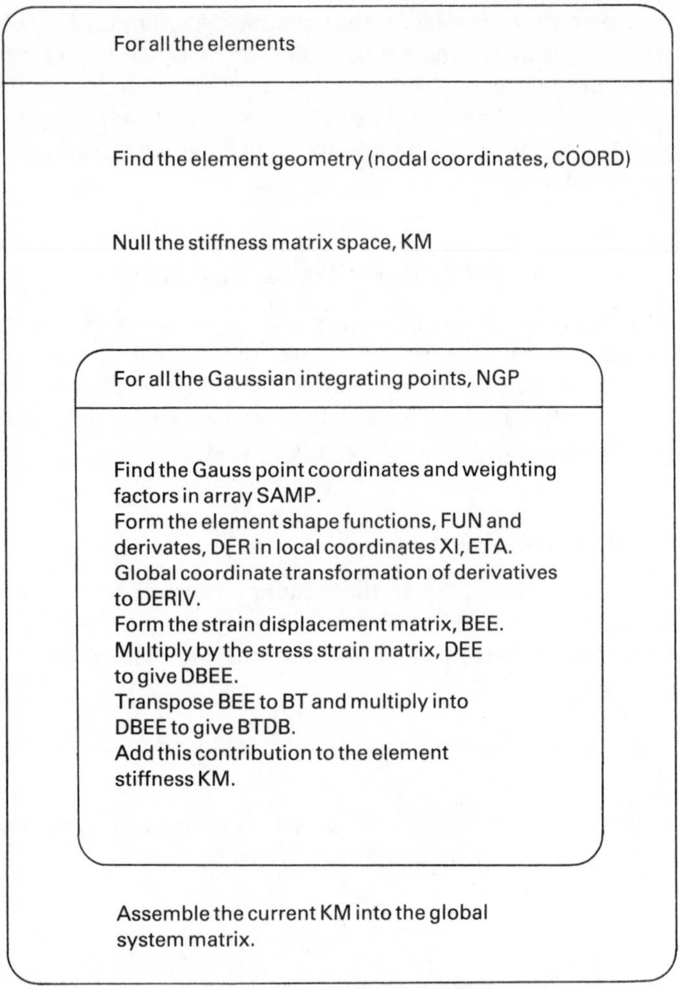

For all the elements
Find the element geometry (nodal coordinates, COORD)
Null the stiffness matrix space, KM
For all the Gaussian integrating points, NGP
Find the Gauss point coordinates and weighting factors in array SAMP.
Form the element shape functions, FUN and derivates, DER in local coordinates XI, ETA.
Global coordinate transformation of derivatives to DERIV.
Form the strain displacement matrix, BEE.
Multiply by the stress strain matrix, DEE to give DBEE.
Transpose BEE to BT and multiply into DBEE to give BTDB.
Add this contribution to the element stiffness KM.
Assemble the current KM into the global system matrix.

Figure 3.8 Structure chart of element matrix assembly

specified in (3.1) by

$$\text{FUN} = \begin{Bmatrix} \frac{1}{4}(1-\text{XI})(1-\text{ETA}) \\ \frac{1}{4}(1-\text{XI})(1+\text{ETA}) \\ \frac{1}{4}(1+\text{XI})(1+\text{ETA}) \\ \frac{1}{4}(1+\text{XI})(1-\text{ETA}) \end{Bmatrix}^{\text{T}} \tag{3.26}$$

The **B** matrix contains derivatives of the shape functions and these are easily computed in the local coordinate system as

$$\text{DER} = \begin{bmatrix} \dfrac{\partial \text{FUN}^{\text{T}}}{\partial \xi} \\ \dfrac{\partial \text{FUN}^{\text{T}}}{\partial \eta} \end{bmatrix} \quad \text{or}$$

$$\text{DER} = \frac{1}{4}\begin{bmatrix} -(1-\text{ETA}) & -(1+\text{ETA}) & (1+\text{ETA}) & (1-\text{ETA}) \\ -(1-\text{XI}) & (1-\text{XI}) & (1+\text{XI}) & -(1+\text{XI}) \end{bmatrix} \tag{3.27}$$

The information in (3.26) and (3.27) for a four-node quadrilateral is formed by the subroutine

FORMLN

for the specific Gaussian integration points $(\text{XI}, \text{ETA})_{\text{I,J}}$ held in the array SAMP where I and J run from 1 to NGP, the number of Gauss points specified in each direction. Figure 3.9(a) shows the typical layout and ordering for two-point Gaussian integration. In all cases SAMP is formed by the subroutine

GAUSS

where NGP can take any value from 1 to 7.

The derivatives DER must then be converted into their counterparts in the (x, y) coordinate system, DERIV, by means of the Jacobian matrix transformation (3.3) or (3.4). From the isoparametric property,

$$\begin{Bmatrix} x \\ y \end{Bmatrix} = \text{COORD}^{\text{T}} * \text{FUN}^{\text{T}} \tag{3.28}$$

and since the Jacobian matrix is given by

$$\begin{bmatrix} \dfrac{\partial x}{\partial \xi} & \dfrac{\partial y}{\partial \xi} \\[2ex] \dfrac{\partial x}{\partial \eta} & \dfrac{\partial y}{\partial \eta} \end{bmatrix} \tag{3.29}$$

it is clear that (3.29) can be obtained by differentiating (3.28) with respect to the local coordinates. In this way

$$\text{JAC} = \text{DER} * \text{COORD} \tag{3.30}$$

In order to compute DERIV we must invert JAC to give JAC1 using TWOBY2

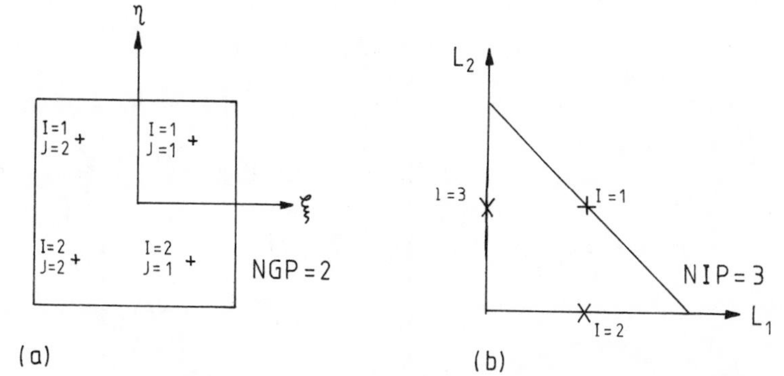

Figure 3.9 Integration schemes for (a) quadrilaterals and (b) triangles

in this two-dimensional case and finally carry out the multiplication

$$\text{DERIV} = \text{JAC1} * \text{DER} \tag{3.31}$$

Thus the sequence of operations

```
CALL FORMLN (DER, IDER, FUN, SAMP, ISAMP, I, J)
CALL MATMUL
  (DER, IDER, COORD, ICOORD, JAC, IJAC, IT, NOD, IT)
CALL TWOBY2 (JAC, IJAC, JAC1, IJAC1, DET)                    (3.32)
CALL MATMUL
  (JAC1, IJAC1, DER, IDER, DERIV, IDERIV, IT, IT, NOD)
```

where NOD is the number of element nodes (four in this case) and IT the number of coordinate dimensions (two in this case), will be found in programs for plane elasticity using four-noded quadrilateral elements. After these operations have been performed, the derivatives of the element shape functions with respect to (x, y) are held in DERIV while DET is the determinant of the Jacobian matrix, required later for the purposes of numerical integration. Following the convention adopted in Chapter 1, IDER is the working size of array DER and so on.

The matrix BEE in (3.24) can now be assembled as it consists of components of DERIV. This assembly is performed by the subroutine

FORMB

for plane problems. Thus the strain-displacement relations are

$$\text{EPS} = \text{BEE} * \text{ELD} \tag{3.33}$$

where in the case of a four-node quadrilateral

$$\text{ELD} = \{u_1\ v_1\ u_2\ v_2\ u_3\ v_3\ u_4\ v_4\}^{\mathrm{T}} \tag{3.34}$$

The variables u and v are simply the nodal displacements in the x and y directions respectively assuming the nodal ordering of Figure 3.5.

The components of the integral of $\text{BEE}^{\mathrm{T}} * \text{DEE} * \text{BEE}$, evaluated at the Gauss points given by all combinations of I and J, can now be computed by transposing BEE to give BT, by forming the stress–strain matrix using the subroutine

FMDEPS (for plane strain)

where

$$\text{DEE} = \frac{\text{E}(1-\text{V})}{(1+\text{V})(1-2\text{V})}\begin{bmatrix} 1 & \dfrac{\text{V}}{1-\text{V}} & 0 \\ \dfrac{\text{V}}{1-\text{V}} & 1 & 0 \\ 0 & 0 & \dfrac{1-2\text{V}}{2(1-\text{V})} \end{bmatrix} \tag{3.35}$$

and by carrying out the multiplications

$$BTDB = BT * DEE * BEE \tag{3.36}$$

A plane stress analysis would be obtained by simply replacing FMDEPS by FMDSIG.

The integral is evaluated numerically by

$$KM = DET * \sum_{I=1}^{NGP} \sum_{J=1}^{NGP} W_I * W_J * BTDB_{I,J} \tag{3.37}$$

where W_I and W_J are the Gaussian weighting coefficients held in array SAMP.

As soon as the element matrix is formed from (3.37) it is assembled into the global system matrix (or matrices) by special subroutines described later in this chapter. Since the assembly process is common to all problems, modifications to the element matrix calculation for different situations will first be described.

3.9.1.2 *Plane strain (stress) analysis of elastic solids using triangular elements*

The previous section showed how the stiffness matrix of a typical four-node quadrilateral could be built up. In order to use triangular elements, very few alterations are required. For example, for a six-node triangular element, the values of the shape functions and their derivatives with respect to local coordinates at a particular location (L_1, L_2, L_3) are formed by the subroutine

FMTRI6

This delivers the shape functions

$$FUN = \begin{Bmatrix} (2L_1-1)L_1 \\ 4L_1L_2 \\ (2L_2-1)L_2 \\ 4L_2L_3 \\ (2L_3-1)L_3 \\ 4L_3L_1 \end{Bmatrix}^T \tag{3.38}$$

and their derivatives with respect to L_1 and L_2

$$DER = \begin{bmatrix} \dfrac{\partial FUN^T}{\partial L_1} \\ \dfrac{\partial FUN^T}{\partial L_2} \end{bmatrix}$$

$$= \begin{bmatrix} (4L_1-1) & 4L_2 & 0 & -4L_2 & -(4L_3-1) & 4(L_3-L_1) \\ 0 & 4L_1 & (4L_2-1) & 4(L_3-L_2) & -(4L_3-1) & -4L_1 \end{bmatrix} \tag{3.39}$$

The nodal numbering and the order in which the integration points are sampled for a typical three-point scheme are shown in Figure 3.9(b).

For integration over triangles, the sampling points in local coordinates (L_1, L_2) are held in the array SAMP and the corresponding weighting coefficients in the vector WT. Both of these items are provided by the subroutine

NUMINT

The version of this subroutine described in Appendix 5 allows the total number of integrating points (NIP) to take the value 1, 3, 4, 6, 7, 12 or 16. The coding should be referred to in order to determine the sequence in which the integrating points are sampled for $NIP \geqslant 3$.

The sequence of operations

```
CALL FMTRI6 (DER, IDER, FUN, SAMP, ISAMP, I)
CALL MATMUL
  (DER, IDER, COORD, ICOORD, JAC, IJAC, IT, NOD, IT)        (3.40)
CALL TWOBY2 (JAC, IJAC, JAC1, IJAC1, DET)
CALL MATMUL
  (JAC1, IJAC1, DER, IDER, DERIV, IDERIV, IT, IT, NOD)
```

(where NOD now equals 6) places the required derivatives with respect to (x, y) in DERIV and finds the Jacobian determinant DET.

Finally, numerical integration is performed by

$$KM = 0.5 * DET * \sum_{I=1}^{NIP} W_I * BTDB_I \tag{3.41}$$

the factor of 0.5 being required because the weights add up to unity whereas the area of the triangle in local coordinates only equals one half.

A higher order triangular element with fifteen nodes is also considered in Chapter 5. The shape function and derivatives for this element are provided by routine FMTR15; otherwise the sequence of operations is virtually identical to those described above for the six-node element.

3.9.1.3 *Axisymmetric strain of elastic solids*

The strain-displacement relations can again be written by (3.33) but in this case BEE must be formed by the subroutine

FMBRAD

where the cylindrical coordinates (r, z) replace their plane strain counterparts (x, y). The stress–strain matrix is still given by an expression similar to (3.35) but the 4×4 DEE matrix is formed by

FMDRAD

In this case the integrated element stiffness is (2.66), namely

$$KM = \iint BT * DEE * BEE * r \, dr \, dz \tag{3.42}$$

Considering a four-node element, the isoparametric property gives

$$r = \text{FUN}\begin{Bmatrix} r_1 \\ r_2 \\ r_3 \\ r_4 \end{Bmatrix}$$

$$= \sum_{\text{K}=1}^{\text{NOD}} \text{FUN(K)} * \text{COORD(K,1)}$$

$$= \text{SUM} \tag{3.43}$$

Hence we have

$$\text{KM} = \text{SUM} * \text{DET} * \sum_{\text{I}=1}^{\text{NGP}} \sum_{\text{J}=1}^{\text{NGP}} \text{W}_\text{I} * \text{W}_\text{J} * \text{BTDB}_{\text{I,J}} \tag{3.44}$$

By comparison with (3.37) it may be seen that when evaluated numerically the algorithms for axisymmetric and plane stiffness formation will be essentially the same, despite the fact that they are algebraically quite different. This is very significant from the points of view of programming effort and of program flexibility.

However (3.44) now involves numerical evaluation of integrals involving $1/r$ which do not have simple polynomial representations. Therefore, unlike planar problems, it will be impossible to evaluate (3.44) exactly by numerical means, especially as r (i.e. SUM) approaches zero. Provided integration points do not lie on the $r = 0$ axis, however, reasonable results are usually achieved using a similar order of quadrature to planar analysis.

3.9.1.4 *Plane steady laminar fluid flow*

It was shown in (2.115) that a fluid element has a 'stiffness' defined by

$$\mathbf{KP} = \iint \mathbf{T}^\text{T} \mathbf{KT} \, \mathrm{d}x \, \mathrm{d}y \tag{3.45}$$

which becomes in program terminology

$$\text{KP} = \iint \text{DERIV}^\text{T} * \text{KAY} * \text{DERIV} \, \mathrm{d}x \, \mathrm{d}y \tag{3.46}$$

and the similarity to (3.24) is obvious. The matrix DERIV simply contains the derivatives of the element shape functions with respect to (x, y) which were previously needed in the analysis of solids and are formed by the sequence (3.32) while KAY contains the permeability properties of the element in the form

$$\text{KAY} = \begin{bmatrix} \text{PX} & 0 \\ 0 & \text{PY} \end{bmatrix} \tag{3.47}$$

In the computations,

$$\text{DTKD} = \text{DERIV}^\text{T} * \text{KAY} * \text{DERIV} \tag{3.48}$$

is formed by appropriate matrix multiplications and the final matrix summation for a quadrilateral element is

$$\mathrm{KP} = \mathrm{DET} * \sum_{\mathrm{I}=1}^{\mathrm{NGP}} \sum_{\mathrm{J}=1}^{\mathrm{NGP}} \mathrm{W_I} * \mathrm{W_J} * \mathrm{DTKD_{I,J}} \tag{3.49}$$

By comparison with (3.37) it will be seen that these physically very different problems are likely to require similar solution algorithms.

3.9.1.5 *Mass matrix formation*

The mass or inertia matrix was shown in Chapter 2, e.g. (2.65), to take the general form

$$\mathbf{MM} = \rho \iint \mathbf{N}^{\mathrm{T}} \mathbf{N} \, \mathrm{d}x \, \mathrm{d}y \tag{3.50}$$

where **N** are just the shape functions. In the case of plane fluid flow, since there is only one degree of freedom per node, the 'mass' matrix is particularly simple in program terminology, namely

$$\mathrm{MM} = \mathrm{RHO} * \iint \mathrm{FUN}^{\mathrm{T}} * \mathrm{FUN} \, \mathrm{d}x \, \mathrm{d}y \tag{3.51}$$

By defining the product of the shape functions

$$\mathrm{FTF} = \mathrm{FUN}^{\mathrm{T}} * \mathrm{FUN} \tag{3.52}$$

we have

$$\mathrm{MM} = \mathrm{RHO} * \mathrm{DET} * \sum_{\mathrm{I}=1}^{\mathrm{NGP}} \sum_{\mathrm{J}=1}^{\mathrm{NGP}} \mathrm{W_I} * \mathrm{W_J} * \mathrm{FTF_{I,J}} \tag{3.53}$$

where RHO is the mass density.

In the case of plane stress or strain of solids, because of the arrangement of the displacement vector in (3.34) it is convenient to use a special subroutine

ECMAT

to form the terms of the mass matrix as ECM before integration. Thereafter

$$\mathrm{MM} = \mathrm{RHO} * \iint \mathrm{ECM} \, \mathrm{d}x \, \mathrm{d}y$$

$$= \mathrm{RHO} * \mathrm{DET} * \sum_{\mathrm{I}=1}^{\mathrm{NGP}} \sum_{\mathrm{J}=1}^{\mathrm{NGP}} \mathrm{W_I} * \mathrm{W_J} * \mathrm{ECM_{I,J}} \tag{3.54}$$

When 'lumped' mass approximations are used **MM** becomes a diagonal matrix. For a four-noded quadrilateral (NOD = 4), for example,

$$\mathrm{MM} = (\mathrm{RHO} * \mathrm{AREA}/\mathrm{NOD}) * \mathrm{I} \tag{3.55}$$

where AREA is the element area and I the unit matrix. For higher order elements, however, all nodes may not receive equal weighting.

3.9.1.6 *Higher order quadrilateral elements*

To emphasise the ease with which element types can be interchanged in programs, consider the next member of the isoparametric quadrilateral group, namely the 'quadratic' quadrilateral with midside nodes shown in Figure 3.10. The same local coordinate system is retained and the coordinate matrix becomes

$$\text{COORD} = \begin{bmatrix} x_1 & y_1 \\ x_2 & y_2 \\ x_3 & y_3 \\ x_4 & y_4 \\ x_5 & y_5 \\ x_6 & y_6 \\ x_7 & y_7 \\ x_8 & y_8 \end{bmatrix} \tag{3.56}$$

The shape functions are now

$$\text{FUN} = \begin{Bmatrix} \frac{1}{4}(1-\text{XI})(1-\text{ETA})(-\text{XI}-\text{ETA}-1) \\ \frac{1}{2}(1-\text{XI})(1-\text{ETA}^2) \\ \frac{1}{4}(1-\text{XI})(1+\text{ETA})(-\text{XI}+\text{ETA}-1) \\ \frac{1}{2}(1-\text{XI}^2)(1+\text{ETA}) \\ \frac{1}{4}(1+\text{XI})(1+\text{ETA})(\text{XI}+\text{ETA}-1) \\ \frac{1}{2}(1+\text{XI})(1-\text{ETA}^2) \\ \frac{1}{4}(1+\text{XI})(1-\text{ETA})(\text{XI}-\text{ETA}-1) \\ \frac{1}{2}(1-\text{XI}^2)(1-\text{ETA}) \end{Bmatrix} \tag{3.57}$$

which, together with their derivatives with respect to local coordinates, DER, are

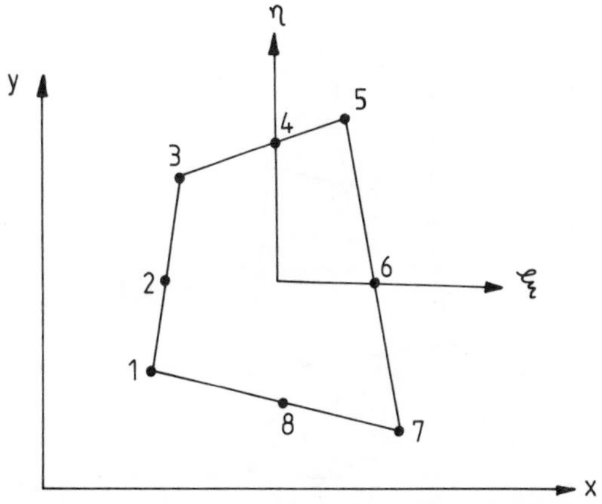

Figure 3.10 General quadratic quadrilateral element

formed by the subroutine

FMQUAD

The sequence of operations:

```
CALL FMQUAD (DER, IDER, FUN, SAMP, ISAMP, I, J)
CALL MATMUL
  (DER, IDER, COORD, ICOORD, JAC, IJAC, IT, NOD, IT)
CALL TWOBY2 (JAC, IJAC, JAC1, IJAC1, DET)                  (3.58)
CALL MATMUL
  (JAC1, IJAC1, DER, IDER, DERIV, IDERIV, IT, IT, NOD)
```

(where NOD now equals 8) places the required derivatives with respect to (x, y) in DERIV and finds the Jacobian determinant DET.

Another plane element used in the programs later in this book is the Lagrangian nine-node element. This element uses 'complete' polynomial interpolation in each direction, but requires a ninth node at its centre (Figure 3.11). The shape functions for this element are given by

$$\text{FUN} = \begin{Bmatrix} \frac{1}{4}(\text{XI})(\text{XI}-1)(\text{ETA})(\text{ETA}-1) \\ -\frac{1}{2}(\text{XI})(\text{XI}-1)(\text{ETA}+1)(\text{ETA}-1) \\ \frac{1}{4}(\text{XI})(\text{XI}-1)(\text{ETA})(\text{ETA}+1) \\ -\frac{1}{2}(\text{XI}+1)(\text{XI}-1)(\text{ETA})(\text{ETA}+1) \\ \frac{1}{4}(\text{XI})(\text{XI}+1)(\text{ETA})(\text{ETA}+1) \\ -\frac{1}{2}(\text{XI})(\text{XI}+1)(\text{ETA}+1)(\text{ETA}-1) \\ \frac{1}{4}(\text{XI})(\text{XI}+1)(\text{ETA})(\text{ETA}-1) \\ -\frac{1}{2}(\text{XI}+1)(\text{XI}-1)(\text{ETA})(\text{ETA}-1) \\ (\text{XI}+1)(\text{XI}-1)(\text{ETA}+1)(\text{ETA}-1) \end{Bmatrix}^{\mathrm{T}} \tag{3.59}$$

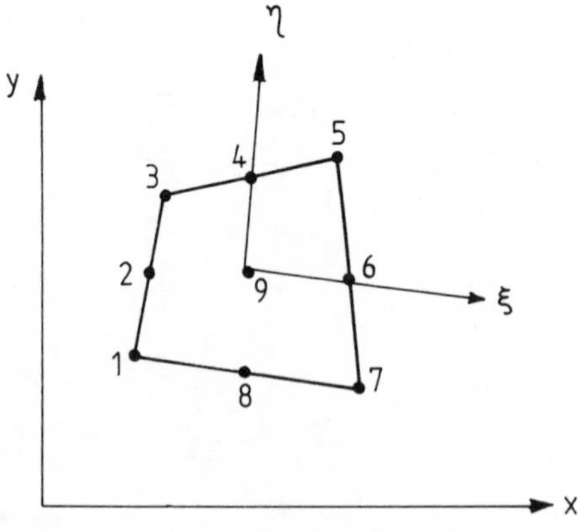

Figure 3.11 The Lagrangian nine-node element

and are formed together with their derivatives with respect to local coordinates by the subroutine

FMLAG9

A comparison of (3.58), (3.40) and (3.32) indicates that programs using different element types will be almost identical, although operating on different sizes of arrays. Indeed, the reader could readily devise a completely general two-dimensional program in which element type selection was accomplished by data (i.e. FORMLN, FMQUAD, FMLAG9, FMTRI6, etc., could be merged into a single routine).

3.9.1.7 *Three-dimensional cubic elements*

As was the case with changes of plane element types, changes of element dimensions are readily made. For example, the eight-node brick element in Figure 3.12 is the three-dimensional extension of the four-noded quadrilateral. Using the local coordinate system (ξ, η, ζ), the coordinate matrix is

$$\text{COORD} = \begin{bmatrix} x_1 & y_1 & z_1 \\ x_2 & y_2 & z_2 \\ x_3 & y_3 & z_3 \\ x_4 & y_4 & z_4 \\ x_5 & y_5 & z_5 \\ x_6 & y_6 & z_6 \\ x_7 & y_7 & z_7 \\ x_8 & y_8 & z_8 \end{bmatrix} \tag{3.60a}$$

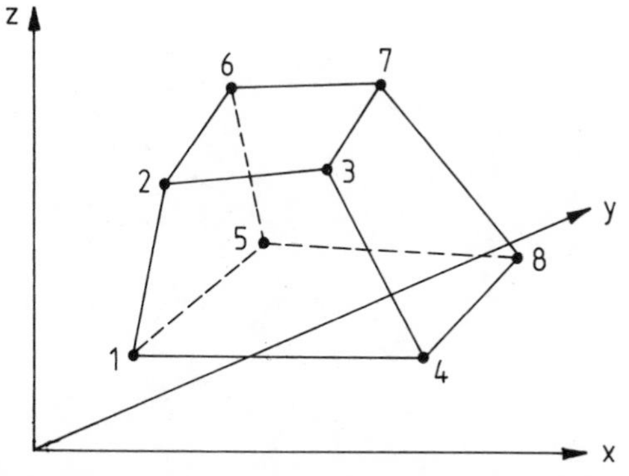

Figure 3.12 General linear brick element

The shape functions are

$$\text{FUN} = \begin{Bmatrix} \frac{1}{8}(1-\text{XI})(1-\text{ETA})(1-\text{ZETA}) \\ \frac{1}{8}(1-\text{XI})(1-\text{ETA})(1+\text{ZETA}) \\ \frac{1}{8}(1+\text{XI})(1-\text{ETA})(1+\text{ZETA}) \\ \frac{1}{8}(1+\text{XI})(1-\text{ETA})(1-\text{ZETA}) \\ \frac{1}{8}(1-\text{XI})(1+\text{ETA})(1-\text{ZETA}) \\ \frac{1}{8}(1-\text{XI})(1+\text{ETA})(1+\text{ZETA}) \\ \frac{1}{8}(1+\text{XI})(1+\text{ETA})(1+\text{ZETA}) \\ \frac{1}{8}(1+\text{XI})(1+\text{ETA})(1-\text{ZETA}) \end{Bmatrix}^{\text{T}} \tag{3.60b}$$

which together with their derivatives with respect to local coordinates are formed by the subroutine

FMLIN3

The sequence of operations:

```
CALL FMLIN3 (DER, IDER, FUN, SAMP, ISAMP, I, J, K)
CALL MATMUL
   (DER, IDER, COORD, ICOORD, JAC, IJAC, IT, NOD, IT)
CALL TREEX3 (JAC, IJAC, JAC1, IJAC1, DET)                    (3.61)
CALL MATMUL
   (JAC1, IJAC1, DER, IDER, DERIV, IDERIV, IT, IT, NOD)
```

(where NOD now equals 8 and IT now equals 3) results in DERIV, the required gradients with respect to (x, y, z) and the Jacobian determinant DET.

A higher order brick element with 20 nodes (Figure 3.13) is also used in

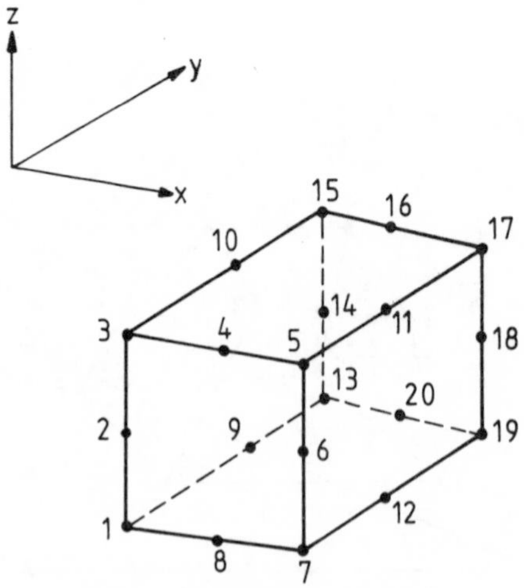

Figure 3.13 A 20-node brick element

programs later in the book. The shape functions for this element are provided by the subroutine

FMQUA3

and are not quoted here, but see the coding in Appendix 5.

For the three-dimensional elastic solid the element stiffness is given by

$$KM = \iiint BT * DEE * BEE \, dx \, dy \, dz \tag{3.62}$$

where BEE and DEE must now be formed by the subroutines

FORMB3
FORMD3

respectively. The final summation becomes

$$KM = DET * \sum_{I=1}^{NGP} \sum_{J=1}^{NGP} \sum_{K=1}^{NGP} W_I * W_J * W_K * BTDB_{I,J,K} \tag{3.63}$$

where W_I, W_J, W_K are as usual Gaussian multipliers obtained from SAMP.

For three-dimensional steady laminar fluid flow the element 'stiffness' is

$$KP = \iiint DERIV^T * KAY * DERIV \, dx \, dy \, dz$$

$$= \iiint DTKD \, dx \, dy \, dz \tag{3.64}$$

where KAY is the principal axes permeability tensor

$$KAY = \begin{bmatrix} PX & 0 & 0 \\ 0 & PY & 0 \\ 0 & 0 & PZ \end{bmatrix} \tag{3.65}$$

Gaussian integration gives

$$KP = DET * \sum_{I=1}^{NGP} \sum_{J=1}^{NGP} \sum_{K=1}^{NGP} W_I * W_J * W_K * DTKD_{I,J,K} \tag{3.66}$$

which is similar to (3.63).

The mass matrix for potential flow is

$$MM = RHO * \iiint FUN^T * FUN \, dx \, dy \, dz$$

$$= RHO * \iiint FTF \, dx \, dy \, dz \tag{3.67}$$

which is replaced by quadrature as

$$MM = RHO * DET * \sum_{I=1}^{NGP} \sum_{J=1}^{NGP} \sum_{K=1}^{NGP} W_I * W_J * W_K * FTF_{I,J,K} \tag{3.68}$$

In solid mechanics, a three-dimensional equivalent of ECMAT would be required. This development is left to the reader.

3.9.1.8 *Three-dimensional tetrahedron elements*

An alternative element for three-dimensional analysis is the tetrahedron, the simplest of which has four corner nodes. The local coordinate system makes use of volume coordinates as shown in Figure 3.14. For example, point P can be identified by four volume coordinates (L_1, L_2, L_3, L_4) where

$$\begin{aligned} L_1 &= \frac{P432}{V_T} \\ L_2 &= \frac{P413}{V_T} \\ L_3 &= \frac{P421}{V_T} \\ L_4 &= \frac{P123}{V_T} \end{aligned} \tag{3.69}$$

As is to be expected, one of these coordinates is redundant due to the identity

$$L_1 + L_2 + L_3 + L_4 = 1 \tag{3.70}$$

The shape functions for the constant strain tetrahedron are

$$\text{FUN} = \begin{Bmatrix} L_1 \\ L_2 \\ L_3 \\ L_4 \end{Bmatrix}^T \tag{3.71}$$

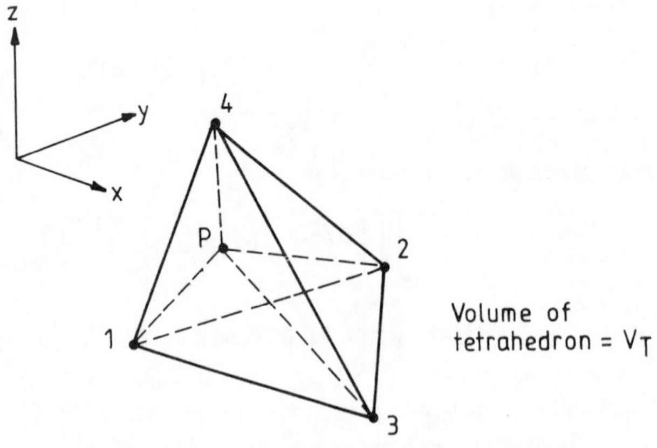

Figure 3.14 A four-node tetrahedron element

and these, together with their derivatives with respect to L_1, L_2 and L_3 are formed by the subroutine

FMTET4

The sequence of operations:

```
CALL FMTET4 (DER, IDER, FUN, SAMP, ISAMP, I)
CALL MATMUL
  (DER, IDER, COORD, ICOORD, JAC, IJAC, IT, NOD, IT)
CALL TREEX3 (JAC, IJAC, JAC1, IJAC1, DET)
CALL MATMUL
  (JAC1, IJAC1, DER, IDER, DERIV, IDERIV, IT, IT, NOD)          (3.72)
```

(where NOD now equals 4) results in DERIV which is needed to form the element matrices and DET which is used in the numerical integrations.

The integration is performed by

$$KM = \tfrac{1}{6} * DET * \sum_{I=1}^{NIP} W_I * BTDB_I \tag{3.73}$$

where W_I is the weighting coefficient corresponding to the particular integrating point. The factor $\frac{1}{6}$ is required because the weighting coefficients provided by the subroutine

NUMIN3

always add up to unity whereas the volume of the tetrahedron in local coordinates equals one-sixth.

The addition of mid-side nodes results in the ten-noded tetrahedron which represents the next member of family. This element could easily be implemented by replacing FMTET4 by an appropriate subroutine.

Transient, coupled poro-elastic transient and elastic–plastic analyses all involve manipulations of the few simple element property matrices described above. Before describing such applications, methods of assembling elements and of solving linear equilibrium and eigenvalue problems must first be described.

3.9.2 Assembly of elements

The nine special purpose subroutines

READNF	FORMKV	FORMKU	FMBIGK
FKDIAG	FORMKB	FORMTB	FORMKC
FSPARV			

are concerned with assembling the individual element matrices to form the system matrix that approximates the desired continuum. Allied to these there must be a specification of the geometrical details, in particular the nodal coordinates of each element and the element's place in some overall node numbering scheme.

Large finite element programs contain mesh generation code which is usually of some considerable complexity. Indeed, in much finite element work, the most expensive and time consuming task is the preparation of the input data for the mesh generation routine. In the present book, this aspect of the computations is essentially ignored and attention is restricted to simple classes of geometry which

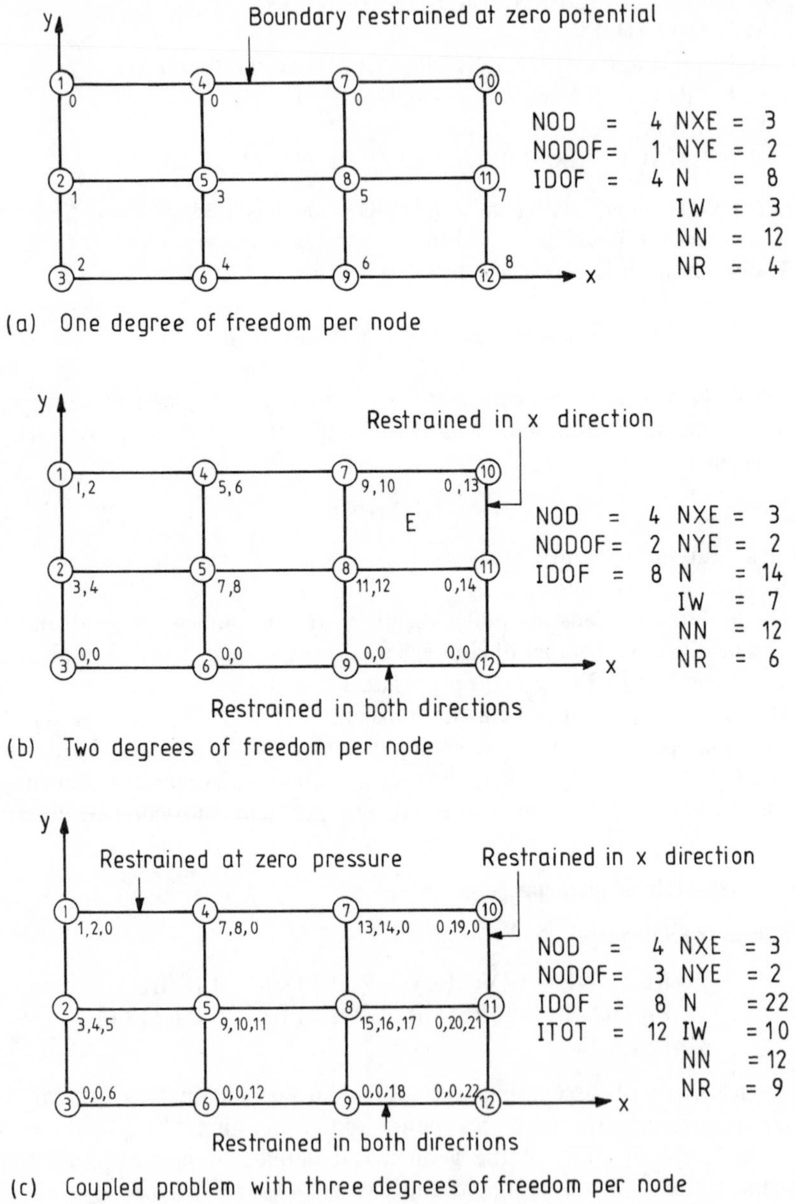

Figure 3.15 Numbering system and data for regular meshes

can be automatically built up by small routines. Examples are the simple meshes made up of four-noded rectangles shown in Figure 3.15.

In the writers' experience, users often want to solve specific classes of problem for which it is the most effective practice to write specialised mesh generation routines. For example, a user interested in the analysis of turbine blades has different specification requirements from one who is interested in analysis of earth dams. In the present work, the majority of programs use plane rectangular elements, so subroutines such as

GEOM4Y (four-node numbering in the y direction)

are provided to generate coordinates and freedom numbering.

A full list of 'geometry' routines is given in Appendix 3.

With reference to Figure 3.15, the nodes of the mesh are first assigned numbers as economically as possible (i.e. always numbering in the 'shorter' direction to minimise the bandwidth). Associated with each node are degrees of freedom (displacements, fluid potentials and so on) which are numbered in the same order as the nodes. However, account is taken at this stage of whether a degree of freedom exists or whether, generally at the boundaries of the region, the freedom is restrained, in which case the freedom number is assigned the value zero.

The variables in Figure 3.15 have the following meaning:

NXE	elements counting in x direction
NYE	elements counting in y direction
N	total number of (non-zero) freedoms in problem
IW	semi-bandwidth
NN	total number of nodes in problem
NR	number of restrained nodes
NOD	nodes per element
NODOF	freedoms per node
IDOF	freedoms per element
ITOT	total freedoms per element (for coupled problems)

In potential problems described by these types of element there is one degree of freedom possible per node, the potential ϕ. In plane or axisymmetric strain there are two, namely the u and v components of displacement specified in that order. In coupled solid–fluid problems the order is u, v, u_w (where u_w = excess pressure) and in three-dimensional displacement problems u, v, w. For Navier–Stokes applications the order u, p, v (where p = pressure) is adopted.

This information about the degrees of freedom present in specific problems is stored in an integer array NF called the 'node freedom array', formed by the building block subroutine

READNF

The node freedom array NF has NN rows, one for each node in the problem analysed, and NODOF columns, one for each degree of freedom per node. Formation of NF is achieved by specifying as data to the subroutine the number of any node whose freedom is restrained in some way, followed by the digit 0 if the node is restrained in that sense and by the digit 1 if it is not. This is the reverse of the convention employed in the first edition of this book. For example, to create NF for the problem shown in Figure 3.15(b) the data specified and the resulting NF are listed in Table 3.4.

Building block subroutines of the GEOM4Y type can then be constructed using NF as input. For example, the rectangular meshes need to be specified by the number of elements in the $x(r)$ and $y(z)$ directions respectively (NXE, NYE), together with their sizes (AA, BB). The subroutine has to work out the nodal coordinates COORD of each element together with a vector G called a 'steering vector', which contains the numbers of the degrees of freedom associated with that particular element in accordance with the nodal order 1234 in Figure 3.1.

For example, element E in Figure 3.15(b) has the steering vector

$$\mathbf{G}^{\mathrm{T}} = [11 \quad 12 \quad 9 \quad 10 \quad 0 \quad 13 \quad 0 \quad 14] \tag{3.74}$$

In its turn G is used to assemble the coefficients of the element property matrices such as KM, KP and MM into the appropriate places in the overall coefficient matrix. This is done according to the following scheme:

Building block subroutine	Banding considered?	Symmetry of coefficients?	Upper or lower triangle?	Storage scheme
FORMKV (real)	Yes	Yes	Upper	Columns
FORMKC (complex)	Yes	Yes	Upper	Columns
FORMKB	Yes	Yes	Lower	Rows
FORMKU	Yes	Yes	Upper	Rows
FORMTB	Yes	No	Both	Rows
FMBIGK	No	No	Both	Rows
FSPARV	Yes	Yes	Lower	Rows (skyline)

A simple three-dimensional mesh is shown in Figure 3.16 and the system coefficients can again be assembled using the same building blocks.

Although the user of these subroutines does not strictly need to know how the storage is carried out, examples are given in Figure 3.17 for the most commonly used assembling routines FORMKV, FORMKB and FSPARV.

Subroutine FORMKV stores the global stiffness matrix as a vector of length N∗(IW + 1) with the diagonal terms occupying the first N positions (Figure 3.17a).

Subroutine FORMKB stores the global stiffness matrix as a rectangular

Table 3.4 Formation of typical nodal freedom array NF

Data read by procedure	Resulting NF array
3 0 0	1 2
6 0 0	3 4
9 0 0	0 0
10 0 1	5 6
11 0 1	7 8
12 0 0	0 0
	9 10
	11 12
	0 0
	0 13
	0 14
	0 0

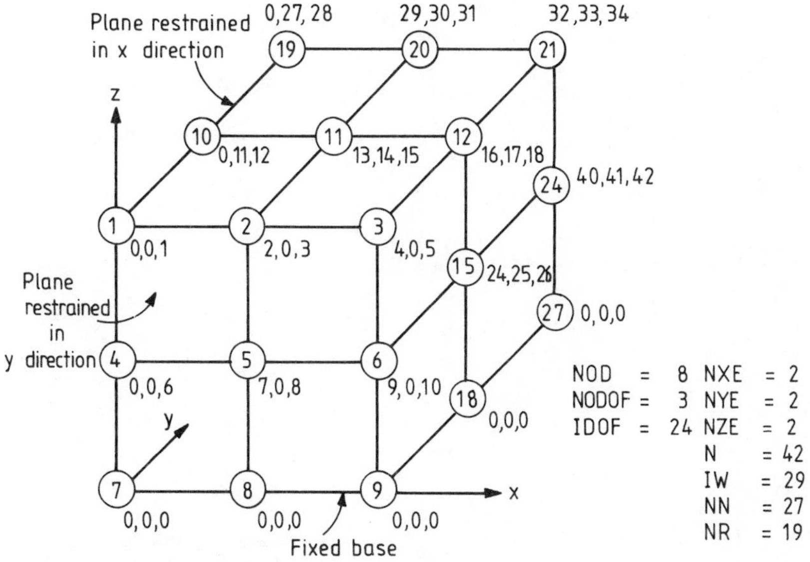

Figure 3.16 A simple three-dimensional mesh

matrix with N rows and (IW + 1) columns. The diagonal terms are held in the IW + 1th column (Figure 3.17b).

Both these strategies include a number of zeros in the storage due to the variability of the bandwidth and the inclusion of 'fictitious' zeros outside the actual array to simplify the subscripting.

Subroutine FSPARV stores only those numbers within the 'skyline' (Figure 3.17c) and can lead to improvements in storage requirements. Inform-

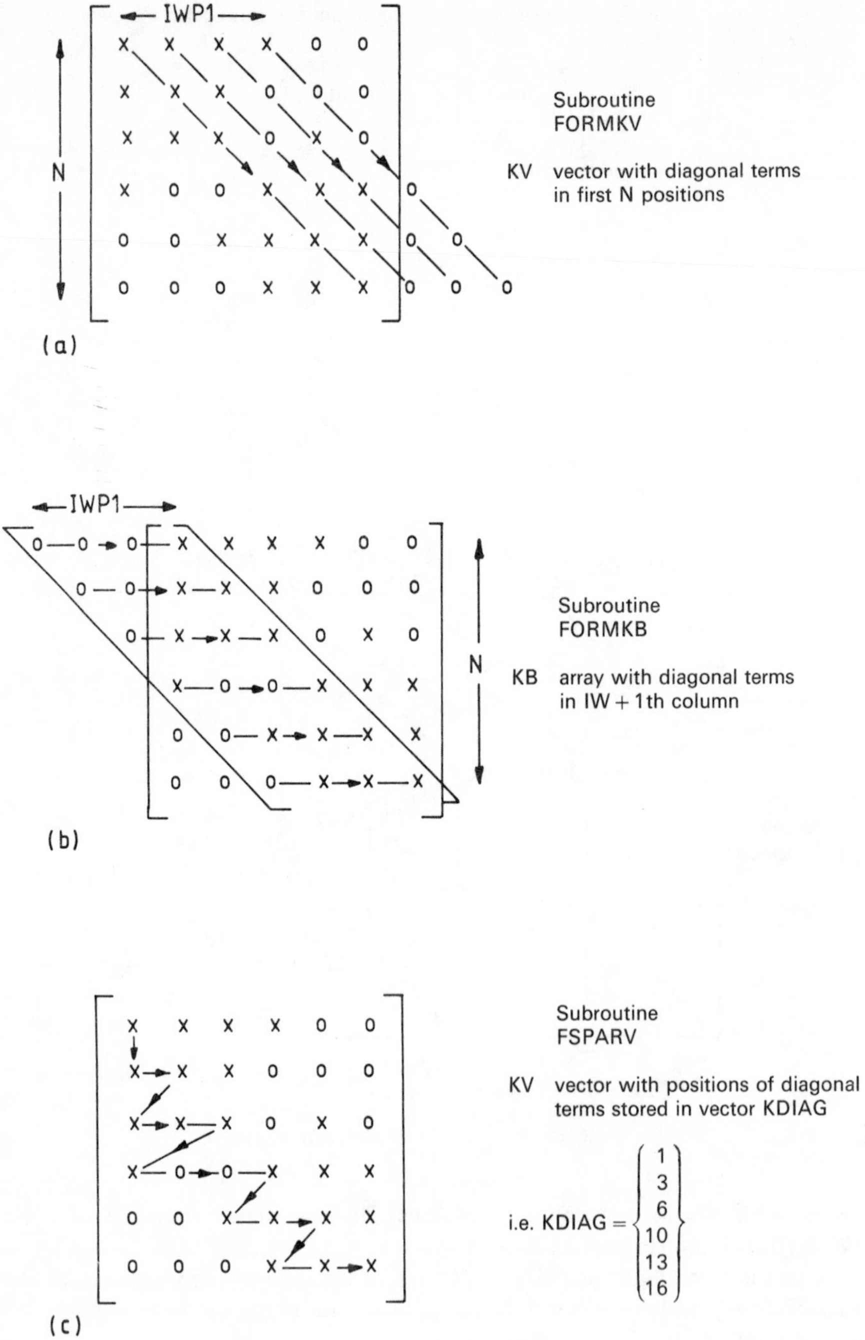

Figure 3.17 Different storage strategies

ation regarding the position of the diagonal terms within the resulting vector must be held in the integer vector KDIAG formed by subroutine FKDIAG. Some examples of this type of storage are given later in the book (e.g. Program 5.10).

3.10 SOLUTION OF EQUILIBRIUM EQUATIONS

After assembly, including specification of boundary conditions, a typical equilibrium equation is

$$\text{KV} * \text{DISPS} = \text{LOADS} \tag{3.75}$$

in which the coefficients of KV have been formed by FORMKV and LOADS are usually just input as data. The black box routines for equation solution depend of course on the method of coefficient storage according to the following scheme:

Solution routines	Coefficients formed by	Matrix/vector product
BANRED BACK1 } BACSUB BACK2 }	FORMKV	LINMUL
CHOLIN CHOBK1 } CHOBAC CHOBK2 }	FORMKB	BANMUL
GAUSBA SOLVBA	FORMTB	BANTML
COMRED COMBK1 } COMBAC COMBK2 }	FORMKC	
SPARIN SPARB1 } SPABAC SPARB2 }	FSPARV	LINMLS

In fact, to save storage, all of the solution routines overwrite the right hand side by the solution. That is, on completion of (3.75) the vector LOADS holds the solution DISPS. The storage strategy adopted by the compiler and the hardware, as described in Chapter 1, strongly influences the solution method that should be adopted. For example, on a virtual store machine in which FORTRAN stores by columns, the CHOLIN, CHOBAC options should be avoided, because of the excessive paging requests that would be involved.

3.11 EVALUATION OF EIGENVALUES AND EIGENVECTORS

Among the black box subroutines provided are TRIDIA which tridiagonalises an $N \times N$ symmetric matrix and EVECTS which finds the eigenvalues and eigenvectors. These same subroutines could be used whether lumped or distributed masses are assumed, it being necessary to reduce the eigenproblem to standard form

$$\mathbf{Ax} = \omega^2 \mathbf{x} \tag{3.76}$$

before solution. This is achieved in different ways for the two mass assumptions.

3.11.1 Lumped masses

The eigenproblem to be solved is

$$\mathrm{BIGK} * \mathrm{X} = \mathrm{OMEGA}^2 * \mathrm{DIAG} * \mathrm{X} \tag{3.77}$$

where BIGK and DIAG are the system stiffness and mass matrices respectively. The stiffness is stored as a complete $N \times N$ matrix having been formed by FMBIGK. The mass is actually a diagonal matrix which can be stored in the vector DIAG.

In order to reduce (3.77) to the required form (3.76) it is necessary to split the mass matrix by forming

$$\mathrm{DIAG} = \mathrm{L} * \mathrm{L}^{\mathrm{T}} \tag{3.78}$$

While this is essentially a Choleski split, it is particularly simple in the case of a diagonal matrix. The non-zero terms in L are simply the square roots of the terms in DIAG and the inverse of L merely consists of the reciprocals of these square roots. Thus

$$\mathrm{BIGK} * \mathrm{X} = \mathrm{OMEGA}^2 * \mathrm{L} * \mathrm{L}^{\mathrm{T}} * \mathrm{X} \tag{3.79}$$

which is then reduced to standard form by making the substitution

$$\mathrm{L}^{\mathrm{T}} * \mathrm{X} = \mathrm{Z} \tag{3.80}$$

Then

$$\mathrm{L}^{-1} * \mathrm{BIGK} * \mathrm{L}^{-\mathrm{T}} * \mathrm{Z} = \mathrm{OMEGA}^2 * \mathrm{Z} \tag{3.81}$$

is of the desired form. Having solved for Z, the required eigenvectors X are readily recovered using (3.80). Subroutine TRIDIA requires an extra vector UDIAG as working space which is then input to routine EVECTS.

In the case where the mass has been lumped, an alternative, more efficient approach is to form the stiffness in band form using FORMKU. The operation $\mathrm{L}^{-1} * \mathrm{KB} * \mathrm{L}^{-\mathrm{T}}$ is easily carried out in this case, and the resulting eigenvalue problem can be solved for the eigenvalues using subroutines BANDRD and BISECT.

3.11.2 Distributed masses

In this case the mass matrix is no longer diagonal, and is stored as a band matrix having the same bandwidth as the stiffness matrix. The eigenproblem is now

$$KB*X = OMEGA^2*MB*X \tag{3.82}$$

where system stiffness KB and mass MB are both formed by FORMKB.
The mass matrix must now be split using CHOLIN so that

$$KB*X = OMEGA^2*L*L^T*X \tag{3.83}$$

Again let

$$L^T*X = Z \tag{3.84}$$

or

$$X = L^{-T}*Z \tag{3.85}$$

so that the symmetrical eigenproblem becomes

$$BIGK*Z = L^{-1}*KB*L^{-T}*Z = OMEGA^2*Z \tag{3.86}$$

In order to avoid computing inverses, the $N \times N$ matrix BIGK is formed in two stages. First the equation

$$L*BIGK = KB \tag{3.87}$$

is solved for BIGK where L and KB are banded using the subroutine NBKBAN. Vhis is equivalent to vhe statement

$$BIGK = N^{-1}*KB \tag{3.88}$$

Secondly, the transpose is taken of the BIGK that has just been found and the equation

$$L*BIGK = (L^{-1}*KB)^T \tag{3.89}$$

is again solved for BIGK using subroutine LBBT hence

$$BIGK = L^{-1}*(L^{-1}*KB)^T \tag{3.90}$$

Finally, using the identity $(\mathbf{AB})^T = \mathbf{B}^T\mathbf{A}^T$ and the fact that for a symmetrical matrix $KB = KB^T$, we get

$$BIGK = L^{-1}*KB*L^{-T} \tag{3.91}$$

which is the required matrix from (3.86) in standard form. After the vectors Z have been found the true eigenvectors X are recovered by solving (3.85) using CHOBK2.

Analysis of this type will be found in Chapter 10.

3.11.3 Lanczos algorithm

The transformation techniques previously described are robust in that they will not fail to find an eigenvalue or to detect multiple roots. However, they are expensive to use on large problems for which, in general, vector iteration methods are preferable. The heart of all of these involves the matrix-by-vector products listed in Section 3.10, using whatever storage strategy is adopted for the global matrices.

For example, the programs described in Chapter 10 use subroutines modified from those described by Parlett and Reid (1981). These subroutines calculate the eigenvalues and eigenvectors of a symmetric matrix, say **A**, requiring the user only to compute the product and sum $\mathbf{Av}+\mathbf{u}$ for $\mathbf{u}$ and $\mathbf{v}$ provided by the subroutines. This would involve, for example, routines BANMUL and VECADD.

There is a slight additional complexity in that, as in (3.82), we often have the generalised eigenvalue problem to solve:

$$\mathrm{KB}*\mathrm{X}=\mathrm{OMEGA}^2*\mathrm{MB}*\mathrm{X} \tag{3.82}$$

If mass factorisation is chosen, a Choleski factorisation gives $\mathrm{MB}=\mathrm{L}*\mathrm{L}^{\mathrm{T}}$ (CHOLIN). Then on each Lanczos iteration, whenever $\mathbf{u}=\mathbf{Av}+\mathbf{u}$ is called for, we compute

(i) $\mathbf{L}^{\mathrm{T}}\mathbf{z}_1=\mathbf{v}$ (CHOBK2)
(ii) $\mathbf{z}_2=\mathbf{KBz}_1$ (BANMUL)
(iii) $\mathbf{Lz}_3=\mathbf{z}_2$ (CHOBK1)
(iv) $\mathbf{u}=\mathbf{u}+\mathbf{z}_3$ (VECADD)

Finally, the true eigenvectors are recovered from the transformed eigenvectors by backward substitution (CHOBK2) (compare equation 3.80).

The Lanczos vector subroutines are LANCZ1 and LANCZ2.

This algorithm has proved to be reasonably robust and is certainly efficient. However, it is possible for clustered eigenvalues to cause difficulties.

3.12 SOLUTION OF FIRST ORDER TIME-DEPENDENT PROBLEMS

A typical equation is given by:

$$\mathbf{KP}\boldsymbol{\phi}+\mathbf{PM}\frac{\mathrm{d}\boldsymbol{\phi}}{\mathrm{d}t}=\mathbf{Q} \tag{3.92}$$

where **Q** may be a function of time. There are many ways of integrating this set of ordinary differential equations, and modern methods for small numbers of equations would probably be based on variable order, variable timestep methods with error control (e.g. 'SPRINT' in the NAG mathematical subroutine library). However, for large engineering systems more primitive methods are still mainly used, involving linear interpolations and fixed time steps Δt. The basic algorithm is written:

$$\mathbf{KP}[\theta\boldsymbol{\phi}_1+(1-\theta)\boldsymbol{\phi}_0]+\mathbf{PM}\left[\theta\frac{\mathrm{d}\boldsymbol{\phi}_1}{\mathrm{d}t}+(1-\theta)\frac{\mathrm{d}\boldsymbol{\phi}_0}{\mathrm{d}t}\right]=\theta\mathbf{Q}_1+(1-\theta)\mathbf{Q}_0 \tag{3.93}$$

leading to the following recurrence relation between timesteps '0' and '1':

$$(\mathbf{PM} + \theta\Delta t\,\mathbf{KP})\boldsymbol{\phi}_1 = [\mathbf{PM} - (1-\theta)\Delta t\,\mathbf{KP}]\boldsymbol{\phi}_0 + \theta\Delta t\,\mathbf{Q}_1 + (1-\theta)\Delta t\,\mathbf{Q}_0 \quad (3.94)$$

This system is only unconditionally 'stable' (i.e. errors will not grow unboundedly) if $\theta \geqslant \frac{1}{2}$. Common choices would be $\theta = \frac{1}{2}$, giving the 'Crank–Nicolson' method (assuming for the moment $\mathbf{Q} = \mathbf{0}$):

$$\left(\mathbf{PM} + \frac{\Delta t}{2}\mathbf{KP}\right)\boldsymbol{\phi}_1 = \left(\mathbf{PM} - \frac{\Delta t}{2}\mathbf{KP}\right)\boldsymbol{\phi}_0 \quad (3.95)$$

and $\theta = 1$ giving the 'fully implicit' method:

$$(\mathbf{PM} + \Delta t\,\mathbf{KP})\boldsymbol{\phi}_1 = \mathbf{PM}\boldsymbol{\phi}_0 \quad (3.96)$$

After assembly of these element equations into system equations using the usual assembly procedures, the solution of problems will clearly involve a marching process from time 0 to time 1. First a matrix by vector multiplication must be carried out on the right hand side using LINMUL (assuming FORMKV has been used for assembly) or its equivalent, depending on the storage strategy, and second, a set of linear equations like the equilibrium equations must be solved on each timestep. If the **KP** and **PM** matrices do not change with time, the reduction of the left hand side coefficients, which is a time consuming operation, need only be performed once, as shown in the structure chart in Figure 3.18.

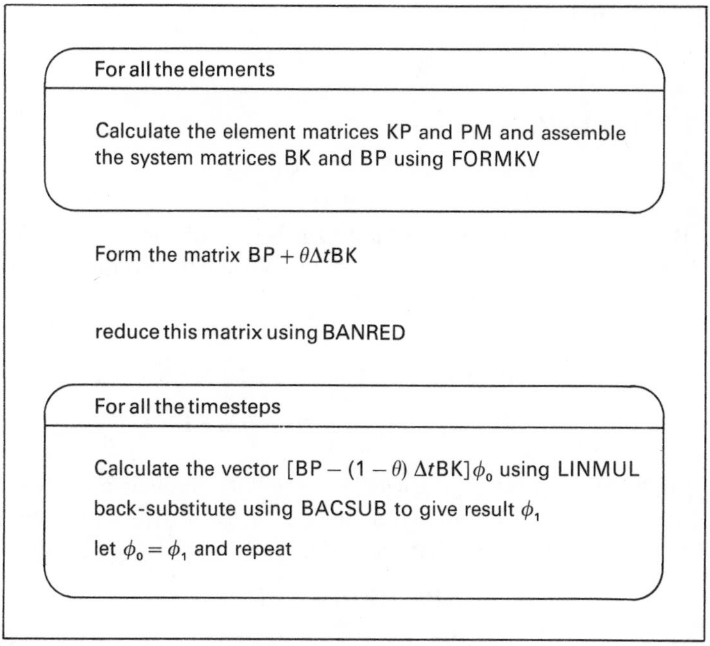
For all the elements
Calculate the element matrices KP and PM and assemble the system matrices BK and BP using FORMKV
Form the matrix BP + θΔtBK
reduce this matrix using BANRED
For all the timesteps
Calculate the vector [BP − (1 − θ) ΔtBK]φ0 using LINMUL
back-substitute using BACSUB to give result φ1
let φ0 = φ1 and repeat

Figure 3.18 Structure chart for first order time dependent problems by implicit methods

The 'implicit' strategies described above are quite effective for linear problems (constant **KP** and **PM**). However, storage requirements can be considerable, and in non-linear problems the necessity to refactorise $\mathbf{BP} + \theta\Delta t\,\mathbf{BK}$ can lead to lengthy calculations.

An alternative, widely used for second order problems (see Section 3.15), is to set $\theta = 0$ and to 'lump' the **PM** matrix (see Section 2.2). In that case the system to be solved is

$$\mathbf{PM}\boldsymbol{\phi}_1 = (\mathbf{PM} - \Delta t\,\mathbf{KP})\boldsymbol{\phi}_0 \tag{3.97}$$

where **PM** is a diagonal matrix and hence $\boldsymbol{\phi}_1$ can be obtained 'explicitly' from $\boldsymbol{\phi}_0$ without resorting to equation solution. Indeed, by appropriate manipulation of element **PM** and **KP** matrices, no large 'global' arrays are necessary, as element-by-element (EBE) summation is possible. The disadvantage is that (3.97) is only stable on condition that Δt is small, and in practice perhaps so small that real times of interest would require an excessive number of steps.

An alternative approach which conserves computer storage while preserving the stability properties of implicit methods involves 'operator splitting' on an element-by-element product basis (Hughes *et al.*, 1983). Although not necessary for the operation of the method, the simplest algorithms result from 'lumping' **PM**. The global equations are, for $\mathbf{Q} = \mathbf{0}$,

$$\boldsymbol{\phi}_1 = (\mathbf{BP} + \theta\Delta t\,\mathbf{BK})^{-1}[\mathbf{BP} - (1-\theta)\Delta t\,\mathbf{BK}]\boldsymbol{\phi}_0 \tag{3.98}$$

Thus **BP** is the global **PM**:

$$\mathbf{BP} = \sum_{\text{elements}} \mathbf{PM} \tag{3.99}$$

and **BK** is the global **KP**

$$\mathbf{BK} = \sum_{\text{elements}} \mathbf{KP} \tag{3.100}$$

The EBE splitting methods are based on binomial theorem expansions of $(\mathbf{BP} + \theta\Delta t\,\mathbf{BK})^{-1}$ which neglect product terms. When **BP** is lumped (diagonal) the method is particularly straightforward because **BP** can be effectively be replaced by **I**.

$$\begin{aligned}(\mathbf{I} + \theta\Delta t\mathbf{BK})^{-1} &= (\mathbf{I} + \theta\Delta t\textstyle\sum\mathbf{KP})^{-1} \\ &\simeq \prod_{\text{elements}} (\mathbf{I} + \theta\Delta t\,\mathbf{KP})^{-1}\end{aligned} \tag{3.101}$$

where $\prod$ indicates a product. As was the case with implicit methods, optimal accuracy consistent with stability is achieved for $\theta = \frac{1}{2}$. It is shown by Hughes *et al.* (1983) that further optimisation is achieved by splitting further to

$$\mathbf{KP} = (\tfrac{1}{2}\mathbf{KP}) + (\tfrac{1}{2}\mathbf{KP}) \tag{3.102}$$

and carrying out the product (3.101) by sweeping twice through the elements. They suggest from first to last and back again, but clearly various choices of sweeps could be employed. It can be shown that as $\Delta t \to 0$, any of these processes converges to the true solution of the global problem (for lumped **PM** of course). A

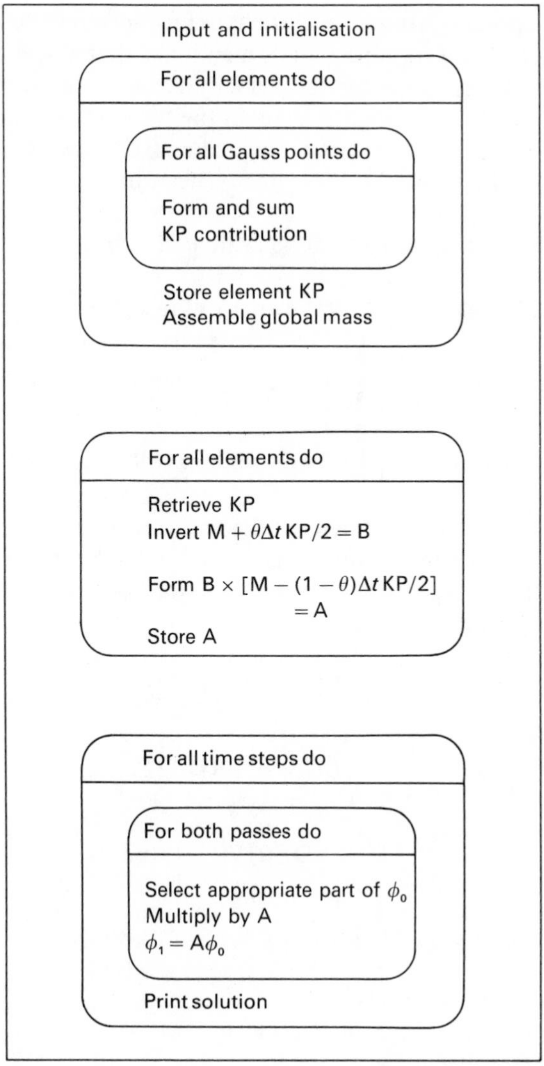

Figure 3.19 Structure chart for the element-by-element product algorithm (2-pass)

structure chart for the process is shown in Figure 3.19 and examples of all the methods described in this section are described in Chapter 8. Consistent mass versions are described by Wong (1987).

3.13 SOLUTION OF COUPLED NAVIER–STOKES PROBLEMS

For steady state conditions, it was shown in Section 2.17 that a non-linear system of algebraic equations has to be solved, involving, at element level, submatrices C_{11}, C_{12}, etc. These element matrices contained velocities $\bar{u}$ and $\bar{v}$ (called UBAR and VBAR in the programs) together with shape functions and their derivatives for the velocity and pressure variables. It was mentioned that it would be possible to use different shape functions for the velocity (vector) quantity and pressure (scalar) quantity and this is what is done in the program in Chapter 9.

The velocity shape functions are designated as FUN and the pressure shape functions as FUNF. Similarly, the velocity derivatives are DERIV and the pressure derivatives DERIVF.

Thus the element integral have to be evaluated numerically, typically

$$\begin{aligned} \text{C11}(=\text{C33}) = &\iint \text{DERIV}^{\text{T}} * \text{KAY} * \text{DERIV}\, \mathrm{d}x\, \mathrm{d}y \\ &+ \iint \text{UBAR} * \text{FUN} * \text{DERIV}(1, -)\, \mathrm{d}x\, \mathrm{d}y \\ &+ \iint \text{VBAR} * \text{FUN} * \text{DERIV}(2, -)\, \mathrm{d}x\, \mathrm{d}y \end{aligned} \tag{3.103}$$

In this equation, DERIV(1, −) signifies the first row of DERIV and so on. The diagonal terms in KAY represent the reciprocal of the Reynolds number. Note the identity of the first term of C11 with (3.46) for uncoupled flow.

The remaining submatrices are

$$\begin{aligned} \text{C12} &= \frac{1}{\text{RHO}} \iint \text{FUN} * \text{DERIVF}(1, -)\, \mathrm{d}x\, \mathrm{d}y \\ \text{C32} &= \frac{1}{\text{RHO}} \iint \text{FUN} * \text{DERIVF}(2, -)\mathrm{d}x\, \mathrm{d}y \\ \text{C21} &= \iint \text{FUNF} * \text{DERIV}(1, -)\mathrm{d}x\, \mathrm{d}y \\ \text{C23} &= \iint \text{FUNF} * \text{DERIV}(2, -)\mathrm{d}x\, \mathrm{d}y \end{aligned} \tag{3.104}$$

where RHO is the mass density.

The (unsymmetrical) matrix built up from these submatrices is formed by a special subroutine called

FRMUPV

and the global, unsymmetrical, band matrix is assembled using FORMTB. The appropriate equation solution routines are GAUSBA and SOLVBA.

3.14 SOLUTION OF COUPLED TRANSIENT PROBLEMS

The element equations for Biot consolidation were shown in Section 2.22 to be given by

$$\begin{aligned} \mathbf{KMr} + \mathbf{Cu}_w &= \mathbf{F} \\ \mathbf{C}^{\text{T}} \frac{\partial \mathbf{r}}{\partial t} - \mathbf{KPu}_w &= \mathbf{0} \end{aligned} \tag{3.105}$$

where **KM** and **KP** are the now familiar solid and fluid stiffnesses. The matrix **C** is the connection matrix which, if it is assumed that the same linear shape functions govern the variation of porewater pressure and displacements within an element, is given for plane strain by

$$\mathrm{C} = \iint \mathrm{VOL} * \mathrm{FUN}\, \mathrm{d}x\, \mathrm{d}y \tag{3.106}$$

The matrix product is called VOLF and C is numerically integrated as

$$\mathrm{C} = \mathrm{DET} * \sum_{\mathrm{I}=1}^{\mathrm{NGP}} \sum_{\mathrm{J}=1}^{\mathrm{NGP}} \mathrm{W_I} * \mathrm{W_J} * \mathrm{VOLF_{I,J}} \tag{3.107}$$

The vector VOL is related to the volumetric strain in an element. In plane strain, for example, the strain-displacement relation is

$$\begin{Bmatrix} \varepsilon_x \\ \varepsilon_y \\ \gamma_{xy} \end{Bmatrix} = \mathrm{BEE} * \mathrm{ELD} \tag{3.108}$$

where ELD are the element displacements and

$$\mathrm{BEE} = \begin{bmatrix} \frac{\partial N_1}{\partial x} & 0 & \frac{\partial N_2}{\partial x} & 0 & \frac{\partial N_3}{\partial x} & 0 & \frac{\partial N_4}{\partial x} & 0 \\ 0 & \frac{\partial N_1}{\partial y} & 0 & \frac{\partial N_2}{\partial y} & 0 & \frac{\partial N_3}{\partial y} & 0 & \frac{\partial N_4}{\partial y} \\ \frac{\partial N_1}{\partial y} & \frac{\partial N_1}{\partial x} & \frac{\partial N_2}{\partial y} & \frac{\partial N_2}{\partial x} & \frac{\partial N_3}{\partial y} & \frac{\partial N_3}{\partial x} & \frac{\partial N_4}{\partial y} & \frac{\partial N_4}{\partial x} \end{bmatrix} \tag{3.109}$$

The element volumetric strain can therefore be written

$$\varepsilon_x + \varepsilon_y = \mathrm{VOL^T} * \mathrm{ELD} \tag{3.110}$$

where

$$\mathrm{VOL^T} = \begin{bmatrix} \frac{\partial N_1}{\partial x} & \frac{\partial N_1}{\partial y} & \frac{\partial N_2}{\partial x} & \frac{\partial N_2}{\partial y} & \frac{\partial N_3}{\partial x} & \frac{\partial N_3}{\partial y} & \frac{\partial N_4}{\partial x} & \frac{\partial N_4}{\partial y} \end{bmatrix} \tag{3.111}$$

and this vector is worked out by the subroutine VOL2D.

To integrate equations (3.105) with respect to time there are again many methods available, but we consider only the simplest linear interpolation in time using finite differences thus:

$$\begin{aligned} \theta \mathbf{KM} \mathbf{r}_1 + \theta \mathbf{C} \boldsymbol{\phi}_1 &= (\theta - 1)\mathbf{KM}\mathbf{r}_0 + (\theta - 1)\mathbf{C}\boldsymbol{\phi}_0 + \mathbf{F} \\ \theta \mathbf{C}^{\mathrm{T}} \mathbf{r}_1 - \theta^2 \Delta t\, \mathbf{KP} \boldsymbol{\phi}_1 &= \theta \mathbf{C}^{\mathrm{T}} \mathbf{r}_0 - \theta(\theta - 1)\Delta t\, \mathbf{KP} \boldsymbol{\phi}_0 \end{aligned} \tag{3.112}$$

where **F** is assumed here to be independent of time.

In the Crank–Nicolson type of approximation, θ is made equal to $\frac{1}{2}$ in both of

(3.112), leading to the recurrence relation:

$$\begin{bmatrix} \mathbf{KM} & \mathbf{C} \\ \mathbf{C}^{\mathrm{T}} & -\dfrac{\Delta t}{2}\mathbf{KP} \end{bmatrix} \begin{Bmatrix} \mathbf{r}_1 \\ \boldsymbol{\phi}_1 \end{Bmatrix} = \begin{bmatrix} -\mathbf{KM} & -\mathbf{C} \\ \mathbf{C}^{\mathrm{T}} & \dfrac{\Delta t}{2}\mathbf{KP} \end{bmatrix} \begin{Bmatrix} \mathbf{r}_0 \\ \boldsymbol{\phi}_0 \end{Bmatrix} + \begin{Bmatrix} 2\mathbf{F} \\ \mathbf{0} \end{Bmatrix} \tag{3.113}$$

It will be shown in Chapter 9 that this approximation can lead to oscillatory results. The oscillations can be smoothed out either by using the fully implicit version of (3.112) with $\theta = 1$, which leads to the recurrence relation

$$\begin{bmatrix} \mathbf{KM} & \mathbf{C} \\ \mathbf{C}^{\mathrm{T}} & -\Delta t\,\mathbf{KP} \end{bmatrix} \begin{Bmatrix} r_1 \\ \boldsymbol{\phi}_1 \end{Bmatrix} = \begin{bmatrix} \mathbf{0} & \mathbf{0} \\ \mathbf{C}^{\mathrm{T}} & \mathbf{0} \end{bmatrix} \begin{Bmatrix} \mathbf{r}_0 \\ \boldsymbol{\phi}_0 \end{Bmatrix} + \begin{Bmatrix} \mathbf{F} \\ \mathbf{0} \end{Bmatrix} \tag{3.114}$$

or by writing the first of (3.112) with $\theta = 1$ and the second with $\theta = \frac{1}{2}$. This hybrid method leads to a recurrence relation

$$\begin{bmatrix} \mathbf{KM} & \mathbf{C} \\ \mathbf{C}^{\mathrm{T}} & -\dfrac{\Delta t}{2}\mathbf{KP} \end{bmatrix} \begin{Bmatrix} \mathbf{r}_1 \\ \boldsymbol{\phi}_1 \end{Bmatrix} = \begin{bmatrix} \mathbf{0} & \mathbf{0} \\ \mathbf{C}^{\mathrm{T}} & \dfrac{\Delta t}{2}\mathbf{KP} \end{bmatrix} \begin{Bmatrix} \mathbf{r}_0 \\ \boldsymbol{\phi}_0 \end{Bmatrix} + \begin{Bmatrix} \mathbf{F} \\ \mathbf{0} \end{Bmatrix} \tag{3.115}$$

Results of calculations using (3.112) are presented in Chapter 9. The algorithms will clearly be of the same form as those described previously for uncoupled equations (3.95, 3.96) and the Structure Chart of Figure 3.18. A right hand side matrix-by-vector multiplication is followed by an equation solution for each timestep. As before a saving in computer time can be achieved if **KM**, **C** and **KP** are independent of time, and constant Δt is used, because the left hand side matrix needs to be reduced only once. Note that the left hand side matrix is always symmetrical whereas the right hand side is not. Therefore FORMKV or FORMKB can be used to assemble the left hand side system equations from (3.113), (3.114) or (3.115) whereas FORMTB must be used to assemble the right hand side system. Element by element summation could be most effective in (3.114) and (3.115) due to sparsity.

3.15 SOLUTION OF SECOND ORDER TIME-DEPENDENT PROBLEMS

The basic second order propagation type of equation was derived in Chapter 2 and takes the form of (2.92), namely

$$\mathbf{KMr} + \mathbf{MM}\frac{\partial^2 \mathbf{r}}{\partial t^2} = \mathbf{F}(t) \tag{3.116}$$

where in the context of solid mechanics, **KM** is the element elastic stiffness and **MM** the element mass. In addition to these elastic and inertial forces, solids in motion experience a third type of force whose action is to dissipate energy. For example, the solid may deform so much that plastic strains result, or may be subjected to internal or external friction. Although these phenomena are non-linear in character and can be treated by the non-linear analysis techniques given in Chapter 6, it is most common to linearise the dissipative forces, for example by assuming that they are proportional to velocity. This allows (3.116) to be modified to

$$\mathbf{KMr} + \mathbf{CM}\frac{\partial \mathbf{r}}{\partial t} + \mathbf{MM}\frac{\partial^2 \mathbf{r}}{\partial t^2} = \mathbf{F}(t) \tag{3.117}$$

where **CM** is assumed to be a constant damping matrix. The three most popular techniques for integrating (3.117) with respect to time will now be described. They are the 'modal superposition', 'direct integration' and 'complex response' methods, the last of which transforms (3.117) into the frequency domain (Clough and Penzien, 1975).

3.15.1 Modal superposition

This method has as its basis the analysis of the free undamped part of (3.117), that is when **CM** and **F** are zero. The reduced equation is

$$\mathbf{KMr} + \mathbf{MM}\frac{\partial^2 \mathbf{r}}{\partial t^2} = \mathbf{0} \tag{3.118}$$

which can of course be converted into an eigenproblem by the assumption of harmonic motion

$$\mathbf{r} = \mathbf{a}\sin(\omega t + \psi) \tag{3.119}$$

to give

$$\mathbf{KMa} - \omega^2\,\mathbf{MMa} = \mathbf{0} \tag{3.120}$$

Solution of this eigenproblem by the techniques previously described results in N modal vectors **a**, where N is the total number of degrees of freedom present. These modes can be considered to be columns of a modal matrix **A** where

$$\mathbf{A} = [\mathbf{a}_i] \; i = 1 \text{ to NMODES} \tag{3.121}$$

Often it is not necessary to include the higher frequency components in an analysis, so that NMODES $\leqslant$ N.

Because of the properties of eigenproblems the mode shapes possess orthogonality one to the other such that

$$\left.\begin{aligned} \mathbf{a}_i^{\mathrm{T}}\mathbf{MMa}_j &= 0 \\ \mathbf{a}_i^{\mathrm{T}}\mathbf{KMa}_j &= 0 \end{aligned}\right\} \quad i \neq j \tag{3.122}$$

$$\left.\begin{aligned} \mathbf{a}_i^{\mathrm{T}}\mathbf{MMa}_j &= \mathbf{MPR}_{ii} \\ \mathbf{a}_i^{\mathrm{T}}\mathbf{KMa}_j &= \mathbf{KPR}_{ii} \end{aligned}\right\} \quad i = j \tag{3.123}$$

where **MPR** and **KPR** are the diagonal 'principal' mass and stiffness matrices. Use of these relationships in (3.118) has the effect of uncoupling the equations in terms of the principal or 'normal' coordinates **ap** in the diagonal matrix equation

$$\mathbf{MPR}\frac{\partial^2 \mathbf{ap}}{\partial t^2} + \mathbf{KPRap} = \mathbf{0} \tag{3.124}$$

The normal coordinates are related to the original coordinates by

$$a_i = \sum_{j=1}^{\text{NMODES}} A_{ij} ap_j \tag{3.125}$$

and the effect of uncoupling has been to reduce the vibration problem to an eigenproblem (3.120), leading to a set of independent second order differential equations, whose number is NMODES (3.124). A final superposition process can be used to recover the actual displacements (3.125).

3.15.1.1 *Inclusion of damping*

Free, damped vibrations, governed by

$$\mathbf{KMr} + \mathbf{CM}\frac{\partial \mathbf{r}}{\partial t} + \mathbf{MM}\frac{\partial^2 \mathbf{r}}{\partial t^2} = \mathbf{0} \tag{3.126}$$

can be handled by the above technique if it is assumed that the undamped mode shapes are also orthogonal with respect to the damping matrix **CM** in the way described by (3.122) and (3.123). This can readily be achieved if **CM** is taken to be a linear combination of **MM** and **KM**,

$$\mathbf{CM} = \alpha\mathbf{MM} + \beta\mathbf{KM} \tag{3.127}$$

where α and β are scalar variables, the so-called 'Rayleigh damping' coefficients.

Because of the orthogonality with respect to **CM** the uncoupled normal coordinate equations are

$$\mathbf{MPR}\frac{\partial^2 \mathbf{ap}}{\partial t^2} + \mathbf{CPR}\frac{\partial \mathbf{ap}}{\partial t} + \mathbf{KPRap} = \mathbf{0} \tag{3.128}$$

where

$$\mathbf{CPR} = \alpha\mathbf{MPR} + \beta\mathbf{KPR} \tag{3.129}$$

The modal matrix is usually normalised with respect to **MM** so that

$$\mathbf{CPR} = (\alpha + \beta\omega^2)\mathbf{I} \tag{3.130}$$

Therefore each ordinary differential equation (3.128) has the form

$$\frac{\partial^2 \mathbf{ap}}{\partial t^2} + (\alpha + \beta\omega^2)\frac{\partial \mathbf{ap}}{\partial t} + \omega^2 \mathbf{ap} = 0 \tag{3.131}$$

and α and β can be related to the more usual 'damping ratio' γ (Timoshenko *et al.*, 1974) by means of

$$\gamma = \frac{\alpha + \beta\omega^2}{2\omega} \tag{3.132}$$

By working in normal coordinates it has become necessary to take a constant γ for the complete mesh being analysed, although γ could be varied from mode to mode. Since many real systems contain areas with markedly different damping properties, this is an undesirable feature of the method in practice (see Chapter 11).

3.15.1.2 *Inclusion of forcing terms*

When $\mathbf{F}(t)$ is non-zero in (3.117), the right hand side of a typical modal equation (3.131) becomes a multiple of $F(t)$ and the transpose of the modal matrix

$$\frac{\partial^2 \mathbf{ap}}{\partial t^2} + (\alpha + \beta\omega^2)\frac{\partial \mathbf{ap}}{\partial t} + \omega^2 \mathbf{ap} = \mathbf{A}^{\mathrm{T}}\mathbf{F}(t) \tag{3.133}$$

For example, suppose that in a specific problem only degrees of freedom 10 and 12 are loaded with forces $\cos\theta t$. A typical member of (3.133) would be

$$\frac{\partial^2 ap_i}{\partial t^2} + (\alpha + \beta\omega_i^2)\frac{\partial ap_i}{\partial t} + \omega_i^2 ap_i = (A_{10,i} + A_{12,i})\cos\theta t \tag{3.134}$$

or

$$\frac{\partial^2 ap_i}{\partial t^2} + 2\gamma\omega_i\frac{\partial ap_i}{\partial t} + \omega_i^2 ap_i = P_i \cos\theta t \tag{3.135}$$

The particular solution to this equation is given by

$$ap_i = \frac{P_i(\omega_i^2 - \theta^2)}{(\omega_i^2 - \theta^2)^2 + 4\gamma^2\omega_i^2\theta^2}\cos\theta t + \frac{P_i 2\gamma\omega_i\theta}{(\omega_i^2 - \theta^2)^2 + 4\gamma^2\omega_i^2\theta^2}\sin\theta t \tag{3.136}$$

The normal coordinates having thus been determined, the displacements can be recovered using (3.125).

For more general forcing functions, (3.133) must be solved by other means, for example by one of the direct integration methods described below.

3.15.2 Direct integration

There are many methods for advancing the solution of (3.117) with respect to time by direct integration, but attention is first focused on two of the simplest popular implicit methods. In both of these, integration is advanced by one time interval Δt, the values of the displacement and its derivatives at one instant in time being sufficient to determine these values at the subsequent instant by means of recurrence relations. Both preserve unconditional stability.

3.15.2.1 *Newmark or Crank–Nicolson method*

If Rayleigh damping is assumed, a class of recurrence relations based on linear interpolation in time can again be constructed, involving the scalar parameter θ which varies between $\frac{1}{2}$ and 1 in the same way as was done for first order problems.

Writing the differential equation (3.117) at both the '0' and '1' stations, thus

$$\mathbf{KMr}_0 + (\alpha\mathbf{MM} + \beta\mathbf{KM})\frac{\partial \mathbf{r}_0}{\partial t} + \mathbf{MM}\frac{\partial^2 \mathbf{r}_0}{\partial t^2} = \mathbf{F}_0$$

$$\mathbf{KMr}_1 + (\alpha\mathbf{MM} + \beta\mathbf{KM})\frac{\partial \mathbf{r}_1}{\partial t} + \mathbf{MM}\frac{\partial^2 \mathbf{r}_1}{\partial t^2} = \mathbf{F}_1 \tag{3.137}$$

and assuming linear interpolation in time giving

$$\begin{aligned} \mathbf{r}_1 &= \mathbf{r}_0 + \Delta t\left[(1-\theta)\frac{\partial \mathbf{r}_0}{\partial t} + \theta\frac{\partial \mathbf{r}_1}{\partial t}\right] \\ \frac{\partial \mathbf{r}_1}{\partial t} &= \frac{\partial \mathbf{r}_0}{\partial t} + \Delta t\left[(1-\theta)\frac{\partial^2 \mathbf{r}_0}{\partial t^2} + \theta\frac{\partial^2 \mathbf{r}_1}{\partial t^2}\right] \end{aligned} \tag{3.138}$$

rearrangement of these equations leads to the following three recurrence relations:

$$\begin{aligned} &\left[\left(\alpha + \frac{1}{\theta\Delta t}\right)\mathbf{MM} + (\beta + \theta\Delta t)\mathbf{KM}\right]\mathbf{r}_1 \\ &\quad = \theta\Delta t\,\mathbf{F}_1 + (1-\theta)\Delta t\,\mathbf{F}_0 + \left(\alpha + \frac{1}{\theta\Delta t}\right)\mathbf{MMr}_0 \\ &\qquad + \frac{1}{\theta}\mathbf{MM}\frac{\partial \mathbf{r}_0}{\partial t} + [\beta - (1-\theta)\Delta t]\,\mathbf{KMr}_0 \end{aligned} \tag{3.139}$$

$$\frac{\partial \mathbf{r}_1}{\partial t} = \frac{1}{\theta\Delta t}(\mathbf{r}_1 - \mathbf{r}_0) - \frac{1-\theta}{\theta}\frac{\partial \mathbf{r}_0}{\partial t} \tag{3.140}$$

$$\frac{\partial^2 \mathbf{r}_1}{\partial t^2} = \frac{1}{\theta\Delta t}\left(\frac{\partial \mathbf{r}_1}{\partial t} - \frac{\partial \mathbf{r}_0}{\partial t}\right) - \frac{1-\theta}{\theta}\frac{\partial^2 \mathbf{r}_0}{\partial t^2} \tag{3.141}$$

In the special case when $\theta = \frac{1}{2}$ this method is Newmark's '$\beta = \frac{1}{4}$' method, which is also the exact equivalent of the Crank–Nicolson method used in first order problems. There are other variants of the Newmark type but this is the most common.

The principal recurrence relation (3.139) is clearly similar to those which arose in first order problems, for example (3.93), (3.112). Although substantially more matrix-by-vector multiplications are involved, together with matrix and vector additions, the recurrence again consists essentially of an equation solution per timestep. Advantage can as usual be taken of a constant left hand side matrix should this occur.

3.15.2.2 *Wilson method*

This method advances the solution of (3.117) from some known state $\mathbf{r}_0$, $\partial \mathbf{r}_0/\partial t$, $\partial^2 \mathbf{r}_0/\partial t^2$ to the new solution $\mathbf{r}_1$, $\partial \mathbf{r}_1/\partial t$, $\partial^2 \mathbf{r}_1/\partial t^2$ an interval Δt later by first linearly extrapolating to a hypothetical solution, say $\mathbf{r}_2$, $\partial \mathbf{r}_2/\partial t$, $\partial^2 \mathbf{r}_2/\partial t^2$ an interval $\delta t = \theta \Delta t$ later where $1.4 \leqslant \theta \leqslant 2$.

If Rayleigh damping is again assumed, $\mathbf{r}_2$ is first computed from

$$\left[\left(\frac{6}{\theta^2 \Delta t^2} + \frac{3\alpha}{\theta \Delta t}\right)\mathbf{MM} + \left(\frac{3\beta}{\theta \Delta t} + 1\right)\mathbf{KM}\right]\mathbf{r}_2$$

$$= \mathbf{F}_2 + \mathbf{MM}\left[\left(\frac{6}{\theta^2 \Delta t^2} + \frac{3\alpha}{\theta \Delta t}\right)\mathbf{r}_0 + \left(\frac{6}{\theta \Delta t} + 2\alpha\right)\frac{\partial \mathbf{r}_0}{\partial t} + \left(2 + \frac{\alpha \theta \Delta t}{2}\right)\frac{\partial^2 \mathbf{r}_0}{\partial t^2}\right]$$

$$+ \mathbf{KM}\left(\frac{3\beta}{\theta \Delta t}\mathbf{r}_0 + 2\beta \frac{\partial \mathbf{r}_0}{\partial t} + \frac{\beta \theta \Delta t}{2}\frac{\partial^2 \mathbf{r}_0}{\partial t^2}\right) \tag{3.142}$$

The acceleration at the hypothetical station can then be computed from

$$\frac{\partial^2 \mathbf{r}_2}{\partial t^2} = \frac{6}{\theta^2 \Delta t^2}(\mathbf{r}_2 - \mathbf{r}_0) - \frac{6}{\theta \Delta t}\frac{\partial \mathbf{r}_0}{\partial t} - 2\frac{\partial^2 \mathbf{r}_0}{\partial t^2} \tag{3.143}$$

and thus the acceleration at the true station can be interpolated or 'averaged' using

$$\frac{\partial^2 \mathbf{r}_1}{\partial t^2} = \frac{\partial^2 \mathbf{r}_0}{\partial t^2} + \frac{1}{\theta}\left(\frac{\partial^2 \mathbf{r}_2}{\partial t^2} - \frac{\partial^2 \mathbf{r}_0}{\partial t^2}\right) \tag{3.144}$$

A Crank–Nicolson equation then gives the desired velocity

$$\frac{\partial \mathbf{r}_1}{\partial t} = \frac{\partial \mathbf{r}_0}{\partial t} + \frac{\Delta t}{2}\left(\frac{\partial^2 \mathbf{r}_0}{\partial t^2} + \frac{\partial^2 \mathbf{r}_1}{\partial t^2}\right) \tag{3.145}$$

and finally the displacement

$$\mathbf{r}_1 = \mathbf{r}_0 + \Delta t \frac{\partial \mathbf{r}_0}{\partial t} + \frac{\Delta t^2}{3}\frac{\partial^2 \mathbf{r}_0}{\partial t^2} + \frac{\Delta t^2}{6}\frac{\partial^2 \mathbf{r}_1}{\partial t^2} \tag{3.146}$$

Finally, $\mathbf{F}_2$ in (3.142) must be replaced by $\mathbf{F}_0 + \theta(\mathbf{F}_1 - \mathbf{F}_0)$ to complete the algorithm.

The principal recurrence relation (3.142) is again of the familiar type for all one-step time integration methods.

3.15.3 Explicit methods and storage-saving strategies

The implicit methods described above are relatively safe to use due to their unconditional stability. However, as was the case for first order problems, storage demands become considerable for large systems, and so can solution times for non-linear problems (even although refactorisation of the left hand side of (3.139) or (3.142) is not usually necessary, the non-linear effects having been transposed to the right hand side).

The simplest option, which at least minimises storage, is the analogue of (3.97), in which θ is set to zero and the mass matrix lumped. In the resulting explicit algorithm, operations are carried out element-wise and no global system storage is necessary. Of course the drawback is potential loss of stability, so that stable timesteps can be very small indeed.

Since stability is governed by the highest natural frequency of the numerical approximation and since such high frequencies are derived from the stiffest elements in the system, it is quite possible to implement hybrid methods in which the very stiff elements are integrated implicitly, but the remainder are integrated explicitly. Equation solution as implied by (3.139), for example, is still necessary, but the bandwidth of the rows in the coefficient matrix associated with freedoms in explicit elements not connected to implicit ones is only one. Thus great savings in storage can be made (Smith, 1984).

Another alternative is to resort to operator splitting, as was done in first order problems.

In Chapter 11, implicit, explicit and mixed implicit/explicit algorithms are described and listed. Although product EBE methods have been programmed (Wong, 1987) they are beyond the scope of the present book.

3.15.4 Complex response

The most troublesome feature of linearised solutions to (3.117) is the proper inclusion of damping. Observations indicate that damping can be frequency-independent and that it varies with material type and location, for example being much higher in soils than in steel or concrete and higher in areas of large (plastic) deformation than in elastic regions.

It has been shown that the modal superposition method is best suited to the analysis of systems where damping is uniformly distributed over the whole system for each mode. When damping is not uniformly distributed, the direct integration methods can be used with variable damping in each element (Idriss *et al.*, 1973) but it is still difficult to preserve the frequency independence of the damping ratio γ. A method which does not suffer from either of these drawbacks is the complex response or complex transfer function method (Timoshenko *et al.*, 1974) which is widely used in seismic analyses of structure–soil interaction (Lysmer *et al.*, 1974).

The method takes as its starting point the undamped equation of motion (3.116) and assumes that $\mathbf{F}(t)$ is harmonic, of the form

$$\mathbf{F}(t) = \mathbf{f}\,e^{i\omega t} \tag{3.147}$$

where $\mathbf{f}$ may be complex. In a linearised system this implies that the response $\mathbf{r}$ is also harmonic:

$$\mathbf{r} = \mathbf{R}\,e^{i\omega t} \tag{3.148}$$

where $\mathbf{R}$ is a (possibly complex) amplitude vector.

When (3.148) are substituted in (3.116) the set of equations

$$(\mathbf{KM} - \omega^2\mathbf{MM})\mathbf{R} = \mathbf{f} \tag{3.149}$$

is obtained which can be solved for $\mathbf{R}$ provided that $\mathbf{KM} - \omega^2\mathbf{MM}$ is nonsingular or, in other words, when ω is not a natural frequency of the undamped system. The real and imaginary parts of the input **f** and output **R** correspond.

An advantage of the method is that damping can be introduced in the form of a complex equivalent of **KM** such that

$$\mathbf{KM}^* = \mathbf{KM}[1 - 2\gamma^2 + 2i\gamma\sqrt{(1-\gamma^2)}] \tag{3.150}$$

This produces the same amplitudes as a modal analysis with damping ratio γ and to a close approximation the same phase. Clearly the equations

$$(\mathbf{KM}^* - \omega^2\mathbf{MM})\mathbf{R} = \mathbf{f} \tag{3.151}$$

can be assembled element by element using varying γ values when required.

For a simple harmonic input with a single frequency ω, (3.151) can be assembled into a set of N complex simultaneous equations in the normal way but using assembly routine FORMKC to form the complex left hand side matrix. Solution routines COMRED and COMBAC would then be used to determine **R** and hence **r** from (3.148).

More usually, $\mathbf{F}(t)$ is a general non-harmonic function, for example a seismic record or a series of ocean wave height measurements. The complex response method can still be used if the random values of $\mathbf{F}(t)$ are converted into equivalent harmonic components.

Examples of the solution of second order time dependent problems by all of the above types of method are presented in Chapter 11.

3.16 CONCLUSIONS

The principles by which finite element computer programs may be constructed from building block subroutines have been outlined in this chapter. In general, local coordinates are used to express the element shape functions, and the element matrices are numerically integrated. However, in the next chapter we begin with programs in which the element stiffness matrix can be explicitly stated.

Element assembly into a global system of equations is done automatically using a nodal numbering system, a 'node freedom array' NF and a 'steering vector' G. Simple boundary conditions are taken care of automatically at this stage. The most common global matrices are the system 'stiffness' and 'mass' matrices.

These matrices are then manipulated to solve three basic types of problem: equilibrium, eigenvalue and propagation. The matrix operations involved are linear equation solution, eigenvalue extraction and linear equation solution with additional matrix-by-vector multiplications and additions.

Extra features will be introduced as they occur in the examples chapters, but these will be found to involve minor adaptations of what has already been described. For example, axisymmetric structures under non-axisymmetric loads will be considered in Chapter 5, while non-linear solid problems take up Chapter 6. However, these are solved by linearising incrementally, so that a minimum of new methodology is necessary.

3.17 REFERENCES

Bathe, K. J., and Wilson, E. L. (1976) *Numerical Methods in Finite Element Analysis.* Prentice-Hall, Englewood Cliffs, New Jersey.

Clough, R. W., and Penzien, J. (1975) *Dynamics of Structures.* McGraw-Hill, New York.

Ergatoudis, J., Irons, B. M., and Zienkiewicz, O. C. (1968) Curved isoparametric quadrilateral elements for finite element analysis. *Int. J. Solids and Structures*, **4**, 31.

Hughes, T. J. R., Levit, I., and Winget, J. (1983) Element by element implicit algorithms for heat conduction. *ASCE J. Eng. Mech.*, **109** (2), 576–585.

Idriss, I. M., Lysmer, J., Hwang, R., and Seed, H. B. (1973) QUAD-4; a computer program for evaluating the seismic response of soil structures by variable damping finite element procedures. University of California, Berkeley, Report EERC 73-16.

Irons, B. M. (1966a) Numerical integration applied to finite element methods. Proc. Conference on Use of Digital Computers in Structural Engineering, University of Newcastle.

Irons, B. M. (1966b) Engineering applications of numerical integration in stiffness method. *J.A.I.A.A*, **14**, 2035.

Irons, B. M., and Ahmad, S. (1980) *Techniques of Finite Elements.* Ellis Horwood Ltd, Chichester.

Kopal, A. (1961) Numerical Analysis, 2nd edition. Chapman and Hall.

Lysmer, J., Udaka, T., Seed, H. B., and Hwang, R. (1974) LUSH: a computer program for complex response analysis of soil-structure systems. University of California, Berkeley, Report EERC 74–4.

Parlett, B. N., and Reid, J. K. (1981) Tracking the progress of the Lanczos algorithm for large symmetric eigenproblems. *IMA J. Num. Anal.*, **1**, 135–55.

Smith, I. M. (1979) Discrete element analysis of pile instability. *Int. J. Num. Anal. Meth. Geomechanics*, **3**, 205–11.

Smith, I. M. (1984) Adaptability of truly modular software. *Engineering Computations*, **1**, No. 1, March, 25–35.

Taig, I. C. (1961) Structural analysis by the matrix displacement method. English Electric Aviation, Report S017.

Timoshenko, S. P., Young, D. H., and Weaver, W. (1974) *Vibration Problems in Engineering*, 4th edition. John Wiley, New York.

Wong, S. W. (1987) Element-by-element methods in geomechanics. Ph.D. Thesis, University of Manchester.

Zienkiewicz, O. C., Irons, B. M., Ergatoudis, S., Ahmad, S., and Scott, F. C. (1969) Isoparametric and associated element families for two and three dimensional analysis. In *Proceedings of Course on Finite Element Methods in Stress Analysis*, ed. I. Holand and K. Bell. Trondheim Technical University.

Zienkiewicz, O. C., Too, J., and Taylor, R. L. (1971) Reduced integration technique in general analysis of plates and shells. *Int. J. Num. Meth. Eng.*, **3**, 275–90.

CHAPTER 4
Static Equilibrium of Structures

4.0 INTRODUCTION

Practical finite element analysis had as its starting point matrix analyses of 'structures', by which engineers usually mean assemblages of elastic, line elements. The matrix displacement (stiffness) method is a special case of finite element analysis and, since many engineers still begin their acquaintance with the finite element method in this way, the opening applications chapter of this book is devoted to 'structural' analysis.

The first application program, Program 4.0, permits the analysis of horizontal uniform beams made up of elements of constant properties (flexural rigidity and length). In this case all elements have the same stiffness matrix which is formed by a single subroutine call. As the chapter unfolds, various additional features will gradually be introduced into the programs. In Program 4.1 stepped beams containing elements of varying properties are considered, and a simple method for imposing 'displacement' boundary conditions is demonstrated.

Anticipating more complex elements used later in the book, the beam element stiffness and 'mass' matrices are formulated numerically using Gaussian quadrature in Program 4.2, in the analysis of a beam on an elastic foundation. Such procedures will be essential for elements whose stiffness matrices are too complex to state explicitly.

Programs 4.3 and 4.4 are for the analysis of two- and three-dimensional frames respectively with arbitrarily inclined members. The programs are quite general and although most joints will be rigid, internal pins are also easily dealt with.

Trusses in which members are connected by pin-joints can be analysed as special cases of the framed structures. It has been decided, however, to give them special treatment in view of their widespread use and Programs 4.5 and 4.6 deal with the two- and three-dimensional cases respectively.

Program 4.7 deals with two-dimensional framed structures which exhibit elastic–perfectly plastic material behaviour. The program enables plastic collapse loads to be computed using a 'constant stiffness' approach. This means that the non-linearity is dealt with iteratively by altering the loads on the structure rather than modifying the stiffness matrix, as is done in more traditional approaches.

Program 4.8 performs elastic stability analyses of two-dimensional framed structures. As loads are increased on the system, axial forces are monitored and the stiffness matrix modified accordingly. A buckling mode is signalled by a change in sign of the determinant of the global stiffness matrix.

Finally, Program 4.9 describes a method for analysing thin plates in bending. This could be considered the first 'genuine' finite element program in the book and anticipates the solid mechanics applications of Chapter 5. However, since one dimension (the thickness) has been reduced to zero, it is still a 'structural' element.

PROGRAM 4.0: EQUILIBRIUM OF UNIFORM BEAMS

```
C
C       PROGRAM 4.0 EQUILIBRIUM OF UNIFORM BEAMS
C
C
C       ALTER NEXT LINE TO CHANGE PROBLEM SIZE
C
      PARAMETER(IKV=400,ILOADS=100,INF=100)
C
      REAL KM(4,4),ELD(4),ACTION(4),KV(IKV),LOADS(ILOADS)
      INTEGER G(4),NF(INF,2)
      DATA IKM,IDOF/2*4/,NODOF/2/,IW/3/
C
C       INPUT SECTION
C
      READ(5,*)NXE,N,NN,NR,EI,ELL
      IR=(IW+1)*N
      CALL NULVEC(KV,IR)
C
C       NODE FREEDOM DATA
C
      CALL READNF(NF,INF,NN,NODOF,NR)
C
C       ELEMENT STIFFNESS MATRIX
C
      CALL BEAMKM(KM,EI,ELL)
C
C       GLOBAL STIFFNESS MATRIX ASSEMBLY
C
      DO 10 IP=1,NXE
      CALL GSTRNG(IP,NODOF,NF,INF,G)
   10 CALL FORMKV(KV,KM,IKM,G,N,IDOF)
C
C       EQUATION SOLUTION
C
      CALL BANRED(KV,N,IW)
      CALL NULVEC(LOADS,N)
      READ(5,*)NL,(K,LOADS(K),I=1,NL)
      CALL BACSUB(KV,LOADS,N,IW)
      CALL PRINTV(LOADS,N)
C
C       RETRIEVE ELEMENT END FORCES AND MOMENTS
```

```
C
      DO 20 IP=1,NXE
      CALL GSTRNG(IP,NODOF,NF,INF,G)
      DO 30 I=1,IDOF
      IF(G(I).EQ.0)ELD(I)=0.
   30 IF(G(I).NE.0)ELD(I)=LOADS(G(I))
      CALL MVMULT(KM,IKM,ELD,IDOF,IDOF,ACTION)
   20 CALL PRINTV(ACTION,IDOF)
      STOP
      END
```

The main features of this program are the elastic beam element stiffness matrix (equation 2.26b) and the global stiffness matrix assembly procedures described in Chapter 3.

The structure chart in Figure 4.1 gives the main sequence of operations. Figure 4.2 shows a simply supported elastic beam of unit length carrying, in case (a), a concentrated unit load at its mid-span and, in case (b), a uniformly distributed unit load. At each node, two degrees of freedom are possible, a vertical translation and a rotation in that order. The global node numbering system reads from the left, as shown in Figure 4.3 and, at the element level, node one is always to the left and node two, to the right. The degrees of freedom within the element are taken in the order $w_1, \theta_1, w_2, \theta_2$ so as to be consistent with the stiffness matrix equation. As explained in Chapter 3, the nodal freedom numbering associated with each element, accounting for any restraints, is contained in the 'steering'

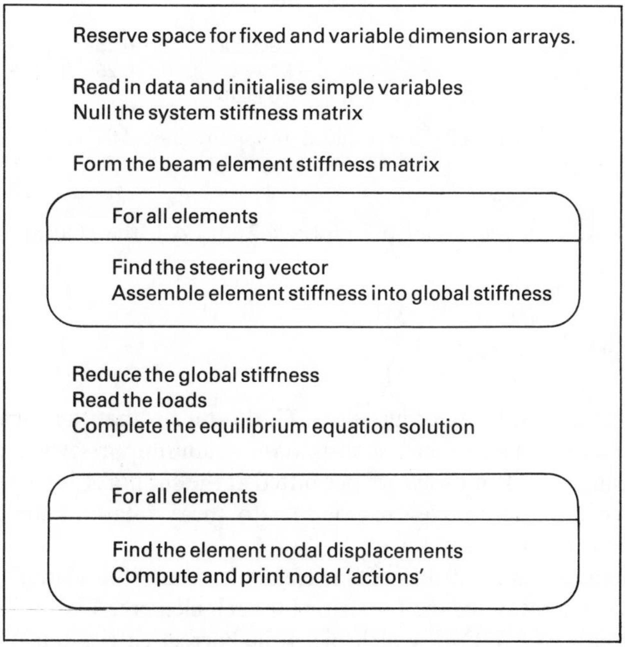
Reserve space for fixed and variable dimension arrays.
Read in data and initialise simple variables
Null the system stiffness matrix
Form the beam element stiffness matrix
For all elements
Find the steering vector
Assemble element stiffness into global stiffness
Reduce the global stiffness
Read the loads
Complete the equilibrium equation solution
For all elements
Find the element nodal displacements
Compute and print nodal 'actions'

Figure 4.1 Structure chart in Program 4.0

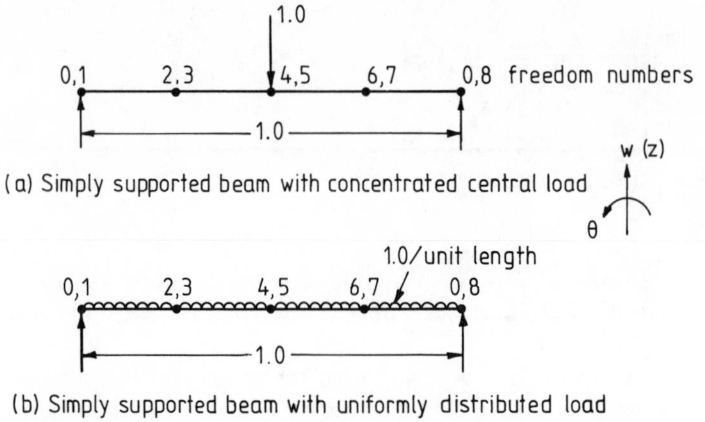

Structure data	NXE	N	NN	NR	
	4	8	5	2	
Element data	EI	ELL			
	1.0	0.25			
Node freedom data	1 0 1	5 0 1			
Loads data	NL	(K_I LOADS (K_I), I = 1, NL)			
	1	4 −1.0			(a)
	5	1 −0.005208	2 −0.25		
		4 −0.25	6 −0.25		(b)
		8 0.005208			

Figure 4.2 Mesh and data for Program 4.0

vector G. Thus, with reference to Figures 4.2 and 4.3, the steering vector for element 1 would be

$$[0 \quad 1 \quad 2 \quad 3]^T$$

and for element 2

$$[2 \quad 3 \quad 4 \quad 5]^T \quad \text{etc.}$$

If a zero appears in the 'steering' vector G, this means that the corresponding displacement has been set to zero, such as at the beam supports where no vertical translation can occur. Rotations are permitted at the supports, however, and the degree of freedom numbers corresponding to these rotations appear in the 'steering' vector in the normal way.

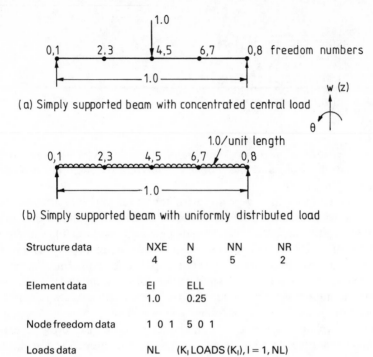

For problems in which strings of line elements occur, it is a simple matter to automate the generation of the G vector for each element. This is done by the library subroutine GSTRNG which picks the correct entries out of the nodal freedom array NF (see Appendix 3 for details). For problems of this type, nodal

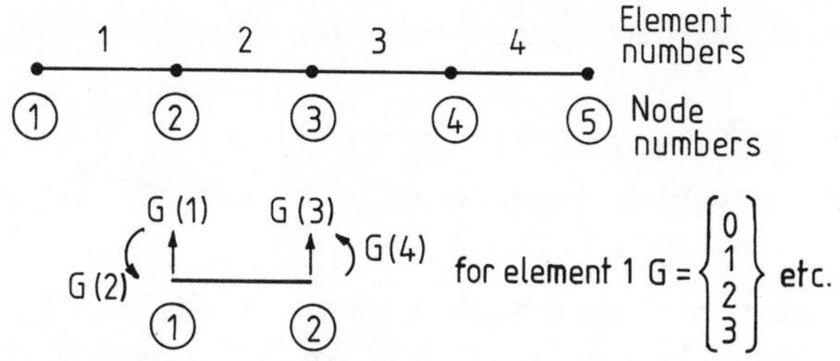

Figure 4.3 Node and element numbering convention

freedom data concerning boundary restraints is read by the subroutine READNF. This data takes the form of the restrained node number followed by either ones or zeros, where the latter implies a restrained degree of freedom and the former an unrestrained degree of freedom. This data must be repeated for every node that has a restraint imposed upon it. In the example of Figure 4.2, there are two restrained nodes (nodes 1 and 5), and in both cases only the translational freedom is suppressed.

Returning to the main program, with the exception of simple integer counters, the meanings of the variable names used with reference to the mesh of Figure 4.2 are listed below:

EI	element flexural rigidity (1.0)
ELL	element length (0.25)
IKM IKV ILOADS INF	working size of arrays KM, KV, LOADS, NF
NXE	number of elements in the x direction (4)
N	number of degrees of freedom in the mesh (8)
NN	number of nodes in the mesh (5)
NR	number of restrained nodes in the mesh (2)
NL	number of loaded freedoms (1(a), 5(b))
IDOF	number of degrees of freedom per element (4)
NODOF	number of degrees of freedom per node (2)
IW	half-bandwidth of mesh (3)
IR	N*(IW + 1)—working length of vector KV

Space is reserved for small fixed length arrays:

KM	element stiffness matrix
ELD	element displacement vector
ACTION	element nodal action vector
G	steering vector

and then for variable length arrays which can be changed to alter the problem size:

KV	global stiffness matrix (IKV $\geqslant$ IR)
LOADS	global load (displacement) vector (ILOADS $\geqslant$ N)
NF	nodal freedom array (INF $\geqslant$ NN)

Data is divided into two types, namely fixed data IKM, IDOF, NODOF and IW and variable data IKV, ILOADS, INF which the user must change as the corresponding arrays are made bigger or smaller. As was pointed out in Chapter 1, the deficiencies of FORTRAN 77 make such changes essential.

In the 'input' section, simple variables are assigned their values from the data and array spaces are nulled where necessary.

The subroutine READNF then reads the 'nodal freedom data' and the element stiffness matrix KM is formed by subroutine BEAMKM.

The global stiffness matrix KV is then assembled element by element by the subroutine FORMKV once the 'steering' vector G has been formed by the subroutine GSTRNG. Gaussian elimination of the system equations is divided into a 'reduction' phase, performed by the subroutine BANRED, and a back-substitution phase performed by the subroutine BACSUB.

Before the back-substitution phase, however, the 'loads data' must be read and this is input as the number of loaded freedoms (NL) followed by the freedom number and the value of the load. Nodal loading in the context of beam analysis can take the form of either point loads or moments. In load case (a), the loading is quite straightforward with a single load of magnitude -1.0 applied to freedom number 4. Load case (b), however, is slightly more involved as the loading is uniformly distributed along the entire length of the beam. In finite element analyses, loading can only be applied at the nodes, so the uniformly distributed load must be simulated by an equivalent set of nodal loads. These arise naturally in the finite element formulation (see equation 2.26a). Alternatively (e.g.

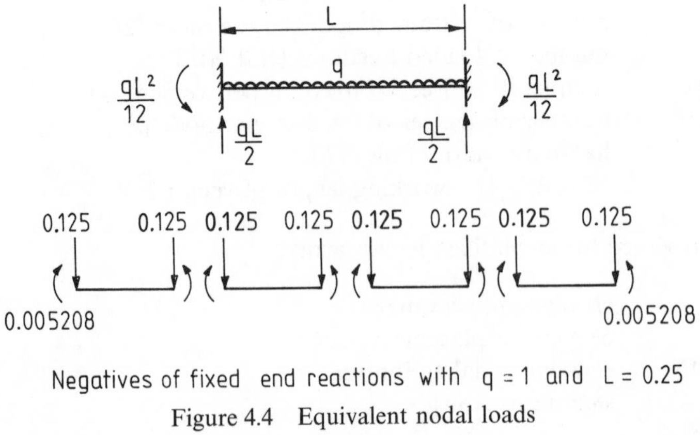

Figure 4.4 Equivalent nodal loads

Przemieniecki, 1968) they may be thought of as the negatives of the end reactions and moments that would apply if each element was fully ‘encastre’ at both ends. Figure 4.4 shows these loads for each of the four elements in Figure 4.2. It is seen that for elements of equal length the internal rotational loads cancel out, leaving couples at the supports and internal point loads to be applied.

Back-substitution solves for the displacements which take the form of vertical translations and rotations. Once the element displacements are known, the element end forces (or ‘actions’) can be retrieved. For this purpose the G vector is formed again to isolate the nodal displacements for each element stored in ELD. Subroutine MVMULT performs the multiplication by the stiffness matrix KM and the end forces and moments are printed.

The computed results for both load cases are reproduced in Figure 4.5. For load case (a), the central deflection is given by freedom number 4 to be 0.02083 which is in exact agreement with the analytical solution $WL^3/48EI$, where W is the central load and L the total length of the beam. The ACTION vector for each element gives what are usually called the ‘shear forces’ and ‘bending moments’ in the beam.

For load case (b), however, although the maximum deflection is again exactly given by 0.01302 or $5qL^4/384EI$, where q is the uniformly distributed load, the fixed end moments and reactions for each element must be added to the ACTION vectors to retrieve the ‘shear forces’ and ‘bending moments’.

For example, for element 1 in load case (b):

		Computed		‘Fixed end’		Actual
Shear	Node	0.3750		0.1250		0.5000
Moment	(1)	−0.0052	+	0.0052	=	0
Shear	Node	−0.3750		0.1250		−0.2500
Moment	(2)	0.0990		−0.0052		0.0938

```
-.6250E-01  -.1432E-01  -.4688E-01  -.2083E-01  -.7216E-15  -.1432E-01 } Displacements
 .4688E-01   .6250E-01
 .5000E+00   .1776E-14  -.5000E+00   .1250E+00 )
 .5000E+00  -.1250E+00  -.5000E+00   .2500E+00 } Actions
-.5000E+00  -.2500E+00   .5000E+00   .1250E+00 }
-.5000E+00  -.1250E+00   .5000E+00   .0000E+00 )

                                                          Case a)

-.4167E-01  -.9277E-02  -.2865E-01  -.1302E-01  -.4441E-15  -.9277E-02 } Displacements
 .2865E-01   .4167E-01
 .3750E+00  -.5208E-02  -.3750E+00   .9896E-01 )
 .1250E+00  -.9896E-01  -.1250E+00   .1302E+00 } Actions
-.1250E+00  -.1302E+00   .1250E+00   .9896E-01 }
-.3750E+00  -.9896E-01   .3750E+00   .5208E-02 )

                                                          Case b)
```

Figure 4.5 Results from Program 4.0

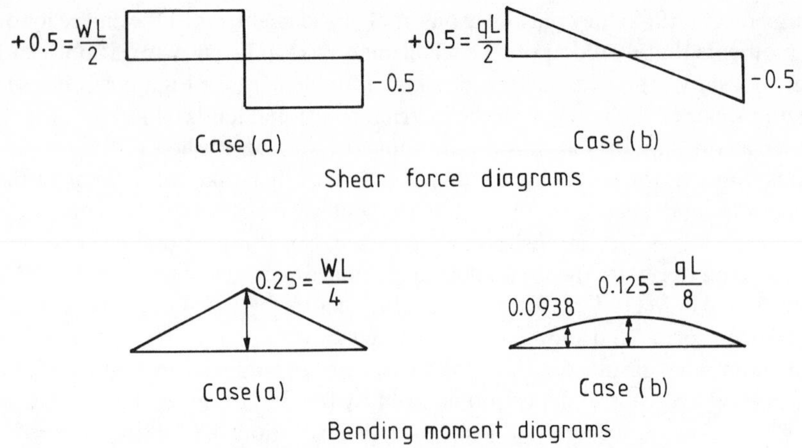

Figure 4.6 Shear force and loading moment diagrams

and for element 2:

Shear	Node	0.1250		0.1250		0.2500
Moment	(2)	−0.0990	+	0.0052	=	−0.0938
Shear	Node	−0.1250		0.1250		0
Moment	(3)	0.1302		−0.0052		0.1250

and so on

The bending moment and shear force diagrams for both cases are summarised in Figure 4.6.

Note that because the deflected shapes are exactly modelled and compatibility and equilibrium are exactly satisfied at the node points, these solutions are exact. Of course, in the problems considered here the same results could have been achieved by using fewer elements and taking account of symmetry.

PROGRAM 4.1: EQUILIBRIUM OF STEPPED BEAMS INCORPORATING PRESCRIBED DISPLACEMENT BOUNDARY CONDITIONS

```
C
C      PROGRAM 4.1  EQUILIBRIUM OF STEPPED BEAMS
C
C
C      ALTER NEXT LINE TO CHANGE PROBLEM SIZE
C
      PARAMETER(IKV=400,ILOADS=100,IPROP=20,INO=20,INF=100)
C
      REAL KM(4,4),ELD(4),ACTION(4),KV(IKV),LOADS(ILOADS),VAL(INO),
     +PROP(IPROP,2)
      INTEGER G(4),NF(INF,2),NO(INO)
      DATA IKM,IDOF/2*4/,NODOF/2/,IW/3/
```

```
C
C        INPUT SECTION
C
      READ(5,*)NXE,N,NN,NR
      IR=(IW+1)*N
      CALL NULVEC(KV,IR)
C
C        NODE FREEDOM DATA
C
      CALL READNF(NF,INF,NN,NODOF,NR)
C
C        GLOBAL STIFFNESS MATRIX ASSEMBLY
C
      DO 10 IP=1,NXE
      READ(5,*)EI,ELL
      PROP(IP,1)=EI
      PROP(IP,2)=ELL
      CALL BEAMKM(KM,EI,ELL)
      CALL GSTRNG(IP,NODOF,NF,INF,G)
   10 CALL FORMKV(KV,KM,IKM,G,N,IDOF)
C
C        EQUATION SOLUTION
C
      CALL NULVEC(LOADS,N)
      READ(5,*)IFIX,(NO(I),VAL(I),I=1,IFIX)
      DO 20 I=1,IFIX
      KV(NO(I))=KV(NO(I))+1.E20
   20 LOADS(NO(I))=KV(NO(I))*VAL(I)
      READ(5,*)NL,(K,LOADS(K),I=1,NL)
      CALL BANRED(KV,N,IW)
      CALL BACSUB(KV,LOADS,N,IW)
      CALL PRINTV(LOADS,N)
C
C        RETRIEVE ELEMENT END FORCES AND MOMENTS
C
      DO 30 IP=1,NXE
      EI=PROP(IP,1)
      ELL=PROP(IP,2)
      CALL BEAMKM(KM,EI,ELL)
      CALL GSTRNG(IP,NODOF,NF,INF,G)
      DO 40 I=1,IDOF
      IF(G(I).EQ.0)ELD(I)=0.
   40 IF(G(I).NE.0)ELD(I)=LOADS(G(I))
      CALL MVMULT(KM,IKM,ELD,IDOF,IDOF,ACTION)
   30 CALL PRINTV(ACTION,IDOF)
      STOP
      END
```

The method used throughout this book is to explore solutions of new problems by making gradual alterations to previously described programs. This program, therefore, is an adaptation of the previous one, the modifications being as follows:

1. Each element can have different properties (flexural rigidity and length).
2. Prescribed displacement boundary conditions are incorporated in addition to conventional loading (see Section 3.6).

Because the problem considered consists of a string of line elements the subroutines which assemble the global stiffness matrix remain unchanged. The only additional data in this program is the integer IFIX which is the number of freedoms at which displacements are to be prescribed and the arrays NO and VAL which hold, respectively, the freedom number to be fixed and its value. Thus, there are IFIX number of entries in each of these vectors. Additional variables required in the PARAMETER statement are IPROP ($\geqslant$ NXE), the working size of array PROP, and INO ($\geqslant$ IFIX), the working size of arrays VAL and NO.

The appropriate structure chart is still essentially that of Figure 4.1, except that the element properties EI and ELL are read inside the element inspection loop and are stored in the properties array PROP. The element stiffness formulation subroutine BEAMKM is thus called once for each element.

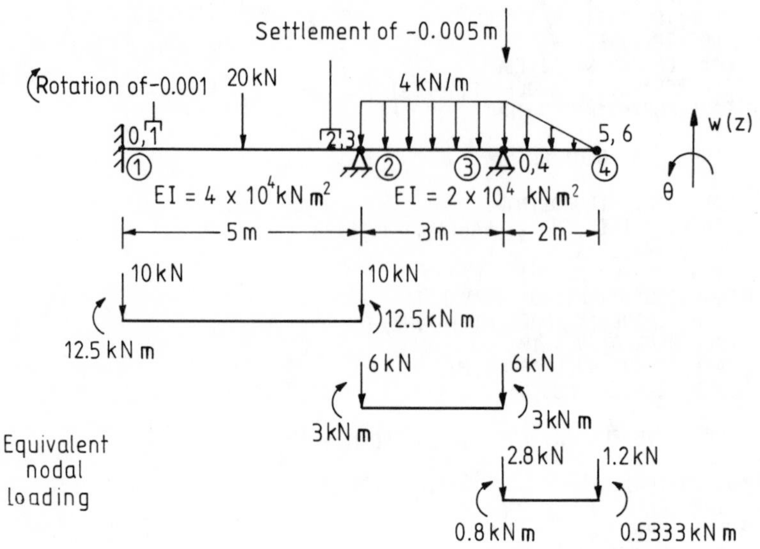

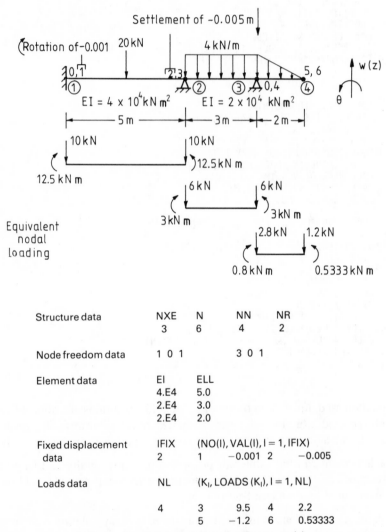

Structure data	NXE	N	NN	NR
	3	6	4	2
Node freedom data	1 0 1		3 0 1	
Element data	EI	ELL		
	4.E4	5.0		
	2.E4	3.0		
	2.E4	2.0		
Fixed displacement data	IFIX	(NO(I), VAL(I), I = 1, IFIX)		
	2	1	−0.001 2	−0.005
Loads data	NL	(K_I, LOADS (K_I), I = 1, NL)		
	4	3	9.5 4	2.2
		5	−1.2 6	0.53333

Figure 4.7 Mesh and data for Program 4.1

```
-.1000E-02  -.5000E-02   .2051E-03   .2410E-02   .4713E-02   .2343E-02 } Displacements
 .1157E+02   .1928E+02  -.1157E+02   .3856E+02 ]
-.9577E+01  -.2906E+02   .9577E+01   .3333E+00 } Actions
 .1200E+01   .1867E+01  -.1200E+01   .5333E+00 ]
```

Figure 4.8 Results from Program 4.1

At the IFIX number of freedoms, held in NO, the large number 1.E20 is added to the appropriate leading diagonal term of the global stiffness matrix KV. The load vector LOADS is then set to this augmented stiffness multiplied by the required value stored in VAL. The prescribed loads are incorporated into the loads vector in the usual way. After solution of the equilibrium equations, the element end forces and moments are retrieved by reforming each of the element stiffness matrices and multiplying it by the corresponding element displacement vector.

The beam shown in Figure 4.7 is subjected to a combination of prescribed displacements and loads. Node (1) is rotated clockwise by 0.001 radians and node (2) is translated by 0.005 m vertically downwards. In addition, a vertical force of 20 kN acts between nodes (1) and (2), a uniform load of 4 kN/m acts between nodes (2) and (3) and a linearly varying load from 4 kN/m to zero acts between nodes (3) and (4). The appropriate nodal loads are computed by incorporating $q(x)$ in (2.25). The computed results are shown in Figure 4.8. The first six numbers printed are the computed displacements, and it is clear that the prescribed values at freedoms 1 and 2 are reproduced correctly. The other displacement values indicate that the rotation at node (2), for example, equals 0.000205 (anti-clockwise).

In order to compute the actual moments and shear forces in the beam, the fixed end moments for each element must be added to the corresponding ACTION vector printed in the output. For example, the moments in the central element at nodes (2) and (3) are given by

$$\begin{aligned} M_2 &= -29.06 + 3 = -26.06\ \text{kN m} \\ M_3 &= \quad 0.33 - 3 = -\ \ 2.67\ \text{kN m} \end{aligned}$$

PROGRAM 4.2: BEAM ON AN ELASTIC FOUNDATION WITH NUMERICALLY INTEGRATED STIFFNESS AND MASS MATRICES

```
C
C      PROGRAM 4.2 NUMERICALLY INTEGRATED BEAM ON ELASTIC FOUNDATION
C
C
C      ALTER NEXT LINE TO CHANGE PROBLEM SIZE
C
      PARAMETER(IKV=400,ILOADS=100,INF=100,IPROP=15)
C
      REAL KM(4,4),ELD(4),ACTION(4),KV(IKV),LOADS(ILOADS),MM(4,4),
     +DTD(4,4),FTF(4,4),DER2(4),FUN(4),SAMP(7,2),MOM(INF),
     +STORKM(IPROP,4,4)
```

```
      INTEGER G(4),NF(INF,2)
      DATA IKM,IDOF/2*4/,NODOF/2/,IW/3/,ISAMP/7/
C
C      INPUT SECTION
C
      READ(5,*)NXE,N,NN,NR,NGP,EI,FS0,FS1,ELL
      IR=(IW+1)*N
      CALL NULVEC(KV,IR)
      CALL GAUSS(SAMP,ISAMP,NGP)
C
C      NODE FREEDOM DATA
C
      CALL READNF(NF,INF,NN,NODOF,NR)
C
C      GLOBAL STIFFNESS AND MASS MATRIX ASSEMBLY
C
      X=0.
      DO 10 IP=1,NXE
      CALL NULL(KM,IKM,IDOF,IDOF)
      CALL NULL(MM,IKM,IDOF,IDOF)
      DO 20 I=1,NGP
      SP=X+ELL*.5*(SAMP(I,1)+1.)
      FS=SP/(ELL*NXE)*(FS1-FS0)+FS0
      WT=SAMP(I,2)
      CALL FMBEAM(DER2,FUN,SAMP,ISAMP,ELL,I)
      DO 30 K=1,IDOF
      DO 30 L=1,IDOF
      FTF(K,L)=FUN(K)*FUN(L)*WT*.5*ELL*FS
   30 DTD(K,L)=DER2(K)*DER2(L)*WT*8.*EI/(ELL*ELL*ELL)
      CALL MATADD(MM,IKM,FTF,IKM,IDOF,IDOF)
   20 CALL MATADD(KM,IKM,DTD,IKM,IDOF,IDOF)
      CALL MATADD(KM,IKM,MM,IKM,IDOF,IDOF)
      DO 40 I=1,IDOF
      DO 40 J=1,IDOF
   40 STORKM(IP,I,J)=KM(I,J)
      CALL GSTRNG(IP,NODOF,NF,INF,G)
      X=X+ELL
   10 CALL FORMKV(KV,KM,IKM,G,N,IDOF)
C
C      EQUATION SOLUTION
C
      CALL BANRED(KV,N,IW)
      CALL NULVEC(LOADS,N)
      READ(5,*)NL,(K,LOADS(K),I=1,NL)
      CALL BACSUB(KV,LOADS,N,IW)
C
C      RETRIEVE ELEMENT END FORCES AND MOMENTS
C
      DO 50 IP=1,NXE
      CALL GSTRNG(IP,NODOF,NF,INF,G)
      DO 60 I=1,IDOF
      DO 60 J=1,IDOF
   60 KM(I,J)=STORKM(IP,I,J)
      DO 70 I=1,IDOF
      IF(G(I).EQ.0)ELD(I)=0.
   70 IF(G(I).NE.0)ELD(I)=LOADS(G(I))
      CALL MVMULT(KM,IKM,ELD,IDOF,IDOF,ACTION)
      MOM(IP)=ACTION(2)
      IF(IP.EQ.NXE)MOM(IP+1)=-ACTION(4)
```

```
50 CONTINUE
   DO 80 I=1,NN
   WRITE(6,'(2E12.4)')LOADS(2*I-1),MOM(I)
80 CONTINUE
   STOP
   END
```

For more complicated elements, or elements whose properties vary along their length, it is often more convenient to form the element matrices numerically. The most efficient way of doing this is to use Gaussian quadrature, so it is necessary to replace the beam coordinate x in the range $(0, L)$ by the local coordinate ξ in the range $(-1, 1)$ using the transformation:

$$x = \frac{L}{2}(\xi + 1) \tag{4.1}$$

The cubic beam shape functions (equation 2.22) may be written thus:

$$\begin{aligned} N_1 &= \tfrac{1}{4}(\xi^3 - 3\xi + 2) \\ N_2 &= \frac{L}{8}(\xi^3 - \xi^2 - \xi + 1) \\ N_3 &= \tfrac{1}{4}(-\xi^3 + 3\xi + 2) \\ N_4 &= \frac{L}{8}(\xi^3 + \xi^2 - \xi - 1) \end{aligned} \tag{4.2}$$

and these (FUN), together with their second derivatives with respect to ξ (DER2), are formed by the subroutine FMBEAM.

As discussed in Chapter 2, the stiffness matrix for a beam can be formed from integrals of the type

$$\mathrm{KM}_{KL} = \mathrm{EI}\int_0^L \frac{\mathrm{d}^2 N_K}{\mathrm{d}x^2}\frac{\mathrm{d}^2 N_L}{\mathrm{d}x^2}\,\mathrm{d}x, \qquad K, L = 1, 2, 3, 4 \tag{4.3}$$

and the mass matrix from integrals of the type

$$\mathrm{MM}_{KL} = \rho A \int_0^L N_K N_L\,\mathrm{d}x, \qquad K, L = 1, 2, 3, 4 \tag{4.4}$$

Converting to local coordinates from equation (4.1) gives

$$\frac{\mathrm{d}^2 N}{\mathrm{d}x^2} = \frac{\mathrm{d}^2 N}{\mathrm{d}\xi^2}\left(\frac{\mathrm{d}\xi}{\mathrm{d}x}\right)^2 = \frac{4}{L^2}\frac{\mathrm{d}^2 N}{\mathrm{d}\xi^2}$$

and (4.5)

$$\mathrm{d}x = \frac{L}{2}\,\mathrm{d}\xi$$

Hence, equations (4.3) and (4.4) become

$$\left.\begin{aligned} \mathrm{KM}_{KL} &= \frac{8}{L^3}\,\mathrm{EI}\int_{-1}^{1}\frac{\mathrm{d}^2N_K}{\mathrm{d}\xi^2}\frac{\mathrm{d}^2N_L}{\mathrm{d}\xi^2}\,\mathrm{d}\xi \qquad (4.6)\\ \mathrm{MM}_{KL} &= \frac{L}{2}\rho A\int_{-1}^{1}N_K N_L\,\mathrm{d}\xi \qquad (4.7)\end{aligned}\right\}\quad K, L = 1, 2, 3, 4$$

In program terminology these expressions become:

$$\mathrm{KM(K,L)} = 8./(\mathrm{ELL}{*}{*}3){*}\sum_{\mathrm{I}=1}^{\mathrm{NGP}}\mathrm{EI}{*}\mathrm{WT_I}{*}\mathrm{DER2(K)}{*}\mathrm{DER2(L)} \qquad (4.8)$$

$$\mathrm{MM(K,L)} = \mathrm{ELL}/2.{*}\sum_{\mathrm{I}=1}^{\mathrm{NGP}}\mathrm{FS}{*}\mathrm{WT_I}{*}\mathrm{FUN(K)}{*}\mathrm{FUN(L)} \qquad (4.9)$$

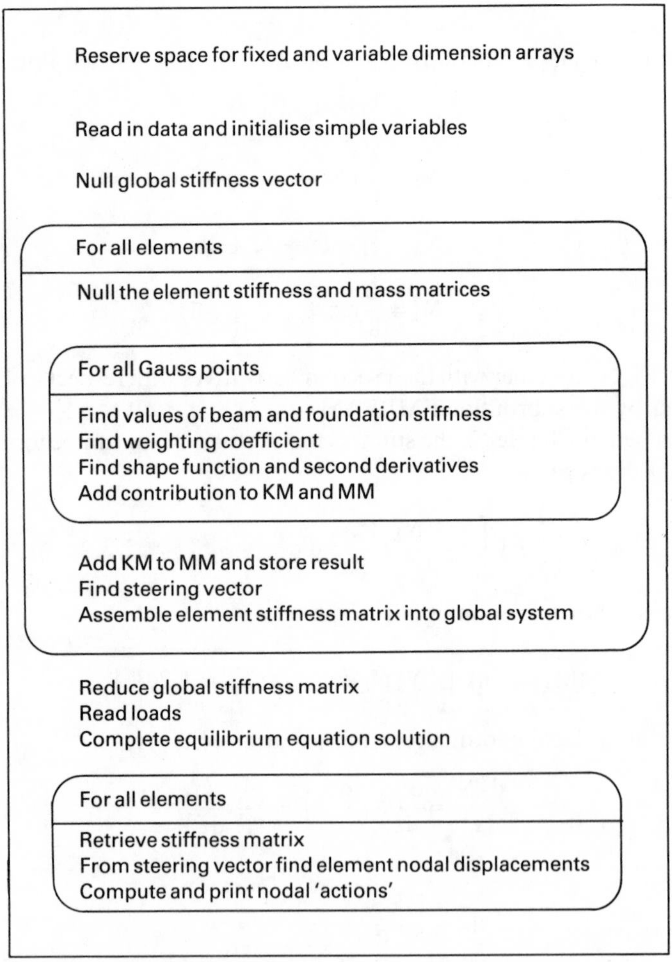
Reserve space for fixed and variable dimension arrays
Read in data and initialise simple variables
Null global stiffness vector
For all elements
Null the element stiffness and mass matrices
For all Gauss points
Find values of beam and foundation stiffness
Find weighting coefficient
Find shape function and second derivatives
Add contribution to KM and MM
Add KM to MM and store result
Find steering vector
Assemble element stiffness matrix into global system
Reduce global stiffness matrix
Read loads
Complete equilibrium equation solution
For all elements
Retrieve stiffness matrix
From steering vector find element nodal displacements
Compute and print nodal 'actions'

Figure 4.9 Structure chart for Program 4.2

where, WT_I is the weighting coefficient at the Ith Gaussian integration point. The weights and sampling points in local coordinates are stored in the array SAMP produced by the library subroutine GAUSS. The required number of Gaussian integration points NGP is read in as data.

Turning to the main program and following the structure chart of Figure 4.9, it can be seen that an additional loop is required where I counts from 1 to NGP. For each of these integrating points the beam stiffness EI and the foundation stiffness FS (analogous to ρA) are found, the weighting coefficient is taken directly from SAMP in the statement WT = SAMP (I, 2) and the library routine FMBEAM is called. This routine computes the vectors of shape functions FUN and second derivatives DER2 at the particular sampling point and multiplications required in equations (4.8) and (4.9) are performed. The resulting arrays DTD and FTF are then added into the accumulating matrices KM and MM respectively using routine MATADD. Once KM and MM have been formed, they are added together to give the net element stiffness matrix (still called KM). This element stiffness matrix is stored in the three-dimensional array STORKM.

The remainder of the program follows a familiar course. The stiffness matrices

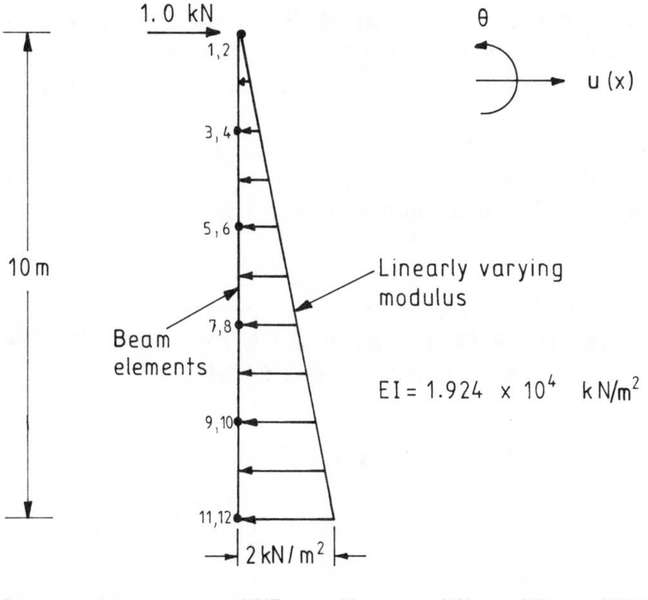

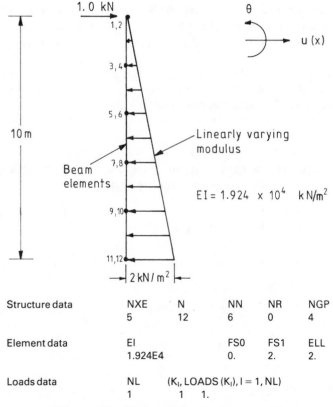

Structure data	NXE	N	NN	NR	NGP
	5	12	6	0	4
Element data	EI		FS0	FS1	ELL
	1.924E4		0.	2.	2.
Loads data	NL	$(K_I, \text{LOADS}(K_I), I = 1, NL)$			
	1	1 1.			

Figure 4.10 Mesh and data for Program 4.2

δ_h	Moment
.9016E+00	.1455E-10
.6605E+00	-.1792E+01
.4199E+00	-.2591E+01
.1797E+00	-.2110E+01
-.6005E-01	-.8311E+00
-.2996E+00	-.2910E-10

Figure 4.11 Results from Program 4.2

are assembled into a global system matrix which is factorised, loads are read in and the system of equations solved to give the displacements. To retrieve the nodal 'actions' a further loop scans each element and obtains the nodal displacements which are multiplied by the element stiffness matrix obtained from STORKM.

The example shown in Figure 4.10 shows a pile made up of five beam elements installed in an elastic soil medium with modulus varying linearly from zero at the ground surface to 2 kN/m^2 at 10 m depth. The beam elements have a constant flexural rigidity EI of 1.924×10^4 kN m^2. This example has no restraints imposed on it so NR equals zero and consequently the routine READNF requires no data. The number of Gauss-points per element NGP is put to 4 because the terms in equation (4.9) are seventh order polynomials; thus the integration will be performed exactly. The data EI, FS0 and FS1 represent the beam flexural rigidity and the foundation stiffness at the top and bottom of the pile respectively and assume a linear variation. A unit load has been applied horizontally at the top of the pile.

The results shown in Figure 4.11 give the horizontal translation and the moment (stored in vector MOM) in the pile at each nodal point. These values agree closely with the exact solution (Hetenyi, 1946).

PROGRAM 4.3: EQUILIBRIUM OF TWO-DIMENSIONAL RIGID-JOINTED FRAMES

```
C
C       PROGRAM 4.3 ANALYSIS OF 2-D FRAMES
C
C
C       ALTER NEXT LINE TO CHANGE PROBLEM SIZE
C
      PARAMETER(IKV=400,ILOADS=100,IPROP=20)
C
      REAL ACTION(6),LOCAL(6),ELD(6),KM(6,6),
     +LOADS(ILOADS),KV(IKV),PROP(IPROP,2),STOREC(IPROP,4)
      INTEGER G(6),STOREG(IPROP,6)
      DATA IKM,IDOF/2*6/,NODOF/3/
C
C       INPUT SECTION
C
      READ(5,*)NXE,N,IW
      IR=N*(IW+1)
      CALL NULVEC(KV,IR)
```

```
C
C       GLOBAL STIFFNESS MATRIX ASSEMBLY
C
      DO 10 IP=1,NXE
      READ(5,*)EA,EI,(STOREC(IP,I),I=1,4),(G(I),I=1,IDOF)
      PROP(IP,1)=EA
      PROP(IP,2)=EI
      CALL BMCOL2(KM,EA,EI,IP,STOREC,IPROP)
      DO 20 I=1,IDOF
   20 STOREG(IP,I)=G(I)
   10 CALL FORMKV(KV,KM,IKM,G,N,IDOF)
C
C       EQUATION SOLUTION
C
      CALL BANRED(KV,N,IW)
      CALL NULVEC(LOADS,N)
      READ(5,*)NL,(K,LOADS(K),I=1,NL)
      CALL BACSUB(KV,LOADS,N,IW)
      CALL PRINTV(LOADS,N)
C
C       RETRIEVE ELEMENT END FORCES AND MOMENTS
C
      DO 30 IP=1,NXE
      EA=PROP(IP,1)
      EI=PROP(IP,2)
      CALL BMCOL2(KM,EA,EI,IP,STOREC,IPROP)
      DO 40 I=1,IDOF
      G(I)=STOREG(IP,I)
      IF(G(I).EQ.0)ELD(I)=0.
   40 IF(G(I).NE.0)ELD(I)=LOADS(G(I))
      CALL MVMULT(KM,IKM,ELD,IDOF,IDOF,ACTION)
      CALL LOC2F(LOCAL,ACTION,IP,STOREC,IPROP)
      CALL PRINTV(ACTION,IDOF)
   30 CONTINUE
      STOP
      END
```

The first three programs in this chapter were concerned only with beam elements, subjected to moments and loading perpendicular to the element axes. It is much more common, of course, to encounter structures made up of members arbitrarily inclined to one another. Loading of such structures results in displacements due to both axial and bending effects, although the former is often ignored in many approximate methods described in texts on structural analysis.

The program described in this section is for the solution of two-dimensional framed structures and uses beam-column elements. These elements each have six degrees of freedom, incorporating two translations and a rotation at each node, with the numbering as shown in Figure 4.12. The element stiffness matrix is obtained by superposing the axial and bending contributions and performing a coordinate transformation to account for the angle of inclination of the element. This generalised stiffness matrix is readily derived and is contained in the library subroutine BMCOL2.

The program is similar to the beam programs described previously but requires more data. For each element, data must be provided to give the stiffness values EA and EI (stored in PROP), the coordinates of the nodes x_1, y_1, x_2, y_2

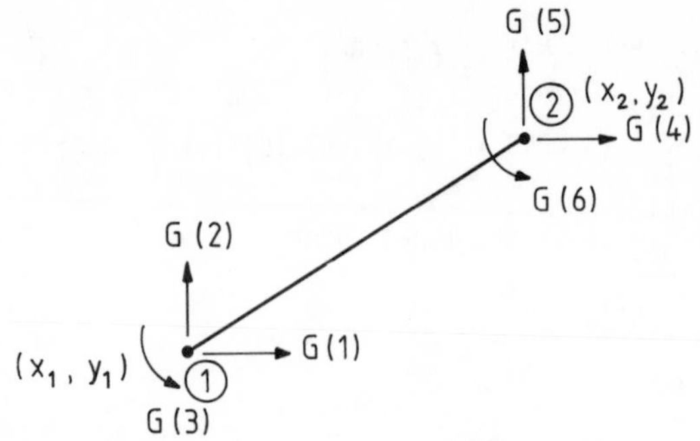

Figure 4.12 Freedom numbering for two-dimensional beam-column element

(stored in STOREC) and the steering vector G in the same order as the coordinates. As the G vector is no longer generated automatically, the variables NN and NR, and the routine READNF, are not needed. To save on storage, the value of the semi-bandwidth IW is read as data and equals the greatest difference between freedom numbers (excluding 'zero' freedoms) occurring for any element in the mesh. The G vector, once read in, is stored in the two-dimensional integer array STOREG.

An additional real array, LOCAL, holds the element end forces and moments in the element local coordinate system. This vector is formed by the routine LOC2F and should be printed out if a shear force diagram is required. In the listing provided here only the global end forces and moments are printed and these are held as usual in the vector ACTION.

The example shown in Figure 4.13 shows a framed structure subjected to a combination of distributed and concentrated loading. The equivalent nodal loading is obtained by independent calculation, although it too could be computed by the program.

The output in Figure 4.14 gives the nodal displacements and the ACTION vector for each element. For example, the rotation at node (1) is found to be – 0.001025. The ACTION vectors for elements 4, 5 and 6 display the correct values, but for elements 1, 2 and 3 the fixed end values must be added. For example, for element 1:

		Computed		'Fixed end'		Actual
F_x		– 30.38		0		– 30.38 kN
F_y	Node	– 19.75	+	60	=	40.25 kN
M	(1)	– 60		60		0 kNm
F_x		30.38		0		30.38 kN
F_y	Node	18.75	+	60	=	79.75 kN
M	(2)	– 58.49		– 60		– 118.49 kN m

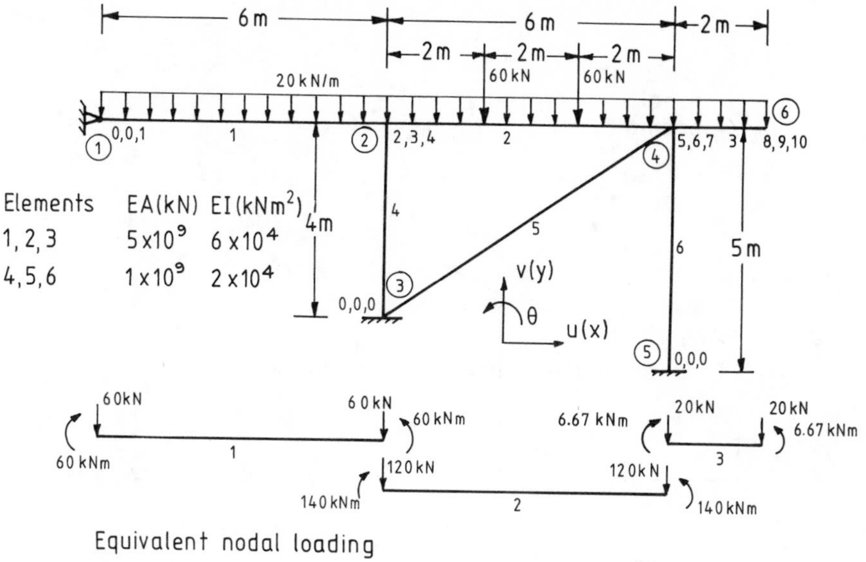

Structure data	NXE	N	IW				
	6	10	5				

Element data	EA	EI	x_1	y_1	x_2	y_2	(G(I), I = 1, 6)
	5.E9	6.E4	0.	0.	6.	0.	0 0 1 2 3 4
	5.E9	6.E4	6.	0.	12.	0.	2 3 4 5 6 7
	5.E9	6.E4	12.	0.	14.	0.	5 6 7 8 9 10
	1.E9	2.E4	6.	−4.	6.	0.	0 0 0 2 3 4
	1.E9	2.E4	6.	−4.	12.	0.	0 0 0 5 6 7
	1.E9	2.E4	12.	−5.	12.	0.	0 0 0 5 6 7

Loads data	NL	(K_I, LOADS (K_I), I = 1, NL)					
	7	1	−60.	3	−180.	4	−80.
		6	−140.	7	133.33	9	−20.
		10	6.67				

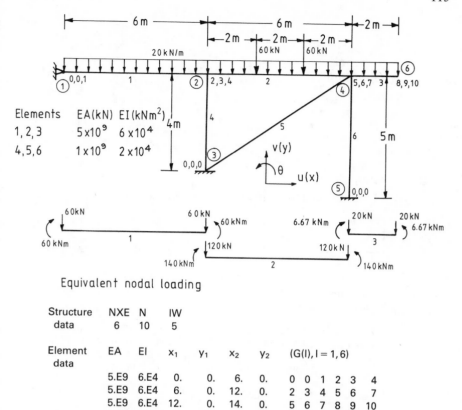

Figure 4.13 Mesh and data for Program 4.3 (first example)

-.1025E-02	.3645E-07	-.8319E-06	-.9497E-03	.6435E-07	-.6283E-06	Displacements
.1774E-02	.6435E-07	.2880E-02	.1329E-02			
-.3038E+02	-.1975E+02	-.6000E+02	.3038E+02	.1975E+02	-.5849E+02	Actions
-.2325E+02	.8238E+01	-.2519E+01	.2325E+02	-.8238E+01	.5195E+02	
.9095E-12	.2000E+02	.3333E+02	-.9095E-12	-.2000E+02	.6670E+01	
.7123E+01	.2080E+03	-.9497E+01	-.7123E+01	-.2080E+03	-.1899E+02	
.3177E+02	.2610E+02	.9839E+01	-.3177E+02	-.2610E+02	.1968E+02	
-.8513E+01	.1257E+03	.1419E+02	.8513E+01	-.1257E+03	.2838E+02	

Figure 4.14 Results from Program 4.3 (first example)

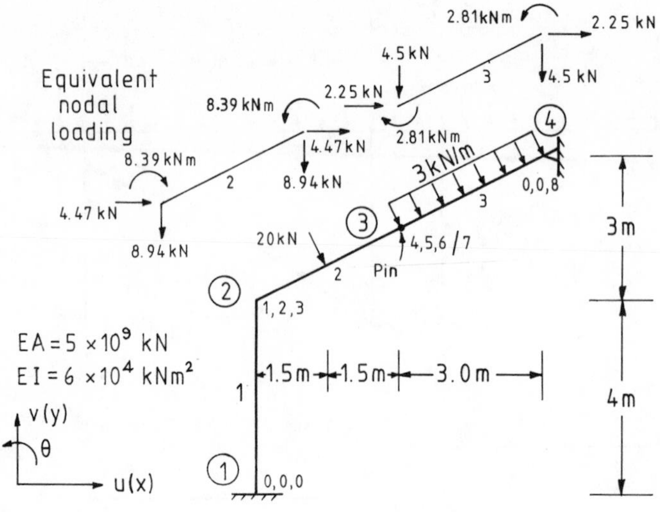

Structure data	NXE	N	IW									
	3	8	5									

							(G(I), I = 1, 6)					
Element data	EA	EI	x_1	y_1	x_2	y_2						
	5.E9	6.E4	0.	0.	0.	4.	0	0	0	1	2	3
	5.E9	6.E4	0.	4.	3.	5.5	1	2	3	4	5	6
	5.E9	6.E4	3.	5.5	6.	7.	4	5	7	0	0	8

Loads data	NL	(K_I, LOADS (K_I), I = 1, NL)					
	8	1	4.47	2	−8.94	3	−8.39
		4	6.72	5	−13.44	6	8.39
		7	−2.81	8	2.81		

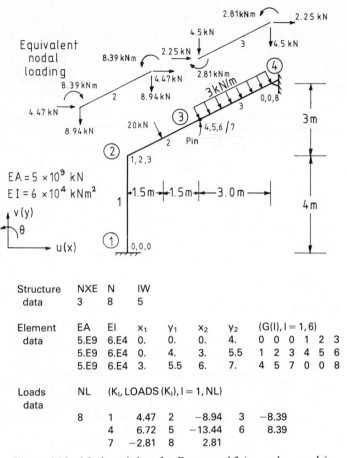

Figure 4.15 Mesh and data for Program 4.3 (second example)

A second example, shown in Figure 4.15, illustrates the generality of the program. The internal pin in the structure is simply dealt with by uncoupling the rotation associated with the element on each side. The nodal loads are computed for elements sustaining loads between nodes in the normal way but in this case the concentrated forces must be resolved into components in the x and y directions before applying them to the structure. The output for this analysis is given in Figure 4.16 and shows that the rotation at node (4) is equal to 0.001623. The true

```
.6543E-07  -.2994E-07  -.8400E-03   .2317E-02  -.4634E-02  -.1780E-02 } Displacement
.1466E-02   .1623E-02
.1890E+02   .3742E+02  -.2520E+02  -.1890E+02  -.3742E+02  -.5040E+02 }
.2337E+02   .2848E+02   .4201E+02  -.2337E+02  -.2848E+02   .8390E+01 } Actions
.3009E+02   .1504E+02  -.2810E+01  -.3009E+02  -.1504E+02   .2810E+01 }
```

Figure 4.16 Results from Program 4.3 (second example)

moments and reactions at the nodes in element 2, for example, are given by:

		Computed		'Fixed end'		Actual
F_x		23.37		−4.47		18.90 kN
F_y	Node	28.48	+	8.94	=	37.42 kN
M	(2)	42.01		8.39		50.40 kN m
F_x		−23.37		−4.47		−27.84 kN
F_y	Node	−28.48	+	8.94	=	−19.54 kN
M	(3)	8.39		−8.39		0 kN m

PROGRAM 4.4: EQUILIBRIUM OF THREE-DIMENSIONAL RIGID-JOINTED FRAMES

```
C
C      PROGRAM 4.4 ANALYSIS OF 3-D FRAMES
C
C
C      ALTER NEXT LINE TO CHANGE PROBLEM SIZE
C
      PARAMETER(IKV=400,ILOADS=100,IPROP=20)
C
      REAL ACTION(12),LOCAL(12),ELD(12),KM(12,12),
     +KV(IKV),LOADS(ILOADS),PROP(IPROP,4),STOREC(IPROP,7)
      INTEGER G(12),STOREG(IPROP,12)
      DATA IKM,IDOF/2*12/,NODOF/6/
C
C      INPUT SECTION
C
      READ(5,*)NXE,N,IW
      IR=N*(IW+1)
      CALL NULVEC(KV,IR)
C
C      GLOBAL STIFFNESS MATRIX ASSEMBLY
C
      DO 10 IP=1,NXE
      READ(5,*)EA,EIY,EIZ,GJ,(STOREC(IP,I),I=1,7),(G(I),I=1,IDOF)
      PROP(IP,1)=EA
      PROP(IP,2)=EIY
      PROP(IP,3)=EIZ
      PROP(IP,4)=GJ
      CALL BMCOL3(KM,EA,EIY,EIZ,GJ,IP,STOREC,IPROP)
      DO 20 I=1,IDOF
   20 STOREG(IP,I)=G(I)
   10 CALL FORMKV(KV,KM,IKM,G,N,IDOF)
C
C      EQUATION SOLUTION
C
      CALL BANRED(KV,N,IW)
      CALL NULVEC(LOADS,N)
      READ(5,*)NL,(K,LOADS(K),I=1,NL)
      CALL BACSUB(KV,LOADS,N,IW)
      CALL PRINTV(LOADS,N)
```

```
C
C      RETRIEVE ELEMENT END FORCES AND MOMENTS
C
      DO 30 IP=1,NXE
      EA=PROP(IP,1)
      EIY=PROP(IP,2)
      EIZ=PROP(IP,3)
      GJ=PROP(IP,4)
      CALL BMCOL3(KM,EA,EIY,EIZ,GJ,IP,STOREC,IPROP)
      DO 40 I=1,IDOF
      G(I)=STOREG(IP,I)
      IF(G(I).EQ.0)ELD(I)=0.
   40 IF(G(I).NE.0)ELD(I)=LOADS(G(I))
      CALL MVMULT(KM,IKM,ELD,IDOF,IDOF,ACTION)
      CALL LOC3F(LOCAL,ACTION,IP,STOREC,IPROP)
      CALL PRINTV(ACTION,IDOF)
   30 CONTINUE
      STOP
      END
```

The extension to problems of three dimensions is relatively simple, but considerably more care is required in the preparation of data and attention to sign conventions. Elements now have three translations and three rotations at each node, resulting in a 12 × 12 stiffness matrix KM, formed by the subroutine BMCOL3. A right-handed global coordinate system has been assumed with positive rotations defined in a clockwise sense as viewed from the origin. The order in which the freedoms are numbered in the G vector is given in Figure 4.17.

Extra data involves the flexural rigidities EIY and EIZ about the element's local y' and z' axes and a torsional stiffness GJ. For each element in turn the coordinates of the nodes and an angle are stored in the vector COORD (I), I = 1, 7 in the order $x_1, y_1, z_1, x_2, y_2, z_2, \gamma$.

For the purposes of data preparation, a 'vertical' element is defined as one which lies parallel to the global y axis. For non-vertical elements the angle γ is

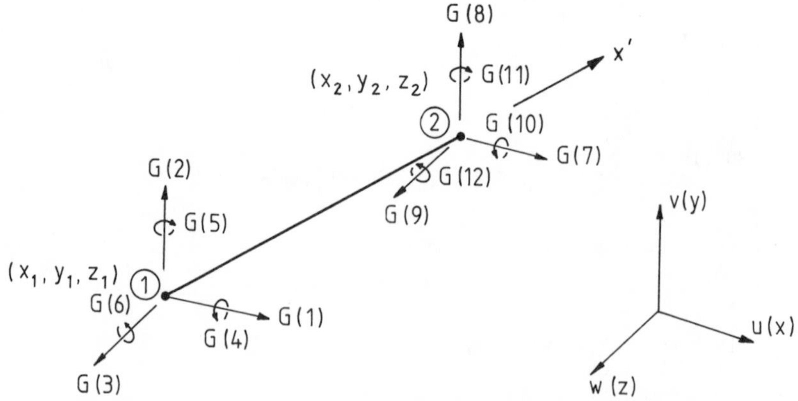

Figure 4.17 Freedom numbering for three-dimensional beam-column element

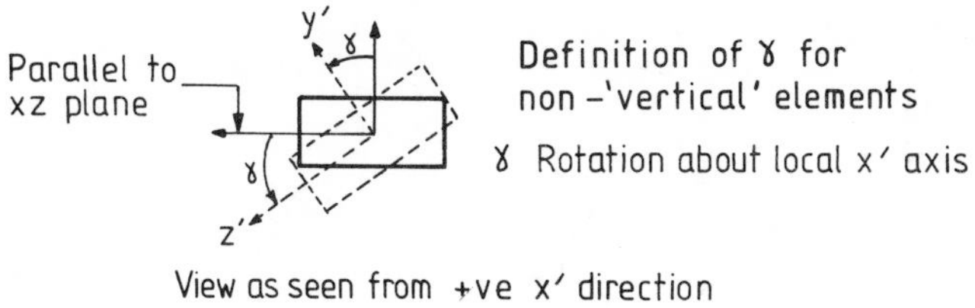

Figure 4.18 Transformation angle for non-vertical elements

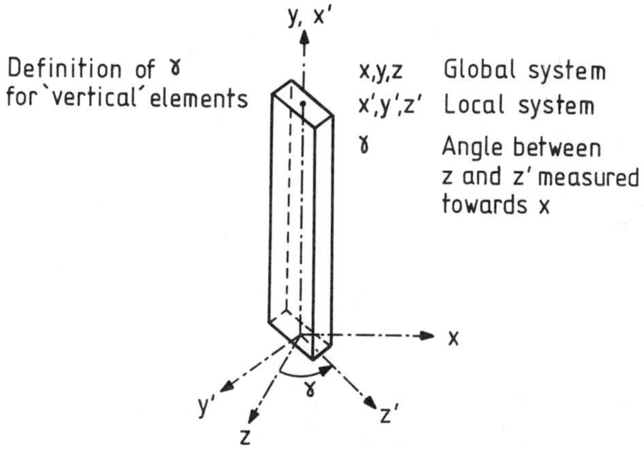

Figure 4.19 Transformation angle for vertical elements

defined as the rotation of the element about its local x' axis (Figure 4.18). For vertical elements, however, γ is defined as the angle between the global z axis and the local z' axis, measured towards the global x axis (Figure 4.19). For vertical elements, it is essential that the local x' axis points in the same direction as the global y axis.

The example shown in Figure 4.20 shows a three-dimensional frame supporting a point load of 100 kN. Before finalising the data, it is necessary to establish the local coordinate system for each element, as shown in Figure 4.21. From Figure 4.17, the positive local x' direction is defined by moving from node (1) to node (2) and this is also the order in which the nodal coordinates and G vector must be read.

In the example of Figure 4.20, elements 1 and 2 both have their local z' axes parallel to the global xz plane; thus γ is set to zero. For vertical element 3, however, γ is set to 90°, which is the angle between the global z and local z' axes measured towards the global x axis.

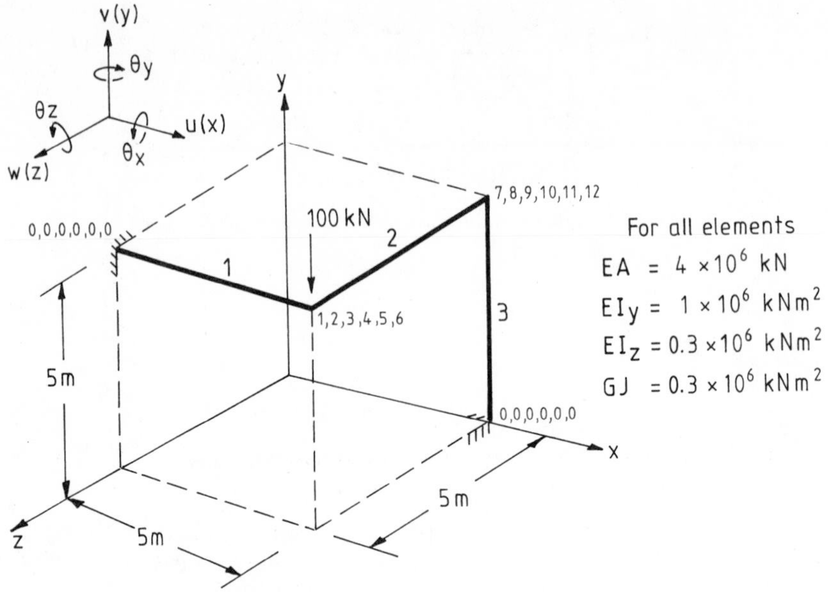

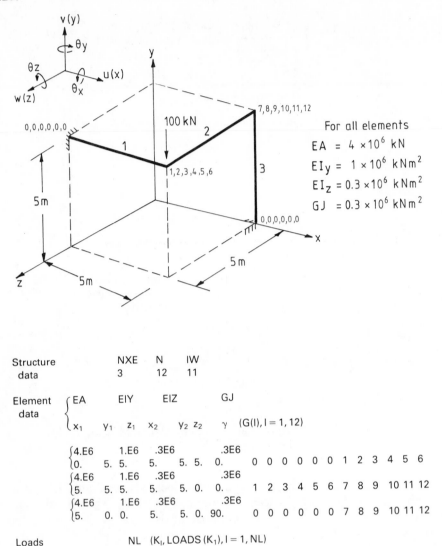

Structure data	NXE	N	IW					
	3	12	11					
Element data	EA		EIY	EIZ			GJ	
	x_1	y_1	z_1	x_2	y_2	z_2	γ	(G(I), I = 1, 12)
	4.E6		1.E6	.3E6			.3E6	
	0.	5.	5.	5.	5.	5.	0.	0 0 0 0 0 0 1 2 3 4 5 6
	4.E6		1.E6	.3E6			.3E6	
	5.	5.	5.	5.	5.	0.	0.	1 2 3 4 5 6 7 8 9 10 11 12
	4.E6		1.E6	.3E6			.3E6	
	5.	0.	0.	5.	5.	0.	90.	0 0 0 0 0 0 7 8 9 10 11 12
Loads data	NL	(K_I, LOADS (K_1), I = 1, NL)						
	1	2	−100.					

Figure 4.20 Mesh and data for Program 4.4

The results given in Figure 4.22 give the displacements followed by the ACTION vector for each element in the order $F_{x1}, F_{y1}, F_{z1}, M_{x1}, M_{y1}, M_{z1}, F_{x2}, F_{y2}, F_{z2}, M_{x2}, M_{y2}, M_{z2}$.

The displacement under the load is given by -0.005997 m and, as a check on equilibrium, F_{y1} for element 1 and F_{y1} for element 3 are seen to equal 63.7 and 36.29 kN respectively.

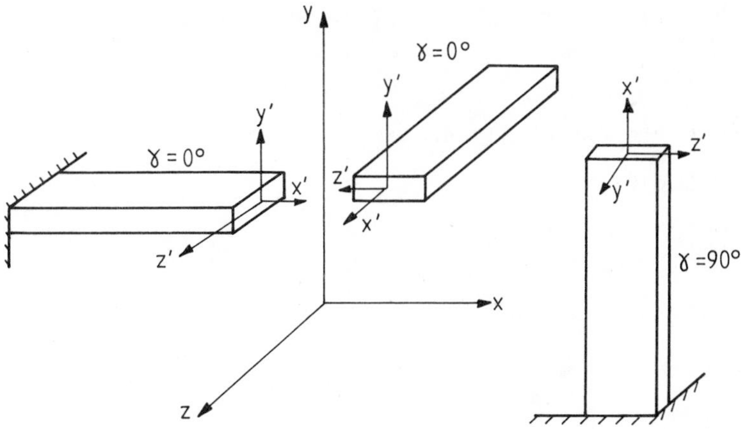

Figure 4.21 Element local coordinate systems

.3039E-05	-.5997E-02	.8769E-03	.1129E-02	-.2360E-03	-.1514E-02	Displacements
.9571E-03	-.4536E-04	.9113E-03	.7470E-03	-.1582E-03	-.3727E-03	
.2431E+01	.6371E+02	-.2754E+02	-.6777E+02	.1160E+03	.2501E+03	Actions
.2431E+01	-.6371E+02	.2754E+02	.6777E+02	.2165E+02	.6846E+02	
.2431E+01	-.3629E+02	-.2754E+02	-.6777E+02	-.2165E+02	-.6846E+02	
.2431E+01	.3629E+02	.2754E+02	-.1137E+03	.9490E+01	.6846E+02	
.2431E+01	.3629E+02	.2754E+02	.2403E+02	.9490E+01	.8062E+02	
.2431E+01	-.3629E+02	-.2754E+02	.1137E+03	-.9490E+01	-.6846E+02	

Figure 4.22 Results from Program 4.4

PROGRAM 4.5: EQUILIBRIUM OF TWO-DIMENSIONAL PIN-JOINTED TRUSSES

```
C
C      PROGRAM 4.5  ANALYSIS OF 2-D TRUSSES
C
C
C       ALTER NEXT LINE TO CHANGE PROBLEM SIZE
C
       PARAMETER(IKV=400,ILOADS=100,IPROP=20)
C
       REAL KM(4,4),ELD(4),ACTION(4),
      +KV(IKV),LOADS(ILOADS),PROP(IPROP),STOREC(IPROP,4)
       INTEGER G(4),STOREG(IPROP,4)
       DATA IKM,IDOF/2*4/,NODOF/2/
C
C       INPUT SECTION
C
       READ(5,*)NXE,N,IW
       IR=N*(IW+1)
       CALL NULVEC(KV,IR)
C
C       GLOBAL STIFFNESS MATRIX ASSEMBLY
C
       DO 10 IP=1,NXE
```

```
      READ(5,*)EA,(STOREC(IP,I),I=1,4),(G(I),I=1,IDOF)
      PROP(IP)=EA
      CALL PINJ2(KM,EA,IP,STOREC,IPROP)
      DO 20 I=1,IDOF
   20 STOREG(IP,I)=G(I)
   10 CALL FORMKV(KV,KM,IKM,G,N,IDOF)
C
C       EQUATION SOLUTION
C
      CALL BANRED(KV,N,IW)
      CALL NULVEC(LOADS,N)
      READ(5,*)NL,(K,LOADS(K),I=1,NL)
      CALL BACSUB(KV,LOADS,N,IW)
      CALL PRINTV(LOADS,N)
C
C      RETRIEVE ELEMENT AXIAL LOADS
C
      DO 30 IP=1,NXE
      EA=PROP(IP)
      CALL PINJ2(KM,EA,IP,STOREC,IPROP)
      DO 40 I=1,IDOF
      G(I)=STOREG(IP,I)
      IF(G(I).EQ.0)ELD(I)=0.
   40 IF(G(I).NE.0)ELD(I)=LOADS(G(I))
      CALL MVMULT(KM,IKM,ELD,IDOF,IDOF,ACTION)
      CALL PRINTV(ACTION,IDOF)
      CALL LOC2T(AXIAL,ACTION,IP,STOREC,IPROP)
   30 WRITE(6,'(E12.4)')AXIAL
      STOP
      END
```

Programs 4.3 and 4.4 for rigid-jointed frames are quite capable of analysing pin-jointed structures in which no moment can develop at the joints. This is done either by putting the flexural rigidity of all members to zero and restraining joint rotations or by uncoupling the rotations of all members meeting at a joint, as was done in the example of Figure 4.15.

The program described in this section, however, applies only to two-dimensional pin-jointed structures. Each member has four degrees of freedom with two translations permitted at each node. The order in which the freedoms are numbered in the G vector is shown in Figure 4.23. The stiffness matrix

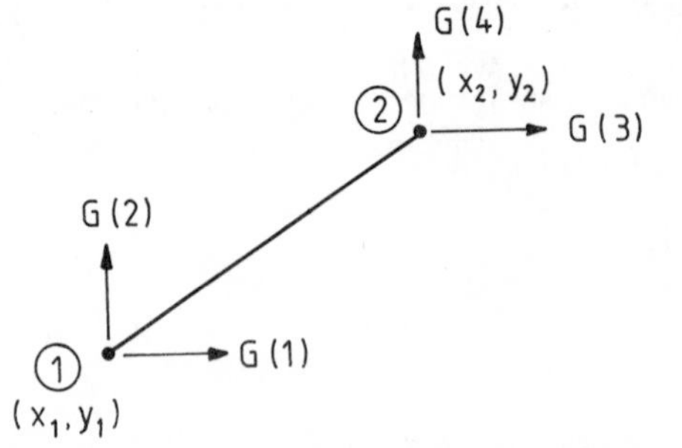

Figure 4.23 Freedom numbering for two-dimensional truss element

provided by the routine PINJ2 is based on the rod element stiffness matrix of equation (2.11) transformed to account for the inclination of the element to the global system. The data for each element involves the axial stiffness EA, followed by the coordinates of each node and the freedom numbering. In the listing provided here, the element ACTION vectors are transformed back into the local system by the routine LOC2T. The resulting variable AXIAL holds the axial force (compression negative) for each element.

The truss shown in the example of Figure 4.24 supports a vertical load of 10 kN. The computed results in Figure 4.25 show that the vertical component of displacement under the load equals −0.007263 m. The axial force in each member is also given; for example the load in member 9 is 16.67 kN (tensile).

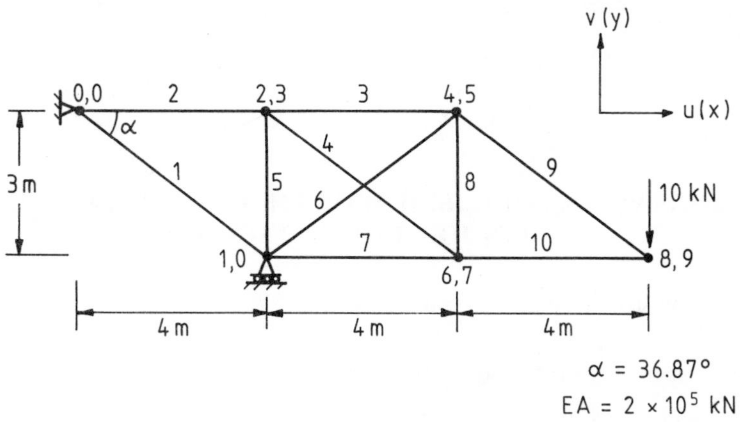

Structure data

NXE	N	IW
10	9	6

Element data

EA	x_1	y_1	x_2	y_2	(G(I), I = 1, 4)
2.E5	0.	3.	4.	0.	0 0 1 0
2.E5	0.	3.	4.	3.	0 0 2 3
2.E5	4.	3.	8.	3.	2 3 4 5
2.E5	4.	3.	8.	0.	2 3 6 7
2.E5	4.	3.	4.	0.	2 3 1 0
2.E5	4.	0.	8.	3.	1 0 4 5
2.E5	4.	0.	8.	0.	1 0 6 7
2.E5	8.	0.	8.	3.	6 7 4 5
2.E5	8.	3.	12	0.	4 5 8 9
2.E5	8.	0.	12.	0.	6 7 8 9

Loads data

NL	(K_I, LOADS (K_I), I = 1, NL)	
1	9	−10.

Figure 4.24 Mesh and data for Program 4.5

```
 -.1042E-02   .5333E-03  -.6562E-04   .9500E-03  -.3046E-02   -.1425E-02 } Displacemen
 -.2981E-02  -.1692E-02  -.7263E-02
  .2667E+02  -.2000E+02  -.2667E+02   .2000E+02 Action } element 1
-.3333E+02   Axial force
 -.2667E+02   .0000E+00   .2667E+02   .0000E+00 Action } element 2
 .2667E+02   Axial force
 -.2083E+02   .0000E+00   .2083E+02   .0000E+00
 .2083E+02                                               etc.
 -.5833E+01   .4375E+01   .5833E+01  -.4375E+01
 .7292E+01
  .0000E+00  -.4375E+01   .0000E+00   .4375E+01
-.4375E+01
  .7500E+01   .5625E+01  -.7500E+01  -.5625E+01
-.9375E+01
  .1917E+02   .0000E+00  -.1917E+02   .0000E+00
-.1917E+02
  .0000E+00   .4375E+01   .0000E+00  -.4375E+01
-.4375E+01
 -.1333E+02   .1000E+02   .1333E+02  -.1000E+02
 .1667E+02
  .1333E+02   .0000E+00  -.1333E+02   .0000E+00
-.1333E+02
```

Figure 4.25 Results from Program 4.5

PROGRAM 4.6: EQUILIBRIUM OF THREE-DIMENSIONAL PIN-JOINTED TRUSSES

```
C
C      PROGRAM 4.6  ANALYSIS OF 3-D TRUSSES
C
C
C       ALTER NEXT LINE TO CHANGE PROBLEM SIZE
C
       PARAMETER(IKV=400,ILOADS=100,IPROP=20)
C
       REAL KM(6,6),ELD(6),ACTION(6),
      +KV(IKV),LOADS(ILOADS),PROP(IPROP),STOREC(IPROP,6)
       INTEGER G(6),STOREG(IPROP,6)
       DATA IKM,IDOF/2*6/,NODOF/3/
C
C       INPUT SECTION
C
       READ(5,*)NXE,N,IW
       IR=N*(IW+1)
       CALL NULVEC(KV,IR)
C
C       GLOBAL STIFFNESS MATRIX ASSEMBLY
C
       DO 10 IP=1,NXE
       READ(5,*)EA,(STOREC(IP,I),I=1,6),(G(I),I=1,IDOF)
       PROP(IP)=EA
       CALL PINJ3(KM,EA,IP,STOREC,IPROP)
       DO 20 I=1,IDOF
    20 STOREG(IP,I)=G(I)
    10 CALL FORMKV(KV,KM,IKM,G,N,IDOF)
C
C       EQUATION SOLUTION
C
```

```
      CALL BANRED(KV,N,IW)
      CALL NULVEC(LOADS,N)
      READ(5,*)NL,(K,LOADS(K),I=1,NL)
      CALL BACSUB(KV,LOADS,N,IW)
      CALL PRINTV(LOADS,N)
C
C      RETRIEVE ELEMENT AXIAL LOADS
C
      DO 30 IP=1,NXE
      EA=PROP(IP)
      CALL PINJ3(KM,EA,IP,STOREC,IPROP)
      DO 40 I=1,IDOF
      G(I)=STOREG(IP,I)
      IF(G(I).EQ.0)ELD(I)=0.
   40 IF(G(I).NE.0)ELD(I)=LOADS(G(I))
      CALL MVMULT(KM,IKM,ELD,IDOF,IDOF,ACTION)
      CALL PRINTV(ACTION,IDOF)
      CALL LOC3T(AXIAL,ACTION,IP,STOREC,IPROP)
   30 WRITE(6,'(E12.4)')AXIAL
      STOP
      END
```

This program is virtually identical to Program 4.5, the only difference being that three coordinates and three freedoms per node are required to define the geometry.

The stiffness matrix is provided by the routine PINJ3 with the freedom numbering shown in Figure 4.26. The transformation to the local system is performed by routine LOC3T.

The truss shown in Figure 4.27 represents a pyramid-like structure loaded by a force Q at its apex which has components of 20, -20, and 30 kN in the x, y and z directions respectively. The computed response in Figure 4.28 indicates that the corresponding displacement components of the loaded node are 0.249×10^{-3}, -0.399×10^{-3} and 0.081×10^{-3} m respectively. Member 3, for example, supports a tensile load of 2.968 kN and, as a check on equilibrium, the x reactions

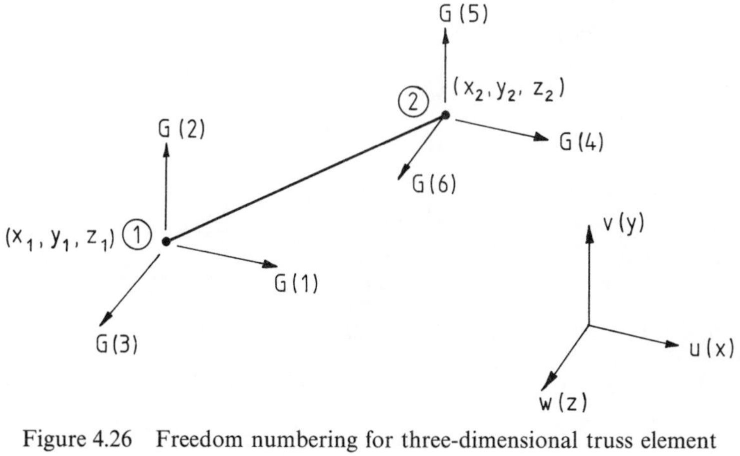

Figure 4.26 Freedom numbering for three-dimensional truss element

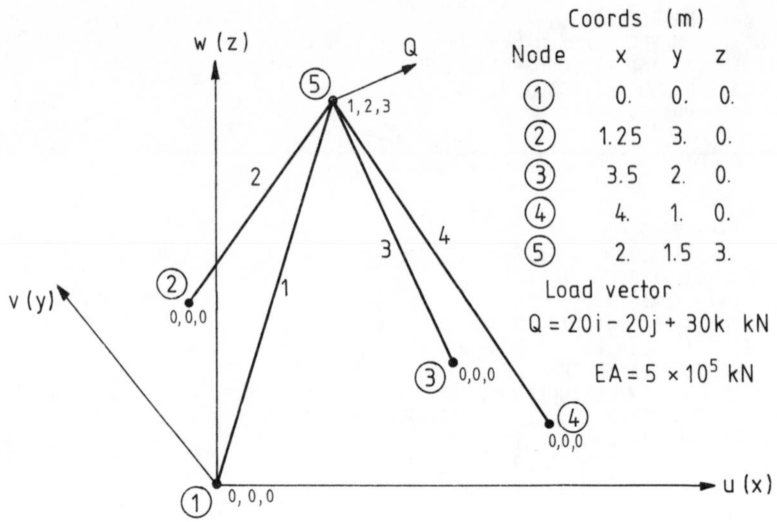

Structure data	NXE	N	IW					
	4	3	2					
Element data	EA	x_1	y_1	z_1	x_2	y_2	z_2	(G(I), I = 1, 6)
	5.E5	0.	0.	0.	2.	1.5	3.	0 0 0 1 2 3
	5.E5	1.25	3.	0.	2.	1.5	3.	0 0 0 1 2 3
	5.E5	3.5	2.	0.	2.	1.5	3.	0 0 0 1 2 3
	5.E5	4.	1.	0.	2.	1.5	3.	0 0 0 1 2 3

Loads data: NL (K$_I$, LOADS (K$_I$), I = 1, NL)

3 1 20. 2 −20. 3 30.

Figure 4.27 Mesh and data for Program 4.6

```
 .2490E-03  -.3991E-03   .8075E-04 Displacements
-.2378E+01  -.1783E+01  -.3567E+01   .2378E+01   .1783E+01   .3567E+01 Actions
 .4643E+01  Axial force
-.9493E+01   .1899E+02  -.3797E+02   .9493E+01  -.1899E+02   .3797E+02 Actions
 .4350E+02  Axial force
 .1313E+01   .4375E+00  -.2625E+01  -.1313E+01  -.4375E+00   .2625E+01
 .2968E+01
-.9442E+01   .2360E+01   .1416E+02   .9442E+01  -.2360E+01  -.1416E+02
-.1718E+02
```

Figure 4.28 Results from Program 4.6

at nodes (1), (2), (3) and (4) are given by the ACTION vectors to equal − 2.378, − 9.493, 1.313 and − 9.442 kN respectively.

PROGRAM 4.7: PLASTIC ANALYSIS OF TWO-DIMENSIONAL FRAMES

```
C
C      PROGRAM 4.7 PLASTIC HINGE ANALYSIS OF RIGID JOINTED FRAMES
C
C      ALTER NEXT LINE TO CHANGE PROBLEM SIZE
C
      PARAMETER(IKV=400,ILOADS=100,IPROP=20,INO=20,ISTEP=20)
C
      REAL OLDSPS(ILOADS),LOADS(ILOADS),BDYLDS(ILOADS),ELDTOT(ILOADS),
     +KV(IKV),HOLDR(IPROP,6),PROP(IPROP,3),VAL(INO),
     +ACTION(6),REACT(6),ELD(6),KM(6,6),STOREC(IPROP,4),DLOAD(ISTEP)
      INTEGER G(6),STOREG(IPROP,6),NO(INO)
      DATA IKM,IDOF/2*6/,NODOF/3/
C
C      INPUT SECTION
C
      READ(5,*)NXE,N,IW,ITS,TOL
      IR=N*(IW+1)
      CALL NULVEC(KV,IR)
      CALL NULVEC(BDYLDS,N)
      CALL NULVEC(ELDTOT,N)
      CALL NULL(HOLDR,IPROP,NXE,IDOF)
      READ(5,*)EA,EI
C
C      GLOBAL STIFFNESS MATRIX ASSEMBLY
C
      DO 10 IP=1,NXE
      READ(5,*)PM,(STOREC(IP,I),I=1,4),(G(I),I=1,6)
      PROP(IP,1)=EA
      PROP(IP,2)=EI
      PROP(IP,3)=PM
      DO 20 I=1,IDOF
   20 STOREG(IP,I)=G(I)
      CALL BMCOL2(KM,EA,EI,IP,STOREC,IPROP)
   10 CALL FORMKV(KV,KM,IKM,G,N,IDOF)
C
C      NODAL LOADING AND STIFFNESS MATRIX REDUCTION
C
      READ(5,*)NL,(NO(I),VAL(I),I=1,NL)
      READ(5,*)INCS,(DLOAD(I),I=1,INCS)
      CALL BANRED(KV,N,IW)
C
C      LOAD INCREMENT LOOP
C
      TOTLO=0.
      DO 30 IY=1,INCS
      TOTLO=TOTLO+DLOAD(IY)
      CALL NULVEC(OLDSPS,N)
      ITERS=0
   40 ITERS=ITERS+1
      CALL NULVEC(LOADS,N)
      DO 50 I=1,NL
   50 LOADS(NO(I))=DLOAD(IY)*VAL(I)
      CALL VECADD(LOADS,BDYLDS,LOADS,N)
```

```
      CALL NULVEC(BDYLDS,N)
      CALL BACSUB(KV,LOADS,N,IW)
C
C     CHECK CONVERGENCE
C
      CALL CHECON(LOADS,OLDSPS,N,TOL,ICON)
C
C     INSPECT MOMENTS IN ALL ELEMENTS
C
      DO 60 IP=1,NXE

      EA=PROP(IP,1)
      EI=PROP(IP,2)
      PM=PROP(IP,3)
      CALL BMCOL2(KM,EA,EI,IP,STOREC,IPROP)
      DO 70 I=1,IDOF
      G(I)=STOREG(IP,I)
      IF(G(I).EQ.0)ELD(I)=0.
   70 IF(G(I).NE.0)ELD(I)=LOADS(G(I))
      CALL MVMULT(KM,IKM,ELD,IDOF,IDOF,ACTION)
      CALL NULVEC(REACT,IDOF)
      IF(ITS.EQ.1)GOTO 80
C
C     IF PM EXCEEDED GENERATE SELF-EQUILIBRATING VECTOR 'REACT'
C     TO SUBTRACT FROM LOADS VECTOR
C
      CALL HING2(IP,HOLDR,STOREC,IPROP,ACTION,REACT,PM)
      DO 90 M=1,IDOF
      IF(G(M).EQ.0)GOTO 90
      BDYLDS(G(M))=BDYLDS(G(M))-REACT(M)
   90 CONTINUE
C
C     AT CONVERGENCE UPDATE ELEMENT REACTIONS, PRINT RESULTS,
C     AND MOVE ON TO NEXT LOAD INCREMENT
C
   80 IF(ITERS.NE.ITS.AND.ICON.NE.1)GOTO 60
      DO 100 M=1,IDOF
  100 HOLDR(IP,M)=HOLDR(IP,M)+REACT(M)+ACTION(M)
   60 CONTINUE
      IF(ITERS.NE.ITS.AND.ICON.NE.1)GOTO 40
      CALL VECADD(LOADS,ELDTOT,ELDTOT,N)
      WRITE(6,'(/,E12.4)')TOTLO
      WRITE(6,'(10E12.4)')(ELDTOT(NO(I)),I=1,NL)
      WRITE(6,'(10I10)')ITERS
   30 CONTINUE
      STOP
      END
```

This program is intended for the analysis of two-dimensional rigid-jointed frames in which a limit is placed on the maximum moment that any member can sustain. As loads on the structure are increased, plastic hinges form progressively and failure occurs when a mechanism develops.

This is the first example in the book of a non-linear analysis in which elastic–perfectly plastic moment–curvature behaviour is assumed as shown in Figure 4.29. Due to the non-linearity, an iterative approach is used to find the nodal displacements and element 'actions' under a given set of applied loads. Moments in excess of the plastic moment are redistributed to other joints.

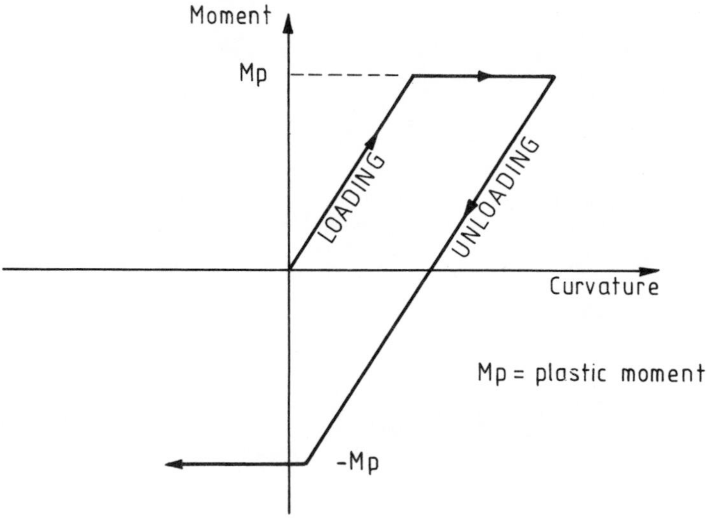

Figure 4.29 Moment–curvature relationship in Program 4.7

Convergence of the iterative process is said to have occurred when, within certain tolerances, moments at the element joints nowhere exceed their limiting values, and the internal forces and moments are in equilibrium with the external loads.

The conventional approach for this type of problem is progressively to modify the global stiffness matrix as joints reach the plastic limit. The modification is necessary because a plastic joint is replaced by a pin-joint with the appropriate plastic moment applied. In the method described by the structure chart in Figure 4.30, the global stiffness matrix is formed once only with the non-linearity introduced by iteratively modifying the applied loads on the structure until convergence is achieved. The process is described in greater detail in Chapter 6. The advantage of splitting the solution of the equilibrium equations into two stages is that the routine BANRED, in which much of the work is performed, is only called once.

Referring to the program, the new constants ISTEP ($\geqslant$ INCS) and INO ($\geqslant$ NL) appear in the PARAMETER statement and additional arrays, together with their functions, are summarised below:

OLDSPS	nodal displacements from previous iteration
BDYLDS	added to external LOADS vector to redistribute moments
ELDTOT	keeps the running total of nodal displacements
HOLDR	keeps the running total of element ACTIONS at convergence
REACT	self-equilibrating 'correction' vector
DLOAD	load increment values

Although EA and EI can vary for each element, to cut down on input they are assumed constant and read once outside the element loop. Variables read within

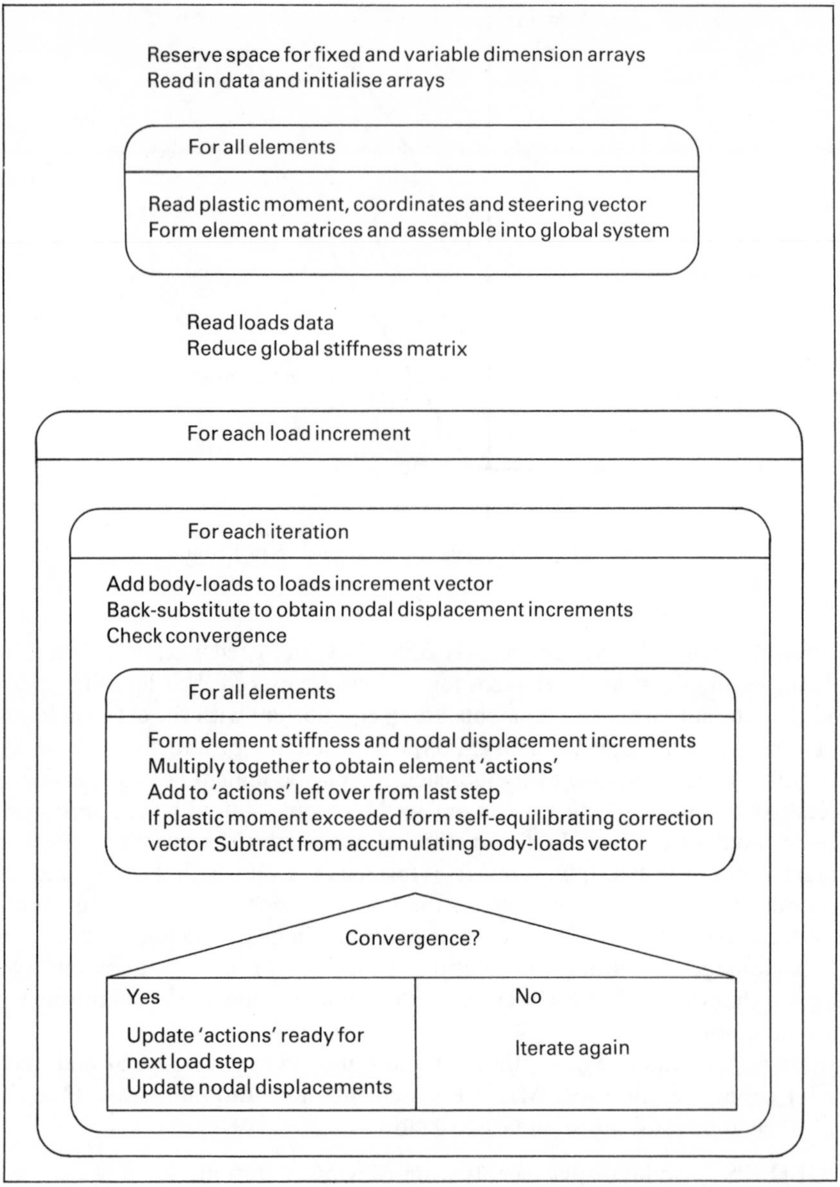

Figure 4.30 Structure chart for Program 4.7

the element loop consist of the plastic moment PM, the COORD array and the G vector.

After assembly of the global stiffness matrix, the number of loads NL and their locations and proportions are read into the vectors NO and VAL. The magnitudes of the INCS load increments are read into the DLOAD vector and the stiffness matrix is reduced (factorised) by the routine BANRED.

The program now enters the load increment loop. For each iteration counted by ITERS, the external load increments are added to the redistribution load vector BDYLDS by the routine VECADD. The equilibrium equations are solved using BACSUB and the resulting nodal displacement increments compared with their values at the previous iteration by the routine CHECON. This routine observes the relative change in displacement increments from one iteration to the next. If the change is less than TOL (read as data) then ICON is set to 1 and convergence has occurred. Alternatively, ICON is set to 0 and another iteration is performed.

Each element is then inspected and its ACTION vector computed by forming

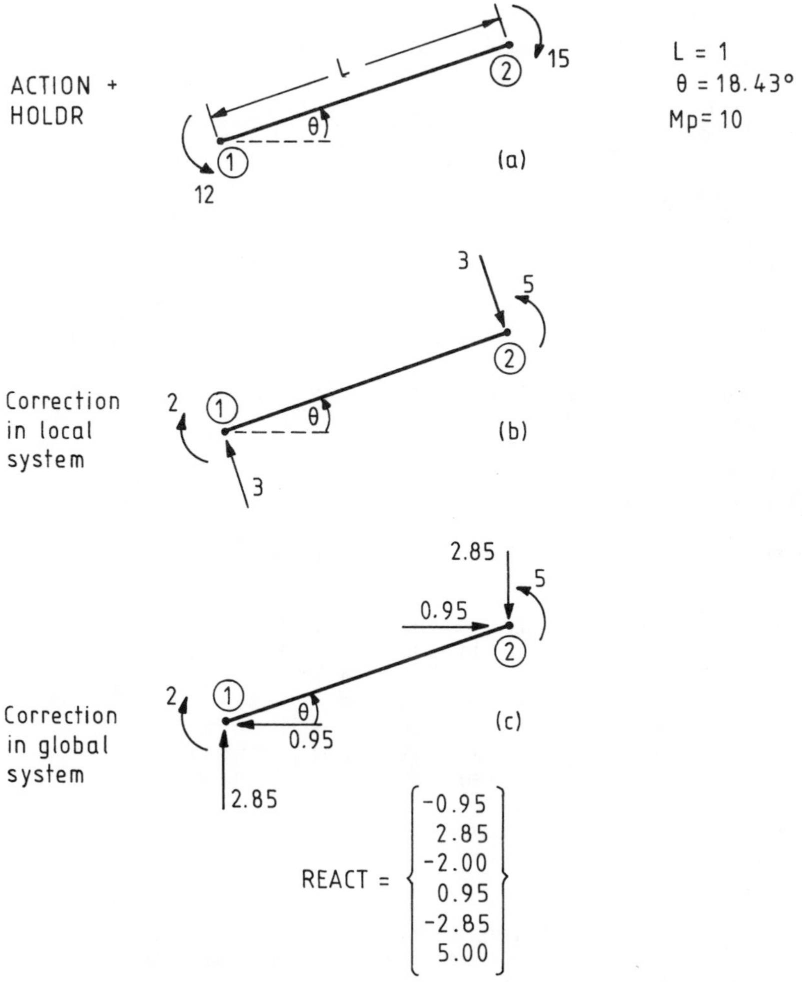

Figure 4.31 Connection vector for 'yielding' element

the product of the nodal displacement increments and the element stiffness matrix. The routine HING2 adds the ACTION vector to the values remaining from the previous load step (held in HOLDR) and checks to see if the plastic moment has been exceeded at either node. If the plastic moment has been exceeded, the self-equilibrating correction vector REACT is formed. In Figure 4.31(a) a typical element is shown in which the moment at both nodes exceeds the plastic limit. The correction vector applies a moment to each end equal to the amount of overshoot. To preserve equilibrium a couple is required, shown in the local coordinate system in Figure 4.31(b). Finally, the couple is transformed into global directions before being accumulated into the BDYLDS vector (Figure 4.31c).

If the algorithm fails to converge in the prescribed maximum number of

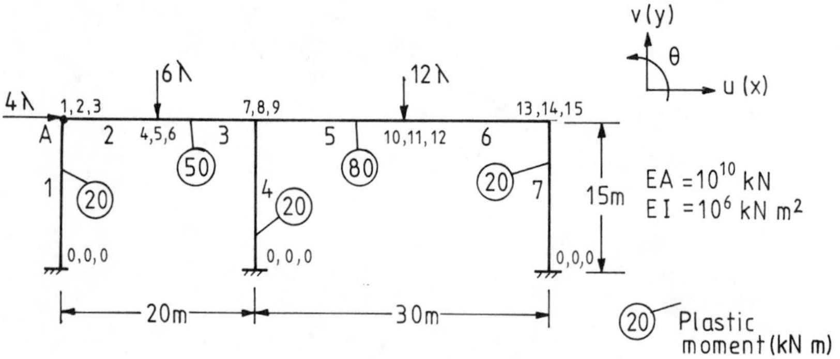

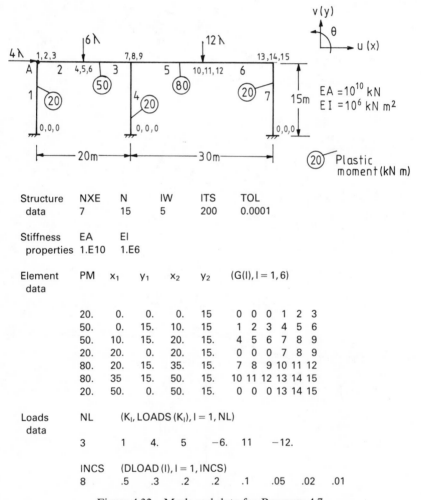

Structure data	NXE	N	IW	ITS	TOL	
	7	15	5	200	0.0001	
Stiffness properties	EA	EI				
	1.E10	1.E6				
Element data	PM	x_1	y_1	x_2	y_2	(G(I), I = 1, 6)
	20.	0.	0.	0.	15	0 0 0 1 2 3
	50.	0.	15.	10.	15	1 2 3 4 5 6
	50.	10.	15.	20.	15.	4 5 6 7 8 9
	20.	20.	0.	20.	15.	0 0 0 7 8 9
	80.	20.	15.	35.	15.	7 8 9 10 11 12
	80.	35	15.	50.	15.	10 11 12 13 14 15
	20.	50.	0.	50.	15.	0 0 0 13 14 15

Loads data   NL   (K_I, LOADS (K_I), I = 1, NL)

3   1   4.   5   −6.   11   −12.

INCS   (DLOAD (I), I = 1, INCS)

8   .5   .3   .2   .2   .1   .05   .02   .01

Figure 4.32 Mesh and data for Program 4.7

```
.5000E+00                                   λ
.2073E-03   -.8478E-04   -.1161E-02    Displacements under loads
     2                                  No. of iterations

.8000E+00
.4912E-03   -.1133E-03   -.2210E-02
    25

.1000E+01
.7737E-03   -.1319E-03   -.2963E-02
    19

.1200E+01
.1085E-02   -.1568E-03   -.3747E-02
    23

.1300E+01
.1457E-02   -.2004E-03   -.4298E-02
    30

.1350E+01
.1786E-02   -.2118E-03   -.4904E-02
    76

.1370E+01
.2043E-02   -.2059E-03   -.5453E-02
    65

.1380E+01
.3203E-02    .2461E-04   -.6727E-02
   200
```

Figure 4.33 Results from Program 4.7

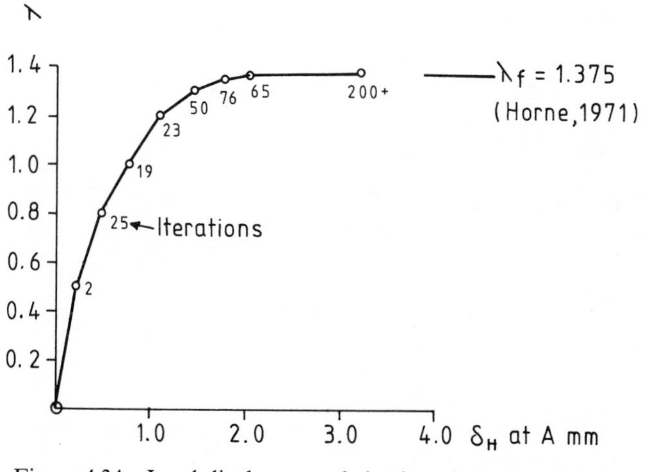

Figure 4.34 Load displacement behaviour from Program 4.7

iterations, ITS, then failure of the structure has probably occurred. Slow convergence, together with rapidly increasing displacements, usually indicates the formation of a mechanism.

The example shown in Figure 4.32 is taken from Horne (1971) and represents a two-bay frame subjected to proportional loading.

After each load increment, the output (Figure 4.33) consists of the load factor λ, the displacements under the three loads and the number of iterations to convergence. The horizontal movement at A is plotted against λ in Figure 4.34 and the peak λ shows close agreement with the analytical solution. These results have been obtained using a rather tight tolerance. More economical, although slightly less accurate solutions, could be obtained by increasing TOL. It is left to users to decide on the best value of TOL for their particular application.

Although beyond the scope of this work, it may be noted that the program is written in a way that allows the effects of unloading on yielding structures to be observed by simply applying negative load increments.

PROGRAM 4.8: STABILITY ANALYSIS OF TWO-DIMENSIONAL FRAMES

```
C
C      PROGRAM 4.8 STABILITY ANALYSIS OF PLANE FRAMES
C
C      ALTER NEXT LINE TO CHANGE PROBLEM SIZE
C
      PARAMETER(IKV=400,ILOADS=100,IPROP=20,INO=20,ISTEP=20)
C
      REAL OLDSPS(ILOADS),ACTION(6),LOCAL(6),ELD(6),KM(6,6),KP(6,6),
     +LOADS(ILOADS),KV(IKV),PROP(IPROP,2),VAL(INO),STOREC(IPROP,4),
     +DISPS(ILOADS),ELDTOT(ILOADS),AXIF(IPROP),AXIP(IPROP),KCOP(IKV),
     +DLOAD(ISTEP)
      INTEGER G(6),STOREG(IPROP,6),NO(INO)
      DATA IKM,IDOF/2*6/,NODOF/3/
C
C      INPUT AND INITIALISATION
C
      READ(5,*)NXE,N,IW,ITS,TOL
      READ(5,*)((PROP(IP,I),I=1,2),(STOREC(IP,I),I=1,4),
     +          (STOREG(IP,I),I=1,6),IP=1,NXE)
      READ(5,*)NL,(NO(I),VAL(I),I=1,NL)
      READ(5,*)INCS,(DLOAD(I),I=1,INCS)
      IR=N*(IW+1)
      CALL NULVEC(AXIP,NXE)
      CALL NULVEC(AXIF,NXE)
      CALL NULVEC(ELDTOT,N)
C
C      LOAD INCREMENT LOOP
C
      TOTLO=0.
      DO 10 IY=1,INCS
      TOTLO=TOTLO+DLOAD(IY)
      CALL NULVEC(LOADS,N)
      DO 20 I=1,NL
   20 LOADS(NO(I))=DLOAD(IY)*VAL(I)
      CALL NULVEC(OLDSPS,N)
```

```
      ITERS=0
   30 ITERS=ITERS+1
      CALL NULVEC(KV,IR)
C
C       GLOBAL STIFFNESS MATRIX ASSEMBLY
C
      DO 40 IP=1,NXE
      DO 50 I=1,6
   50 G(I)=STOREG(IP,I)
      EA=PROP(IP,1)
      EI=PROP(IP,2)
      PAX=AXIF(IP)
      CALL STAB2D(KM,EA,EI,IP,STOREC,IPROP,PAX)
   40 CALL FORMKV(KV,KM,IKM,G,N,IDOF)
      CALL VECCOP(KV,KCOP,IR)
      CALL KVDET(KCOP,N,IW,DET,KSC)
C
C       EQUATION SOLUTION
C
      CALL VECCOP(LOADS,DISPS,N)
      CALL BANRED(KV,N,IW)
      CALL BACSUB(KV,DISPS,N,IW)
C
C       CHECK CONVERGENCE
C
      CALL CHECON(DISPS,OLDSPS,N,TOL,ICON)
C
C       RETRIEVE ELEMENT END FORCES AND MOMENTS
C
      DO 60 IP=1,NXE
      DO 70 I=1,6
   70 G(I)=STOREG(IP,I)
      EA=PROP(IP,1)
      EI=PROP(IP,2)
      PAX=AXIF(IP)
      CALL STAB2D(KM,EA,EI,IP,STOREC,IPROP,PAX)
      DO 80 I=1,IDOF
      IF(G(I).EQ.0)ELD(I)=0.
   80 IF(G(I).NE.0)ELD(I)=DISPS(G(I))
      CALL MVMULT(KM,IKM,ELD,IDOF,IDOF,ACTION)
      CALL LOC2F(LOCAL,ACTION,IP,STOREC,IPROP)
      AXIF(IP)=AXIP(IP)+LOCAL(4)
   60 CONTINUE
      IF(ITERS.NE.ITS.AND.ICON.EQ.0)GOTO 30
C
C       AT CONVERGENCE UPDATE DISPLACEMENTS AND AXIAL FORCES
C
      CALL VECCOP(AXIF,AXIP,NXE)
      CALL VECADD(ELDTOT,DISPS,ELDTOT,N)
      WRITE(6,'(/,E12.4)')TOTLO
      WRITE(6,'(10E12.4)')(ELDTOT(NO(I)),I=1,NL)
      WRITE(6,'(I10,E12.4)')ITERS,DET
   10 CONTINUE
      STOP
      END
```

This program enables the effects of axial loading on members to be accounted for in the form of a modified element stiffness matrix. The stiffness matrix for each element is formed by routine STAB2D, which uses a popular form of the stability function method (see Section 2.6) involving s and c functions (e.g. Horne and Merchant, 1965). The simplest example of buckling is a pinned strut which reaches its first buckling mode when the compressive axial loading equals the Euler load π^2 EI/L^2. In a similar way frames made up of several members connected at their joints may become unstable when the loading reaches certain critical levels. The onset of instability can be observed numerically in several ways. The easiest approach for small structures is to monitor the determinant of the global stiffness matrix as loads are increased on the structure. Instability corresponds to a singular system in which the determinant equals zero. Physically, this implies that a combination of loading has been reached which

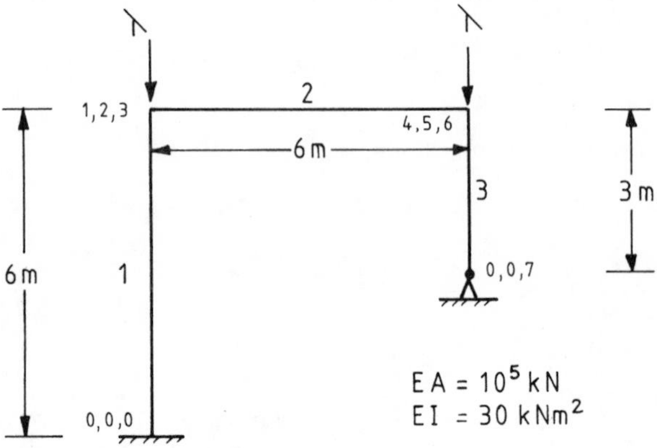

Structure data	NXE	N	IW	ITS	TOL		
	3	7	5	50	0.0001		
Element data	EA	EI	x_1	y_1	x_2	y_2	(G(I), I = 1, 6)
	1.E5	30.	0.	0.	0.	6.	0 0 0 1 2 3
	1.E5	30.	0.	6.	6.	6.	1 2 3 4 5 6
	1.E5	30.	6.	6.	6.	3.	4 5 6 0 0 7
Loads data	NL	(K_I, LOADS (K_I), I = 1, NL)					
	2	2 −1. 5 −1.					
	INCS	(DLOAD (I), I = 1, INCS)					
	10	3.0	1.0	0.5	0.2	0.1	
		0.1	0.1	0.1	0.1	0.1	

Figure 4.35 Mesh and data for Program 4.8

results in a state of neutral equilibrium and a non-unique displacement field. Alternatively, if the critical buckling mode can be predicted in advance (e.g. sway) then a disturbing force can be applied in the appropriate direction. As instability is approached the displacement caused by the disturbing force should increase significantly.

Referring to the program, the following new real arrays are declared:

AXIF	holds axial forces during iterations
AXIP	holds axial forces at convergence
KCOP	holds a copy of the global stiffness matrix stored as vector

and the following subroutines appear for the first time:

VECCOP copies one vector to another

```
  .3000E+01                     λ
 -.1800E-03  -.9000E-04         Displacements under loads
        3   .7043E+18           No. of iterations, Determinant

  .4000E+01
 -.2400E-03  -.1200E-03
        3   .3554E+18

  .4500E+01
 -.2700E-03  -.1350E-03
        3   .1929E+18

  .4700E+01
 -.2820E-03  -.1410E-03
        3   .1301E+18

  .4800E+01
 -.2880E-03  -.1440E-03
        3   .9921E+17

  .4900E+01
 -.2940E-03  -.1470E-03
        3   .6862E+17

  .5000E+01
 -.3000E-03  -.1500E-03
        3   .3836E+17

  .5100E+01
 -.3060E-03  -.1530E-03
        4   .8427E+16

  .5200E+01
 -.3120E-03  -.1560E-03
        4  -.2122E+17

  .5300E+01
 -.3180E-03  -.1590E-03
        3  -.5054E+17
```

Figure 4.36 Results from Program 4.8

KVDET forms the determinant of a symmetric matrix stored as a vector (formed by FORMKV)

STAB2D forms the modified element stiffness matrix accounting for tensile (positive) or compressive (negative) axial forces

As in the previous non-linear problem, iterations are carried out within each load increment to reach a converged solution. The main difference, however, is that the global stiffness matrix is modified during each iteration; thus coding describing its assembly is to be found inside the iteration loop. Once the global system is assembled its determinant is computed using the routine KVDET.

For every load increment, the first iteration takes the axial forces left over from the previous step to compute the stiffness matrices. The resulting displacement increments after solution of the equilibrium equations enable the axial forces to be corrected and the stiffness matrices duly modified. Convergence is said to occur when compatibility is reached between displacements, axial forces and hence stiffness matrices. Naturally, equilibrium between internal actions and external loads must also be satisfied.

The example shown in Figure 4.35 represents a single-bay frame with one pinned support. The critical value of the forces applied to the columns to cause instability is required. The data and output are very similar to the previous program, the difference being that the input consists of EA and EI for all members and the output (Figure 4.36) includes the determinant of the global stiffness matrix after convergence for each load step. The values of the determinant suggest that a 'sway' mode is reached when $5.1 \leqslant \lambda \leqslant 5.2$.

PROGRAM 4.9: EQUILIBRIUM OF RECTANGULAR PLATES IN BENDING

```
C
C      PROGRAM 4.9 EQUILIBRIUM OF TRANSVERSLY LOADED
C      RECTANGULAR PLATES
C
C      ALTER NEXT LINE TO CHANGE PROBLEM SIZE
C
      PARAMETER(IKV=400,ILOADS=100,INF=100)
C
      REAL KV(IKV),LOADS(ILOADS),KM(16,16),DTD(16,16),FUN(16),
     +D1X(16),D2X(16),D1Y(16),D2Y(16),D2XY(16),SAMP(7,2)
      INTEGER NF(INF,4),G(16)
      DATA NODOF/4/,ISAMP/7/,IDOF,IKM,IDTD/3*16/
C
C      INPUT SECTION
C
      READ(5,*)NXE,NYE,N,IW,NN,NR,NGP,E,V,TH,AA,BB
      CALL READNF(NF,INF,NN,NODOF,NR)
      D=E*TH**3/(12.*(1.-V*V))
      IR=N*(IW+1)
      CALL NULVEC(KV,IR)
      CALL NULL(KM,IKM,IDOF,IDOF)
      CALL GAUSS(SAMP,ISAMP,NGP)
```

```
C
C      FORM ELEMENT STIFFNESS MATRIX BY NUMERICAL INTEGRATION
C
      DO 10 I=1,NGP
      DO 10 J=1,NGP
      CALL FMPLAT(FUN,D1X,D1Y,D2X,D2Y,D2XY,SAMP,ISAMP,AA,BB,I,J)
      DO 20 K=1,IDOF
      DO 20 L=1,IDOF
   20 DTD(K,L)=4.*AA*BB*D*SAMP(I,2)*SAMP(J,2)*(D2X(K)*D2X(L)/(AA**4)+
     +          D2Y(K)*D2Y(L)/(BB**4)+(V*D2X(K)*D2Y(L)+
     +          V*D2X(L)*D2Y(K)+2.*(1.-V)*D2XY(K)*D2XY(L))/(AA**2*BB**2)
     +)
   10 CALL MATADD(KM,IKM,DTD,IDTD,IDOF,IDOF)
C
C      GLOBAL STIFFNESS MATRIX ASSEMBLY
C
      DO 30 IP=1,NXE
      DO 30 IQ=1,NYE
      CALL FORMGP(IP,IQ,NYE,G,NF,INF)
   30 CALL FORMKV(KV,KM,IKM,G,N,IDOF)
C
C      EQUATION SOLUTION
C
      CALL BANRED(KV,N,IW)
      CALL NULVEC(LOADS,N)
      READ(5,*)NL,(K,LOADS(K),I=1,NL)
      CALL BACSUB(KV,LOADS,N,IW)
      CALL PRINTV(LOADS,N)
      STOP
      END
```

The previous examples have illustrated the principles of finite element analysis applied to 'structures' made up of one-dimensional elements. Solutions to these idealised problems were not usually dependent upon the number of elements, which was chosen conveniently to reflect positions of load application and changes in geometry. This example attempts to model a two-dimensional thin plate structure using a genuine finite element approximation. The number of elements used to model the plate is decided by the user, but as the number increases, so the solution should improve. The success of a finite element analysis rests on 'close enough' solutions being found using a reasonable number of elements.

The formulation described here enforces complete compatibility of displacements between elements and equilibrium at the nodes, but there will in general be some loss of equilibrium between nodes. Figure 4.37 illustrates a typical element and gives the freedom numbering of the G vector. It can be seen that there are sixteen degrees of freedom per element comprising a vertical translation with two ordinary rotations, θ_x, θ_y, and a 'twist' rotation, θ_{xy}, at each node.

The structure of the program is very similar to Program 4.2 except that the element stiffness matrix is calculated numerically using a double integral in the x and y directions. Additional data involves E (Young's modulus), V (Poisson's ratio), TH (plate thickness) and AA and BB (the dimensions of the elements in the x and y directions respectively). The flexural rigidity of the plate D is calculated

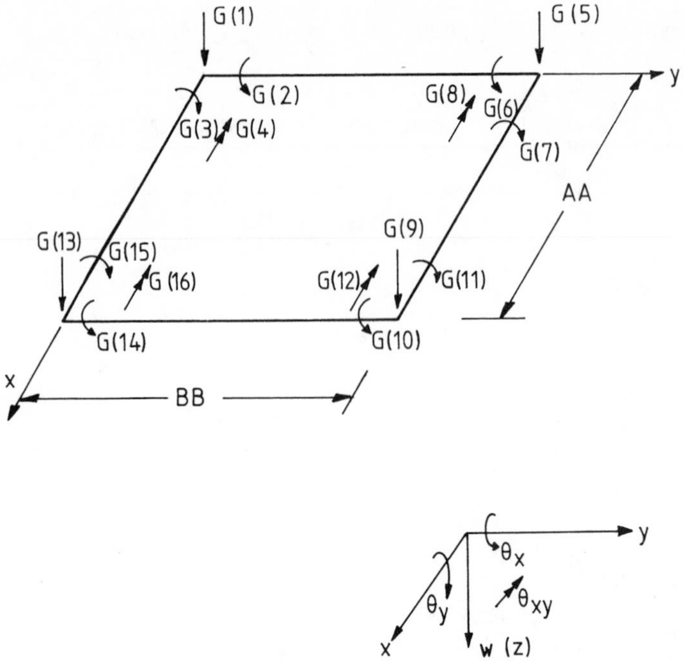

Figure 4.37 Freedom numbering for plate element

from the plate thickness and elastic properties. Additional arrays required are D1X and D2X which hold the first and second derivatives of the shape functions with respect to ξ in the local coordinate system. Similarly D1Y and D2Y hold the derivatives with respect to η and D2XY holds the 'mixed' derivative. All these arrays are delivered by the library routine FMPLAT which evaluates the quantities at each Gaussian integration point.

The array DTD is then formed in accordance with equation (2.86) and represents the contribution to the element stiffness matrix made by each Gauss point. These contributions are added into the accumulating matrix KM. Thereafter the program takes a familiar course.

The subroutine FORMGP generates the G vector for each element assuming a rectangular assembly of elements and performs the same task that GSTRNG did for strings of line elements.

The example in Figure 4.38 illustrates a symmetrical quadrant of a square plate simply supported at its edges and represented by four elements. The plate supports a central unit load. It may be noted that the stiffness matrix requires integration of sixth order polynomials. NGP has been set to 4 ensuring exact integration over the elements.

The results in Figure 4.39 show the central deflection of the plate (freedom 16) to be 0.01147 whereas the exact solution is known to be 0.01160 (Timoshenko and Woinowsky-Krieger, 1959). By increasing the number of elements, better approximations to the exact value could be obtained.

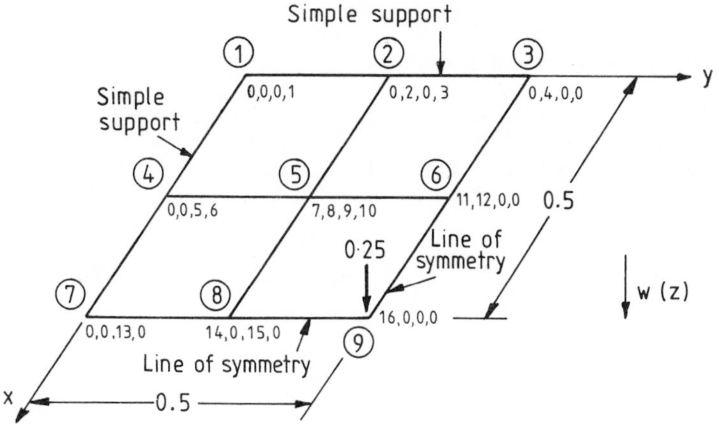

Structure data	NXE	NYE	N	IW	NN	NR	NGP
	2	2	16	10	9	8	4

Element data	E	V	TH	AA	BB
	10.92	0.3	1.0	0.25	0.25

Node freedom data	1 0 0 0 1	2 0 1 0 1	3 0 1 0 0
	4 0 0 1 1	6 1 1 0 0	7 0 0 1 0
	8 1 0 1 0	9 1 0 0 0	

Loads data	NL	(K_I, LOADS (KI), I = 1, NL)
	1	16 0.25

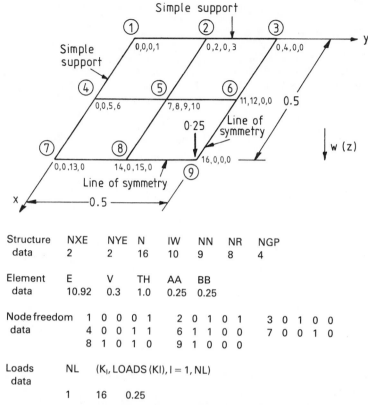

Figure 4.38 Mesh and data for Program 4.9

```
8743E-01   .2021E-01   .6814E-01   .2956E-01   .2021E-01   .6814E-01 )
4772E-02   .1661E-01   .1661E-01   .6460E-01   .7125E-02   .2594E-01 } Displacements
2956E-01   .7125E-02   .2594E-01   .1147E-01                          )
```

Figure 4.39 Results from Program 4.9

4.1 CONCLUSIONS

It has been shown how sample programs can be built up from the library of subroutines described in Chapter 3. The method of exposition is that used throughout this book, namely the introduction of a basis program in some detail followed by the gradual modification of the code to suit new problem conditions.

A central feature of the programs has been their brevity. A typical program has about 50 lines and can easily be compiled on a 'micro' computer. The total program can be listed on one page, which makes for ready comprehension. In subsequent chapters programs of greater complexity are introduced but the central theme of conciseness is adhered to.

4.2 REFERENCES

Hetenyi, M. (1946) *Beams on Elastic Foundations.* University of Michigan Press, ann arbour.

Horne, M. R. (1971) *Plastic Theory of Structures.* Nelson.

Horne, M. R., and Merchant, W. (1965) *The Stability of Frames.* Pergamon Press.

Przemieniecki, J. S. (1968) *Theory of Matrix Structural Analysis.* McGraw-Hill.

Timoshenko, S. P., and Woinowsky-Krieger, S. (1959) *Theory of Plates and Shells.* McGraw-Hill.

CHAPTER 5
STATIC EQUILIBRIUM OF LINEAR ELASTIC SOLIDS

5.0 INTRODUCTION

This chapter describes eleven programs which can be used to solve equilibrium problems in small strain solid elasticity. The programs differ only slightly from each other and, following the method adopted in Chapter 4, the first is described in some detail with changes gradually introduced into the later programs. The first program deals with a plane stress analysis using constant-strain triangular elements. Later programs incorporate plane strain, axisymmetric strain, non-axisymmetric strain of axisymmetric bodies and three-dimensional strain conditions. Various types of two- and three-dimensional finite elements are also considered. The majority of examples in this chapter consider problems involving a regular (usually rectangular) geometry. This has been done to simplify the presentation and minimise the volume of data required. The simple geometries enable the nodal coordinates and freedom numbers to be generated automatically. This is done by a series of geometry routines (GEOM3X, GEOM4Y, GEOM8X, etc.) which depend on the element type and numbering system adopted (see Appendix 3). The programs are all quite capable of analysing geometrically more complex problems, in which case the user must replace the simple geometry routines by some other means of generating nodal coordinates and freedom numbers.

Many users may prefer to generate this information by means of a pre-processing mesh generator program affording graphical checks.

Post-processing of results using graphics packages is highly recommended. Plots of deformed meshes and contours of stresses enable the overall properties of the solution to be taken in at a glance. Machine dependence of such programs, however, has precluded their use in the present book, although in Chapter 6 some typical plots are reproduced for demonstration.

PROGRAM 5.0: PLANE STRESS ANALYSIS USING THREE-NODE TRIANGLES

```
C
C      PROGRAM 5.0 PLANE STRESS OF AN ELASTIC
C      SOLID USING 3-NODE TRIANGULAR ELEMENTS
C
C
C      ALTER NEXT LINE TO CHANGE PROBLEM SIZE
C
      PARAMETER(IKV=2400,ILOADS=200,INF=100)
C
      REAL DEE(3,3),SAMP(16,2),COORD(3,2),JAC(2,2),JAC1(2,2),DER(2,3),
     +DERIV(2,3),BEE(3,6),DBEE(3,6),BTDB(6,6),KM(6,6),ELD(6),
     +EPS(3),SIGMA(3),BT(6,3),FUN(3),WT(16),KV(IKV),LOADS(ILOADS)
      INTEGER NF(INF,2),G(6)
      DATA ISAMP/16/,IBTDB,IKM,IBT,IDOF/4*6/
      DATA IDEE,ICOORD,IBEE,IDBEE,IH,NOD/6*3/
      DATA IJAC,IJAC1,IDER,IDERIV,NODOF,IT/6*2/
C
C      INPUT AND INITIALISATION
C
      READ(5,*)NCE,NYE,N,IW,NN,NR,NIP,AA,BB,E,V
      CALL READNF(NF,INF,NN,NODOF,NR)
      IR=(IW+1)*N
      CALL NULVEC(KV,IR)
      CALL FMDSIG(DEE,IDEE,E,V)
      CALL NUMINT(SAMP,ISAMP,WT,NIP)
C
C      ELEMENT STIFFNESS INTEGRATION AND ASSEMBLY
C
      DO 10 IP=1,NCE
      DO 10 IQ=1,NYE
      CALL GEOM3X(IP,IQ,NCE,AA,BB,COORD,ICOORD,G,NF,INF)
      CALL NULL(KM,IKM,IDOF,IDOF)
      DO 20 I=1,NIP
      CALL FMTRI3(DER,IDER,FUN,SAMP,ISAMP,I)
      CALL MATMUL(DER,IDER,COORD,ICOORD,JAC,IJAC,IT,NOD,IT)
      CALL TWOBY2(JAC,IJAC,JAC1,IJAC1,DET)
      CALL MATMUL(JAC1,IJAC1,DER,IDER,DERIV,IDERIV,IT,IT,NOD)
      CALL NULL(BEE,IBEE,IH,IDOF)
      CALL FORMB(BEE,IBEE,DERIV,IDERIV,NOD)
      CALL MATMUL(DEE,IDEE,BEE,IBEE,DBEE,IDBEE,IH,IH,IDOF)
      CALL MATRAN(BT,IBT,BEE,IBEE,IH,IDOF)
      CALL MATMUL(BT,IBT,DBEE,IDBEE,BTDB,IBTDB,IDOF,IH,IDOF)
      QUOT=.5*DET*WT(I)
      CALL MSMULT(BTDB,IBTDB,QUOT,IDOF,IDOF)
   20 CALL MATADD(KM,IKM,BTDB,IBTDB,IDOF,IDOF)
   10 CALL FORMKV(KV,KM,IKM,G,N,IDOF)
C
C      EQUATION SOLUTION
C
      CALL BANRED(KV,N,IW)
      CALL NULVEC(LOADS,N)
      READ(5,*)NL,(K,LOADS(K),I=1,NL)
      CALL BACSUB(KV,LOADS,N,IW)
      CALL PRINTV(LOADS,N)
C
C      RECOVER STRESSES AT TRIANGLE CENTRES
C
      DO 30 IP=1,NCE
      DO 30 IQ=1,NYE
```

```
      CALL GEOM3X(IP,IQ,NCE,AA,BB,COORD,ICOORD,G,NF,INF)
      DO 40 M=1,IDOF
      IF(G(M).EQ.0)ELD(M)=0.
   40 IF(G(M).NE.0)ELD(M)=LOADS(G(M))
      DO 30 I=1,NIP
      CALL FMTRI6(DER,IDER,FUN,SAMP,ISAMP,I)
      CALL MATMUL(DER,IDER,COORD,ICOORD,JAC,IJAC,IT,NOD,IT)
      CALL TWOBY2(JAC,IJAC,JAC1,IJAC1,DET)
      CALL MATMUL(JAC1,IJAC1,DER,IDER,DERIV,IDERIV,IT,IT,NOD)
      CALL NULL(BEE,IBEE,IH,IDOF)
      CALL FORMB(BEE,IBEE,DERIV,IDERIV,NOD)
      CALL MVMULT(BEE,IBEE,ELD,IH,IDOF,EPS)
      CALL MVMULT(DEE,IDEE,EPS,IH,IH,SIGMA)
      IF(IP.EQ.1.AND.IQ/2*2.NE.IQ)WRITE(6,'(E12.4)')SIGMA(2)
   30 CONTINUE
      STOP
      END
```

The structure chart in Figure 5.1 illustrates the sequence of calculations for this program. In fact the same chart is essentially valid for all programs in this chapter. The three-node (constant-strain) triangle is a poor element and is not recommended for use in practice. In view of its simplicity, however, the first program in this chapter is devoted to it.

As shown in the structure chart, the element stiffness matrices are formed numerically following the procedures described in Chapter 3, equations (3.11), (3.23) and (3.24). For such a simple element only one integrating point is required at each element's centroid.

Figure 5.2 shows a square block of material of unit side length and unit thickness subjected to an equivalent axial stress of 1 kN/m^2. The boundary conditions imply that two planes of symmetry exist and that only one quarter of the problem is being considered. The freedom numbers at each node represent possible displacements in the x and y directions respectively. Figure 5.3 shows the nodal numbering system adopted for this example and although it does not matter in which direction nodes are numbered for a case such as this, the most efficient numbering system for general rectangular shapes will count in the direction with the least nodes. The particular geometry assumed in this case identifies elements by counters IP and IQ running from 1 to NCE and NYE respectively. The simple geometry generated by routine GEOM3X assumes that all elements are right-angled, congruent and formed by diagonal lines drawn from the bottom left hand corner to the top right hand corner of rectangles. The rectangles are assumed to be a constant size with width AA (x direction) and depth BB (y direction). Figure 5.4 shows the order of node and freedom numbering at the element level. Node (1) can be any corner, but subsequent corners and freedoms must follow in an anticlockwise sense. Thus, the top left element (IP = 1, IQ = 1) in Figure 5.2 has a steering vector

$$0 \quad 1 \quad 0 \quad 6 \quad 2 \quad 3$$

and its neighbour (IP = 1, IQ = 2) has a steering vector

$$7 \quad 8 \quad 2 \quad 3 \quad 0 \quad 6$$

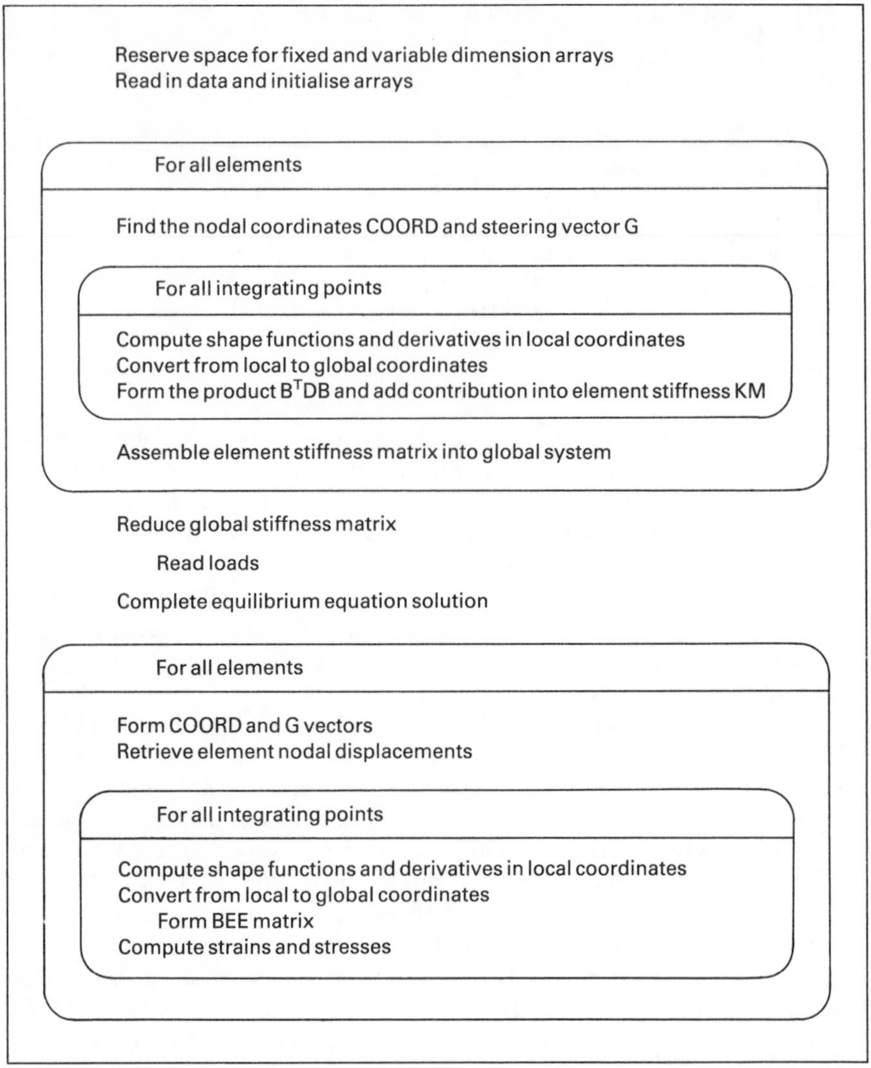
Reserve space for fixed and variable dimension arrays
Read in data and initialise arrays
For all elements
Find the nodal coordinates COORD and steering vector G
For all integrating points
Compute shape functions and derivatives in local coordinates
Convert from local to global coordinates
Form the product B^TDB and add contribution into element stiffness KM
Assemble element stiffness matrix into global system
Reduce global stiffness matrix
Read loads
Complete equilibrium equation solution
For all elements
Form COORD and G vectors
Retrieve element nodal displacements
For all integrating points
Compute shape functions and derivatives in local coordinates
Convert from local to global coordinates
Form BEE matrix
Compute strains and stresses

Figure 5.1 Structure chart for all Chapter 5 programs

It is expected that, where necessary, users will replace the simple geometry routine (GEOM3X) by more sophisticated versions. It need only be ensured that the coordinates and freedom numbers are generated consistently.

Returning to the main program, with the exception of simple integer counters, the meanings of the variable names with reference to the mesh of Figure 5.2 are given below:

NCE	number of element columns in x direction (2)
NYE	number of elements in y direction (4)

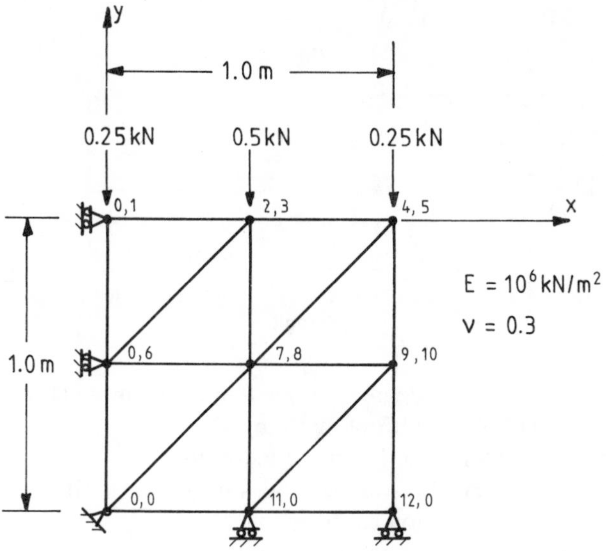

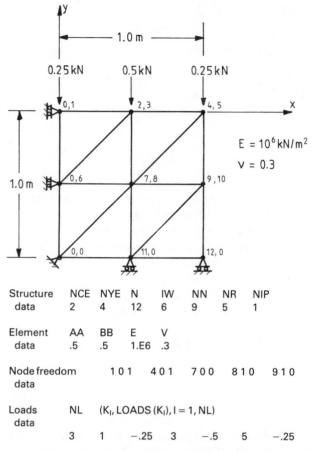

Structure data	NCE	NYE	N	IW	NN	NR	NIP
	2	4	12	6	9	5	1

Element data	AA	BB	E	V
	.5	.5	1.E6	.3

Node freedom data	1 0 1	4 0 1	7 0 0	8 1 0	9 1 0

Loads data	NL	(K_I, LOADS (K_I), I = 1, NL)					
	3	1	−.25	3	−.5	5	−.25

Figure 5.2 Mesh and data for Program 5.0

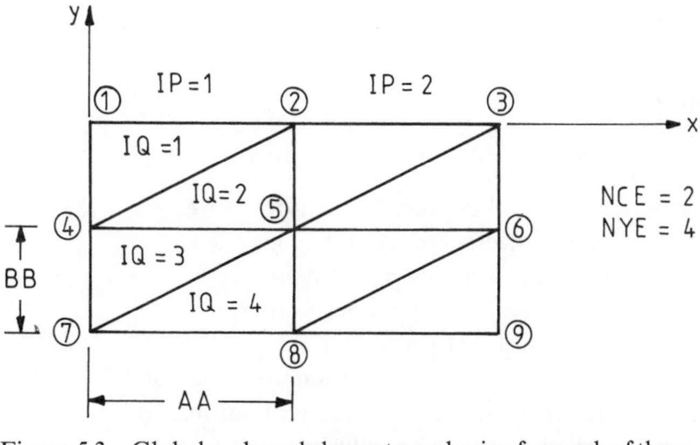

Figure 5.3 Global node and element numbering for mesh of three-node triangle

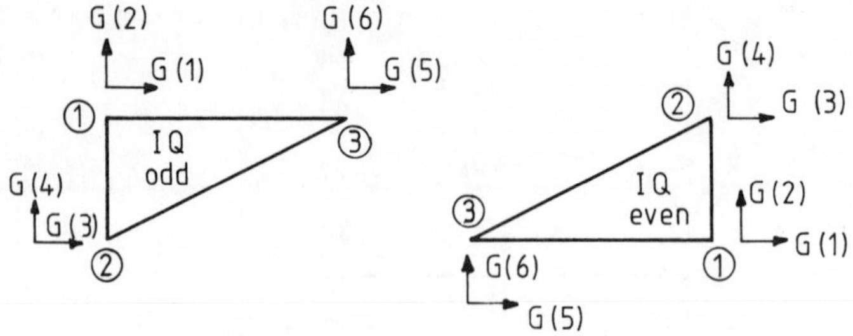

Figure 5.4 Local node and freedom numbering for three-node triangle

N	number of degrees of freedom in the mesh (12)
IW	half-bandwidth of mesh (6)
NN	number of nodes in the mesh (9)
NR	number of restrained nodes in the mesh (5)
NL	number of loaded freedoms (3)
NIP	number of integrating points (1)
AA	dimension of elements in x direction (0.5)
BB	dimension of elements in y direction (0.5)
E	Young's modulus (10^6)
V	Poisson's ratio (0.3)
IBEE, IBT, IBTDB, ICOORD, IDEE, IDER, IDERIV, IJAC, IJAC1, IKM, IKV, ILOADS, INF, ISAMP	working size of matrices BEE, BT, BTDB, COORD, DBEE DEE, DER, DERIV, JAC, JAC1, KM, KV, LOADS, NF, SAMP initialised using DATA and PARAMETER statements
IDOF	number of degrees of freedom per element (6)
IH	size of stress–strain matrix (3)
NOD	number of nodes per element (3)
NODOF	number of freedoms per node (2)
IT	dimensions of problem (2)
IR	N*(IW + 1)—working length of vector KV (84)
DET	determinant of element Jacobian matrix
QUOT	scaled weighting coefficient

Space is reserved for small fixed length arrays:

DEE	stress–strain matrix
SAMP	quadrature abscissae
COORD	element nodal coordinates
JAC	Jacobian matrix
JAC1	inverse of Jacobian matrix
DER	derivatives of shape functions in local coordinates
DERIV	derivatives of shape function in global coordinates
BEE	strain-displacement matrix
DBEE	product of DEE and BEE
BTDB	product of transpose of BEE and DBEE
KM	element stiffness matrix
ELD	element displacement vector
EPS	element strain vector
SIGMA	element stress vector
BT	transpose of BEE
FUN	element shape functions in local coordinates
WT	quadrature weights
G	element steering vector

and then for variable length arrays which can be changed to alter the problem size:

KV	global stiffness matrix (IKV $\geqslant$ IR)
LOADS	global load (displacement) vector (ILOADS $\geqslant$ N)
NF	nodal freedom array (INF $\geqslant$ NN)

After declaration of arrays and DATA statements the program enters the 'input and initialisation' stage. Data concerning the mesh and its properties is now presented and the routine READNF reads the nodal freedom data as given in Figure 5.2. The global system matrix KV is nulled in preparation for the assembly phase, and calls to the routines FMDSIG and NUMINT which form the plane stress stress–strain matrix and the numerical quadrature weights and abscissae respectively.

In the next section of the program, labelled 'element stiffness integration and assembly', the elements are inspected one by one using GEOM3X. This delivers the nodal coordinates COORD and the steering vector G for each element.

After the stiffness matrix KM has been nulled, the integration loop is entered. The local coordinates of each integrating point (only 1 in this case) are extracted from SAMP, and the shape functions (FUN) and their derivatives with respect to those coordinates (DER) are provided for the three-node element by the routine FMTRI3. The conversion of these derivatives to the global system (DERIV) requires a sequence of routine calls similar to those described by equations (3.40). The BEE matrix is then nulled and formed by the routine FORMB. The next three routine calls manipulate the BEE and DEE matrices to form the product

```
1000E-05   .1500E-06  -.1000E-05   .3000E-06  -.1000E-05  -.5000E-06 } Displacements
1500E-06  -.5000E-06   .3000E-06  -.5000E-06   .1500E-06   .3000E-06 }
1776E-14  -.1000E+01   .0000E+00 }
5329E-14  -.1000E+01   .2606E-14 }
7105E-14  -.1000E+01   .0000E+00 }
1066E-13  -.1000E+01   .1303E-14 } Centroid stresses
5329E-14  -.1000E+01  -.1043E-13 } σx σy τxy
1066E-13  -.1000E+01  -.2606E-14 }
3553E-14  -.1000E+01  -.2606E-14 }
1776E-14  -.1000E+01   .0000E+00 }
```

Figure 5.5 Results from Program 5.0

$\mathbf{B}^T\mathbf{DB}$. The contribution from each integration point is scaled by the weighting factor QUOT and added into the element stiffness matrix KM. Eventually, the completed KM is assembled into the global stiffness KV using the routine FORMKV which was used extensively in Chapter 4.

When all element stiffnesses have been assembled, the program enters the 'equation solution' stage. The global matrix KV is reduced by the routine BANRED and the required loads vector is read as data. The final stage of the equation solution is performed by the routine BACSUB and the resulting displacement vector (still called LOADS) is printed.

If required, the stresses and strains within the elements can now be computed. Clearly, these could be found anywhere in the elements, but it is convenient and often more accurate to use the integrating points that were used in the stiffness formulation. In this example only one integrating point (the centroid) was employed for each element, so it is at these locations that stresses and strains will be calculated. Each element is scanned once more and its nodal displacement (ELD) retrieved from the global displacement vector. The BEE matrix for each integrating point is recalculated and the product of BEE and ELD yields the strains EPS from equation (3.33). Multiplication by the stress–strain matrix DEE gives the stresses SIGMA which are printed.

The computed results for the example shown in Figure 5.2 are given in Figure 5.5. For this simple case the results are seen to be exact. The vertical displacements under the loads (freedoms 1, 3 and 5) all equal 10^{-6} m and the Poisson's ratio effect has caused horizontal movement at freedoms 4, 9 and 12 equal to 0.3×10^{-6} m. The stress components, printed in the order $\sigma_x, \sigma_y, \tau_{xy}$ are constant and give $\sigma_y = 1\,\text{kN/m}^2$ with $\sigma_x = \tau_{xy} = 0$.

This simple element does not perform well for problems involving less uniform loading distributions, so the remaining sections of this chapter are therefore devoted to more reliable higher order elements. The program structure, however, will remain essentially unchanged.

PROGRAM 5.1: PLANE STRESS ANALYSIS USING SIX-NODE TRIANGLES

```
C
C      PROGRAM 5.1 PLANE STRESS OF AN ELASTIC
C      SOLID USING 6-NODE TRIANGULAR ELEMENTS
```

```
C
C      ALTER NEXT LINE TO CHANGE PROBLEM SIZE
C
      PARAMETER(IKB1=200,IKB2=22,ILOADS=200,INF=100)
C
      REAL DEE(3,3),SAMP(16,2),COORD(6,2),JAC(2,2),JAC1(2,2),DER(2,6),
     +DERIV(2,6),BEE(3,12),DBEE(3,12),BTDB(12,12),KM(12,12),ELD(12),
     +EPS(3),SIGMA(3),BT(12,3),FUN(6),WT(16),KB(IKB1,IKB2),LOADS(ILOADS)
      INTEGER NF(INF,2),G(12)
      DATA ISAMP/16/,IBTDB,IKM,IBT,IDOF/4*12/
      DATA IDEE,IBEE,IDBEE,IH/4*3/,ICOORD,NOD/2*6/
      DATA IJAC,IJAC1,IDER,IDERIV,NODOF,IT/6*2/
C
C      INPUT AND INITIALISATION
C
      READ(5,*)NCE,NYE,N,IW,NN,NR,NIP,AA,BB,E,V
      CALL READNF(NF,INF,NN,NODOF,NR)
      IWP1=IW+1
      CALL NULL(KB,IKB1,N,IWP1)
      CALL FMDSIG(DEE,IDEE,E,V)
      CALL NUMINT(SAMP,ISAMP,WT,NIP)
C
C      ELEMENT STIFFNESS INTEGRATION AND ASSEMBLY
C
      DO 10 IP=1,NCE
      DO 10 IQ=1,NYE
      CALL GEOM6X(IP,IQ,NCE,AA,BB,COORD,ICOORD,G,NF,INF)
      CALL NULL(KM,IKM,IDOF,IDOF)
      DO 20 I=1,NIP
      CALL FMTRI6(DER,IDER,FUN,SAMP,ISAMP,I)
      CALL MATMUL(DER,IDER,COORD,ICOORD,JAC,IJAC,IT,NOD,IT)
      CALL TWOBY2(JAC,IJAC,JAC1,IJAC1,DET)
      CALL MATMUL(JAC1,IJAC1,DER,IDER,DERIV,IDERIV,IT,IT,NOD)
      CALL NULL(BEE,IBEE,IH,IDOF)
      CALL FORMB(BEE,IBEE,DERIV,IDERIV,NOD)
      CALL MATMUL(DEE,IDEE,BEE,IBEE,DBEE,IDBEE,IH,IH,IDOF)
      CALL MATRAN(BT,IBT,BEE,IBEE,IH,IDOF)
      CALL MATMUL(BT,IBT,DBEE,IDBEE,BTDB,IBTDB,IDOF,IH,IDOF)
      QUOT=.5*DET*WT(I)
      CALL MSMULT(BTDB,IBTDB,QUOT,IDOF,IDOF)
   20 CALL MATADD(KM,IKM,BTDB,IBTDB,IDOF,IDOF)
   10 CALL FORMKB(KB,IKB1,KM,IKM,G,IW,IDOF)
C
C      EQUATION SOLUTION
C
      CALL CHOLIN(KB,IKB1,N,IW)
      CALL NULVEC(LOADS,N)
      READ(5,*)NL,(K,LOADS(K),I=1,NL)
      CALL CHOBAC(KB,IKB1,LOADS,N,IW)
      WRITE(6,'(10E12.4)')(LOADS(I),I=1,81,20)
C
C      RECOVER STRESSES AT TRIANGLE CENTRES
C
      NIP=1
      CALL NUMINT(SAMP,ISAMP,WT,NIP)
      DO 30 IP=1,NCE
      DO 30 IQ=1,NYE
      CALL GEOM6X(IP,IQ,NCE,AA,BB,COORD,ICOORD,G,NF,INF)
      DO 40 M=1,IDOF
      IF(G(M).EQ.0)ELD(M)=0.
   40 IF(G(M).NE.0)ELD(M)=LOADS(G(M))
      DO 30 I=1,NIP
```

```
      CALL FMTRI6(DER,IDER,FUN,SAMP,ISAMP,I)
      CALL MATMUL(DER,IDER,COORD,ICOORD,JAC,IJAC,IT,NOD,IT)
      CALL TWOBY2(JAC,IJAC,JAC1,IJAC1,DET)
      CALL MATMUL(JAC1,IJAC1,DER,IDER,DERIV,IDERIV,IT,IT,NOD)
      CALL NULL(BEE,IBEE,IH,IDOF)
      CALL FORMB(BEE,IBEE,DERIV,IDERIV,NOD)
      CALL MVMULT(BEE,IBEE,ELD,IH,IDOF,EPS)
      CALL MVMULT(DEE,IDEE,EPS,IH,IH,SIGMA)
      IF(IP.EQ.1.AND.IQ/2*2.NE.IQ)WRITE(6,'(E12.4)')SIGMA(2)
   30 CONTINUE
      STOP
      END
```

A more versatile and useful element is the linear-strain triangle which has six nodes. The program is virtually identical to its predecessor. The sizes of some of the fixed arrays and DATA statements have been increased to account for the fact that each element now has twelve freedoms instead of six. The geometry subroutine GEOM6X has been inserted to generate the COORD and G arrays for this particular element, and the routine FMTRI6 computes shape functions and derivatives at each integrating point. The geometry routine GEOM6X generates an identical layout of elements to that assumed in the previous program. The node numbering system at the element level starts at a corner, and counts around the element perimeter in an anticlockwise sense, as shown in Figure 5.6.

A further modification that has been made to this program is the introduction of Choleski's method of solving linear equilibrium equations rather than Gauss's method. In preparation for this, the system equation coefficients are assembled into the rectangular array KB rather than the vector KV as was previously the case. The dimensions of KB are assigned in the PARAMETER statement where IKB1 $\geqslant$ N and IKB2 $\geqslant$ IW + 1. This is done by the routine FORMKB (instead of FORMKV) and the reduction and back-substitution phases of the equilibrium equation solution by CHOLIN (instead of BANRED) and CHOBAC (instead of BACSUB). Thereafter the programs follow a familiar course.

It may be noted that exact integration of the six-node element requires three integration points per element. If the stresses are still required at the centroid,

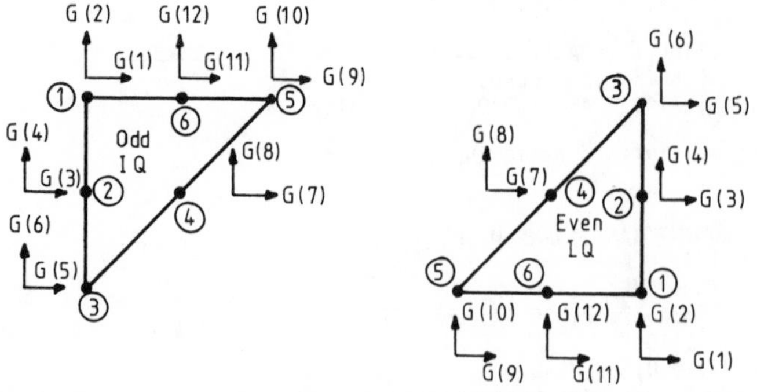

Figure 5.6 Local node and freedom numbering for six-node triangle

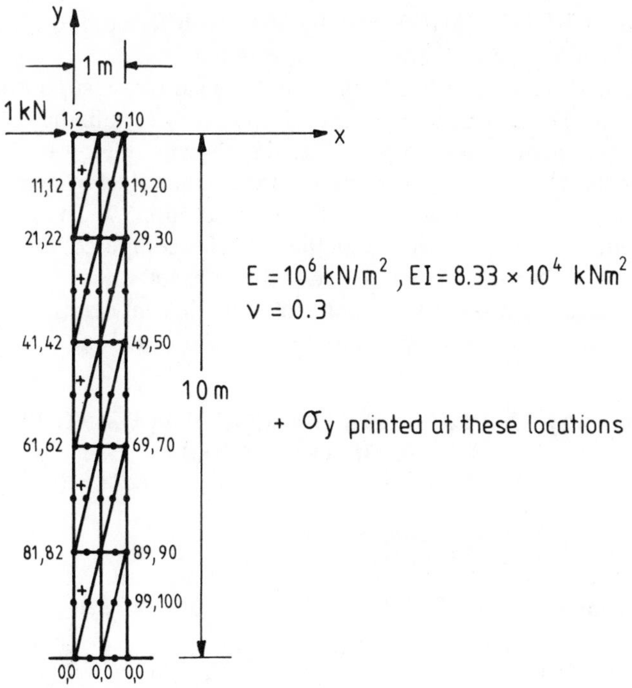

Structure data	NCE	NYE	N	IW	NN	NR	NIP
	2	10	100	21	55	5	3
Element data	AA	BB	E	V			
	.5	2.	1.E6	.3			
Node freedom data	51 0 0	52 0 0	53 0 0	54 0 0	55 0 0		
Loads data	NL	$(K_I, LOADS(K_I), I = 1, NL)$					
	1	1	1.				

Figure 5.7 Mesh and data for Program 5.1

```
.3974E-02   .2795E-02   .1714E-02   .8241E-03   .2226E-03 Chosen displacements
.3414E+01
.1137E+02
.1937E+02  Chosen σy
.2740E+02
.3507E+02
```

Figure 5.8 Results from Program 5.1

however, the routine NUMINT must be called with NIP put to 1 just before the loops in which strains and stresses are computed.

The example and data in Figure 5.7 relate to a cantilever supporting a load of 1 kN at its tip. To reduce the volume of output, only certain displacements and stresses are printed. The computed results shown in Figure 5.8 give the x displacements at freedoms 1, 21, 41, 61 and 81 and the values of σ_y at the centroids of elements on the tensile side, as indicated in Figure 5.7. The results indicate close agreement with slender beam theory. The computed tip displacement (freedom 1) is 3.974×10^{-3} m compared with the 'exact' value of 4×10^{-3} m. The computed value of stress σ_y at the centroid of the element closest to the fixed end of the cantilever is 35.07 kN/m^2, compared with the 'exact' value of 34.67 kN/m^2.

PROGRAM 5.2: PLANE STRAIN ANALYSIS USING FIFTEEN-NODE TRIANGLES

```
C
C       PROGRAM 5.2 PLANE STRAIN OF AN ELASTIC
C       SOLID USING 15-NODE TRIANGULAR ELEMENTS
C
C       ALTER NEXT LINE TO CHANGE PROBLEM SIZE
C
      PARAMETER(IKB1=200,IKB2=42,ILOADS=200,INF=100,
     +INC=20,INY=20)
C
      REAL DEE(3,3),SAMP(16,2),COORD(15,2),JAC(2,2),JAC1(2,2),DER(2,15),
     +DERIV(2,15),BEE(3,30),DBEE(3,30),BTDB(30,30),KM(30,30),ELD(30),
     +EPS(3),SIGMA(3),BT(30,3),FUN(15),WT(16),KB(IKB1,IKB2),
     +LOADS(ILOADS),WID(INC),DEP(INY)
      INTEGER NF(INF,2),G(30)
      DATA ISAMP/16/,IBTDB,IKM,IBT,IDOF/4*30/
      DATA IDEE,IBEE,IDBEE,IH/4*3/,ICOORD,NOD/2*15/
      DATA IJAC,IJAC1,IDER,IDERIV,NODOF,IT/6*2/
C
C       INPUT AND INITIALISATION
C
      READ(5,*)NCE,NYE,N,IW,NN,NR,NIP,E,V
      READ(5,*)(WID(I),I=1,NCE+1)
      READ(5,*)(DEP(I),I=1,(NYE+2)/2)
      CALL READNF(NF,INF,NN,NODOF,NR)
      IWP1=IW+1
      CALL NULL(KB,IKB1,N,IWP1)
      CALL FMDEPS(DEE,IDEE,E,V)
      CALL NUMINT(SAMP,ISAMP,WT,NIP)
C
C       ELEMENT STIFFNESS INTEGRATION AND ASSEMBLY
C
      DO 10 IP=1,NCE
      DO 10 IQ=1,NYE
      CALL GEV15Y(IP,IQ,NYE,WID,DEP,COORD,ICOORD,G,NF,INF)
      CALL NULL(KM,IKM,IDOF,IDOF)
      DO 20 I=1,NIP
      CALL FMTR15(DER,IDER,FUN,SAMP,ISAMP,I)
      CALL MATMUL(DER,IDER,COORD,ICOORD,JAC,IJAC,IT,NOD,IT)
      CALL TWOBY2(JAC,IJAC,JAC1,IJAC1,DET)
      CALL MATMUL(JAC1,IJAC1,DER,IDER,DERIV,IDERIV,IT,IT,NOD)
      CALL NULL(BEE,IBEE,IH,IDOF)
```

```
      CALL FORMB(BEE,IBEE,DERIV,IDERIV,NOD)
      CALL MATMUL(DEE,IDEE,BEE,IBEE,DBEE,IDBEE,IH,IH,IDOF)
      CALL MATRAN(BT,IBT,BEE,IBEE,IH,IDOF)
      CALL MATMUL(BT,IBT,DBEE,IDBEE,BTDB,IBTDB,IDOF,IH,IDOF)
      QUOT=.5*DET*WT(I)
      CALL MSMULT(BTDB,IBTDB,QUOT,IDOF,IDOF)
   20 CALL MATADD(KM,IKM,BTDB,IBTDB,IDOF,IDOF)
   10 CALL FORMKB(KB,IKB1,KM,IKM,G,IW,IDOF)
C
C       EQUATION SOLUTION
C
      CALL CHOLIN(KB,IKB1,N,IW)
      CALL NULVEC(LOADS,N)
      READ(5,*)NL,(K,LOADS(K),I=1,NL)
      CALL CHOBAC(KB,IKB1,LOADS,N,IW)
      CALL PRINTV(LOADS,N)
C
C       RECOVER STRESSES AT TRIANGLE CENTRES
C
      NIP=1
      CALL NUMINT(SAMP,ISAMP,WT,NIP)
      DO 30 IP=1,NCE
      DO 30 IQ=1,NYE
      CALL GEV15Y(IP,IQ,NYE,WID,DEP,COORD,ICOORD,G,NF,INF)
      DO 40 M=1,IDOF
      IF(G(M).EQ.0)ELD(M)=0.
   40 IF(G(M).NE.0)ELD(M)=LOADS(G(M))
      DO 30 I=1,NIP
      CALL FMTR15(DER,IDER,FUN,SAMP,ISAMP,I)
      CALL MATMUL(DER,IDER,COORD,ICOORD,JAC,IJAC,IT,NOD,IT)
      CALL TWOBY2(JAC,IJAC,JAC1,IJAC1,DET)
      CALL MATMUL(JAC1,IJAC1,DER,IDER,DERIV,IDERIV,IT,IT,NOD)
      CALL NULL(BEE,IBEE,IH,IDOF)
      CALL FORMB(BEE,IBEE,DERIV,IDERIV,NOD)
      CALL MVMULT(BEE,IBEE,ELD,IH,IDOF,EPS)
      CALL MVMULT(DEE,IDEE,EPS,IH,IH,SIGMA)
      CALL PRINTV(SIGMA,IH)
   30 CONTINUE
      STOP
      END
```

The last of the triangular elements to be considered here is the fifteen-noded 'cubic strain' triangle. Plane strain conditions are introduced at this stage, and this is achieved by replacing the routine FMDSIG (for plane stress) by FMDEPS. Although the triangles are still formed by drawing diagonals of rectangles, a variation is introduced in that the rectangles can vary in size. Thus, instead of constants AA and BB, the widths and depths are read by the arrays WID and DEP. The dimensions of these arrays appear in the PARAMETER statement where:

$$\text{INC} \geqslant \text{NCE} + 1$$

and

$$\text{INY} \geqslant (\text{NYE} + 2)/2$$

Nodal coordinates COORD and the steering vector G are generated by the 'variable' geometry routine GEV15Y which assumes nodes are numbered in the

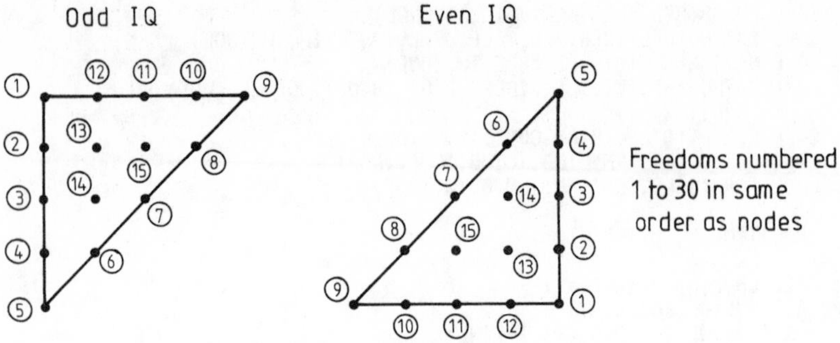

Figure 5.9 Local node numbering for fifteen-node triangle

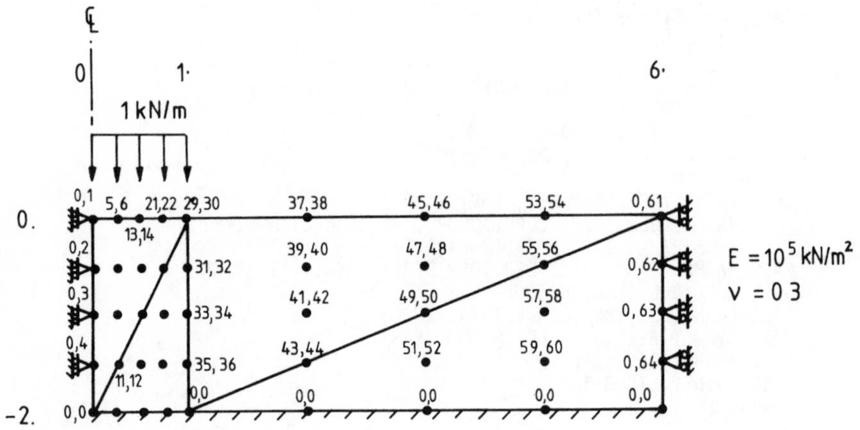

Structure data	NCE	NYE	N	IW	NN	NR	NIP
	2	2	64	32	45	17	12

Material properties	E	V
	1.E5	.2

Widths (WID(I), I = 1, NCE + 1) 0. 1. 6.
Depths (DEP(I), I = 1, (NYE + 2)/2) 0. −2.

Node freedom data					
	1 0 1	2 0 1	3 0 1	4 0 1	5 0 0
	10 0 0	15 0 0	20 0 0	25 0 0	30 0 0
	35 0 0	40 0 0	41 0 1	42 0 1	43 0 1
	44 0 1	45 0 0			

Loads data	NL	(LOADS (K_I), I = 1, NL)					
	5	1	−.0778	6	−.3556	14	−.1333
		22	−.3556	30	−.0778		

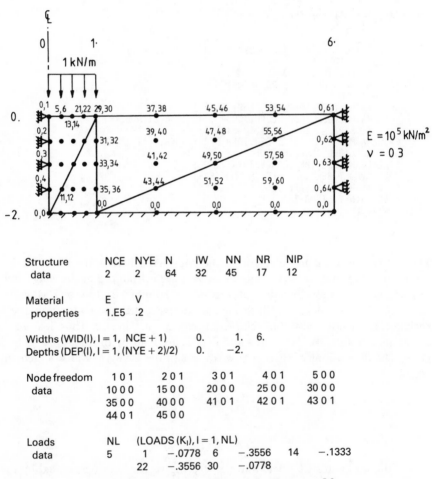

Figure 5.10 Mesh and data for Program 5.2

Displacements:

```
-.1591E-04  -.1158E-04  -.7226E-05  -.3354E-05  -.9321E-06  -.1559E-04
 .1493E-06  -.1128E-04   .4540E-06  -.7019E-05   .3347E-06  -.3255E-05
-.1888E-05  -.1463E-04   .3231E-06  -.1040E-04   .8542E-06  -.6504E-05
 .6511E-06  -.3028E-05  -.2689E-05  -.1259E-04   .4988E-06  -.8864E-05
 .1189E-05  -.5660E-05   .9051E-06  -.2698E-05  -.2962E-05  -.8438E-05
 .5009E-06  -.6729E-05   .1371E-05  -.4643E-05   .1075E-05  -.2333E-05
-.1164E-05  -.4884E-06   .9373E-07  -.6201E-06   .6987E-06  -.6133E-06
 .6450E-06  -.4335E-06  -.1638E-06   .2388E-06  -.5387E-07   .1893E-06
 .1279E-06   .1101E-06   .1656E-06   .2727E-07   .1471E-06  -.9032E-07
 .2513E-07   .8103E-07  -.3083E-08   .7027E-07  -.4914E-08   .3575E-07
 .4022E-07  -.3260E-07  -.3931E-07  -.1906E-07
```

Centroid stresses σ_x σ_y τ_{xy}:

```
-.8302E-01  -.9098E+00   .7671E-01
-.4434E-01  -.6555E+00   .1123E+00
-.2042E-01   .3240E-01  -.1323E-01
-.7382E-02   .1345E-01  -.3256E-02
```

Figure 5.11 Results from Program 5.2

'depth' direction. The shape functions and derivatives at each integrating point are provided for this element by the routine FMTR15. To form the stiffness matrix for this element exactly in plane strain, NIP is set to 12.

The node numbering at the element level follows the pattern of the lower order triangular elements described previously and is given in Figure 5.9.

The example and data in Figure 5.10 show half of a flexible footing resting on a uniform elastic layer. The nodal loads imply a uniform stress of 1 kN/m^2 (see Appendix 1). It is seen from the data that due to the number of nodes associated with each element, the bandwidths can become quite large. The high order of the interpolating polynomials, however, suggests that less elements would be required for a typical boundary value problem.

The computed results for this example, given in Figure 5.11, indicate a centreline displacement of -1.591×10^{-3} m. This is in good agreement with the closed form solution of -1.53×10^{-3} m given by Poulos and Davis (1974).

PROGRAM 5.3: PLANE STRAIN ANALYSIS USING FOUR-NODE QUADRILATERALS

```
C
C      PROGRAM 5.3 PLANE STRAIN OF AN ELASTIC
C      SOLID USING 4-NODE QUADRILATERAL ELEMENTS
C
C
C      ALTER NEXT LINE TO CHANGE PROBLEM SIZE
C
      PARAMETER(IKV=3500,ILOADS=100,INF=100)
C
      REAL DEE(3,3),SAMP(4,2),COORD(4,2),JAC(2,2),JAC1(2,2),
     +DER(2,4),DERIV(2,4),BEE(3,8),DBEE(3,8),
     +BTDB(8,8),KM(8,8),ELD(8),EPS(3),SIGMA(3),
     +BT(8,3),FUN(4),KV(IKV),LOADS(ILOADS)
      INTEGER NF(INF,2),G(8)
      DATA IDEE,IBEE,IDBEE,IH/4*3/,IDOF,IBTDB,IBT,IKM/4*8/
      DATA IJAC,IJAC1,NODOF,IT,IDER,IDERIV/6*2/,ICOORD,NOD/2*4/
      DATA ISAMP/4/
```

```
C
C      INPUT AND INITIALISATION
C
      READ(5,*)NXE,NYE,N,IW,NN,NR,NGP,AA,BB,E,V
      CALL READNF(NF,INF,NN,NODOF,NR)
      IR=N*(IW+1)
      CALL NULVEC(KV,IR)
      CALL FMDEPS(DEE,IDEE,E,V)
      CALL GAUSS(SAMP,ISAMP,NGP)
C
C      ELEMENT STIFFNESS INTEGRATION AND ASSEMBLY
C
      DO 10 IP=1,NXE
      DO 10 IQ=1,NYE
      CALL GEOM4Y(IP,IQ,NYE,AA,BB,COORD,ICOORD,G,NF,INF)
      CALL NULL(KM,IKM,IDOF,IDOF)
      DO 20 I=1,NGP
      DO 20 J=1,NGP
      CALL FORMLN(DER,IDER,FUN,SAMP,ISAMP,I,J)
      CALL MATMUL(DER,IDER,COORD,ICOORD,JAC,IJAC,IT,NOD,IT)
      CALL TWOBY2(JAC,IJAC,JAC1,IJAC1,DET)
      CALL MATMUL(JAC1,IJAC1,DER,IDER,DERIV,IDERIV,IT,IT,NOD)
      CALL NULL(BEE,IBEE,IH,IDOF)
      CALL FORMB(BEE,IBEE,DERIV,IDERIV,NOD)
      CALL MATMUL(DEE,IDEE,BEE,IBEE,DBEE,IDBEE,IH,IH,IDOF)
      CALL MATRAN(BT,IBT,BEE,IBEE,IH,IDOF)
      CALL MATMUL(BT,IBT,DBEE,IDBEE,BTDB,IBTDB,IDOF,IH,IDOF)
      QUOT=DET*SAMP(I,2)*SAMP(J,2)
      CALL MSMULT(BTDB,IBTDB,QUOT,IDOF,IDOF)
   20 CALL MATADD(KM,IKM,BTDB,IBTDB,IDOF,IDOF)
   10 CALL FORMKV(KV,KM,IKM,G,N,IDOF)
C
C      EQUATION SOLUTION
C
      CALL BANRED(KV,N,IW)
      CALL NULVEC(LOADS,N)
      READ(5,*)NL,(K,LOADS(K),I=1,NL)
      CALL BACSUB(KV,LOADS,N,IW)
      CALL PRINTV(LOADS,N)
C
C      RECOVER STRESSES AT ELEMENT GAUSS-POINTS
C
      DO 30 IP=1,NXE
      DO 30 IQ=1,NYE
      CALL GEOM4Y(IP,IQ,NYE,AA,BB,COORD,ICOORD,G,NF,INF)
      DO 40 M=1,IDOF
      IF(G(M).EQ.0)ELD(M)=0.
   40 IF(G(M).NE.0)ELD(M)=LOADS(G(M))
      DO 30 I=1,NGP
      DO 30 J=1,NGP
      CALL FORMLN(DER,IDER,FUN,SAMP,ISAMP,I,J)
      CALL MATMUL(DER,IDER,COORD,ICOORD,JAC,IJAC,IT,NOD,IT)
      CALL TWOBY2(JAC,IJAC,JAC1,IJAC1,DET)
      CALL MATMUL(JAC1,IJAC1,DER,IDER,DERIV,IDERIV,IT,IT,NOD)
      CALL NULL(BEE,IBEE,IH,IDOF)
      CALL FORMB(BEE,IBEE,DERIV,IDERIV,NOD)
      CALL MVMULT(BEE,IBEE,ELD,IH,IDOF,EPS)
      CALL MVMULT(DEE,IDEE,EPS,IH,IH,SIGMA)
      IF(IQ.EQ.1)CALL PRINTV(SIGMA,IH)
   30 CONTINUE
      STOP
      END
```

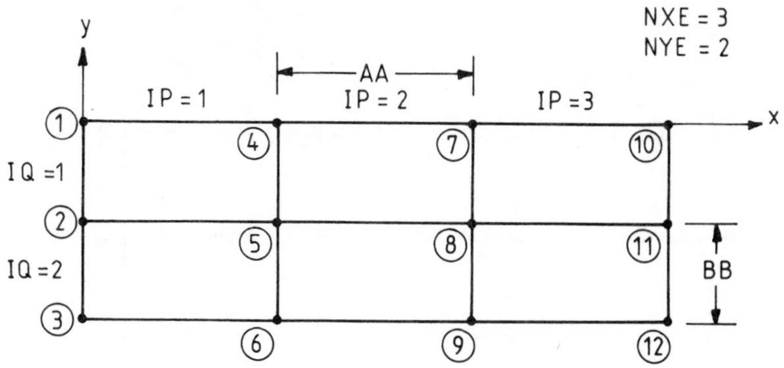

Figure 5.12 Global node and element numbering for mesh of four-node quadrilaterals

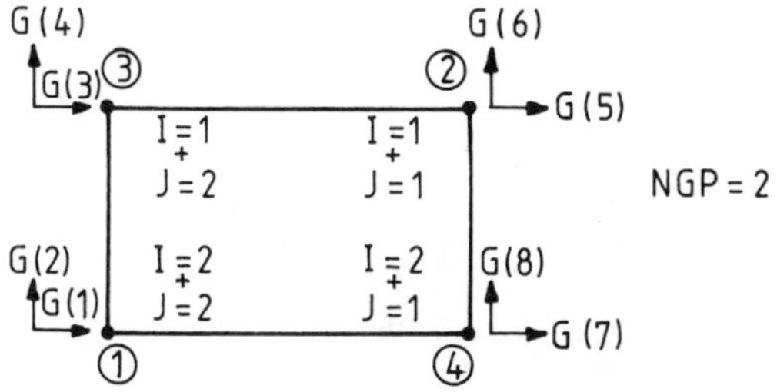

Figure 5.13 Local node, freedom and gauss point numbering for four-node quadrilateral

The next element considered is the four-node 'linear strain' quadrilateral.

Numerical integration is performed using Gaussian quadrature, with weights and abscissae provided by the routine GAUSS (which replaces NUMINT for triangles). Due to the symmetry of Gaussian sampling points in the element local coordinate system, NGP for quadrilaterals is defined as the number of sampling points required in each direction. Thus, it is convenient to scan the NGP^2 sampling points using two DO loops.

The geometry routine GEOM4Y counts in the *y* direction and assumes that all elements are rectangular and equal in size. Figure 5.12 shows a typical mesh of elements together with the global node and element numbering system. Figure 5.13 indicates the sequence of node and freedom numbering at the element level and also the order in which Gauss points are sampled. It may be noted that the four-node element is exactly integrated by putting NGP equal to 2. The element shape functions and derivatives with respect to the local coordinates for this element are provided by the routine FORMLN.

In the particular listing provided here, the strains and stresses are computed at the same Gauss points that were used in the stiffness integrations.

The example of Figure 5.14 shows the mesh and data for a strip load bearing on

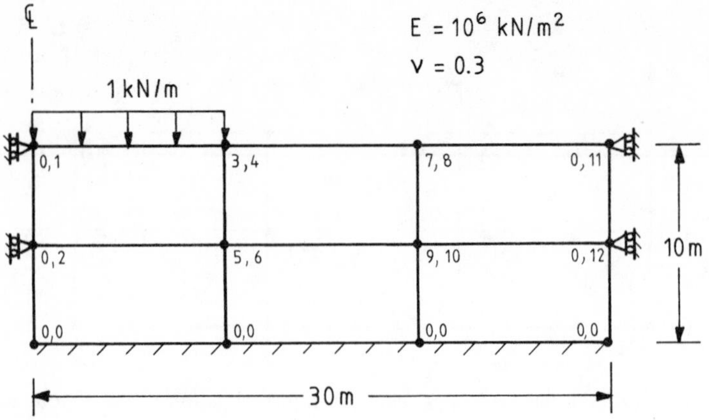

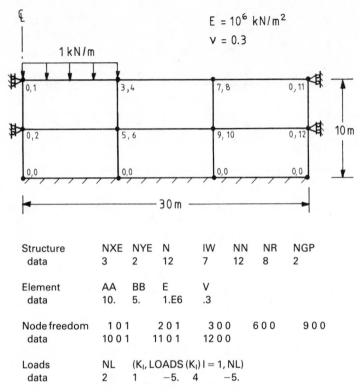

Structure data	NXE	NYE	N	IW	NN	NR	NGP
	3	2	12	7	12	8	2

Element data	AA	BB	E	V
	10.	5.	1.E6	.3

Node freedom data	1 0 1	2 0 1	3 0 0	6 0 0	9 0 0
	10 0 1	11 0 1	12 0 0		

Loads data	NL	$(K_I, \text{LOADS}(K_I)\ I = 1, NL)$			
	2	1	−5.	4	−5.

Figure 5.14 Mesh and data for Program 5.3

Displacements:

```
-.8601E-05  -.4056E-05  -.5472E-07  -.3771E-05   .1225E-05  -.1910E-05
 .1792E-06   .5860E-06   .6280E-06   .1707E-06   .1129E-06   .1066E-06
```

Gauss-point (NGP = 2) stresses σ_x σ_y τ_{xy} for IQ = 1:

```
-.2511E+00  -.6411E+00   .8629E-01
-.4299E+00  -.1058E+01   .1431E+00
-.1516E+00  -.5985E+00   .2667E-01
-.3304E+00  -.1016E+01   .8352E-01
 .2765E-03  -.1429E-01   .1010E+00
-.1513E+00  -.3680E+00   .6412E-01
-.6432E-01  -.4197E-01   .5050E-01
-.2159E+00  -.3957E+00   .1359E-01
-.2619E-01   .9144E-02  -.2217E-01
 .1061E-02   .7274E-01  -.4210E-01
-.6108E-01  -.5806E-02  -.1308E-01
-.3382E-01   .5779E-01  -.3302E-01
```

Figure 5.15 Results from Program 5.3

a uniform elastic layer. Because of symmetry, only half of the layer need be analysed and the width has been arbitrarily terminated at a roller boundary at three times the load width. The computed results in Figure 5.15 give the vertical displacement at freedoms 1 and 4 to be 8.601×10^{-6} and 3.771×10^{-6} respectively. Comparison with closed form or other numerical solutions will show that, with such a coarse mesh of these elements, the results can be quite inaccurate. To cut down on the volume of output, stresses have only been printed at the Gauss points in the first row (IQ = 1) of elements. The vertical stresses in the element immediately beneath the load are 1.058 kN/m^2 at Gauss point I = 1, J = 2 and 0.641 kN/m^2 at Gauss point I = 1, J = 1. Thus, locally, the crude solution indicates some decay between the Gauss point stresses and the applied stress. Such discretisation errors are inevitable in finite element work, and it is the user's responsibility to experiment with mesh designs to help discover whether the numerical solution is adequate.

PROGRAM 5.4: PLANE STRAIN ANALYSIS USING EIGHT-NODE QUADRILATERALS

```
C
C      PROGRAM 5.4 PLANE STRAIN OF AN ELASTIC
C      SOLID USING 8-NODE QUADRILATERAL ELEMENTS
C
C
C      ALTER NEXT LINE TO CHANGE PROBLEM SIZE
C
      PARAMETER(IKB1=100,IKB2=35,ILOADS=100,INF=100)
C
      REAL DEE(3,3),SAMP(4,2),COORD(8,2),JAC(2,2),JAC1(2,2),
     +DER(2,8),DERIV(2,8),BEE(3,16),DBEE(3,16),
     +BTDB(16,16),KM(16,16),ELD(16),EPS(3),SIGMA(3),
     +BT(16,3),FUN(8),KB(IKB1,IKB2),LOADS(ILOADS)
      INTEGER NF(INF,2),G(16)
      DATA IDEE,IBEE,IDBEE,IH/4*3/,IDOF,IBTDB,IBT,IKM/4*16/
      DATA IJAC,IJAC1,NODOF,IT,IDER,IDERIV/6*2/,ICOORD,NOD/2*8/
      DATA ISAMP/4/
C
C      INPUT AND INITIALISATION
C
      READ(5,*)NXE,NYE,N,IW,NN,NR,NGP,AA,BB,E,V
      CALL READNF(NF,INF,NN,NODOF,NR)
      IWP1=IW+1
      CALL NULL(KB,IKB1,N,IWP1)
      CALL FMDEPS(DEE,IDEE,E,V)
      CALL GAUSS(SAMP,ISAMP,NGP)
C
C      ELEMENT STIFFNESS INTEGRATION AND ASSEMBLY
C
      DO 10 IP=1,NXE
      DO 10 IQ=1,NYE
      CALL GEOM8X(IP,IQ,NXE,AA,BB,COORD,ICOORD,G,NF,INF)
      CALL NULL(KM,IKM,IDOF,IDOF)
      DO 20 I=1,NGP
      DO 20 J=1,NGP
      CALL FMQUAD(DER,IDER,FUN,SAMP,ISAMP,I,J)
      CALL MATMUL(DER,IDER,COORD,ICOORD,JAC,IJAC,IT,NOD,IT)
      CALL TWOBY2(JAC,IJAC,JAC1,IJAC1,DET)
      CALL MATMUL(JAC1,IJAC1,DER,IDER,DERIV,IDERIV,IT,IT,NOD)
      CALL NULL(BEE,IBEE,IH,IDOF)
      CALL FORMB(BEE,IBEE,DERIV,IDERIV,NOD)
      CALL MATMUL(DEE,IDEE,BEE,IBEE,DBEE,IDBEE,IH,IH,IDOF)
      CALL MATRAN(BT,IBT,BEE,IBEE,IH,IDOF)
      CALL MATMUL(BT,IBT,DBEE,IDBEE,BTDB,IBTDB,IDOF,IH,IDOF)
```

```
      QUOT=DET*SAMP(I,2)*SAMP(J,2)
      CALL MSMULT(BTDB,IBTDB,QUOT,IDOF,IDOF)
   20 CALL MATADD(KM,IKM,BTDB,IBTDB,IDOF,IDOF)
   10 CALL FORMKB(KB,IKB1,KM,IKM,G,IW,IDOF)
C
C      EQUATION SOLUTION
C
      CALL CHOLIN(KB,IKB1,N,IW)
      CALL NULVEC(LOADS,N)
      READ(5,*)NL,(K,LOADS(K),I=1,NL)
      CALL CHOBAC(KB,IKB1,LOADS,N,IW)
      CALL PRINTV(LOADS,N)
C
C      RECOVER STRAINS AND STRESSES AT ELEMENT CENTRES
C
      NGP=1
      CALL GAUSS(SAMP,ISAMP,NGP)
      DO 30 IP=1,NXE
      DO 30 IQ=1,NYE
      CALL GEOM8X(IP,IQ,NXE,AA,BB,COORD,ICOORD,G,NF,INF)
      DO 40 M=1,IDOF
      IF(G(M).EQ.0)ELD(M)=0.
   40 IF(G(M).NE.0)ELD(M)=LOADS(G(M))
      DO 30 I=1,NGP
      DO 30 J=1,NGP
      CALL FMQUAD(DER,IDER,FUN,SAMP,ISAMP,I,J)
      CALL MATMUL(DER,IDER,COORD,ICOORD,JAC,IJAC,IT,NOD,IT)
      CALL TWOBY2(JAC,IJAC,JAC1,IJAC1,DET)
      CALL MATMUL(JAC1,IJAC1,DER,IDER,DERIV,IDERIV,IT,IT,NOD)
      CALL NULL(BEE,IBEE,IH,IDOF)
      CALL FORMB(BEE,IBEE,DERIV,IDERIV,NOD)
      CALL MVMULT(BEE,IBEE,ELD,IH,IDOF,EPS)
      CALL MVMULT(DEE,IDEE,EPS,IH,IH,SIGMA)
      CALL PRINTV(EPS,IH)
      CALL PRINTV(SIGMA,IH)
   30 CONTINUE
      STOP
      END
```

This program illustrates the use of a higher order element, namely the eight-noded quadrilateral. Two changes to the previous program have been made apart from the necessary increases in fixed array dimensions. The routine FMQUAD has replaced FORMLN to perform calculations of shape function values and their derivatives at the Gauss points, and GEOM8X has replaced GEOM4Y to generate the arrays COORD and G. As the name of the geometry routine suggests, nodes are assumed to be numbered in the x direction for maximum storage efficiency, as shown in Figure 5.16. The local node and freedom numbering for this element (Figure 5.17) indicates that node 1 occurs at a corner and the rest follow in a clockwise sense. The Gauss point numbering follows the pattern indicated by Figure 5.9. The general eight-node quadrilateral element stiffness matrix contains fourth order polynomial terms and thus requires NGP equal to 3 for 'exact' integration. It is often the case, however, that the use of 'reduced' integration, by putting NGP equal to 2, improves the performance of

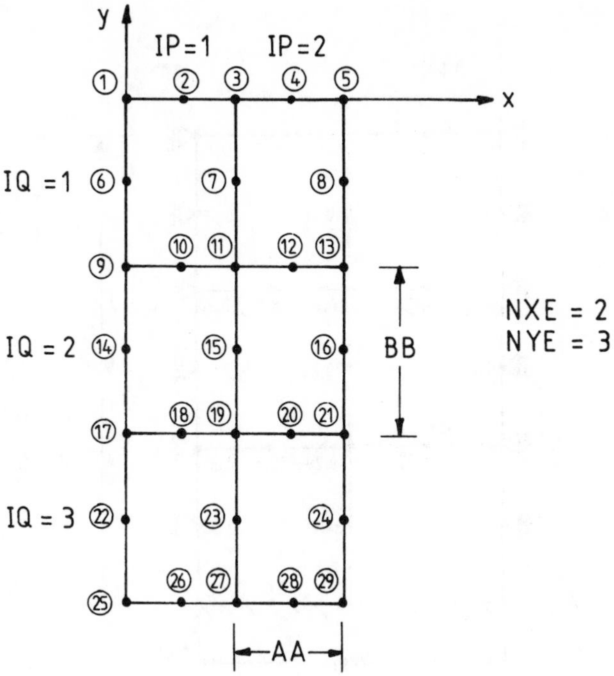

Figure 5.16 Global node and element numbering for mesh of eight-node quadrilaterals

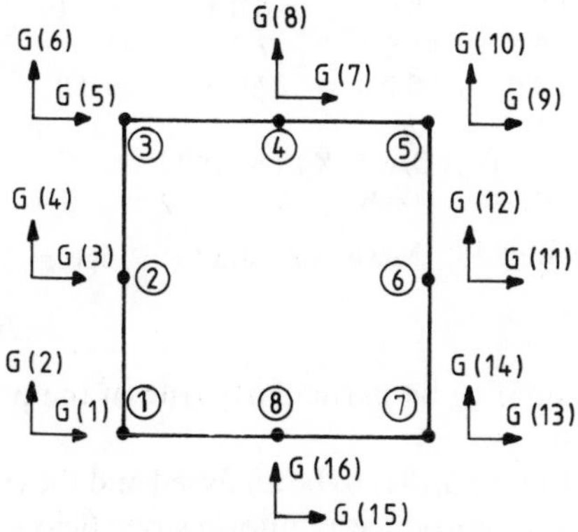

Figure 5.17 Local node and freedom numbering for eight-node quadrilateral

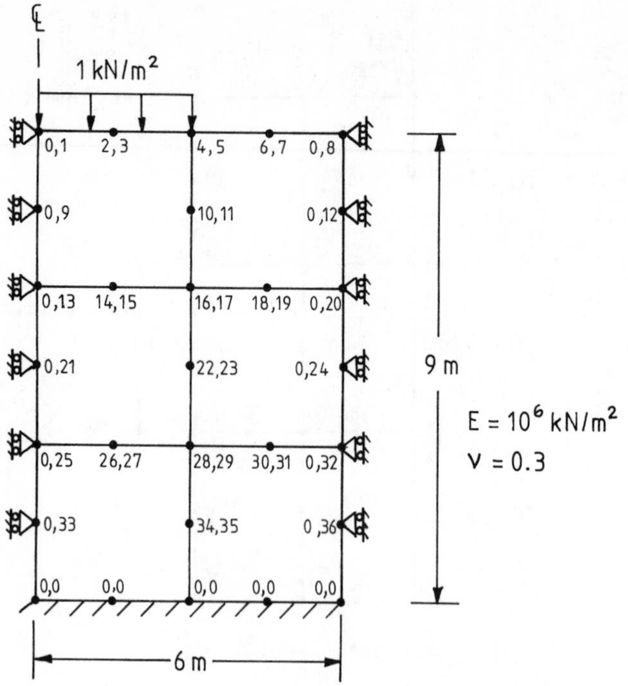

Structure data	NXE	NYE	N	IW	NN	NR	NGP
	2	3	36	16	29	17	2
Element data	AA	BB	E	V			
	3.	3.	1.E6	.3			
Node freedom data	1 0 1	5 0 1	6 0 1	8 0 1	9 0 1	13 0 1	
	14 0 1	16 0 1	17 0 1	21 0 1	22 0 1	24 0 1	
	25 0 0	26 0 0	27 0 0	28 0 0	29 0 0		
Loads data	NL	$(K_I, LOADS(K_I), I = 1, NL)$					
	3	1 −.5	3 −2.	5 −.5			

Figure 5.18 Mesh and data for Program 5.4

this element. This is found to be particularly true of the plasticity applications described in Chapter 6.

The simple mesh in Figure 5.18 is to be analysed and the consistent nodal loads (Appendix 1) necessary to reproduce a uniform stress field should be noted in the data. The listing provided here prints the nodal displacements and also the strains and stresses at the element centroids. The computed results are given in Figure 5.19.

```
-.5311E-05  -.4211E-06  -.5041E-05  -.7222E-06  -.3343E-05  -.4211E-06 ⎫
-.1644E-05  -.1375E-05  -.4288E-05   .3774E-06  -.2786E-05  -.1283E-05 ⎪
-.3243E-05   .2708E-06  -.2873E-05   .3670E-06  -.2229E-05   .2708E-06 ⎪
-.1584E-05  -.1214E-05  -.2217E-05   .2996E-06  -.1671E-05  -.1125E-05 ⎬ Displacements
-.1379E-05   .1370E-06  -.1299E-05   .1912E-06  -.1114E-05   .1370E-06 ⎪
-.9299E-06  -.8498E-06  -.6454E-06   .1122E-06  -.5571E-06  -.4689E-06 ⎭
 .1258E-06  -.7227E-06   .2703E-06 ⎫
-.2476E+00  -.9003E+00   .1040E+00 ⎪
 .9988E-07  -.5248E-06   .2266E-06 ⎪
-.1683E+00  -.6489E+00   .8714E-01 ⎪
 .3740E-07  -.4329E-06   .7509E-07 ⎪ Centroid
-.1994E+00  -.5612E+00   .2888E-01 ⎬ strains εx εy γxy
-.1258E-06  -.2017E-07   .2703E-06 ⎪ and
-.1810E+00  -.9973E-01   .1040E+00 ⎪ stresses σx σy τxy
-.9988E-07  -.2180E-06   .2266E-06 ⎪
-.2602E+00  -.3511E+00   .8714E-01 ⎪
-.3740E-07  -.3100E-06   .7509E-07 ⎪
-.2292E+00  -.4388E+00   .2888E-01 ⎭
```

Figure 5.19 Results from Program 5.4

PROGRAM 5.5: PLANE STRAIN ANALYSIS USING NINE-NODE QUADRILATERALS

```
C
C      PROGRAM 5.5 PLANE STRAIN OF AN ELASTIC
C      SOLID USING 9-NODE QUADRILATERAL ELEMENTS
C
C
C      ALTER NEXT LINE TO CHANGE PROBLEM SIZE
C
      PARAMETER(IKB1=100,IKB2=35,ILOADS=100,INF=100)
C
      REAL DEE(3,3),SAMP(4,2),COORD(9,2),JAC(2,2),JAC1(2,2),
     +DER(2,9),DERIV(2,9),BEE(3,18),DBEE(3,18),
     +BTDB(18,18),KM(18,18),ELD(18),EPS(3),SIGMA(3),
     +BT(18,3),FUN(9),KB(IKB1,IKB2),LOADS(ILOADS)
      INTEGER G(18),NF(INF,2)
      DATA IDEE,IBEE,IDBEE,IH/4*3/,IDOF,IBTDB,IBT,IKM/4*18/
      DATA IJAC,IJAC1,NODOF,IT,IDER,IDERIV/6*2/,ICOORD,NOD/2*9/
      DATA ISAMP/4/
C
C      INPUT AND INITIALISATION
C
      READ(5,*)NXE,NYE,N,IW,NN,NR,NGP,AA,BB,E,V
      CALL READNF(NF,INF,NN,NODOF,NR)
      IWP1=IW+1
      CALL NULL(KB,IKB1,N,IWP1)
      CALL FMDEPS(DEE,IDEE,E,V)
      CALL GAUSS(SAMP,ISAMP,NGP)
C
C      ELEMENT STIFFNESS INTEGRATION AND ASSEMBLY
C
      DO 10 IP=1,NXE
      DO 10 IQ=1,NYE
      CALL GEOM9X(IP,IQ,NXE,AA,BB,COORD,ICOORD,G,NF,INF)
      CALL NULL(KM,IKM,IDOF,IDOF)
      DO 20 I=1,NGP
      DO 20 J=1,NGP
      CALL FMLAG9(DER,IDER,FUN,SAMP,ISAMP,I,J)
```

```
      CALL MATMUL(DER,IDER,COORD,ICOORD,JAC,IJAC,IT,NOD,IT)
      CALL TWOBY2(JAC,IJAC,JAC1,IJAC1,DET)
      CALL MATMUL(JAC1,IJAC1,DER,IDER,DERIV,IDERIV,IT,IT,NOD)
      CALL NULL(BEE,IBEE,IH,IDOF)
      CALL FORMB(BEE,IBEE,DERIV,IDERIV,NOD)
      CALL MATMUL(DEE,IDEE,BEE,IBEE,DBEE,IDBEE,IH,IH,IDOF)
      CALL MATRAN(BT,IBT,BEE,IBEE,IH,IDOF)
      CALL MATMUL(BT,IBT,DBEE,IDBEE,BTDB,IBTDB,IDOF,IH,IDOF)
      QUOT=DET*SAMP(I,2)*SAMP(J,2)
      CALL MSMULT(BTDB,IBTDB,QUOT,IDOF,IDOF)
   20 CALL MATADD(KM,IKM,BTDB,IBTDB,IDOF,IDOF)
   10 CALL FORMKB(KB,IKB1,KM,IKM,G,IW,IDOF)
C
C       EQUATION SOLUTION
C
      CALL CHOLIN(KB,IKB1,N,IW)
      CALL NULVEC(LOADS,N)
      READ(5,*)NL,(K,LOADS(K),I=1,NL)
      CALL CHOBAC(KB,IKB1,LOADS,N,IW)
      CALL PRINTV(LOADS,N)
C
C       RECOVER STRAINS AND STRESSES AT ELEMENT CENTRES
C
      NGP=1
      CALL GAUSS(SAMP,ISAMP,NGP)
      DO 30 IP=1,NXE
      DO 30 IQ=1,NYE
      CALL GEOM9X(IP,IQ,NXE,AA,BB,COORD,ICOORD,G,NF,INF)
      DO 40 M=1,IDOF
      IF(G(M).EQ.0)ELD(M)=0.
   40 IF(G(M).NE.0)ELD(M)=LOADS(G(M))
      DO 30 I=1,NGP
      DO 30 J=1,NGP
      CALL FMLAG9(DER,IDER,FUN,SAMP,ISAMP,I,J)
      CALL MATMUL(DER,IDER,COORD,ICOORD,JAC,IJAC,IT,NOD,IT)
      CALL TWOBY2(JAC,IJAC,JAC1,IJAC1,DET)
      CALL MATMUL(JAC1,IJAC1,DER,IDER,DERIV,IDERIV,IT,IT,NOD)
      CALL NULL(BEE,IBEE,IH,IDOF)
      CALL FORMB(BEE,IBEE,DERIV,IDERIV,NOD)
      CALL MVMULT(BEE,IBEE,ELD,IH,IDOF,EPS)
      CALL MVMULT(DEE,IDEE,EPS,IH,IH,SIGMA)
      CALL PRINTV(EPS,IH)
      CALL PRINTV(SIGMA,IH)
   30 CONTINUE
      STOP
      END
```

The nine-noded quadrilateral is in many ways similar to the eight-noded element described in the previous program. The addition of an extra node at the element centroid, however, gives the shape functions of this element a greater degree of symmetry. This symmetry occurs due to the use of Lagrangian polynomials to form the shape functions; hence this element is often referred to as a Lagrangian element.

Apart from the obvious array bound changes necessary to accommodate the extra freedoms associated with this element, the only other alterations involve the following subroutines:

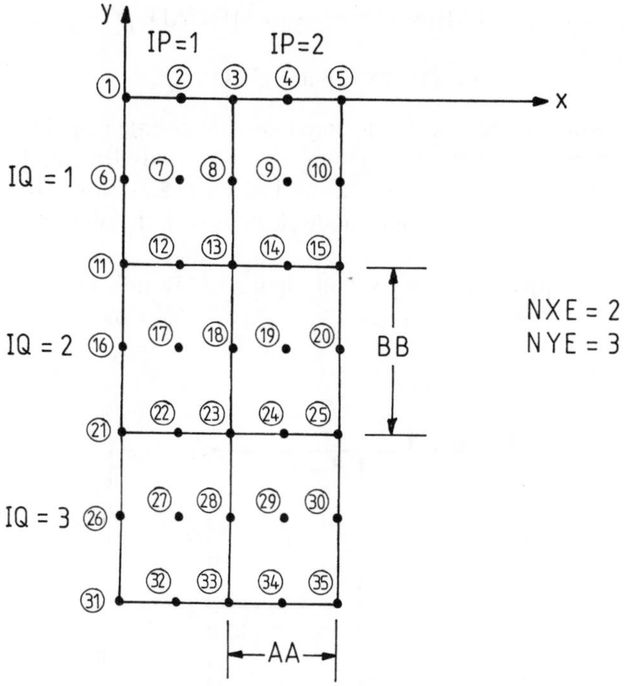

Figure 5.20 Global node and element numbering for mesh of nine-node quadrilaterals

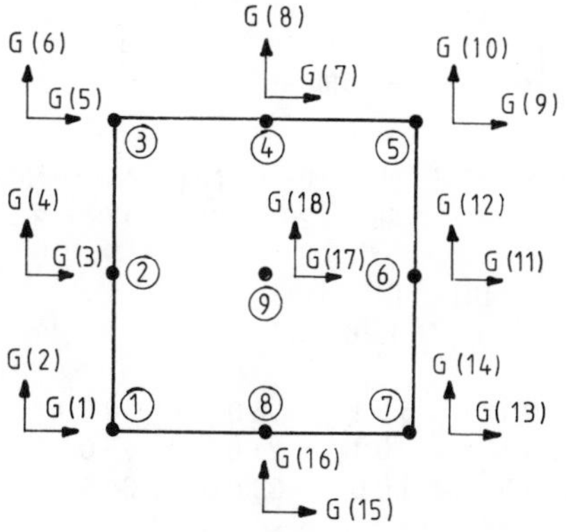

Figure 5.21 Local node and freedom numbering for nine-node quadrilateral

FMLAG9 replaces FMQUAD

and

GEOM9X replace GEOM8X

Once more, assuming nodes and freedoms are counted in the x direction, Figure 5.20 gives the numbering system for nodes and elements for a typical mesh. The local numbering system given in Figure 5.21 shows that the node numbering follows that of the eight-node element with the ninth node placed at the centroid.

The simple example and data given in Figure 5.22 is the same boundary value

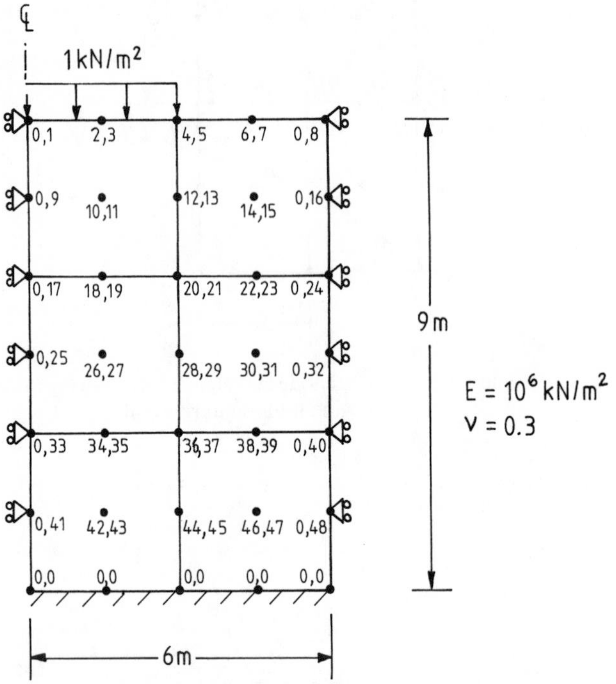

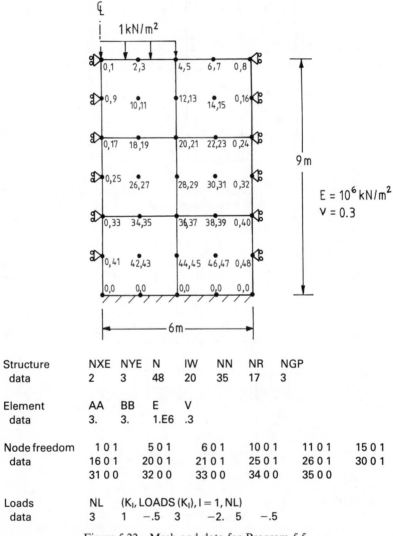

Structure data	NXE	NYE	N	IW	NN	NR	NGP
	2	3	48	20	35	17	3

Element data	AA	BB	E	V
	3.	3.	1.E6	.3

Node freedom data						
	1 0 1	5 0 1	6 0 1	10 0 1	11 0 1	15 0 1
	16 0 1	20 0 1	21 0 1	25 0 1	26 0 1	30 0 1
	31 0 0	32 0 0	33 0 0	34 0 0	35 0 0	

Loads data	NL	(K_I, LOADS (K_I), I = 1, NL)					
	3	1	−.5	3	−2.	5	−.5

Figure 5.22 Mesh and data for Program 5.5

```
.5299E-05  -.4004E-06  -.4988E-05  -.6190E-06  -.3343E-05  -.4004E-06
.1697E-05  -.1387E-05  -.4307E-05   .1856E-06  -.3911E-05   .3167E-06
.2786E-05   .1856E-06  -.1661E-05  -.1264E-05  -.3170E-05   .2748E-06
.2899E-05   .3777E-06  -.2229E-05   .2748E-06  -.1558E-05  -.1287E-05
.2195E-05   .2109E-06  -.2047E-05   .3051E-06  -.1671E-05   .2109E-06
.1296E-05  -.1148E-05  -.1373E-05   .1369E-06  -.1299E-05   .1948E-06
.1114E-05   .1369E-06  -.9293E-06  -.8555E-06  -.6492E-06   .7643E-07
.6245E-06   .1105E-06  -.5571E-06   .7643E-07  -.4898E-06  -.4651E-06
.1056E-06  -.6964E-06   .2822E-06
.2597E+00  -.8766E+00   .1085E+00
.1017E-06  -.5333E-06   .2206E-06
.1707E+00  -.6592E+00   .8484E-01
.3682E-07  -.4331E-06   .7633E-07
.2003E+00  -.5618E+00   .2936E-01
.1056E-06  -.4644E-07   .2822E-06
.1689E+00  -.1234E+00   .1085E+00
.1017E-06  -.2096E-06   .2206E-06
.2578E+00  -.3408E+00   .8484E-01
.3682E-07  -.3098E-06   .7633E-07
.2283E+00  -.4382E+00   .2936E-01
```

(First eight rows: Displacements. Remaining rows: Centroid strains ε_x ε_y γ_{xy} and stresses σ_x σ_y τ_{xy})

Figure 5.23 Results from Program 5.5

problem as that solved using the eight-node element. An exact integration scheme (NGP = 3) has been used to obtain the computed results given in Figure 5.23, which are directly comparable with those in Figure 5.19.

PROGRAM 5.6: AXISYMMETRIC STRAIN ANALYSIS USING FOUR-NODE QUADRILATERALS

```
C
C       PROGRAM 5.6 AXISYMMETRIC STRAIN OF AN ELASTIC
C       SOLID USING 4-NODE QUADRILATERAL ELEMENTS
C
C
C       ALTER NEXT LINE TO CHANGE PROBLEM SIZE
C
      PARAMETER(IKV=3500,ILOADS=100,INF=100,INR=20,IND=20,IPROP=20)
C
      REAL DEE(4,4),SAMP(4,2),COORD(4,2),JAC(2,2),JAC1(2,2),
     +DER(2,4),DERIV(2,4),BEE(4,8),DBEE(4,8),KV(IKV),LOADS(ILOADS),
     +BTDB(8,8),KM(8,8),ELD(8),EPS(4),SIGMA(4),
     +BT(8,4),FUN(4),RAD(INR),DEP(IND),PROP(IPROP,2)
      INTEGER G(8),NF(INF,2)
      DATA IDEE,IBEE,IDBEE,IH/4*4/,IDOF,IBTDB,IBT,IKM/4*8/
      DATA IJAC,IJAC1,NODOF,IT,IDER,IDERIV/6*2/,ICOORD,NOD/2*4/
      DATA ISAMP/4/
C
C       INPUT AND INITIALISATION
C
      READ(5,*)NRE,NDE,N,IW,NN,NR,NGP
      READ(5,*)(RAD(I),I=1,NRE+1)
      READ(5,*)(DEP(I),I=1,NDE+1)
      READ(5,*)((PROP(I,J),I=1,NDE),J=1,2)
      CALL READNF(NF,INF,NN,NODOF,NR)
      IR=N*(IW+1)
      CALL NULVEC(KV,IR)
      CALL GAUSS(SAMP,ISAMP,NGP)
```

```
C
C      ELEMENT STIFFNESS INTEGRATION AND ASSEMBLY
C
      DO 10 IP=1,NRE
      DO 10 IQ=1,NDE
      E=PROP(IQ,1)
      V=PROP(IQ,2)
      CALL FMDRAD(DEE,IDEE,E,V)
      CALL GEOV4Y(IP,IQ,NDE,RAD,DEP,COORD,ICOORD,G,NF,INF)
      CALL NULL(KM,IKM,IDOF,IDOF)
      DO 20 I=1,NGP
      DO 20 J=1,NGP
      CALL FORMLN(DER,IDER,FUN,SAMP,ISAMP,I,J)
      CALL MATMUL(DER,IDER,COORD,ICOORD,JAC,IJAC,IT,NOD,IT)
      CALL TWOBY2(JAC,IJAC,JAC1,IJAC1,DET)
      CALL MATMUL(JAC1,IJAC1,DER,IDER,DERIV,IDERIV,IT,IT,NOD)
      CALL NULL(BEE,IBEE,IH,IDOF)
      CALL FMBRAD(BEE,IBEE,DERIV,IDERIV,FUN,COORD,ICOORD,SUM,NOD)
      CALL MATMUL(DEE,IDEE,BEE,IBEE,DBEE,IDBEE,IH,IH,IDOF)
      CALL MATRAN(BT,IBT,BEE,IBEE,IH,IDOF)
      CALL MATMUL(BT,IBT,DBEE,IDBEE,BTDB,IBTDB,IDOF,IH,IDOF)
      QUOT=SUM*DET*SAMP(I,2)*SAMP(J,2)
      CALL MSMULT(BTDB,IBTDB,QUOT,IDOF,IDOF)
   20 CALL MATADD(KM,IKM,BTDB,IBTDB,IDOF,IDOF)
   10 CALL FORMKV(KV,KM,IKM,G,N,IDOF)
C
C      EQUATION SOLUTION
C
      CALL BANRED(KV,N,IW)
      CALL NULVEC(LOADS,N)
      READ(5,*)NL,(K,LOADS(K),I=1,NL)
      CALL BACSUB(KV,LOADS,N,IW)
      CALL PRINTV(LOADS,N)
C
C      RECOVER STRESSES AT ELEMENT CENTRES
C
      NGP=1
      CALL GAUSS(SAMP,ISAMP,NGP)
      DO 30 IP=1,NRE
      DO 30 IQ=1,NDE
      E=PROP(IQ,1)
      V=PROP(IQ,2)
      CALL FMDRAD(DEE,IDEE,E,V)
      CALL GEOV4Y(IP,IQ,NDE,RAD,DEP,COORD,ICOORD,G,NF,INF)
      DO 40 M=1,IDOF
      IF(G(M).EQ.0)ELD(M)=0.
   40 IF(G(M).NE.0)ELD(M)=LOADS(G(M))
      DO 30 I=1,NGP
      DO 30 J=1,NGP
      CALL FORMLN(DER,IDER,FUN,SAMP,ISAMP,I,J)
      CALL MATMUL(DER,IDER,COORD,ICOORD,JAC,IJAC,IT,NOD,IT)
      CALL TWOBY2(JAC,IJAC,JAC1,IJAC1,DET)
      CALL MATMUL(JAC1,IJAC1,DER,IDER,DERIV,IDERIV,IT,IT,NOD)
      CALL NULL(BEE,IBEE,IH,IDOF)
      CALL FMBRAD(BEE,IBEE,DERIV,IDERIV,FUN,COORD,ICOORD,SUM,NOD)
      CALL MVMULT(BEE,IBEE,ELD,IH,IDOF,EPS)
      CALL MVMULT(DEE,IDEE,EPS,IH,IH,SIGMA)
      CALL PRINTV(EPS,IH)
      CALL PRINTV(SIGMA,IH)
   30 CONTINUE
      STOP
      END
```

This program is the axisymmetric counterpart of Program 5.3, which dealt with plane strain conditions. Slight variations are introduced in that the mesh is now irregular in a similar way to that described in Program 5.2 (although still 'rectangular') while the elastic properties E and V are allowed to assume different values in each horizontal layer of elements. These properties are stored in the array PROP and the dimension of this array appears in the PARAMETER statements, where

$$\text{IPROP} \geqslant \text{NDE}$$

NRE and NDE for axisymmetric meshes are the counterparts of NXE and NYE in planar problems.

Nodal coordinates COORD and the steering vector G are generated by the 'variable' geometry subroutine GEOV4Y, which assumes that nodes are numbered in the 'depth' direction. Two other subroutines appear for the first time

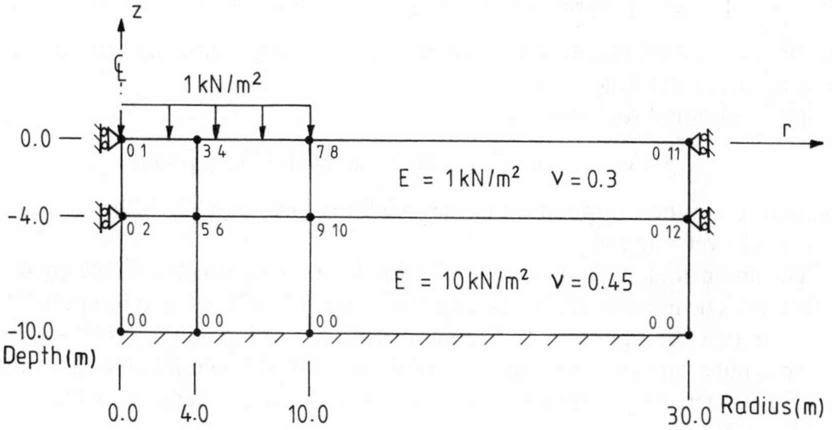

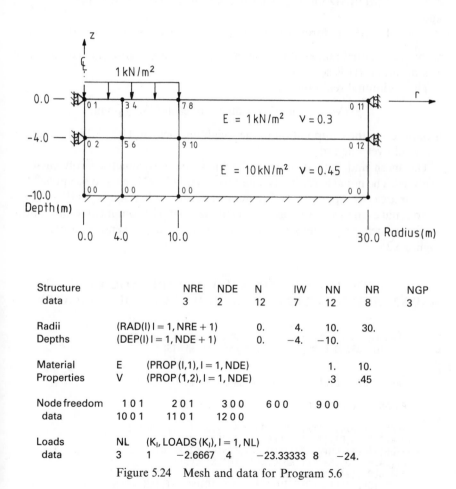

Structure data		NRE	NDE	N	IW	NN	NR	NGP
		3	2	12	7	12	8	3
Radii	(RAD(I) I = 1, NRE + 1)			0.	4.	10.	30.	
Depths	(DEP(I) I = 1, NDE + 1)			0.	−4.	−10.		
Material Properties	E (PROP (I,1), I = 1, NDE)					1.	10.	
	V (PROP (1,2), I = 1, NDE)					.3	.45	
Node freedom data	1 0 1	2 0 1	3 0 0	6 0 0	9 0 0			
	10 0 1	11 0 1	12 0 0					
Loads data	NL (KI, LOADS (K_I), I = 1, NL)							
	3 1 −2.6667 4 −23.33333 8 −24.							

Figure 5.24 Mesh and data for Program 5.6

```
-.3176E+01  -.3231E+00   .1395E+00  -.3991E+01   .1165E+00  -.2498E+00 } Displacements
 .1704E+00  -.6046E+00   .1330E+00  -.4421E-01   .2588E+00   .3091E-01 }
 .3200E-01  -.8242E+00  -.8977E-01   .3200E-01
-.4140E+00  -.1073E+01  -.3453E-01  -.4140E+00
 .1456E-01  -.4774E-01   .1887E-01   .1456E-01
-.4776E+00  -.9072E+00   .6508E-01  -.4776E+00
 .3949E-02  -.5376E+00   .3068E+00   .1998E-01
-.2933E+00  -.7098E+00   .1180E+00  -.2810E+00
 .1378E-02  -.2450E-01   .3792E-01   .8911E-02
-.4316E+00  -.6101E+00   .1308E+00  -.3796E+00
-.7584E-02  -.4157E-01   .2813E-01   .3792E-02
-.3200E-01  -.5814E-01   .1082E-01  -.2325E-01
-.3325E-02  -.1109E-02   .1296E-01   .1663E-02
-.1090E+00  -.9367E-01   .4470E-01  -.7455E-01
```

Centroid strains ε_r ε_z γ_{rz} ε_θ and stresses σ_r σ_z τ_{rz} σ_θ

Figure 5.25 Results from Program 5.6

for axisymmetric conditions:

FMDRAD forms the 4 × 4 stress–strain DEE matrix (2.68)

and

FMBRAD forms the 4 × 8 strain displacement BEE matrix (2.67)

It should be noted that as four components of strain and stress are computed by this program, IH is set to 4.

The additional real number:

SUM, the radial coordinate of each Gauss point,

is required in the computation of the weighting factor QUOT which integrates the solid over one radian.

The mesh and data in Figure 5.24 show the axisymmetric problem to be analysed. The nodal loads imply a uniform stress of 1 kN/m^2 is to be applied to a circular area (see Appendix 1). The number of integrating points, NGP, has been set to 3, but even this is not exact, especially as $r \to 0$. The computed results for this problem, including strains and stresses at the element 'centres', are given in Figure 5.25.

PROGRAM 5.7: NON-AXISYMMETRIC STRAIN OF AXISYMMETRIC SOLIDS USING EIGHT-NODE QUADRILATERALS

```
C
C      PROGRAM 5.7 NON-AXISYMMETRIC STRAIN OF AXISYMMETRIC
C      SOLIDS USING 8-NODE QUADRILATERAL ELEMENTS
C
C
C      ALTER NEXT LINE TO CHANGE PROBLEM SIZE
C
      PARAMETER(IKB1=100,IKB2=35,ILOADS=100,INF=100)
C
      REAL DEE(6,6),SAMP(4,2),COORD(8,2),JAC(2,2),JAC1(2,2),
     +DER(2,8),DERIV(2,8),BEE(6,24),DBEE(6,24),
     +BTDB(24,24),KM(24,24),ELD(24),EPS(6),SIGMA(6),
     +BT(24,6),FUN(8),KB(IKB1,IKB2),LOADS(ILOADS)
      INTEGER G(24),NF(INF,3)
      DATA IDEE,IBEE,IDBEE,IH/4*6/,IDOF,IBTDB,IBT,IKM/4*24/
```

```
      DATA IJAC,IJAC1,IT,IDER,IDERIV/5*2/,ICOORD,NOD/2*8/
      DATA ISAMP/4/,NODOF/3/
C
C       INPUT AND INITIALISATION
C
      READ(5,*)NRE,NDE,N,IW,NN,NR,NGP,AA,BB,E,V
      READ(5,*)LTH,IFLAG,CHI
      CALL READNF(NF,INF,NN,NODOF,NR)
      PI=4.*ATAN(1.)
      CHI=CHI*PI/180.
      CA=COS(CHI)
      SA=SIN(CHI)
      IWP1=IW+1
      CALL NULL(KB,IKB1,N,IWP1)
      CALL FORMD3(DEE,IDEE,E,V)
      CALL GAUSS(SAMP,ISAMP,NGP)
C
C       ELEMENT STIFFNESS INTEGRATION AND ASSEMBLY
C
      DO 10 IP=1,NRE
      DO 10 IQ=1,NDE
      CALL GENA8X(IP,IQ,NRE,AA,BB,COORD,ICOORD,G,NF,INF)
      CALL NULL(KM,IKM,IDOF,IDOF)
      DO 20 I=1,NGP
      DO 20 J=1,NGP
      CALL FMQUAD(DER,IDER,FUN,SAMP,ISAMP,I,J)
      CALL MATMUL(DER,IDER,COORD,ICOORD,JAC,IJAC,IT,NOD,IT)
      CALL TWOBY2(JAC,IJAC,JAC1,IJAC1,DET)
      CALL MATMUL(JAC1,IJAC1,DER,IDER,DERIV,IDERIV,IT,IT,NOD)
      CALL NULL(BEE,IBEE,IH,IDOF)
      CALL BNONAX(BEE,IBEE,DERIV,IDERIV,FUN,COORD,ICOORD,
     +            SUM,NOD,IFLAG,LTH)
      CALL MATMUL(DEE,IDEE,BEE,IBEE,DBEE,IDBEE,IH,IH,IDOF)
      CALL MATRAN(BT,IBT,BEE,IBEE,IH,IDOF)
      CALL MATMUL(BT,IBT,DBEE,IDBEE,BTDB,IBTDB,IDOF,IH,IDOF)
      QUOT=SUM*DET*SAMP(I,2)*SAMP(J,2)
      CALL MSMULT(BTDB,IBTDB,QUOT,IDOF,IDOF)
   20 CALL MATADD(KM,IKM,BTDB,IBTDB,IDOF,IDOF)
   10 CALL FORMKB(KB,IKB1,KM,IKM,G,IW,IDOF)
C
C       EQUATION SOLUTION
C
      CALL CHOLIN(KB,IKB1,N,IW)
      CALL NULVEC(LOADS,N)
      READ(5,*)NL,(K,LOADS(K),I=1,NL)
      CALL CHOBAC(KB,IKB1,LOADS,N,IW)
      CALL PRINTV(LOADS,N)
C
C       RECOVER STRAINS AND STRESSES AT ELEMENT CENTRES
C
      NGP=1
      CALL GAUSS(SAMP,ISAMP,NGP)
      DO 30 IP=1,NRE
      DO 30 IQ=1,NDE
      CALL GENA8X(IP,IQ,NRE,AA,BB,COORD,ICOORD,G,NF,INF)
      DO 40 M=1,IDOF
      IF(G(M).EQ.0)ELD(M)=0.
   40 IF(G(M).NE.0)ELD(M)=LOADS(G(M))
      DO 30 I=1,NGP
      DO 30 J=1,NGP
      CALL FMQUAD(DER,IDER,FUN,SAMP,ISAMP,I,J)
      CALL MATMUL(DER,IDER,COORD,ICOORD,JAC,IJAC,IT,NOD,IT)
      CALL TWOBY2(JAC,IJAC,JAC1,IJAC1,DET)
```

```
      CALL MATMUL(JAC1,IJAC1,DER,IDER,DERIV,IDERIV,IT,IT,NOD)
      CALL NULL(BEE,IBEE,IH,IDOF)
      CALL BNONAX(BEE,IBEE,DERIV,IDERIV,FUN,COORD,ICOORD,
     +             SUM,NOD,IFLAG,LTH)
      DO 50 L=1,IDOF
      DO 60 K=1,4
   60 BEE(K,L)=BEE(K,L)*CA
      DO 50 K=5,6
   50 BEE(K,L)=BEE(K,L)*SA
      CALL MVMULT(BEE,IBEE,ELD,IH,IDOF,EPS)
      CALL MVMULT(DEE,IDEE,EPS,IH,IH,SIGMA)
      CALL PRINTV(SIGMA,IH)
   30 CONTINUE
      STOP
      END
```

This program allows the analysis of axisymmetric solids subjected to non-axisymmetric loads. Variations in displacements, and hence strains and stresses, tangentially are described by Fourier series (e.g. Wilson, 1965; Zienkiewicz, 1977). Although the analysis is genuinely three dimensional, with three degrees of freedom at each node, it is only necessary to discretise the problem in a radial plane. The integrals in radial planes are performed using Gaussian quadrature in the usual way. Orthogonality relationships between typical terms in the tangential direction enable the integrals in the third direction to be stated explicitly. The problem therefore takes on the 'appearance' of a two-dimensional analysis with the obvious benefits in terms of storage requirements. The disadvantage of the method over conventional three-dimensional finite element analyses is that, for complicated loading distributions, several loading harmonic terms may be required and a global stiffness matrix must be stored for each. Several harmonic terms may be required for elasto-plastic analyses (Griffiths, 1986), but for most elastic analyses such as the one described here, one harmonic will often be sufficient.

It is important to realise that the basic stiffness relationships relate amplitudes of load to amplitudes of displacement. Once the amplitudes of a displacement are known, the actual displacement at a particular circumferential location is easily found.

For simplicity, consider only the components of nodal load which are symmetric about the $\theta = 0$ axis of the axisymmetric body. In this case a general loading distribution may be given by

$$\begin{aligned} R &= \tfrac{1}{2}\bar{R}^0 + \bar{R}^1 \cos\theta + \bar{R}^2 \cos 2\theta + \cdots \\ Z &= \tfrac{1}{2}\bar{Z}^0 + \bar{Z}^1 \cos\theta + \bar{Z}^1 \cos 2\theta + \cdots \\ T &= \qquad\quad \bar{T}^1 \sin\theta + \bar{T}^2 \sin 2\theta + \cdots \end{aligned} \tag{5.1}$$

where R, Z and T represent the load per radian in the radial, depth and tangential directions. The bar terms represent amplitudes of these quantities.

For antisymmetric loading, symmetrical about $\theta = \pi/2$, these expressions become

$$\begin{aligned} R &= \qquad\quad \bar{R}^1 \sin\theta + \bar{R}^2 \sin 2\theta + \cdots \\ Z &= \qquad\quad \bar{Z}^1 \sin\theta + \bar{Z}^2 \sin 2\theta + \cdots \\ T &= \tfrac{1}{2}\bar{T}^0 + \bar{T}^1 \cos\theta + \bar{T}^2 \cos 2\theta + \cdots \end{aligned} \tag{5.2}$$

Corresponding to these quantities are amplitudes of displacement in the radial, depth and tengential directions.

Since there are now three displacements per node, there are six strains at any point taken in the order

$$\mathrm{EPS}^{\mathrm{T}} = [\varepsilon_r\, \varepsilon_z\, \varepsilon_\theta\, \gamma_{rz}\, \gamma_{z\theta}\, \gamma_{\theta r}] \tag{5.3}$$

and six corresponding stresses. The 6 × 6 stress–strain matrix DEE is formed by the subroutine FORMD3. With reference to equation (2.67), the **A** matrix now becomes

$$\mathbf{A} = \begin{bmatrix} \frac{\partial}{\partial r} & 0 & 0 \\ 0 & \frac{\partial}{\partial z} & 0 \\ \frac{1}{r} & 0 & \frac{1}{r}\frac{\partial}{\partial \theta} \\ \frac{\partial}{\partial z} & \frac{\partial}{\partial r} & 0 \\ 0 & \frac{1}{r}\frac{\partial}{\partial \theta} & \frac{\partial}{\partial z} \\ \frac{1}{r}\frac{\partial}{\partial \theta} & 0 & \frac{\partial}{\partial r} - \frac{1}{r} \end{bmatrix} \tag{5.4}$$

For each harmonic i, the strain-displacement relationship provided by the subroutine BNONAX is of the form

$$\mathbf{B}^i = [B_1^i\, B_2^i\, B_3^i\, B_4^i \cdots B_j^i \cdots B_{\mathrm{NOD}}^i] \tag{5.5}$$

where NOD is the number of nodes in an element.

A typical submatrix from the above expression for symmetric loading is given by

$$B_j^i = \begin{bmatrix} \frac{\partial N_j}{\partial r}\cos i\theta & 0 & 0 \\ 0 & \frac{\partial N_j}{\partial z}\cos i\theta & 0 \\ \frac{N_j}{r}\cos i\theta & 0 & \frac{iN_j}{r}\cos i\theta \\ \frac{\partial N_j}{\partial z}\cos i\theta & \frac{\partial N_j}{\partial r}\cos i\theta & 0 \\ 0 & \frac{iN_j}{r}\sin i\theta & \frac{\partial N_j}{\partial z}\sin i\theta \\ -\frac{iN_j}{r}\sin i\theta & 0 & \left(\frac{\partial N_j}{\partial r} - \frac{N_j}{r}\right)\sin i\theta \end{bmatrix} \tag{5.6}$$

The equivalent expression for antisymmetry is similar to equation (5.6) but with the sine and cosine terms interchanged and the signs of elements (3, 3), (5, 2) and (6, 1) reversed. Additional integer variables required by this subroutine are IFLAG and LTH. The variable IFLAG is set to 1 or -1 for symmetry or antisymmetry respectively, and the variable LTH gives the harmonic on which loads are to be applied.

It should be noted that if LTH = 0 and IFLAG = 1, the analysis reduces to ordinary axisymmetry as described by Program 5.6.

An additional variable input to this program is the angle CHI (in degrees in the range 0 to 360). This is the angle at which stresses are evaluated and printed. Naturally, stresses could be printed at other locations if required.

The program uses eight-node quadrilateral elements and can be considered a variant of Program 5.4. The geometry routine GENA8X assumes constant size elements with nodes numbered in the x (or radial) direction. Each element has 24 degrees of freedom, as shown in Figure 5.26.

The example shown in Figure 5.27 represents a cylindrical cantilever subjected to a transverse force of 1 kN at its tip. The nature of harmonic loading is such that a radial load amplitude of 1 unit on the first harmonic (LTH = 1) in symmetry (IFLAG = 1) results in a net thrust in the 0° direction of π. Thus the load amplitude applied at freedom 6 equals $1/\pi$. The nodal freedom data takes account of the fact that there can be no vertical movement along the centreline; hence these freedoms are restrained. The computed displacements in Figure 5.28 give the end deflection of the cantilever to be 6.755×10^{-2} m, compared with the slender beam value of 6.791×10^{-2} m. If the same load amplitude was applied to freedom 7, it would correspond to a net moment of 0.5 kN m. The computed

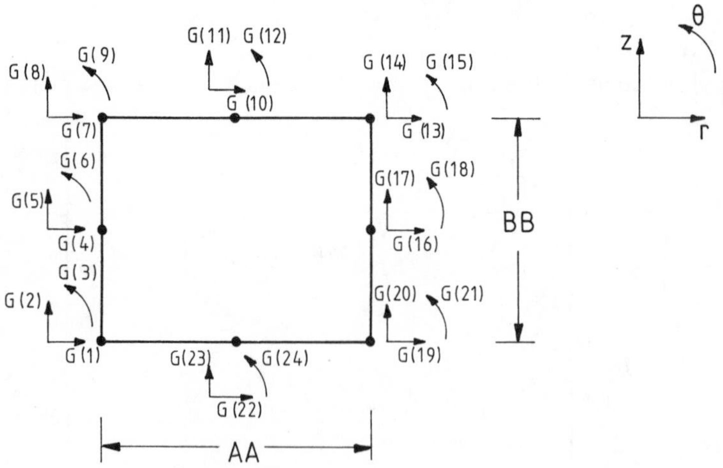

Figure 5.26 Local node and freedom numbering for eight-node quadrilateral (three freedoms per node)

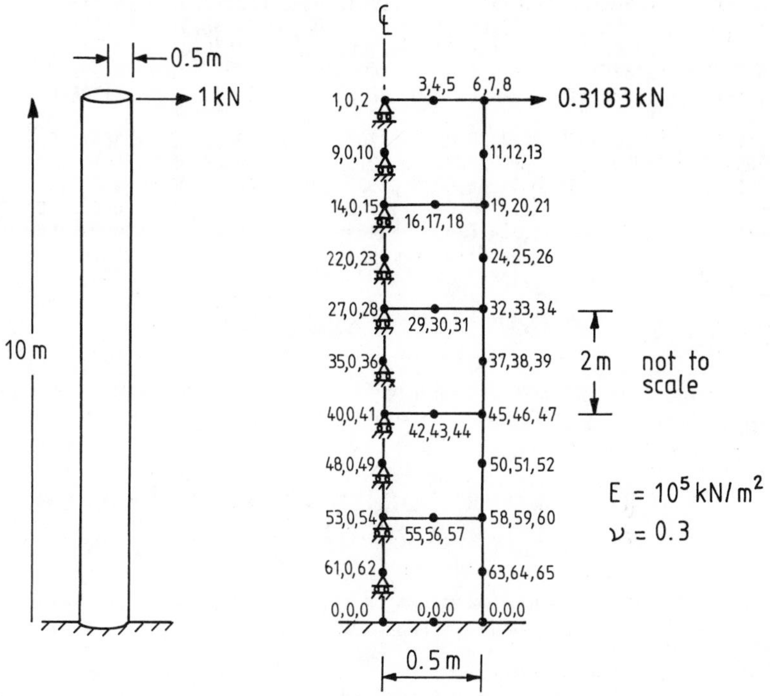

Structure	NRE	NDE	N	IW	NN	NR	NGP
data	1	5	65	20	28	13	2
Element	AA	BB	E	V			
data	.5	2.	1.E5	.3			
Harmonic	LTH	IFLAG	CHI				
data	1	1	0.				
Node freedom	1 1 0 1	4 1 0 1	6 1 0 1	9 1 0 1	11 1 0 1		
data	14 1 0 1	16 1 0 1	19 1 0 1	21 1 0 1	24 1 0 1		
	26 0 0 0	27 0 0 0	28 0 0 0				
Loads	NL,	$(K_I, LOADS(K_I), I = 1, NL)$					
data	1	6	.3183				

Figure 5.27 Mesh and data for Program 5.7

.6755E-01	-.6755E-01	.6755E-01	-.2528E-02	-.6755E-01	.6755E-01	Displacements
-.5063E-02	-.6754E-01	.5743E-01	-.5743E-01	.5744E-01	-.5006E-02	
-.5743E-01	.4753E-01	-.4752E-01	.4753E-01	-.2426E-02	-.4752E-01	
.4753E-01	-.4858E-02	-.4750E-01	.3801E-01	-.3801E-01	.3804E-01	
-.4598E-02	-.3799E-01	.2914E-01	-.2913E-01	.2913E-01	-.2121E-02	
-.2912E-01	.2914E-01	-.4248E-02	-.2909E-01	.2103E-01	-.2104E-01	
.2109E-01	-.3782E-02	-.2101E-01	.1400E-01	-.1399E-01	.1399E-01	
-.1614E-02	-.1398E-01	.1401E-01	-.3233E-02	-.1394E-01	.8132E-02	
-.8137E-02	.8212E-02	-.2559E-02	-.8085E-02	.3764E-02	-.3736E-02	
.3739E-02	-.9010E-03	-.3732E-02	.3754E-02	-.1809E-02	-.3687E-02	
.9378E-03	-.9620E-03	.1052E-02	-.9219E-03	-.8806E-03		
.6441E+00	-.5036E+01	-.4726E+00	.8638E-01	.0000E+00	.0000E+00	Centroid stresses σ_r σ_z σ_θ τ_{rz} $\tau_{z\theta}$ $\tau_{\theta r}$
.2661E+01	-.1413E+02	.1144E+01	.6800E-01	.0000E+00	.0000E+00	
.5441E+01	-.2274E+02	.3341E+01	.1484E+00	.0000E+00	.0000E+00	
.8040E+01	-.3164E+02	.5250E+01	.3627E+00	.0000E+00	.0000E+00	
.1700E+02	-.3521E+02	.1580E+02	.9873E+00	.0000E+00	.0000E+00	

Figure 5.28 Results from Program 5.7

displacement in this case at freedom 6 would be -5.063×10^{-3} m, compared with the slender beam value of -5.093×10^{-3} m.

PROGRAM 5.8: THREE-DIMENSIONAL ANALYSIS USING FOUR-NODED TETRAHEDRA

```
C
C      PROGRAM 5.8 THREE-DIMENSIONAL ELASTIC
C      ANALYSIS USING 4-NODE TETRAHEDRON ELEMENTS
C
C
C      ALTER NEXT LINE TO CHANGE PROBLEM SIZE
C
      PARAMETER(IKB1=320,IKB2=12,ILOADS=320,ISTOR=20)
C
      REAL DEE(6,6),SAMP(16,3),COORD(4,3),JAC(3,3),JAC1(3,3),
     +DER(3,4),DERIV(3,4),BEE(6,12),DBEE(6,12),BTDB(12,12),KM(12,12),
     +ELD(12),EPS(6),SIGMA(6),BT(12,6),FUN(4),WT(16),
     +KB(IKB1,IKB2),LOADS(ILOADS),STOREC(ISTOR,4,3)
      INTEGER G(12),STOREG(ISTOR,12)
      DATA ISAMP/16/,IBTDB,IKM,IBT,IDOF/4*12/,IDEE,IBEE,IDBEE,IH/4*6/
      DATA ICOORD,NOD/2*4/,IJAC,IJAC1,IDER,IDERIV,NODOF,IT/6*3/
C
C       INPUT AND INITIALISATION
C
      READ(5,*)NEL,N,IW,NIP,E,V
      IWP1=IW+1
      CALL NULL(KB,IKB1,N,IWP1)
      CALL FORMD3(DEE,IDEE,E,V)
      CALL NUMIN3(SAMP,ISAMP,WT,NIP)
C
C       ELEMENT STIFFNESS INTEGRATION AND ASSEMBLY
C
      DO 10 IP=1,NEL
      READ(5,*)((COORD(I,J),J=1,3),I=1,4)
      DO 20 I=1,NOD
      DO 20 J=1,IT
   20 STOREC(IP,I,J)=COORD(I,J)
      READ(5,*)(G(I),I=1,IDOF)
      DO 30 I=1,IDOF
```

```
   30 STOREG(IP,I)=G(I)
      CALL NULL(KM,IKM,IDOF,IDOF)
      DO 40 I=1,NIP
      CALL FMTET4(DER,IDER,FUN,SAMP,ISAMP,I)
      CALL MATMUL(DER,IDER,COORD,ICOORD,JAC,IJAC,IT,NOD,IT)
      CALL TREEX3(JAC,IJAC,JAC1,IJAC1,DET)
      CALL MATMUL(JAC1,IJAC1,DER,IDER,DERIV,IDERIV,IT,IT,NOD)
      CALL NULL(BEE,IBEE,IH,IDOF)
      CALL FORMB3(BEE,IBEE,DERIV,IDERIV,NOD)
      CALL MATMUL(DEE,IDEE,BEE,IBEE,DBEE,IDBEE,IH,IH,IDOF)
      CALL MATRAN(BT,IBT,BEE,IBEE,IH,IDOF)
      CALL MATMUL(BT,IBT,DBEE,IDBEE,BTDB,IBTDB,IDOF,IH,IDOF)
      QUOT=DET*WT(I)/6.
      CALL MSMULT(BTDB,IBTDB,QUOT,IDOF,IDOF)
      CALL MATADD(KM,IKM,BTDB,IBTDB,IDOF,IDOF)
   40 CONTINUE
      CALL FORMKB(KB,IKB1,KM,IKM,G,IW,IDOF)
   10 CONTINUE
C
C       EQUATION SOLUTION
C
      CALL CHOLIN(KB,IKB1,N,IW)
      CALL NULVEC(LOADS,N)
      READ(5,*)NL,(K,LOADS(K),I=1,NL)
      CALL CHOBAC(KB,IKB1,LOADS,N,IW)
      CALL PRINTV(LOADS,N)
C
C       RECOVER ELEMENT STRAINS AND STRESSES
C       AT TETRAHEDRON CENTROIDS
C
      DO 50 IP=1,NEL
      DO 60 I=1,NOD
      DO 60 J=1,IT
   60 COORD(I,J)=STOREC(IP,I,J)
      DO 70 I=1,IDOF
   70 G(I)=STOREG(IP,I)
      DO 50 I=1,NIP
      CALL FMTET4(DER,IDER,FUN,SAMP,ISAMP,I)
      CALL MATMUL(DER,IDER,COORD,ICOORD,JAC,IJAC,IT,NOD,IT)
      CALL TREEX3(JAC,IJAC,JAC1,IJAC1,DET)
      CALL MATMUL(JAC1,IJAC1,DER,IDER,DERIV,IDERIV,IT,IT,NOD)
      CALL NULL(BEE,IBEE,IH,IDOF)
      CALL FORMB3(BEE,IBEE,DERIV,IDERIV,NOD)
      DO 80 M=1,IDOF
      IF(G(M).EQ.0)ELD(M)=0.0
   80 IF(G(M).NE.0)ELD(M)=LOADS(G(M))
      CALL MVMULT(BEE,IBEE,ELD,IH,IDOF,EPS)
      CALL MVMULT(DEE,IDEE,EPS,IH,IH,SIGMA)
      CALL PRINTV(SIGMA,IH)
   50 CONTINUE
      STOP
      END
```

In cases where many Fourier harmonics are required to define a loading pattern it becomes more efficient to solve fully three-dimensional problems. The simplest three-dimensional element is the four-noded tetrahedron. This 'constant-strain' element is analogous to the three-noded triangle for planar problems described in Program 5.0 and, like the triangle, is not recommended for practical calculations. Due to its simplicity, however, this element is a convenient starting point for three-dimensional applications, and the program described here can easily be modified to include more sophisticated tetrahedron elements.

Comparing with Program 5.0, few changes are required. The subroutine NUMIN3 replaces NUMINT and provides the integrating points in volume coordinates and weighting coefficients for integration over tetrahedra. It may be noted that this simple element only requires one integrating point at its centroid. Three-dimensional **B** and **D** matrices are formed by the routines FORMB3 and FORMD3 and the Jacobian matrix, being now of dimensions 3×3, is inverted by the routine TREEX3.

For the simple example demonstrated here, it has been decided that the COORD and G arrays for each of the NEL elements should be read in as data. This of course could have been done in any of the previous programs if these quantities could not have been conveniently generated automatically. These extra READ statements mean that no geometry routine is required for this program. Extra arrays are declared, however, to store COORD and G for each element after they are read, and these are called STOREC and STOREG respectively. The variable dimension of these arrays, ISTOR ($\geqslant$ NEL), is defined in the PARAMETER statement. The element shape functions and derivatives are provided by the routine FMTET4. The only other obvious difference between this program and Program 5.0 is that the former uses Gaussian elimination to solve the equilibrium equations, whereas this program uses Choleski's method.

The local element node and freedom numbering is indicated in Figure 5.29 and it is important that the nodal coordinates and freedoms are read in a consistent manner. The system adopted here is that nodes (1), (2) and (3) go clockwise as viewed from node (4).

The example and data given in Figure 5.30 represent a cube made up of six tetrahedra. One corner of the cube is fixed and the three adjacent faces are restrained to move only in their own planes. The four nodal forces applied are equivalent to a uniform vertical compressive stress of 1 kN/m^2.

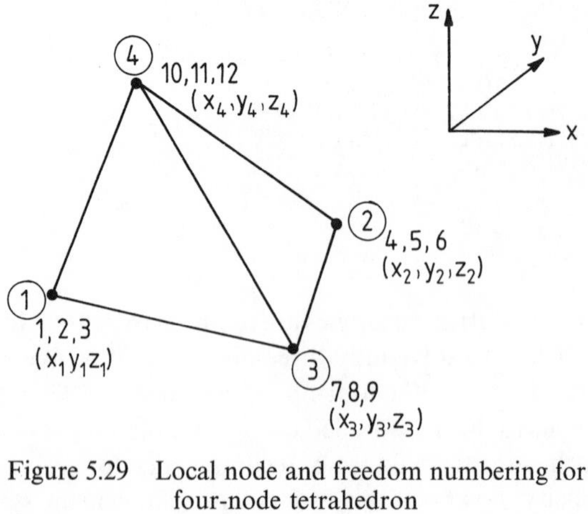

Figure 5.29 Local node and freedom numbering for four-node tetrahedron

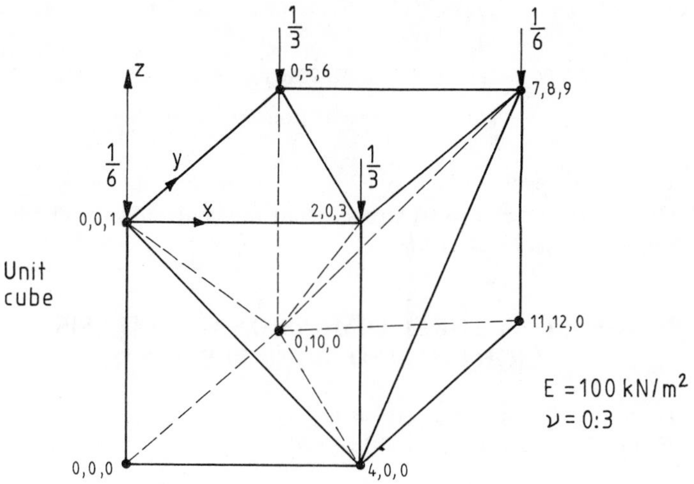

Structure data	NEL	N	IW	NGP
	6	12	9	1

Element data
E V
100 .3
((COORD (I, J), J = 1, 3), I = 1, 4) for each element
(G(I), I = 1, 12)

0.	0.	0.	0.	0.	−1.	1.	0.	−1.	0.	1.	−1.
0	0	1	0	0	0	4	0	0	0	10	0
0.	0.	0.	1.	0.	−1.	1.	0.	0.	0.	1.	−1.
0	0	1	4	0	0	2	0	3	0	10	0
0.	0.	0.	1.	0.	0.	0.	1.	0.	0.	1.	−1.
0	0	1	2	0	3	0	5	6	0	10	0
1.	1.	0.	1.	0.	−1.	1.	1.	−1.	0.	1.	−1.
7	8	9	4	0	0	11	12	0	0	10	0
1.	1.	0.	1.	0.	0.	1.	0.	−1.	0.	1.	−1.
7	8	9	2	0	3	4	0	0	0	10	0
1.	1.	0.	0.	1.	0.	1.	0.	0.	0.	1.	−1.
7	8	9	0	5	6	2	0	3	0	10	0

Loads data
NL (K$_I$, LOADS (K$_I$), I = 1, NL)
4 1 −.1667 3 −.3333 6 −.3333 9 −.1667

Figure 5.30 Mesh and data for Program 5.8

-.1000E-01	.3000E-02	-.1000E-01	.3000E-02	.3000E-02	-.9999E-02	Displacements
.3000E-02	.3000E-02	-.1000E-01	.3000E-02	.3000E-02	.3000E-02	
-.1965E-04	-.2100E-04	-.1000E+01	.0000E+00	.0000E+00	.0000E+00	Centroid stresses
.2302E-05	.1389E-04	-.1000E+01	-.6475E-05	.2974E-04	.2326E-04	$\sigma_x \sigma_y \sigma_z$
.1230E-04	.2075E-05	-.9999E+00	.0000E+00	.3641E-04	.2974E-04	$\tau_{xy} \tau_{yz} \tau_{zx}$
-.1087E-04	-.9023E-05	-.1000E+01	.1556E-05	-.7467E-05	-.9316E-05	
.6102E-05	-.1297E-05	-.1000E+01	-.8752E-05	-.2541E-04	-.3189E-04	
.9809E-05	.1536E-04	-.9999E+00	.2158E-05	-.3632E-04	-.4299E-04	

Figure 5.31 Results from Program 5.8

The computed results given in Figure 5.31 show that the cube compresses uniformly and that the vertical stress, σ_z, at the centroid of each element is in equilibrium with the applied loads.

PROGRAM 5.9: THREE-DIMENSIONAL ANALYSIS USING EIGHT-NODED BRICK ELEMENTS

```
C
C       PROGRAM 5.9 THREE-DIMENSIONAL ELASTIC
C       ANALYSIS USING 8-NODE BRICK ELEMENTS
C
C
C       ALTER NEXT LINE TO CHANGE PROBLEM SIZE
C
      PARAMETER(IKV=1000,ILOADS=100,INF=50)
C
      REAL DEE(6,6),SAMP(4,2),COORD(8,3),JAC(3,3),JAC1(3,3),
     +DER(3,8),DERIV(3,8),BEE(6,24),DBEE(6,24),BTDB(24,24),KM(24,24),
     +ELD(24),EPS(6),SIGMA(6),BT(24,6),FUN(8),KV(IKV),LOADS(ILOACS)
      INTEGER G(24),NF(INF,3)
      DATA ISAMP/4/,IBTDB,IKM,IBT,IDOF/4*24/,IDEE,IBEE,IDBEE,IH/4*6/
      DATA ICOORD,NOD/2*8/,IJAC,IJAC1,IDER,IDERIV,NODOF,IT/6*3/
C
C       INPUT AND INITIALISATION
C
      READ(5,*)NXE,NYE,NZE,N,IW,NN,NR,NGP,AA,BB,CC,E,V
      IR=N*(IW+1)
      CALL NULVEC(KV,IR)
      CALL FORMD3(DEE,IDEE,E,V)
      CALL GAUSS(SAMP,ISAMP,NGP)
      CALL READNF(NF,INF,NN,NODOF,NR)
C
C       ELEMENT STIFFNESS INTEGRATION AND ASSEMBLY
C
      DO 10 IP=1,NXE
      DO 10 IQ=1,NYE
      DO 10 IS=1,NZE
      CALL GEO83D(IP,IQ,IS,NXE,NZE,AA,BB,CC,COORD,ICOORD,G,NF,INF)
      CALL NULL(KM,IKM,IDOF,IDOF)
      DO 20 I=1,NGP
      DO 20 J=1,NGP
      DO 20 K=1,NGP
      CALL FMLIN3(DER,IDER,FUN,SAMP,ISAMP,I,J,K)
      CALL MATMUL(DER,IDER,COORD,ICOORD,JAC,IJAC,IT,NOD,IT)
      CALL TREEX3(JAC,IJAC,JAC1,IJAC1,DET)
      CALL MATMUL(JAC1,IJAC1,DER,IDER,DERIV,IDERIV,IT,IT,NOD)
      CALL NULL(BEE,IBEE,IH,IDOF)
      CALL FORMB3(BEE,IBEE,DERIV,IDERIV,NOD)
```

```
      CALL MATMUL(DEE,IDEE,BEE,IBEE,DBEE,IDBEE,IH,IH,IDOF)
      CALL MATRAN(BT,IBT,BEE,IBEE,IH,IDOF)
      CALL MATMUL(BT,IBT,DBEE,IDBEE,BTDB,IBTDB,IDOF,IH,IDOF)
      QUOT=DET*SAMP(I,2)*SAMP(J,2)*SAMP(K,2)
      CALL MSMULT(BTDB,IBTDB,QUOT,IDOF,IDOF)
      CALL MATADD(KM,IKM,BTDB,IBTDB,IDOF,IDOF)
   20 CONTINUE
      CALL FORMKV(KV,KM,IKM,G,N,IDOF)
   10 CONTINUE
C
C      EQUATION SOLUTION
C
      CALL BANRED(KV,N,IW)
      CALL NULVEC(LOADS,N)
      READ(5,*)NL,(K,LOADS(K),I=1,NL)
      CALL BACSUB(KV,LOADS,N,IW)
      CALL PRINTV(LOADS,N)
C
C      RECOVER ELEMENT STRAINS AND STRESSES
C      AT BRICK CENTROIDS
C
      NGP=1

      CALL GAUSS(SAMP,ISAMP,NGP)
      DO 30 IP=1,NXE
      DO 30 IQ=1,NYE
      DO 30 IS=1,NZE
      CALL GEO83D(IP,IQ,IS,NXE,NZE,AA,BB,CC,COORD,ICOORD,G,NF,INF)
      DO 40 M=1,IDOF
      IF(G(M).EQ.0)ELD(M)=0.0
   40 IF(G(M).NE.0)ELD(M)=LOADS(G(M))
      DO 30 I=1,NGP
      DO 30 J=1,NGP
      DO 30 K=1,NGP
      CALL FMLIN3(DER,IDER,FUN,SAMP,ISAMP,I,J,K)
      CALL MATMUL(DER,IDER,COORD,ICOORD,JAC,IJAC,IT,NOD,IT)
      CALL TREEX3(JAC,IJAC,JAC1,IJAC1,DET)
      CALL MATMUL(JAC1,IJAC1,DER,IDER,DERIV,IDERIV,IT,IT,NOD)
      CALL NULL(BEE,IBEE,IH,IDOF)
      CALL FORMB3(BEE,IBEE,DERIV,IDERIV,NOD)
      CALL MVMULT(BEE,IBEE,ELD,IH,IDOF,EPS)
      CALL MVMULT(DEE,IDEE,EPS,IH,IH,SIGMA)
      CALL PRINTV(SIGMA,IH)
   30 CONTINUE
      STOP
      END
```

This element is the three-dimensional equivalent of the four-noded quadrilateral element for planar problems described in Program 5.3. Comparing with this earlier program, the only additional input required are NZE, the number of elements in the z direction, and CC, the z dimension of these elements. The geometry routine GEO83D generates the COORD and G arrays, assuming a cubical mesh made up of constant size brick elements. The assumed nodal numbering system is described in Figure 5.32 and counts in the y direction in xz planes. Once one xz plane has been numbered, the next one in the y direction follows.

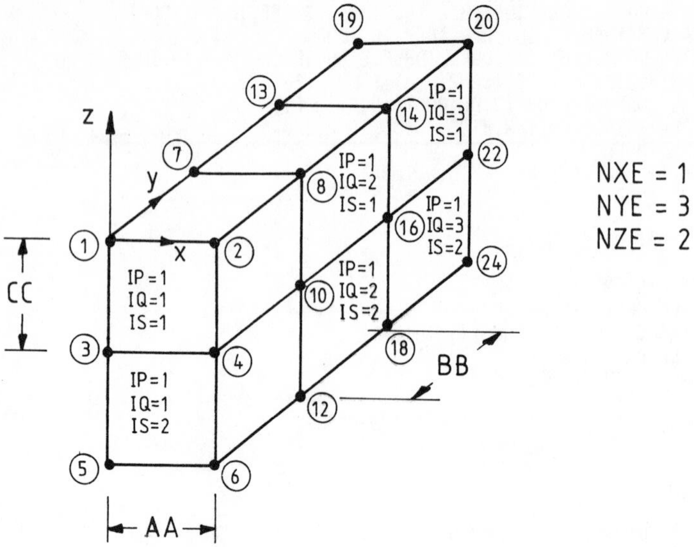

Figure 5.32 Global node and element numbering for eight-node bricks

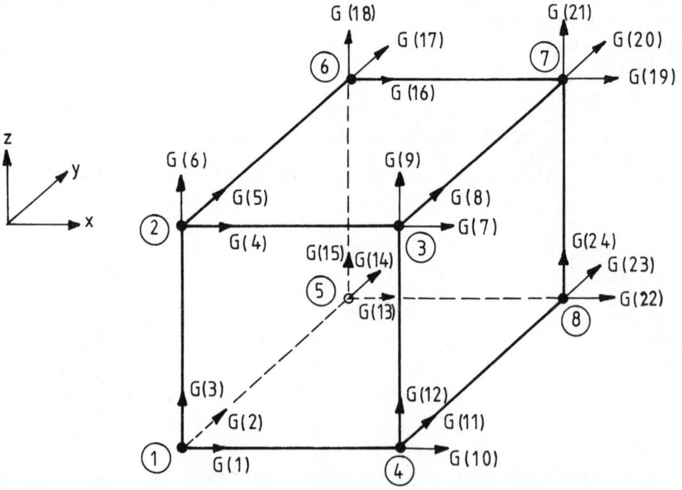

Figure 5.33 Local node and freedom numbering for eight-node brick

Due to the symmetry of Gaussian sampling points in the local coordinate system a third loop, in which K counts from 1 to NGP, tracks Gauss points across elements in the z direction. The local node and freedom numbering system for the eight-node element is demonstrated in Figure 5.33. The shape functions and their derivatives for this element are provided by routine FMLIN3. The element

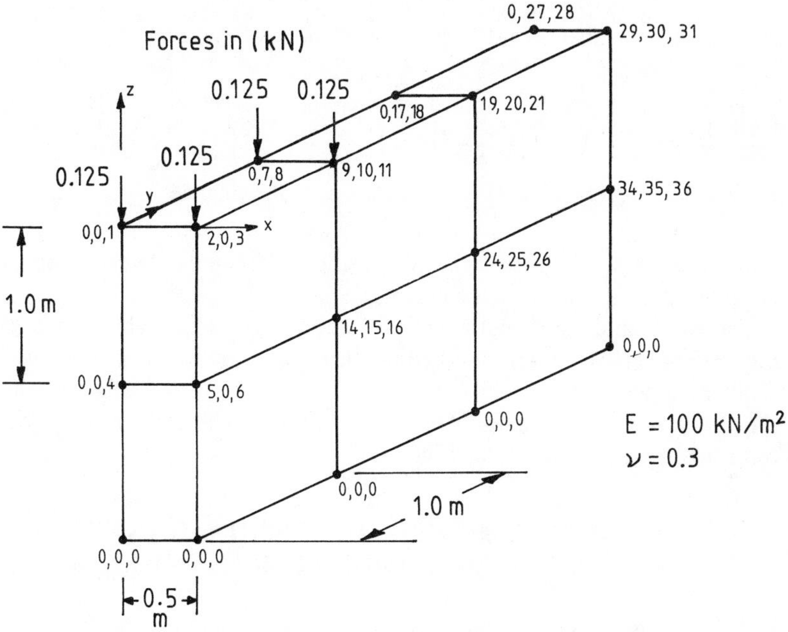

Structure data	NXE 1	NYE 3	NZE 2	N 36	IW 19	NN 24	NR 18	NGP 2

Element data	AA .5	BB 1.	CC 1.	E 100.	V .3

Node freedom data	1 0 0 1	2 1 0 1	3 0 0 1	4 1 0 1	5 0 0 0
	6 0 0 0	7 0 1 1	9 0 1 1	11 0 0 0	12 0 0 0
	13 0 1 1	15 0 1 1	17 0 0 0	18 0 0 0	19 0 1 1
	21 0 1 1	23 0 0 0	24 0 0 0		

Loads	NL	(K_I, LOADS (K_I), I = 1, NL)							
data	4	1	−.125	3	−.125	8	−.125	11	−.125

Figure 5.34 Mesh and data for Program 5.9

-.1595E-01	.2297E-02	-.1612E-01	-.6534E-02	.1550E-02	-.6842E-02	Displacements
-.1530E-02	-.9034E-02	.1056E-02	-.1197E-02	-.9011E-02	.1632E-02	
-.4272E-02	.1030E-02	.1820E-02	-.4437E-02	-.1193E-02	-.1107E-02	
-.1366E-03	-.1050E-02	-.9665E-03	.1431E-02	-.9641E-03	.3100E-03	
.1557E-02	-.9722E-03	-.6390E-03	.5760E-03	-.1735E-03	-.6629E-03	
.6552E-03	.8487E-03	.2654E-03	.3694E-04	.8924E-03	.2818E-03	
.5554E-02	-.2087E+00	-.7617E+00	-.6923E-02	.1203E+00	-.4432E-02	Centroid stresses
-.9508E-01	-.1279E+00	-.6190E+00	-.1389E-02	.7806E-01	.1570E-01	
.1573E-01	-.7076E-01	-.2534E+00	-.3187E-02	.1090E+00	-.4232E-02	
-.7003E-01	-.1305E+00	-.3263E+00	-.8784E-03	.1270E+00	.9555E-02	σ_x σ_y σ_z
.5950E-02	-.1346E-02	.1505E-01	.2583E-02	-.2391E-01	-.1930E-02	τ_{xy} τ_{yz} τ_{zx}
-.1468E-01	-.5201E-01	-.5473E-01	.6397E-03	.6936E-01	.3496E-02	

Figure 5.35 Results from Program 5.9

stiffness matrix is integrated exactly by using NGP equal to two, resulting in eight sampling points within each element.

The mesh and data for a simple example using this element is given in Figure 5.34. The symmetry implied by the freedom numbering indicates that the problem under analysis represents one quarter of a flexible footing bearing on an elastic layer supporting a uniform load of 1 kN/m^2. The nodal displacements and centroid stresses are given in Figure 5.35.

PROGRAM 5.10: THREE-DIMENSIONAL ANALYSIS USING TWENTY-NODED BRICK ELEMENTS

```
C
C       PROGRAM 5.10 THREE-DIMENSIONAL ELASTIC
C       ANALYSIS USING 20-NODE BRICK ELEMENTS
C
C
C       ALTER NEXT LINE TO CHANGE PROBLEM SIZE
C
      PARAMETER(IKV=4388,ILOADS=124,INF=70)
C
      REAL DEE(6,6),SAMP(4,2),COORD(20,3),JAC(3,3),JAC1(3,3),
     +DER(3,20),DERIV(3,20),BEE(6,60),DBEE(6,60),BTDB(60,60),KM(60,60),
     +ELD(60),EPS(6),SIGMA(6),BT(60,6),FUN(20),KV(IKV),LOADS(ILOADS)
      INTEGER G(60),NF(INF,3),KDIAG(ILOADS)
      DATA ISAMP/4/,IBTDB,IKM,IBT,IDOF/4*60/,IDEE,IBEE,IDBEE,IH/4*6/
      DATA ICOORD,NOD/2*20/,IJAC,IJAC1,IDER,IDERIV,NODOF,IT/6*3/
C
C       INPUT AND INITIALISATION
C
      READ(5,*)NXE,NYE,NZE,N,NN,NR,NGP,AA,BB,CC,E,V
      CALL READNF(NF,INF,NN,NODOF,NR)
      DO 10 I=1,N
   10 KDIAG(I)=0
      DO 20 IP=1,NXE
      DO 20 IQ=1,NYE
      DO 20 IS=1,NZE
      CALL GE203D(IP,IQ,IS,NXE,NZE,AA,BB,CC,COORD,ICOORD,G,NF,INF)
   20 CALL FKDIAG(KDIAG,G,IDOF)
      KDIAG(1)=1
      DO 30 I=2,N
   30 KDIAG(I)=KDIAG(I)+KDIAG(I-1)
      IR=KDIAG(N)
```

```
      CALL NULVEC(KV,IR)
      CALL FORMD3(DEE,IDEE,E,V)
      CALL GAUSS(SAMP,ISAMP,NGP)
C
C       ELEMENT STIFFNESS INTEGRATION AND ASSEMBLY
C
      DO 40 IP=1,NXE
      DO 40 IQ=1,NYE
      DO 40 IS=1,NZE
      CALL GE203D(IP,IQ,IS,NXE,NZE,AA,BB,CC,COORD,ICOORD,G,NF,INF)
      CALL NULL(KM,IKM,IDOF,IDOF)
      DO 50 I=1,NGP
      DO 50 J=1,NGP
      DO 50 K=1,NGP
      CALL FMQUA3(DER,IDER,FUN,SAMP,ISAMP,I,J,K)
      CALL MATMUL(DER,IDER,COORD,ICOORD,JAC,IJAC,IT,NOD,IT)
      CALL TREEX3(JAC,IJAC,JAC1,IJAC1,DET)
      CALL MATMUL(JAC1,IJAC1,DER,IDER,DERIV,IDERIV,IT,IT,NOD)
      CALL NULL(BEE,IBEE,IH,IDOF)
      CALL FORMB3(BEE,IBEE,DERIV,IDERIV,NOD)
      CALL MATMUL(DEE,IDEE,BEE,IBEE,DBEE,IDBEE,IH,IH,IDOF)
      CALL MATRAN(BT,IBT,BEE,IBEE,IH,IDOF)
      CALL MATMUL(BT,IBT,DBEE,IDBEE,BTDB,IBTDB,IDOF,IH,IDOF)
      QUOT=DET*SAMP(I,2)*SAMP(J,2)*SAMP(K,2)
      CALL MSMULT(BTDB,IBTDB,QUOT,IDOF,IDOF)
      CALL MATADD(KM,IKM,BTDB,IBTDB,IDOF,IDOF)
   50 CONTINUE
      CALL FSPARV(KV,KM,IKM,G,KDIAG,IDOF)
   40 CONTINUE
C
C       EQUATION SOLUTION
C
      CALL SPARIN(KV,N,KDIAG)
      CALL NULVEC(LOADS,N)
      READ(5,*)NL,(K,LOADS(K),I=1,NL)
      CALL SPABAC(KV,LOADS,N,KDIAG)
      CALL PRINTV(LOADS,34)
C
C       RECOVER ELEMENT STRAINS AND STRESSES
C       AT BRICK CENTROIDS
C
      NGP=1
      CALL GAUSS(SAMP,ISAMP,NGP)
      DO 60 IP=1,NXE
      DO 60 IQ=1,NYE
      DO 60 IS=1,NZE
      CALL GE203D(IP,IQ,IS,NXE,NZE,AA,BB,CC,COORD,ICOORD,G,NF,INF)
      DO 70 M=1,IDOF
      IF(G(M).EQ.0)ELD(M)=0.0
   70 IF(G(M).NE.0)ELD(M)=LOADS(G(M))
      DO 60 I=1,NGP
      DO 60 J=1,NGP
      DO 60 K=1,NGP
      CALL FMQUA3(DER,IDER,FUN,SAMP,ISAMP,I,J,K)
      CALL MATMUL(DER,IDER,COORD,ICOORD,JAC,IJAC,IT,NOD,IT)
      CALL TREEX3(JAC,IJAC,JAC1,IJAC1,DET)
      CALL MATMUL(JAC1,IJAC1,DER,IDER,DERIV,IDERIV,IT,IT,NOD)
      CALL NULL(BEE,IBEE,IH,IDOF)
      CALL FORMB3(BEE,IBEE,DERIV,IDERIV,NOD)
      CALL MVMULT(BEE,IBEE,ELD,IH,IDOF,EPS)
      CALL MVMULT(DEE,IDEE,EPS,IH,IH,SIGMA)
      CALL PRINTV(SIGMA,IH)
```

```
60 CONTINUE
   STOP
   END
```

A more sophisticated three-dimensional element, the twenty-node brick, is the subject of the final program in this chapter. The element is the three-dimensional analogue of the eight-noded quadrilateral used in planar problems and described in Program 5.4. Although this element is recommended for three-dimensional analysis, especially in non-linear problems, storage requirements rapidly become substantial on even the most powerful computers.

To reduce storage requirements, a 'skyline' storage strategy is adopted. This has the effect of reducing the number of zero terms that would be stored in the vector KV due to bandwidth variability using conventional storage methods.

In programming terms, an additional integer column array KDIAG is required. A preliminary scan of all elements is performed in the 'input and initialisation' section of the program. For each element, the G vector is formed and routine FKDIAG finds the greatest bandwidth associated with each freedom. This variable bandwidth is accumulated in the vector KDIAG as described for a simple case in Figure 3.16. Finally, the total storage requirement for the stiffness matrix is given by KDIAG(N), where N is the total number of freedoms. This figure should be considerably less than N*(IW + 1) which is the length of the KV vector assuming a constant bandwidth. It may be noted that IW is no longer required as data. In the authors' experience a saving of the order of 30% is often achieved using this method.

Due to this different storage strategy, new subroutines are required to assemble the global stiffness matrix in this pattern and also to solve the equilibrium equations. Assembly of the global stiffness matrix is performed by the routine FSPARV and Choleski's reduction and back-substitution performed by SPARIN and SPABAC respectively.

Other changes to the program to account for the new element may be summarised as follows:

routine FMQUA3 replaces FMLIN3 to form shape functions and derivatives

and

routine GE203D replaces GEO83D to generate the COORD and G arrays

As in the previous program, the geometry routine assumes that nodes are numbered in xz planes moving in the y direction, as illustrated in Figure 5.36. The node numbering at the element level is given in Figure 5.37 and freedoms are assumed to be numbered in the same sense.

The example and data of Figure 5.38 is the same boundary value problem as that analysed previously using eight-noded elements in Figure 5.34. An exact integration scheme has been used (NGP = 3) in the present analysis. The nodal forces to simulate a uniform stress field involve corner loads which act in the opposite direction to the mid-side loads (Appendix 1).

The computed displacements and centroid stresses are given in Figure 5.39. It

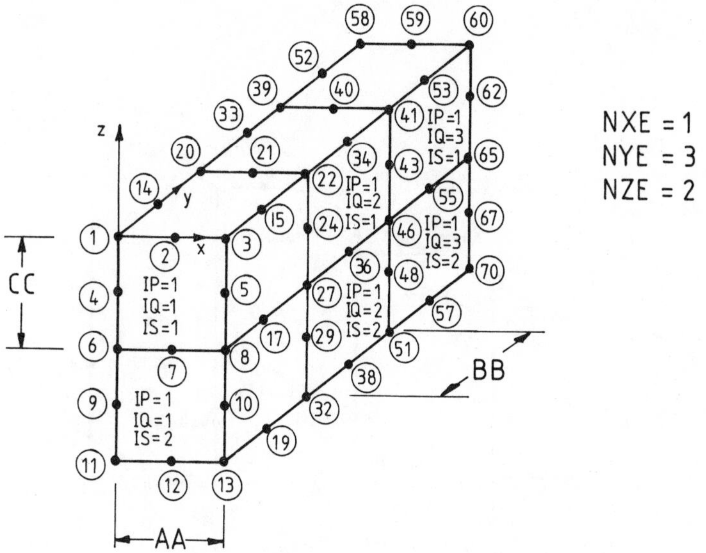

Figure 5.36 Global node and element numbering for 20-node brick

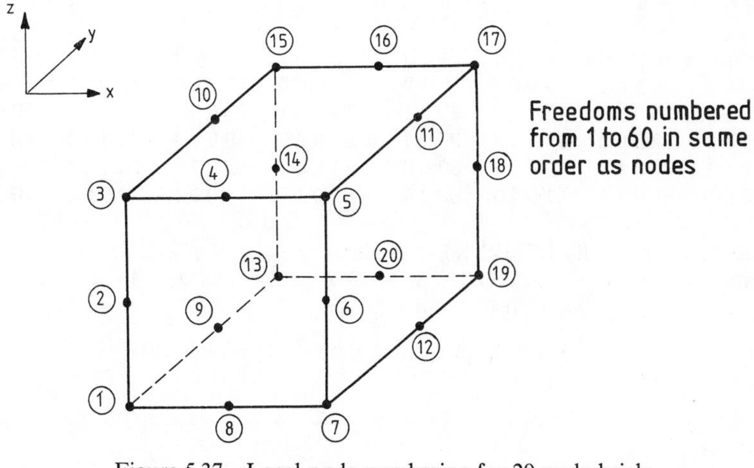

Figure 5.37 Local node numbering for 20-node brick

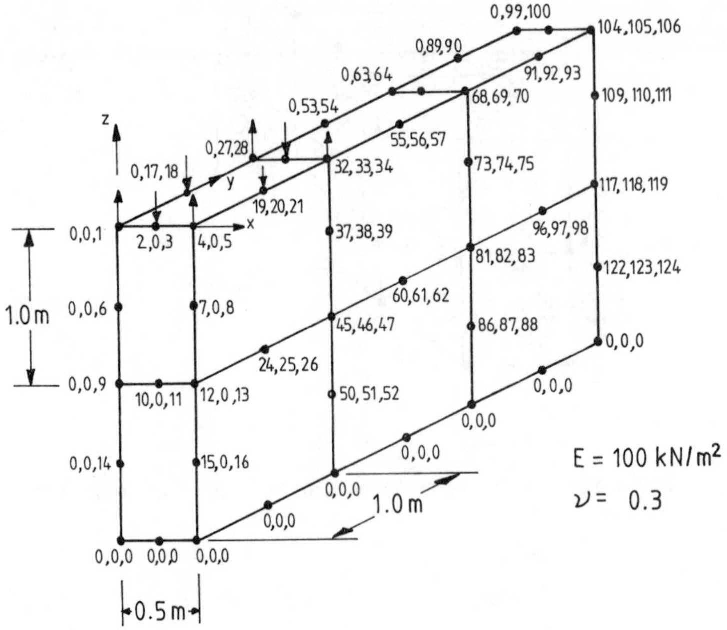

Structure data	NXE	NYE	NZE	N	NN	NR	NGP
	1	3	2	124	70	46	3

Element data	AA	BB	CC	E	V
	.5	1.	1.	100.	.3

Nodal freedom data

		1 0 0 1	2 1 0 1	3 1 0 1	4 0 0 1	5 1 0 1	6 0 0 1
7 1 0 1	8 1 0 1	9 0 0 1	10 1 0 1	11 0 0 0	12 0 0 0	13 0 0 0	14 0 1 1
16 0 1 1	18 0 0 0	19 0 0 0	20 0 1 1	23 0 1 1	25 0 1 1	28 0 1 1	30 0 0 0
31 0 0 0	32 0 0 0	33 0 1 1	35 0 1 1	37 0 0 0	38 0 0 0	39 0 1 1	42 0 1 1
44 0 1 1	47 0 1 1	49 0 0 0	50 0 0 0	51 0 0 0	52 0 1 1	54 0 1 1	56 0 0 0
57 0 0 0	58 0 1 1	61 0 1 1	63 0 1 1	66 0 1 1	68 0 0 0	69 0 0 0	70 0 0 0

Loads data	NL	(K$_I$, LOADS (K$_I$), I = 1, NL)							
	8	1	.0417	3	−.1667	5	.0417	18	−.1667
		21	−.1667	28	.0417	31	−.1667	34	.0417

Figure 5.38 Mesh and data for Program 5.10

-.1606E-01	.1344E-02	-.1616E-01	.2687E-02	-.1658E-01	-.1168E-01	Displacements
.1703E-02	-.1197E-01	-.7423E-02	.6783E-03	-.7406E-02	.1345E-02	
-.7490E-02	-.3078E-02	.1220E-02	-.3328E-02	-.1678E-02	-.1475E-01	
.2249E-02	-.1379E-02	-.1510E-01	.1049E-02	-.6449E-02	.1249E-02	
.1161E-02	-.6516E-02	-.2881E-02	-.9123E-02	.5036E-03	-.2800E-02	
-.9143E-02	.1012E-02	-.2388E-02	-.9150E-02			
.4610E-02	-.1648E+00	-.8883E+00	.2472E-02	.8305E-01	.2310E-02	Centroid stresses
.2584E-02	-.6851E-01	-.6645E+00	-.6375E-03	.8298E-01	.6887E-02	
-.6057E-02	-.9576E-01	-.1171E+00	.1028E-02	.8193E-01	-.2192E-02	
.2349E-02	-.9292E-01	-.2653E+00	-.1321E-02	.1252E+00	.3128E-02	
-.5789E-03	-.1013E-01	-.2536E-02	-.1240E-02	-.5978E-02	.6601E-03	
.2175E-02	-.4532E-01	-.2886E-01	-.1753E-03	.5598E-01	.3086E-04	

Figure 5.39 Results from Program 5.10

was noted that the length of the vector KV was reduced from 7068 to 4388 by the use of the 'skyline' storage strategy—an improvement of approximately 38%.

5.1 REFERENCES

Griffiths, D. V. (1986) HARMONY—a program for analysing elasto-plastic axisymmetric bodies subjected to non-axisymmetric loads. *Internal Report*, Engineering Dept., University of Manchester.

Poulos, H. G., and Davis, E. H. (1974) *Elastic Solutions for soil and Rock Mechanics*. John Wiley, New York.

Wilson, E. L. (1965) Structural analysis of axisymmetric solids. *J.A.I.A.A.*, **3**, 2269–74.

Zienkiewicz, O. C. (1977) *The Finite Element Method*, 3rd edition, p. 388. McGraw-Hill, New York.

CHAPTER 6
Material Non-Linearity

6.0 INTRODUCTION

Non-linear processes pose very much greater analytical problems than do the linear processes so far considered in this book. The non-linearity may be found in the dependence of the equation coefficients on the solution itself or in the appearance of powers and products of the unknowns or their derivatives.

Two main types of non-linearity can manifest themselves in finite element analysis of solids: material non-linearity, in which the relationship between stresses and strains (or other material properties) are complicated functions which result in the equation coefficients depending on the solution, and geometric non-linearity (otherwise known as 'large strain' or 'large displacement' analysis), which leads to products of the unknowns in the equations.

In order to keep the present book to a manageable size, the five programs described in this chapter deal only with material non-linearity. As far as the organisation of computer programs is concerned, material non-linearity is simpler to implement than geometric non-linearity. However, it is hoped that readers may appreciate how programs could be adapted to cope with geometric non-linearity as well (see, for example, Teng, 1981).

In practical finite element analysis two main types of solution procedure can be adopted to model material non-linearity. The first approach involves 'constant stiffness' iterations in which non-linearity is introduced by iteratively modifying the right hand side 'loads' vector. The (usually elastic) global stiffness matrix in such an analysis is formed once only. Each iteration thus represents an elastic analysis of the type described in Chapter 5. Convergence is said to occur when stresses generated by the loads satisfy some yield or failure criterion within prescribed tolerances. The loads vector at each iteration consists of externally applied loads and self-equilibrating 'body-loads'. The body-loads have the effect of redistributing stresses within the system, but as they are self-equilibrating, they do not alter the net loading on the system. The 'constant stiffness' method is shown diagrammatically in Figure 6.1. For load controlled problems, many iterations may be required as failure is approached, because the elastic (constant) global stiffness matrix starts to seriously overestimate the actual material

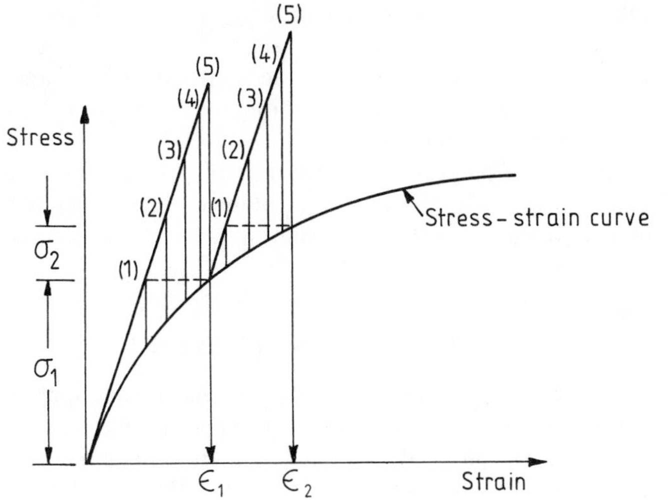

Figure 6.1 Constant stiffness method

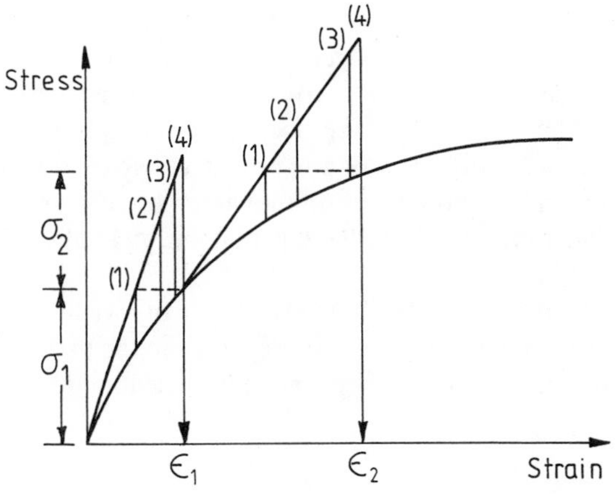

Figure 6.2 Variable stiffness method

stiffness. The numbers in parentheses on the figure indicate the number of iterations that might be required to reach convergence.

Less iterations per load step are required if the second approach, the 'variable or tangent stiffness' method is adopted. This method, shown in Figure 6.2, takes account of the reduction in stiffness of the material as failure is approached. If small enough load steps are taken, the method becomes equivalent to a simple Euler method. In practice, the global stiffness matrix may be updated periodically and iterations employed to achieve convergence. In contrasting the two methods, the extra cost of reforming and reducing the global stiffness matrix in the 'variable stiffness' method is offset by reduced numbers of iterations, especially as failure is approached.

All the programs in this chapter employ the 'constant stiffness' approach for the sake of simplicity, and are similar in structure to Program 4.7 described previously for plastic analysis of frames.

Before describing the programs, some discussion is necessary regarding the form of the stress–strain laws that are to be adopted. In addition, two popular methods of generating body-loads for 'constant stiffness' methods, namely 'visco-plasticity' and 'initial stress' are described.

6.1 STRESS–STRAIN BEHAVIOUR

Although non-linear elastic constitutive relations have been applied in finite element analyses and especially soil mechanics applications (e.g. Duncan and Chang, 1970), the main physical feature of non-linear material behaviour is usually the irrecoverability of strain. A convenient mathematical framework for describing this phenomenon is to be found in the theory of plasticity (e.g. Hill, 1950). The simplest stress–strain law of this type that could be implemented in a finite element analysis involves elastic–perfectly plastic material behaviour (Figure 6.3). Although a simple law of this type was described in Chapter 4 (Figure 4.29), it is convenient in solid mechanics to introduce a 'yield' surface in principal stress space which separates stress states that give rise to elastic and plastic (irrecoverable) strains. To take account of complicated processes like cyclic loading, the yield surface may move in stress space 'kinematically' (e.g. Prevost, 1977) but in this book only immovable surfaces are considered. An additional simplification introduced here is that the yield and ultimate 'failure' surfaces are identical.

Algebraically, the surfaces are expressed in terms of a yield or failure function F. This function, which has units of stress, depends on the material strength and invariant combinations of the stress components. The function is designed such

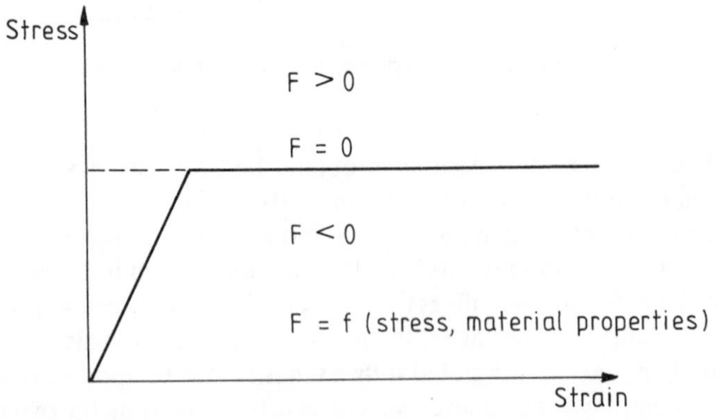

Figure 6.3 Elastic–perfectly plastic stress–strain behaviour

that it is negative within the failure surface and zero on the failure surface. Positive values of F imply stresses lying outside the failure surface which are illegal and which must be redistributed via the iterative process described previously.

During plastic straining, the material may flow in an 'associated' manner, that is the vector of plastic strain increment may be normal to the yield or failure surface. Alternatively, normality may not exist and the flow may be 'non-associated'. Associated flow leads to various mathematically attractive simplifications and, when allied to the von Mises or Tresca failure criterion, accurately predicts zero plastic volume change during yield for undrained clays. For frictional materials, whose ultimate state is described by the Mohr–Coulomb criterion, associated flow leads to physically unrealistic volumetric expansion or dilation during yield. In such cases, non-associated flow rules may be preferred in which plastic straining is described by a plastic potential function Q. This function is often geometrically similar to the failure function F but with the friction angle ϕ replaced by a dilation angle ψ. The implementation of the plastic potential function will be described further in a later section.

Before describing some commonly used failure criteria and their representations in principal stress space, some useful stress invariant expressions are briefly reviewed.

6.2 STRESS INVARIANTS

The Cartesian stress tensor defining the stress conditions at a point within a loaded body is given by

$$\{\sigma_x \sigma_y \sigma_z \tau_{xy} \tau_{yz} \tau_{zx}\} \tag{6.1}$$

which can be shown to be equivalent to three principal stresses acting on orthogonal planes

$$\{\sigma_1 \sigma_2 \sigma_3\} \tag{6.2}$$

Principal stress space is obtained by treating the principal stresses as three-dimensional coordinates and such a plot represents a useful means of defining the stresses acting at a point. It may be noted that although principal stress space defines the magnitude of the principal stresses, no indication is given of their orientation in physical space.

Instead of defining a point in principal stress space with coordinates $(\sigma_1, \sigma_2, \sigma_3)$ it is often more convenient to use invariants (s, t, θ), where

$$s = \frac{1}{\sqrt{3}}(\sigma_x + \sigma_y + \sigma_z)$$

$$t = \frac{1}{\sqrt{3}}[(\sigma_x - \sigma_y)^2 + (\sigma_y - \sigma_z)^2 + (\sigma_z - \sigma_x)^2 + 6\tau_{xy}^2 + 6\tau_{yz}^2 + 6\tau_{zx}^2]^{1/2} \tag{6.3}$$

$$\theta = \tfrac{1}{3}\arcsin\left(\frac{-3\sqrt{6}J_3}{t^3}\right)$$

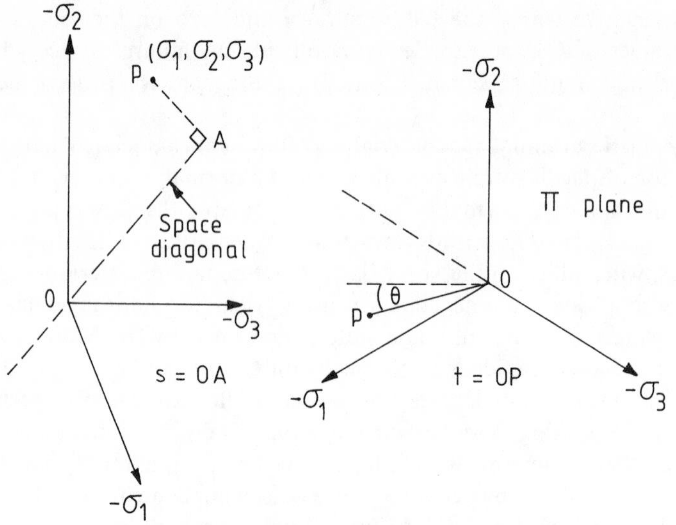

Figure 6.4 Principal stress space

where

$$J_3 = s_x s_y s_z - s_x \tau_{yz}^2 - s_y \tau_{zx}^2 - s_z \tau_{xy}^2 + 2\tau_{xy}\tau_{yz}\tau_{zx}$$

and

$$s_x = (2\sigma_x - \sigma_y - \sigma_z)/3, \text{ etc.}$$

As shown in Figure 6.4, s gives the distance from the origin to the π plane in which the stress point lies and t represents the perpendicular distance of the stress point from the space diagonal. The Lode angle θ is a measure of the angular position of the stress point within the π plane.

It may be noted that in many soil mechanics applications, plane strain conditions apply and equations (6.3) are simplified because $\tau_{yz} = \tau_{zx} = 0$.

In the programs described later in this chapter, the invariants that are used are slightly different to these defined in (6.3) whence

$$\begin{aligned} \sigma_m &= s/\sqrt{3} \\ \bar{\sigma} &= t\sqrt{(3/2)} \end{aligned} \tag{6.4}$$

These expressions, called SIGM and DSBAR in program terminology, have more physical meaning than s and t in that they represent respectively the 'mean stress' and 'deviator stress' in a triaxial test. The relationship between principal stresses and invariants is given as follows:

$$\begin{aligned} \sigma_1 &= \sigma_m + \sqrt{\tfrac{2}{3}}\bar{\sigma}\sin\left(\theta - \frac{2\pi}{3}\right) \\ \sigma_2 &= \sigma_m + \sqrt{\tfrac{2}{3}}\bar{\sigma}\sin\theta \\ \sigma_3 &= \sigma_m + \sqrt{\tfrac{2}{3}}\bar{\sigma}\sin\left(\theta + \frac{2\pi}{3}\right) \end{aligned} \tag{6.5}$$

6.3 FAILURE CRITERIA

Several failure criteria have been proposed as suitable for representing the strength of soils as engineering materials. For soils possessing both frictional and cohesive components of shear strength, the best known criterion is undoubtedly that due to Mohr–Coulomb and takes the form of an irregular hexagonal cone in principal stress space.

For metals or undrained clays which behave in a 'frictionless' ($\phi_u = 0$) manner, cylindrical failure criteria are appropriate. These are the simplest criteria which do not depend on the first invariant s (or σ_m). The Tresca criterion in fact does not require separate treatment mathematically because it is a special case of the Mohr–Coulomb criterion, as will be shown. Alternatively, the von Mises criterion can be used. The difference in strengths predicted by the two criteria does not exceed about 15%.

6.3.1 Von Mises

As shown in Figure 6.5, this criterion takes the form of a right circular cylinder lying along the space diagonal. Only one of the three invariants, namely t (or $\bar{\sigma}$), is of any significance when determining whether a stress state has reached the limit of elastic behaviour. The onset of yield in a von Mises material is not dependent upon invariants s or θ.

The symmetry of the von Mises criterion when viewed in the π plane indicates why it is not ideally suited to correlations with traditional soil mechanics concepts of strength. The criterion gives equal weighting to all three principal stresses, so if it is to be used to model undrained clay behaviour, consideration must be given to the value of the intermediate principal stress σ_2 at failure.

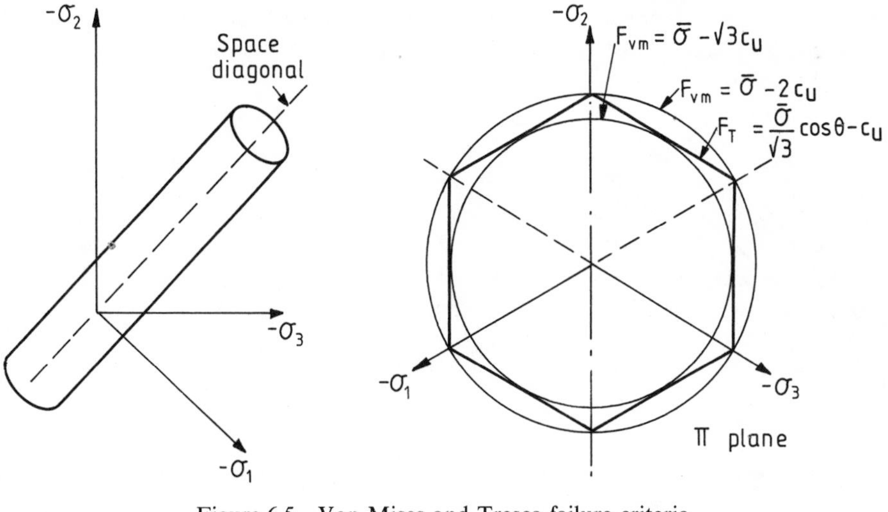

Figure 6.5 Von Mises and Tresca failure criteria

For plane strain applications it can be shown that at failure

$$\sigma_2 = \frac{\sigma_1 + \sigma_3}{2} \tag{6.6}$$

Hence, the failure criterion is given by

$$F = \bar{\sigma} - \sqrt{3}c_u \tag{6.7}$$

where c_u = undrained 'cohesion' of the soil.

Under triaxial conditions, where at all times

$$\sigma_2 = \sigma_3 \tag{6.8}$$

the criterion is given by

$$F = \bar{\sigma} - 2c_u \tag{6.9}$$

Both of these expressions ensure that at failure

$$c_u = \frac{\sigma_1 - \sigma_3}{2} \tag{6.10}$$

6.3.2 Mohr–Coulomb

In principal stress space, this criterion takes the form of an irregular hexagonal cone, as shown in Figure 6.6. The irregularity is due to the fact that σ_2 is not taken into account. In order to derive the invariant form of this criterion, it should first

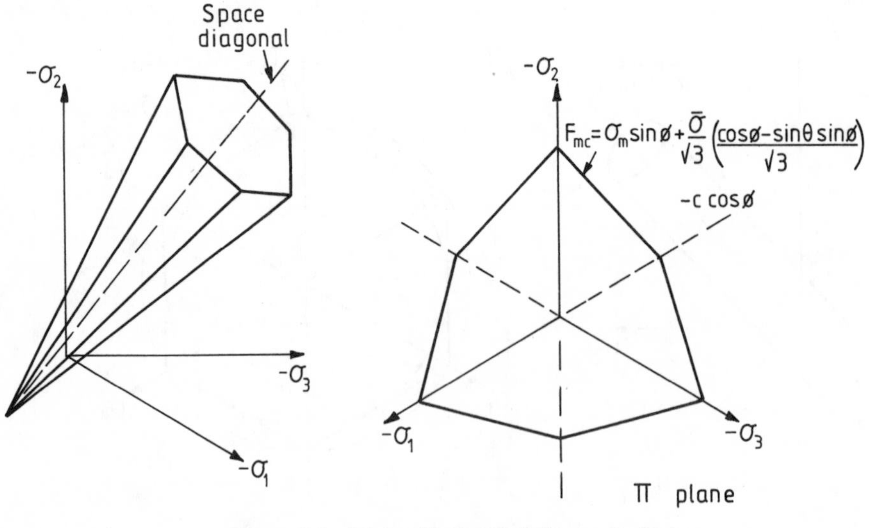

Figure 6.6 Mohr–Coulomb failure criterion

be written in terms of principal stresses from the geometry of Mohr's circle; thus

$$\frac{\sigma_1+\sigma_3}{2}\sin\phi-\frac{\sigma_1-\sigma_3}{2}-c\cos\phi=0 \tag{6.11}$$

Substituting for σ_1 and σ_3 from equations (6.5) gives the function

$$F=\sigma_{\mathrm{m}}\sin\phi+\bar{\sigma}\left(\frac{\cos\phi}{\sqrt{3}}-\frac{\sin\theta\sin\phi}{3}\right)-c\cos\phi \tag{6.12}$$

which is clearly dependent upon all three invariants (σ_{m}, $\bar{\sigma}$, θ).

Tresca's criterion is obtained from (6.12) by putting $\phi=0$ to give

$$F=\frac{\bar{\sigma}\cos\theta}{\sqrt{3}}-c_{\mathrm{u}} \tag{6.13}$$

This criterion is preferred to von Mises for applications involving undrained clays because (6.10) is always satisfied at failure, regardless of the value of σ_2. In principal stress space Tresca's criterion is a regular hexagonal cylinder tangential to the von Mises cylinder defined by equation (6.7), as shown in Figure 6.5.

6.4 GENERATION OF BODY-LOADS

Constant stiffness methods of the type described in this chapter use repeated elastic solutions to achieve convergence by iteratively varying the loads on the system. Within each load increment, the system of equations

$$\mathbf{K}\boldsymbol{\delta}^i=\mathbf{P}^i \tag{6.14}$$

must be solved for the displacement increments $\boldsymbol{\delta}^i$, where i represents the iteration number.

The element displacement increments $\mathbf{u}^i$ are extracted from $\boldsymbol{\delta}^i$, and these lead to total strain increments via the element strain–displacement relationships:

$$\boldsymbol{\Delta\varepsilon}^i=\mathbf{B}\mathbf{u}^i \tag{6.15}$$

Assuming the material is yielding, the strains will contain both elastic and (visco) plastic components; thus

$$\boldsymbol{\Delta\varepsilon}^i=(\boldsymbol{\Delta\varepsilon}^{\mathrm{e}}+\boldsymbol{\Delta\varepsilon}^{\mathrm{p}})^i \tag{6.16}$$

It is only the elastic strain increments $\boldsymbol{\Delta\varepsilon}^{\mathrm{e}}$ that generate stresses through the elastic stress–strain matrix; hence

$$\boldsymbol{\Delta\sigma}^i=\mathbf{D}^{\mathrm{e}}(\Delta\varepsilon^{\mathrm{e}})^i \tag{6.17}$$

These stress increments are added to stresses already existing from the previous load step and the updated stresses substituted into the failure criterion.

If stress redistribution is necessary, this is done by altering the load increment vector $\mathbf{P}^i$ in equation (6.14). In general, this vector holds two types of load, as given by

$$\mathbf{P}^i=\mathbf{P}_{\mathrm{a}}+\mathbf{P}_{\mathrm{b}}^i \tag{6.18}$$

where $\mathbf{P}_a$ is the actual applied load increment that is required and $\mathbf{P}_b^i$ is the body-loads vector that varies from one iteration to the next. The $\mathbf{P}_b^i$ vector must be self-equilibrating so that the net loading on the system is not affected by it. Two methods for generating body-loads are now described briefly.

6.5 VISCO-PLASTICITY

In this method (Zienkiewicz and Cormeau, 1974), the material is allowed to sustain stresses outside the failure criterion for finite 'periods'. Overshoot of the failure criterion, as signified by a positive value of F, is an integral part of the method and is actually used to drive the algorithm.

Instead of plastic strains, we now refer to visco-plastic strains and these are generated at a rate that is related to the amount by which yield has been violated through the expression

$$\dot{\boldsymbol{\varepsilon}}^{\mathrm{VP}} = F\frac{\partial Q}{\partial \boldsymbol{\sigma}} \tag{6.19}$$

It should be noted that a pseudo viscosity property equal to unity is implied on the right hand side of equation (6.19) from dimensional considerations.

Multiplication of the visco-plastic strain rate by a pseudo time step gives an increment of visco-plastic strain which is accumulated from one 'time step' or iteration to the next; thus

$$(\boldsymbol{\delta\varepsilon}^{\mathrm{VP}})^i = \Delta t(\dot{\boldsymbol{\varepsilon}}^{\mathrm{VP}})^i \tag{6.20}$$

and

$$(\boldsymbol{\Delta\varepsilon}^{\mathrm{VP}})^i = (\boldsymbol{\Delta\varepsilon}^{\mathrm{VP}})^{i-1} + (\boldsymbol{\delta\varepsilon}^{\mathrm{VP}})^i \tag{6.21}$$

The 'time step' for unconditional numerical stability has been derived by Cormeau (1975) and depends on the assumed failure criterion. Thus, for von Mises materials:

$$\Delta t = \frac{4(1+v)}{3E} \tag{6.22}$$

and for Mohr–Coulomb materials:

$$\Delta t = \frac{4(1+v)(1-2v)}{E(1-2v+\sin^2\phi)} \tag{6.23}$$

The derivatives of the plastic potential function Q with respect to stresses are conveniently expressed through the Chain Rule; thus

$$\frac{\partial Q}{\partial \boldsymbol{\sigma}} = \frac{\partial Q}{\partial \sigma_{\mathrm{m}}}\frac{\partial \sigma_{\mathrm{m}}}{\partial \boldsymbol{\sigma}} + \frac{\partial Q}{\partial J_2}\frac{\partial J_2}{\partial \boldsymbol{\sigma}} + \frac{\partial Q}{\partial J_3}\frac{\partial J_3}{\partial \boldsymbol{\sigma}} \tag{6.24}$$

where $J_2 = \frac{1}{2}t^2$ and the visco-plastic strain rate given by equation (6.19) is evaluated numerically by an expression of the form

$$\dot{\boldsymbol{\varepsilon}}^{\mathrm{VP}} = F(\mathrm{DQ1}\,\mathbf{M}^1 + \mathrm{DQ2}\,\mathbf{M}^2 + \mathrm{DQ3}\,\mathbf{M}^3)\boldsymbol{\sigma} \tag{6.25}$$

where DQ1, DQ2 and DQ3 are scalars equal to $\partial Q/\partial \sigma_m$, $\partial Q/\partial J_2$ and $\partial Q/\partial J_3$ respectively, and $\mathbf{M}^1\boldsymbol{\sigma}$, $\mathbf{M}^2\boldsymbol{\sigma}$ and $\mathbf{M}^3\boldsymbol{\sigma}$ are vectors representing $\partial\sigma_m/\partial\boldsymbol{\sigma}$, $\partial J_2/\partial\boldsymbol{\sigma}$ and $\partial J_3/\partial\boldsymbol{\sigma}$ respectively. This is essentially the notation used by Zienkiewicz (1977) and these quantities are given in more detail in Appendix 2.

The body-loads $\mathbf{P}_b^i$ are accumulated at each 'time step' within each load step by summing the following integrals for all elements containing a yielding Gauss point:

$$\mathbf{P}_b^i = \mathbf{P}_b^{i-1} + \sum_{\text{elements}}^{\text{all}} \iint \mathbf{B}^T \mathbf{D}^e (\delta\varepsilon^{VP})^i \mathrm{d}x\,\mathrm{d}y \tag{6.26}$$

This process is repeated at each 'time step' iteration until no Gauss point stresses violate the failure criterion to within certain tolerances. The convergence criterion is based on a dimensionless measure of the amount by which the displacement increment vector $\boldsymbol{\delta}^i$ changes from one iteration to the next. The convergence checking process is identical to that used in Program 4.7. A more detailed description of the whole algorithm can be found in Griffiths (1980).

6.6 INITIAL STRESS

The viscoplastic algorithm is often referred to as an 'initial strain' method to distinguish it from the more widely used 'initial stress' approaches (e.g. Zienkiewicz *et al.*, 1969).

Initial stress methods involve an explicit relationship between increments of stress and increments of strain. Thus, whereas linear elasticity was described by

$$\boldsymbol{\Delta\sigma} = \mathbf{D}^e \boldsymbol{\Delta\varepsilon} \tag{6.27}$$

elasto-plasticity is described by

$$\boldsymbol{\Delta\sigma} = \mathbf{D}^{PL} \boldsymbol{\Delta\varepsilon} \tag{6.28}$$

where

$$\mathbf{D}^{PL} = \mathbf{D}^e - \mathbf{D}^P \tag{6.29}$$

For perfect plasticity in the absence of hardening or softening it is assumed that once a stress state reaches a failure surface, subsequent changes in stress may shift the stress state to a different position on the failure surface, but not outside it; thus

$$\frac{\partial F}{\partial \boldsymbol{\sigma}} \boldsymbol{\Delta\sigma} = 0 \tag{6.30}$$

Allowing for the possibility of non-associated flow, plastic strain increments occur normal to a plastic potential surface; thus

$$\boldsymbol{\Delta\varepsilon}^P = \lambda \frac{\partial Q}{\partial \boldsymbol{\sigma}} \tag{6.31}$$

Assuming stress changes are generated by elastic strain components only gives

$$\boldsymbol{\Delta\sigma} = \mathbf{D}^e \left(\boldsymbol{\Delta\varepsilon} - \lambda \frac{\partial Q}{\partial \boldsymbol{\sigma}} \right) \tag{6.32}$$

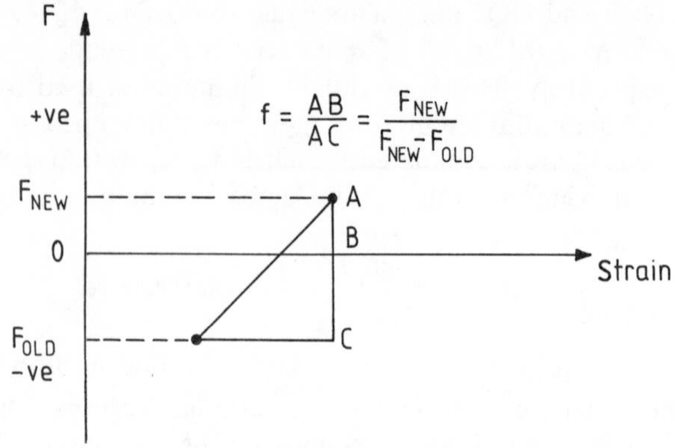

Figure 6.7 Factoring process for just yielding elements

Substitution of equation (6.32) into (6.30) leads to

$$\mathbf{D}^{\mathrm{P}} = \frac{\mathbf{D}^{\mathrm{e}} \dfrac{\partial Q}{\partial \boldsymbol{\sigma}} \left(\dfrac{\partial F}{\partial \boldsymbol{\sigma}} \right)^{\mathrm{T}} \mathbf{D}^{\mathrm{e}}}{\left(\dfrac{\partial F}{\partial \boldsymbol{\sigma}} \right)^{\mathrm{T}} \mathbf{D}^{\mathrm{e}} \dfrac{\partial Q}{\partial \boldsymbol{\sigma}}} \tag{6.33}$$

Explicit versions of $\mathbf{D}^{\mathrm{P}}$ may be obtained for simple failure and potential functions and these are given for von Mises (Yamada *et al.*, 1968) and Mohr–Coulomb (Griffiths and Willson, 1986) in Appendix 2.

The body-loads $\mathbf{P}_{\mathrm{b}}^{i}$ in the stress redistribution process are reformed at each iteration by summing the following integral for all elements that possess yielding Gauss points; thus

$$\mathbf{P}_{\mathrm{b}}^{i} = \sum_{\text{elements}}^{\text{all}} \iint \mathbf{B}^{\mathrm{T}} (\mathbf{D}^{\mathrm{P}} \Delta \boldsymbol{\varepsilon})^{i} \, \mathrm{d}x \, \mathrm{d}y \tag{6.34}$$

In the event of a loading increment causing a Gauss point to go plastic for the first time, it may be necessary to factor the matrix $\mathbf{D}^{\mathrm{P}}$ in (6.34). A linear interpolation can be used as indicated in Figure 6.7. Thus, instead of using $\mathbf{D}^{\mathrm{P}}$ we use $f\mathbf{D}^{\mathrm{P}}$ where

$$f = \frac{F_{\mathrm{NEW}}}{F_{\mathrm{NEW}} - F_{\mathrm{OLD}}} \tag{6.35}$$

It may be noted that the visco-plastic algorithm requires no such scaling factor because overshoot of the failure criterion is an integral part of the algorithm.

6.7 CORNERS ON THE FAILURE AND POTENTIAL SURFACES

For non-continuous F and Q surfaces, the derivatives in equations (6.19) and (6.33) become indeterminate. In the case of the Mohr–Coulomb (or Tresca) surface, this occurs when the angular invariant $\theta = \pm 30°$. The method used in the programs to overcome this difficulty is to replace the hexagonal surface by a smooth conical surface if

$$|\sin\theta| > 0.49 \tag{6.36}$$

The conical surfaces are those obtained by substituting either $\theta = 30°$ or $\theta = -30°$ into (6.12), depending upon the sign of θ as it approaches $\pm 30°$. It should be noted that in the initial stress approach, both F and Q functions must be approximated in this way due to the inclusion of both $\partial F/\partial \boldsymbol{\sigma}$ and $\partial Q/\partial \boldsymbol{\sigma}$ terms in (6.33). The visco-plastic algorithm, however, only requires the potential function Q to be smoothed from (6.19), as derivatives of F with respect to stresses are not required.

All the two-dimensional programs in this chapter have been constructed using the eight-node quadrilateral element, together with reduced integration (four Gauss points per element). This particular combination has been chosen for its simplicity, and also its well-known ability to compute collapse loads accurately (e.g. Zienkiewicz, 1977; Griffiths, 1980). Of course, other element types could be used if required, and this would involve making similar subroutine changes to those illustrated in Chapter 5.

The first four programs involve computation of collapse loads in problems which have known closed form solutions. In addition to the different boundary value problems tackled, both the visco-plastic method and the initial stress method are implemented. The fifth program shows how the constant stiffness method is extended to deal with three-dimensional strain conditions.

PROGRAM 6.0: PLANE STRAIN BEARING CAPACITY ANALYSIS—VISCO-PLASTIC METHOD USING THE VON MISES CRITERION

```
C
C       PROGRAM 6.0 PLANE STRAIN OF AN ELASTIC-PLASTIC
C       (VON-MISES) SOLID USING 8-NODE QUADRILATERAL ELEMENTS
C       VISCOPLASTIC STRAIN METHOD
C
C
C       ALTER NEXT LINE TO CHANGE PROBLEM SIZE
C
      PARAMETER(IKB1=184,IKB2=30,ILOADS=184,INF=130,
     +          IEVPT=512,INX=20,INY=20,INO=10,IQINC=20)
C
      REAL DEE(4,4),SAMP(4,2),COORD(8,2),JAC(2,2),JAC1(2,2),
     +DER(2,8),DERIV(2,8),BEE(4,16),DBEE(4,16),WIDTH(INX),DEPTH(INY),
     +BTDB(16,16),KM(16,16),ELD(16),EPS(4),SIGMA(4),BLOAD(16),
     +BT(16,4),FUN(8),KB(IKB1,IKB2),LOADS(ILOADS),ELOAD(16),
     +TOTD(ILOADS),BDYLDS(ILOADS),EVPT(IEVPT),OLDIS(ILOADS),
     +SX(INX,INY,4),SY(INX,INY,4),TXY(INX,INY,4),SZ(INX,INY,4),
     +VAL(INO),ERATE(4),EVP(4),DEVP(4),M1(4,4),M2(4,4),M3(4,4),
     +FLOW(4,4),STRESS(4),QINC(IQINC)
      INTEGER NF(INF,2),G(16),NO(INO)
      DATA IDEE,IBEE,IDBEE,IH,IFLOW/5*4/,IDOF,IBTDB,IBT,IKM/4*16/
```

```
      DATA IJAC,IJAC1,NODOF,IT,IDER,IDERIV/6*2/,ICOORD,NOD/2*8/
      DATA ISAMP/4/
C
C      INPUT AND INITIALISATION
C
      READ(5,*)CU,E,V,NXE,NYE,N,IW,NN,NR,NGP,ITS
      CALL READNF(NF,INF,NN,NODOF,NR)
      READ(5,*)(WIDTH(I),I=1,NXE+1)
      READ(5,*)(DEPTH(I),I=1,NYE+1)
      IWP1=IW+1
      CALL NULL(KB,IKB1,N,IWP1)
      CALL NULVEC(OLDIS,N)
      CALL NULVEC(TOTD,N)
      CALL FMDRAD(DEE,IDEE,E,V)
      CALL GAUSS(SAMP,ISAMP,NGP)
      DT=4.*(1.+V)/(3.*E)
C
C      ELEMENT STIFFNESS INTEGRATION AND ASSEMBLY
C
      DO 10 IP=1,NXE
      DO 10 IQ=1,NYE
      CALL GEOV8Y(IP,IQ,NYE,WIDTH,DEPTH,COORD,ICOORD,G,NF,INF)
      CALL NULL(KM,IKM,IDOF,IDOF)
      IG=0
      DO 20 I=1,NGP
      DO 20 J=1,NGP
      IG=IG+1
      SX(IP,IQ,IG)=0.
      SY(IP,IQ,IG)=0.
      TXY(IP,IQ,IG)=0.
      SZ(IP,IQ,IG)=0.
      CALL FMQUAD(DER,IDER,FUN,SAMP,ISAMP,I,J)
      CALL MATMUL(DER,IDER,COORD,ICOORD,JAC,IJAC,IT,NOD,IT)
      CALL TWOBY2(JAC,IJAC,JAC1,IJAC1,DET)
      CALL MATMUL(JAC1,IJAC1,DER,IDER,DERIV,IDERIV,IT,IT,NOD)
      CALL NULL(BEE,IBEE,IH,IDOF)
      CALL FORMB(BEE,IBEE,DERIV,IDERIV,NOD)
      CALL MATMUL(DEE,IDEE,BEE,IBEE,DBEE,IDBEE,IH,IH,IDOF)
      CALL MATRAN(BT,IBT,BEE,IBEE,IH,IDOF)
      CALL MATMUL(BT,IBT,DBEE,IDBEE,BTDB,IBTDB,IDOF,IH,IDOF)
      QUOT=DET*SAMP(I,2)*SAMP(J,2)
      CALL MSMULT(BTDB,IBTDB,QUOT,IDOF,IDOF)
   20 CALL MATADD(KM,IKM,BTDB,IBTDB,IDOF,IDOF)
   10 CALL FORMKB(KB,IKB1,KM,IKM,G,IW,IDOF)
C
C      READ LOAD WEIGHTINGS AND REDUCE EQUATIONS
C
      READ(5,*)NL,(NO(I),VAL(I),I=1,NL)
      CALL CHOLIN(KB,IKB1,N,IW)
C
C      LOAD INCREMENT LOOP
C
      READ(5,*)INCS,(QINC(I),I=1,INCS)
      PTOT=0.
      DO 30 IY=1,INCS
      PTOT=PTOT+QINC(IY)
      ITERS=0
      CALL NULVEC(BDYLDS,N)
      CALL NULVEC(EVPT,NXE*NYE*IH*NGP*NGP)
C
C      ITERATION LOOP
C
```

```
   40 ITERS=ITERS+1
      CALL NULVEC(LOADS,N)
      DO 50 I=1,NL
   50 LOADS(NO(I))=VAL(I)*QINC(IY)
      CALL VECADD(LOADS,BDYLDS,LOADS,N)
      CALL CHOBAC(KB,IKB1,LOADS,N,IW)
C
C      CHECK CONVERGENCE
C
      CALL CHECON(LOADS,OLDIS,N,0.001,ICON)
      IF(ITERS.EQ.1)ICON=0
      IF(ICON.EQ.1.OR.ITERS.EQ.ITS)CALL NULVEC(BDYLDS,N)
C
C      INSPECT ALL GAUSS POINTS
C
      NM=0
      DO 60 IP=1,NXE
      DO 60 IQ=1,NYE
      NM=NM+1
      CALL NULVEC(BLOAD,IDOF)
      CALL GEOV8Y(IP,IQ,NYE,WIDTH,DEPTH,COORD,ICOORD,G,NF,INF)
      DO 70 M=1,IDOF
      IF(G(M).EQ.0)ELD(M)=0.
   70 IF(G(M).NE.0)ELD(M)=LOADS(G(M))
      IG=0
      DO 80 I=1,NGP
      DO 80 J=1,NGP
      IG=IG+1
      IN=NGP*NGP*IH*(NM-1)+IH*(IG-1)
      CALL FMQUAD(DER,IDER,FUN,SAMP,ISAMP,I,J)
      CALL MATMUL(DER,IDER,COORD,ICOORD,JAC,IJAC,IT,NOD,IT)
      CALL TWOBY2(JAC,IJAC,JAC1,IJAC1,DET)
      CALL MATMUL(JAC1,IJAC1,DER,IDER,DERIV,IDERIV,IT,IT,NOD)
      CALL NULL(BEE,IBEE,IH,IDOF)
      CALL FORMB(BEE,IBEE,DERIV,IDERIV,NOD)
      CALL MATRAN(BT,IBT,BEE,IBEE,IH,IDOF)
      CALL MVMULT(BEE,IBEE,ELD,IH,IDOF,EPS)
      DO 90 K=1,IH
   90 EPS(K)=EPS(K)-EVPT(IN+K)
      CALL MVMULT(DEE,IDEE,EPS,IH,IH,SIGMA)
      STRESS(1)=SIGMA(1)+SX(IP,IQ,IG)
      STRESS(2)=SIGMA(2)+SY(IP,IQ,IG)
      STRESS(3)=SIGMA(3)+TXY(IP,IQ,IG)
      STRESS(4)=SIGMA(4)+SZ(IP,IQ,IG)
      CALL INVAR(STRESS,SIGM,DSBAR,THETA)
C
C      CHECK WHETHER YIELD IS VIOLATED
C
      F=DSBAR-SQRT(3.)*CU
      IF(ICON.EQ.1.OR.ITERS.EQ.ITS)GOTO 100
      IF(F.LT.0.)GOTO 110
      DQ1=0.
      DQ2=1.5/DSBAR
      DQ3=0.
      CALL FORMM(STRESS,M1,M2,M3)
      DO 120 L=1,IH
      DO 120 M=1,IH
  120 FLOW(L,M)=F*(M1(L,M)*DQ1+M2(L,M)*DQ2+M3(L,M)*DQ3)
      CALL MVMULT(FLOW,IFLOW,STRESS,IH,IH,ERATE)
      DO 130 K=1,IH
      EVP(K)=ERATE(K)*DT
  130 EVPT(IN+K)=EVPT(IN+K)+EVP(K)
```

```
      CALL MVMULT(DEE,IDEE,EVP,IH,IH,DEVP)
      GOTO 140
  100 CALL VECCOP(STRESS,DEVP,IH)
  140 CALL MVMULT(BT,IBT,DEVP,IDOF,IH,ELOAD)
      QUOT=DET*SAMP(I,2)*SAMP(J,2)
      DO 150 K=1,IDOF
  150 BLOAD(K)=BLOAD(K)+ELOAD(K)*QUOT
  110 IF(ICON.NE.1.AND.ITERS.NE.ITS)GOTO 80
C
C       UPDATE GAUSS POINT STRESSES
C
      SX(IP,IQ,IG)=STRESS(1)
      SY(IP,IQ,IG)=STRESS(2)
      TXY(IP,IQ,IG)=STRESS(3)
      SZ(IP,IQ,IG)=STRESS(4)
   80 CONTINUE
C
C       COMPUTE TOTAL BODYLOADS VECTOR
C
      DO 160 M=1,IDOF
      IF(G(M).EQ.0)GOTO 160
      BDYLDS(G(M))=BDYLDS(G(M))+BLOAD(M)
  160 CONTINUE
   60 CONTINUE
      IF(ICON.NE.1.AND.ITERS.NE.ITS)GOTO 40
      CALL VECADD(TOTD,LOADS,TOTD,N)
      WRITE(6,1000)PTOT
      WRITE(6,1000)(TOTD(NO(I)),I=1,NL)
      WRITE(6,2000)ITERS
      IF(ITERS.EQ.ITS)GOTO 170
   30 CONTINUE
  170 CONTINUE
 1000 FORMAT(10E12.4)
 2000 FORMAT(10I12)
      STOP
      END
```

The first four programs in this chapter are derived from Program 5.4 for linear elastic analysis, which used eight-node quadrilateral elements. Program 6.0 employs the visco-plastic method to predict the response to loading of an elastic–perfectly plastic (von Mises) material. Plane strain conditions are enforced and, in order to monitor the load-displacement response, the loads are applied incrementally.

In the first part of the program, however, the global stiffness matrix is formed in the usual way and the resulting matrix, KB, is then 'reduced' using (in this case) Choleski's method. An obvious advantage of constant stiffness methods is that the time-consuming reduction process is only performed once. An outline of the visco-plastic algorithm which follows the stiffness matrix formation is given in the structure chart in Figure 6.8.

New variable and array names that have not already been defined in Chapter 5 are listed below:

CU	undrained 'cohesion'
INCS	number of load increments
ITS	maximum number of iterations
DT	critical time step
ITERS	iteration counter

PTOT	keeps running total of loading
ICON	checks convergence (1 = convergence)
SIGM	mean stress invariant σ_m
DSBAR	deviatoric stress invariant $\bar{\sigma}$
THETA	Lode angle θ
F	value of failure function
DQ1 DQ2 DQ3	$\partial Q/\partial \sigma_m$, $\partial Q/\partial J_2$, $\partial Q/\partial J_3$
IQINC ⩾ INCS INX ⩾ NXE + 1 INY ⩾ NYE + 1 IEVP ⩾ NXE*NYE*NGP^2*IH INO ⩾ NL IFLOW	working size of matrices QINC, WIDTH, DEPTH, EVPT, FLOW VAL, NO
ELOAD	product $\mathbf{B}^T\mathbf{D}\delta\boldsymbol{\varepsilon}^{VP}$ at each Gauss point
BLOAD	accumulates ELOAD from each Gauss point
BDYLDS	accumulates BLOAD from each element
OLDIS	displacements from previous iteration
QINC	load increments
WIDTH	x coordinates of elements
DEPTH	y coordinates of elements
TOTD	accumulated displacements
SX SY TXY SZ	accumulates σ_x, σ_y, τ_{xy} and σ_z at each Gauss point
STRESS	vector holding σ_x, σ_y, τ_{xy} and σ_z
NO	loaded freedom numbers
VAL	load weighting at each freedom
ERATE	visco-plastic strain rate $\dot{\boldsymbol{\varepsilon}}^{VP}$
EVP	visco-plastic strain increment $\delta\boldsymbol{\varepsilon}^{VP}$
DEVP	product $\mathbf{D}^e\delta\boldsymbol{\varepsilon}^{VP}$
EVPT	accumulated visco-plastic strain at each Gauss point
M1, M2, M3	arrays holding $\partial\sigma_m/\partial\boldsymbol{\sigma}$, $\partial J_2/\partial\boldsymbol{\sigma}$, $\partial J_3/\partial\boldsymbol{\sigma}$
FLOW	used to calculate $\partial Q/\partial\boldsymbol{\sigma}$

Several subroutines appear for the first time in Program 6.0. All elements are rectangular with nodes counted in the y direction; thus the nodal coordinates and steering vector are provided by the eight-node variable geometry routine GEOV8Y (Appendix 3). Routine INVAR forms the three invariants σ_m, $\bar{\sigma}$ and θ from the four Cartesian stress components held in the one-dimensional array STRESS. It should be noted that in plane strain plasticity applications, it is necessary to retain four components of stress and strain. Although, by definition, one of the strains (ε_z) must equal zero, the elastic strain in that direction may be non-zero, provided

$$\varepsilon_z^e = -\varepsilon_z^{VP} \tag{6.37}$$

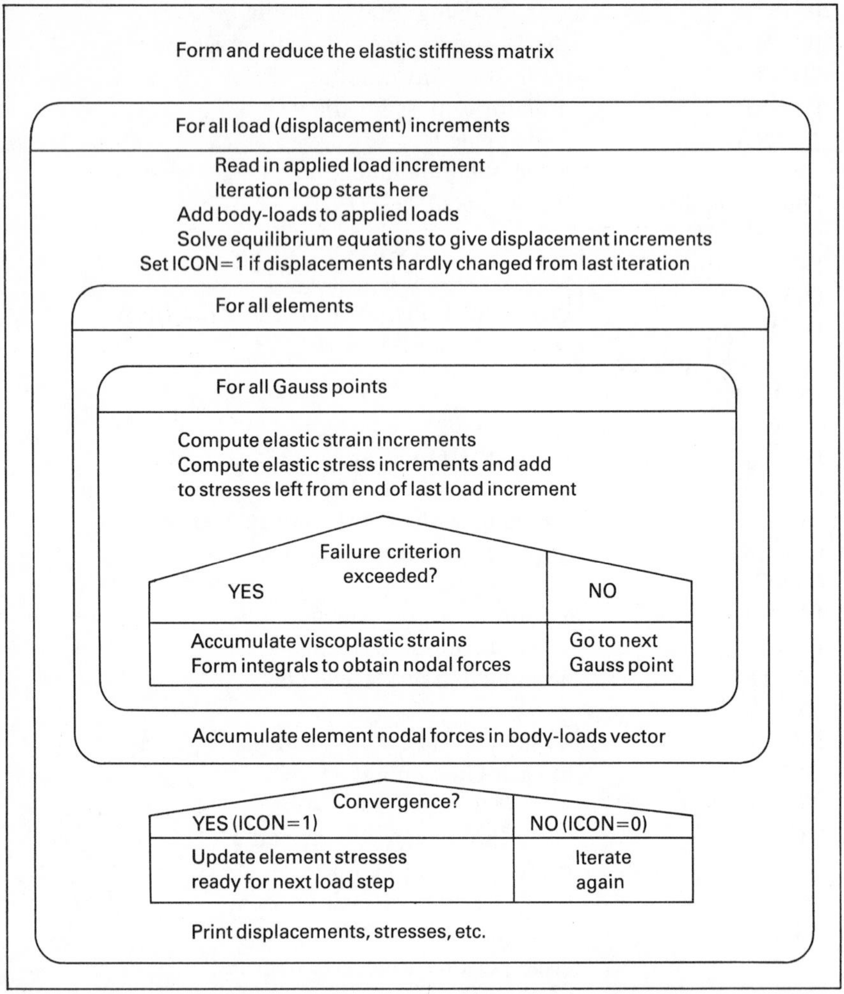

Figure 6.8 Structure chart for visco-plastic algorithm

is always true. For this reason, the 4×4 elastic stress–strain matrix is provided by the routine FMDRAD.

The only other subroutine not encountered before is FORMM, which creates arrays M1, M2 and M3 used in the calculation of the visco-plastic strain rate from equation (6.25).

The example shown in Figure 6.9 is of a flexible strip footing at the surface of a layer of uniform undrained clay. The footing supports a uniform stress, q, which is increased incrementally to failure. The elasto-plastic soil is described by three parameters, namely the elastic properties, E, ν and the undrained 'cohesion' c_u. Bearing failure in this problem occurs when q reaches the 'Prandtl' load given by

$$q_{\text{ULT}} = (2 + \pi)c_u \tag{6.38}$$

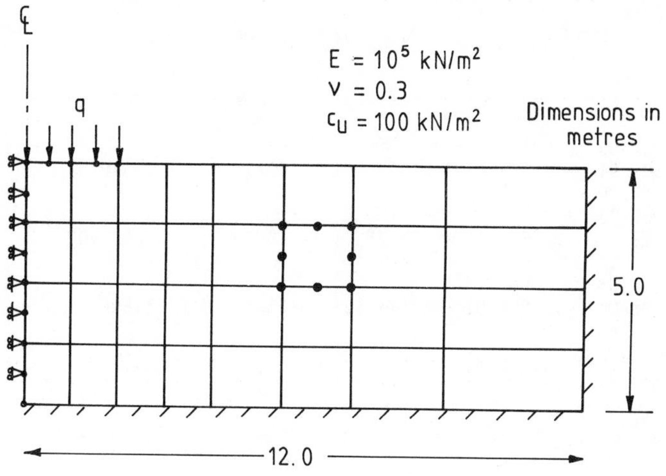

Soil	CU	E	V					
properties	100.	1.E5	0.3					
Structure	NXE	NYE	N	IW	NN	NR	NGP	ITS
data	8	4	184	29	121	33	2	250

Node freedom data

	1 0 1	2 0 1	3 0 1	4 0 1	5 0 1	6 0 1
7 0 1	8 0 1	9 0 0	14 0 0	23 0 0	28 0 0	37 0 0
42 0 0	51 0 0	56 0 0	65 0 0	70 0 0	79 0 0	84 0 0
93 0 0	98 0 0	107 0 0	112 0 0	113 0 0	114 0 0	115 0 0
116 0 0	117 0 0	118 0 0	119 0 0	120 0 0	121 0 0	

Width data	0.	1.	2.	3.	4.	5.5	7.	9.	12
Depth data	0.	−1.25	−2.5	−3.75	−5.				

Load weightings NL (NO(I), VAL(I), I = 1, NL)

5	1	−.166667	10	−.666667	18	−.333333
	34	−.666667	42	−.166667		

Load increments INCS QINC(I), I = 1, INCS

10	200.	100.	50.	50.	50.	30.	20.	10.	5.	5.

Figure 6.9 Mesh and data for Program 6.0

The data follows the familiar pattern established in Chapter 5. The 'load weightings' provide a uniform stress of 1 kN/m^2 across the footing semi-width of 2 m (Appendix 1). These 'weightings' are then increased proportionally by the load increment values held in the vector QINC. It is usual in problems of this type to make the load increments smaller as the failure load is approached. At load levels well below failure, convergence should occur in relatively few iterations. In the data provided in this example ITS is set to 250, and this represents the maximum number of iterations that will be allowed within any load increment. If

```
 .2000E+03  q
-.6592E-02  -.6487E-02  -.6116E-02  -.5418E-02  -.3849E-02  displacements
         2  iterations                                      under load
 .3000E+03
-.1155E-01  -.1128E-01  -.1051E-01  -.9099E-02  -.6005E-02
        11
 .3500E+03
-.1630E-01  -.1596E-01  -.1512E-01  -.1327E-01  -.7557E-02
        20
 .4000E+03
-.2316E-01  -.2283E-01  -.2217E-01  -.1986E-01  -.9360E-02
        33
 .4500E+03
-.3317E-01  -.3285E-01  -.3242E-01  -.2958E-01  -.1150E-01
        45
 .4800E+03
-.4227E-01  -.4193E-01  -.4173E-01  -.3852E-01  -.1314E-01
        65
 .5000E+03
-.5084E-01  -.5041E-01  -.5034E-01  -.4677E-01  -.1455E-01
        81
 .5100E+03
-.5665E-01  -.5611E-01  -.5600E-01  -.5209E-01  -.1543E-01
        99
 .5150E+03
-.6094E-01  -.6026E-01  -.5998E-01  -.5569E-01  -.1594E-01
       159
 .5200E+03
-.7460E-01  -.7318E-01  -.7123E-01  -.6463E-01  -.1654E-01
       250
```

Figure 6.10 Results from Program 6.0

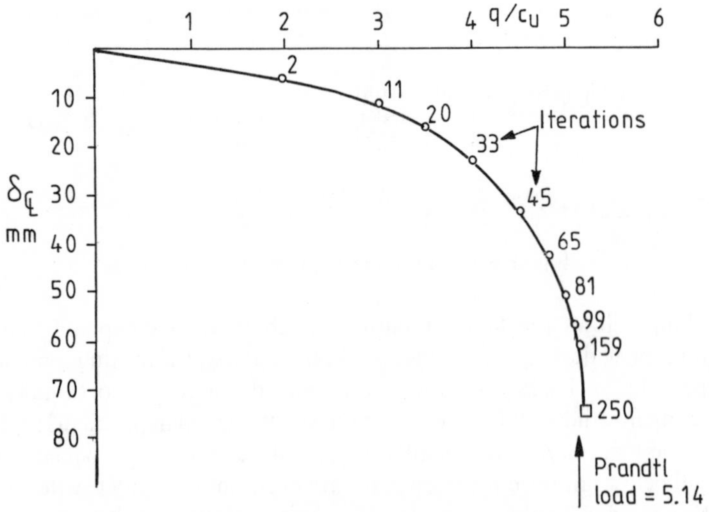

Figure 6.11 Plot of bearing stress versus centreline displacement

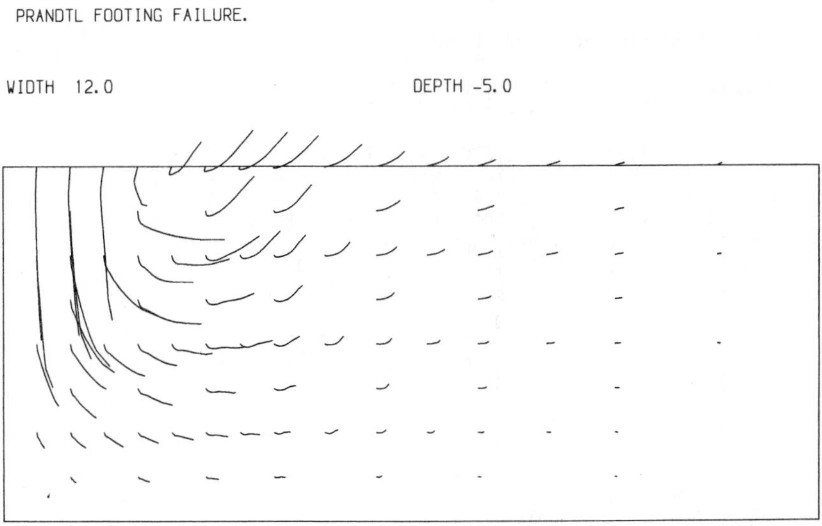

Figure 6.12 Displacement vectors at failure from Program 6.0

ITERS ever becomes equal to ITS, the algorithm stops and no more load increments are applied.

The computed results for this example are given in Figure 6.10 and show the applied stress, the vertical displacement under the loaded nodes and the number of iterations at each stage of the calculation. These results have been plotted in Figure 6.11 in the form of a dimensionless bearing capacity factor q/c_u versus centreline displacement. The number of iterations to achieve convergence for each load increment is also shown. It is seen that convergence was achieved in 159 iterations when $q/c_u = 5.1$, but convergence could not be achieved within the upper limit of 250 when $q/c_u = 5.2$. In addition, the displacements are also increasing rapidly at this level of loading, indicating that bearing failure is taking place at a value very close to the 'Prandtl' load of 5.14.

Although Program 6.0 does not incorporate any graphical output capability, it is a simple matter for users to write their own from commercially available libraries. Figure 6.12 shows a plot of the displacement vectors for this example at failure. Although the finite element mesh is constrained to remain a continuum, the vectors are still able to give an indication of the form of the failure mechanism.

PROGRAM 6.1: PLANE STRAIN SLOPE STABILITY ANALYSIS—VISCO-PLASTIC METHOD USING THE MOHR–COULOMB CRITERION

```
C
C      PROGRAM 6.1 PLANE STRAIN OF AN ELASTIC-PLASTIC
C      (MOHR-COULOMB) SOLID USING 8-NODE QUADRILATERAL ELEMENTS
C      VISCOPLASTIC STRAIN METHOD (GRAVITY LOADING)
C
```

```
C
C      ALTER NEXT LINE TO CHANGE PROBLEM SIZE
C
      PARAMETER(IKB1=184,IKB2=38,ILOADS=184,INF=96,
     +          IEVPT=512,INX=20,INY=20,IFOS=10)
C
      REAL DEE(4,4),SAMP(4,2),COORD(8,2),JAC(2,2),JAC1(2,2),DER(2,8),
     +DERIV(2,8),BEE(4,16),DBEE(4,16),DEPTH(INY),GRAVLO(ILOADS),
     +BTDB(16,16),KM(16,16),ELD(16),EPS(4),SIGMA(4),BLOAD(16),
     +BT(16,4),FUN(8),KB(IKB1,IKB2),LOADS(ILOADS),ELOAD(16),
     +BDYLDS(ILOADS),EVPT(IEVPT),OLDIS(ILOADS),
     +ERATE(4),EVP(4),DEVP(4),M1(4,4),M2(4,4),M3(4,4),
     +FOS(IFOS),FLOW(4,4),STRESS(4),TOP(INX),BOT(INX)
      INTEGER NF(INF,2),G(16)
      DATA IDEE,IBEE,IDBEE,IH,IFLOW/5*4/,IDOF,IBTDB,IBT,IKM/4*16/
      DATA IJAC,IJAC1,NODOF,IT,IDER,IDERIV/6*2/,ICOORD,NOD/2*8/
      DATA ISAMP/4/
C
C      INPUT AND INITIALISATION
C
      READ(5,*)PHI,C,PSI,GAMA,E,V,NXE,NYE,N,IW,NN,NR,NGP,ITS
      CALL READNF(NF,INF,NN,NODOF,NR)
      READ(5,*)(TOP(I),I=1,NXE+1)
      READ(5,*)(BOT(I),I=1,NXE+1)
      READ(5,*)(DEPTH(I),I=1,NYE+1)
      IWP1=IW+1
      CALL NULL(KB,IKB1,N,IWP1)
      CALL NULVEC(OLDIS,N)
      CALL NULVEC(GRAVLO,N)
      CALL FMDRAD(DEE,IDEE,E,V)
      CALL GAUSS(SAMP,ISAMP,NGP)
      PI=ACOS(-1.)
      TNPH=TAN(PHI*PI/180.)
C
C      ELEMENT STIFFNESS INTEGRATION AND ASSEMBLY
C
      DO 10 IP=1,NXE
      DO 10 IQ=1,NYE
      CALL SLOGEO(IP,IQ,NYE,TOP,BOT,DEPTH,COORD,ICOORD,G,NF,INF)
      CALL NULL(KM,IKM,IDOF,IDOF)
      CALL NULVEC(ELD,IDOF)
      DO 20 I=1,NGP
      DO 20 J=1,NGP
      CALL FMQUAD(DER,IDER,FUN,SAMP,ISAMP,I,J)
      CALL MATMUL(DER,IDER,COORD,ICOORD,JAC,IJAC,IT,NOD,IT)
      CALL TWOBY2(JAC,IJAC,JAC1,IJAC1,DET)
      CALL MATMUL(JAC1,IJAC1,DER,IDER,DERIV,IDERIV,IT,IT,NOD)
      CALL NULL(BEE,IBEE,IH,IDOF)
      CALL FORMB(BEE,IBEE,DERIV,IDERIV,NOD)
      CALL MATMUL(DEE,IDEE,BEE,IBEE,DBEE,IDBEE,IH,IH,IDOF)
      CALL MATRAN(BT,IBT,BEE,IBEE,IH,IDOF)
      CALL MATMUL(BT,IBT,DBEE,IDBEE,BTDB,IBTDB,IDOF,IH,IDOF)
      QUOT=DET*SAMP(I,2)*SAMP(J,2)
      DO 30 K=2,IDOF,2
   30 ELD(K)=ELD(K)+FUN(K/2)*QUOT
      CALL MSMULT(BTDB,IBTDB,QUOT,IDOF,IDOF)
   20 CALL MATADD(KM,IKM,BTDB,IBTDB,IDOF,IDOF)
      CALL FORMKB(KB,IKB1,KM,IKM,G,IW,IDOF)
      DO 40 K=1,IDOF
   40 IF(G(K).NE.0)GRAVLO(G(K))=GRAVLO(G(K))-ELD(K)*GAMA
   10 CONTINUE
C
```

```
C      REDUCE EQUATIONS
C
      CALL CHOLIN(KB,IKB1,N,IW)
C
C      TRIAL FACTOR OF SAFETY LOOP
C
      READ(5,*)INCS,(FOS(I),I=1,INCS)
      DO 50 IY=1,INCS
      PHIF=ATAN(TNPH/FOS(IY))*180./PI
      SNPH=SIN(PHIF*PI/180.)
      DT=4.*(1.+V)*(1.-2.*V)/(E*(1.-2.*V+SNPH**2.))
      CF=C/FOS(IY)
      ITERS=0
      CALL NULVEC(BDYLDS,N)
      CALL NULVEC(EVPT,NXE*NYE*IH*NGP*NGP)
C
C      ITERATION LOOP
C
   60 ITERS=ITERS+1
      CALL VECADD(GRAVLO,BDYLDS,LOADS,N)
      CALL CHOBAC(KB,IKB1,LOADS,N,IW)
C
C      CHECK CONVERGENCE
C
      CALL CHECON(LOADS,OLDIS,N,0.0001,ICON)
      IF(ITERS.EQ.1)ICON=0
      IF(ICON.EQ.1.OR.ITERS.EQ.ITS)CALL NULVEC(BDYLDS,N)
C
C      INSPECT ALL GAUSS POINTS
C
      NM=0
      DO 70 IP=1,NXE
      DO 70 IQ=1,NYE
      NM=NM+1
      CALL NULVEC(BLOAD,IDOF)
      CALL SLOGEO(IP,IQ,NYE,TOP,BOT,DEPTH,COORD,ICOORD,G,NF,INF)
      DO 80 M=1,IDOF
      IF(G(M).EQ.0)ELD(M)=0.
   80 IF(G(M).NE.0)ELD(M)=LOADS(G(M))
      IG=0
      DO 90 I=1,NGP
      DO 90 J=1,NGP
      IG=IG+1
      IN=NGP*NGP*IH*(NM-1)+IH*(IG-1)
      CALL FMQUAD(DER,IDER,FUN,SAMP,ISAMP,I,J)
      CALL MATMUL(DER,IDER,COORD,ICOORD,JAC,IJAC,IT,NOD,IT)
      CALL TWOBY2(JAC,IJAC,JAC1,IJAC1,DET)
      CALL MATMUL(JAC1,IJAC1,DER,IDER,DERIV,IDERIV,IT,IT,NOD)
      CALL NULL(BEE,IBEE,IH,IDOF)
      CALL FORMB(BEE,IBEE,DERIV,IDERIV,NOD)
      CALL MATRAN(BT,IBT,BEE,IBEE,IH,IDOF)
      CALL MVMULT(BEE,IBEE,ELD,IH,IDOF,EPS)
      DO 100 K=1,IH
  100 EPS(K)=EPS(K)-EVPT(IN+K)
      CALL MVMULT(DEE,IDEE,EPS,IH,IH,SIGMA)
      CALL INVAR(SIGMA,SIGM,DSBAR,THETA)
C
C      CHECK WHETHER YIELD IS VIOLATED
C
      CALL MOCOUF(PHIF,CF,SIGM,DSBAR,THETA,F)
      IF(ICON.EQ.1.OR.ITERS.EQ.ITS)GOTO 110
      IF(F.LT.0.)GOTO 90
```

```
      CALL MOCOUQ(PSI,DSBAR,THETA,DQ1,DQ2,DQ3)
      CALL FORMM(SIGMA,M1,M2,M3)
      DO 120 L=1,IH
      DO 120 M=1,IH
  120 FLOW(L,M)=F*(M1(L,M)*DQ1+M2(L,M)*DQ2+M3(L,M)*DQ3)
      CALL MVMULT(FLOW,IFLOW,SIGMA,IH,IH,ERATE)
      DO 130 K=1,IH
      EVP(K)=ERATE(K)*DT
  130 EVPT(IN+K)=EVPT(IN+K)+EVP(K)
      CALL MVMULT(DEE,IDEE,EVP,IH,IH,DEVP)
      GOTO 140
  110 CALL VECCOP(SIGMA,DEVP,IH)
  140 CALL MVMULT(BT,IBT,DEVP,IDOF,IH,ELOAD)
      QUOT=DET*SAMP(I,2)*SAMP(J,2)
      DO 150 K=1,IDOF
  150 BLOAD(K)=BLOAD(K)+ELOAD(K)*QUOT
   90 CONTINUE
C
C      COMPUTE TOTAL BODYLOADS VECTOR
C
      DO 160 M=1,IDOF
      IF(G(M).EQ.0)GOTO 160
      BDYLDS(G(M))=BDYLDS(G(M))+BLOAD(M)
  160 CONTINUE
   70 CONTINUE
      IF(ICON.NE.1.AND.ITERS.NE.ITS)GOTO 60
      BIG=0.
      DO 170 I=1,N
  170 IF(ABS(LOADS(I)).GT.BIG)BIG=ABS(LOADS(I))
      WRITE(6,1000)FOS(IY),BIG
      WRITE(6,2000)ITERS
      IF(ITERS.EQ.ITS)GOTO 180
   50 CONTINUE
  180 CONTINUE
 1000 FORMAT(1H ,10E12.4)
 2000 FORMAT(1H ,10I12)
      STOP
      END
```

The program is, in many ways, similar to its predecessor. The problem to be analysed is a slope of Mohr–Coulomb material subjected to gravity loading. The factor of safety (FOS) of the slope is to be assessed, and this quantity is defined as the proportion by which $\tan\phi$ and c must be reduced in order to cause failure. This is in contrast to the previous program in which failure was induced by increasing the loads with the material properties remaining constant.

The gravity loading vector $\mathbf{P}_a$ is accumulated element by element from integrals of the type

$$\mathbf{P}_a = \gamma \sum_{\text{elements}}^{\text{all}} \iint \mathbf{N}^T \,dx\,dy \tag{6.39}$$

and these calculations are performed in the same part of the program that forms the global stiffness matrix. It may be noted that only those freedoms corresponding to vertical movement are incorporated in the integrals.

At the element level, the one-dimensional array ELD is used to gather the contributions from each Gauss point. The global gravity loads vector GRAVLO accumulates ELD from each element after multiplication by the unit weight GAMA (γ).

The gravity loads vector in this program is applied to the slope in a single increment, and what was previously called the 'load increment loop' is now called the 'trial factor of safety loop'. Each entry into this loop corresponds to a different factor of safety (FOS) on the soil strength parameters. The factored soil strength parameters that go into the elasto-plastic analysis are obtained from

$$\begin{aligned} \phi_f &= \arctan(\tan\phi/\mathrm{FOS}) \\ c_f &= c/\mathrm{FOS} \end{aligned} \tag{6.40}$$

Keeping the loads constant, several (usually increasing) values of the factor of safety are attempted until the algorithm fails to converge. The actual factor of safety of the slope is the value to cause failure.

The following variables that are new to this program are now defined:

PHI	friction angle (ϕ)
C	cohesion(c)
PSI	dilation angle (ψ)
PHIF	factored friction angle (ϕ_f)
CF	factored cohesion (c_f)
TNPH	$\tan\phi_f$
SNPH	$\sin\phi_f$
FOS	one-dimensional array holding trial factors of safety
IFOS $\geqslant$ INCS	working size of array FOS
BIG	largest displacement

The first subroutine that is new to this program is the geometry routine SLOGEO. This subroutine generates a mesh shaped like a trapezium with the restriction that the top and bottom boundaries are parallel to the x axis. Hence, instead of a single one-dimensional array to hold the x coordinates of the element boundaries, the following two one-dimensional arrays must be read in as data:

TOP	element x coordinates at top of mesh
BOT	element x coordinates at bottom of mesh

The routine assumes that these coordinates are connected by straight lines. The element y coordinates are read into the array DEPTH, as was done in Program 6.0.

The routine MOCOUF forms the Mohr–Coulomb failure function F from the current stress state and the operating shear strength parameters. The routine MOCOUQ forms the derivatives of the Mohr–Coulomb potential function Q with respect to three stress invariants and these values are held in DQ1, DQ2 and DQ3.

In Program 6.0, similar subroutines corresponding to the von Mises criterion could have been used (VMF, VMQ, etc.), but the required expressions were so trivial that they were written directly into the main program.

Figure 6.13 shows the mesh and data for a typical slope stability analysis. The

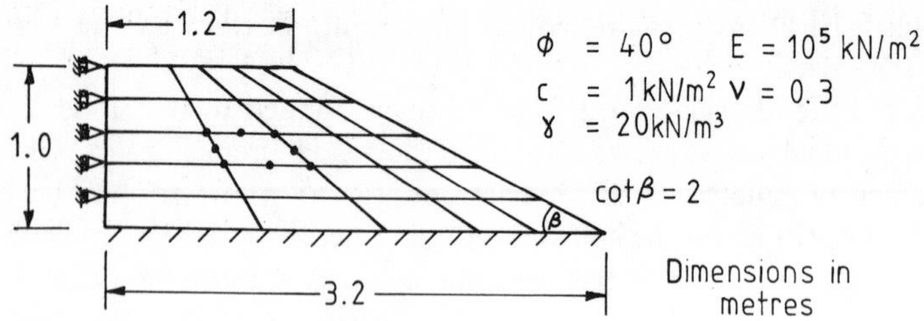

Soil properties	PHI	C	PSI	GAMA	E	V		
	40.	1.	0.	20.	1.E5	0.3		
Structure data	NXE	NYE	N	IW	NN	NR	NGP	ITS
	5	5	160	35	96	21	2	250

Node freedom data	1 0 1	2 0 1	3 0 1	4 0 1	5 0 1	6 0 1
	7 0 1	8 0 1	9 0 1	10 0 1	11 0 0	17 0 0
	28 0 0	34 0 0	45 0 0	51 0 0	62 0 0	68 0 0
	79 0 0	85 0 0	96 0 0			

Top width data	0.	0.4	0.6	0.8	1.0	1.2
Bottom width data	0.	1.0	1.8	2.4	2.8	3.2
Depth data	0.	−0.2	−0.4	−0.6	−0.8	−1.0

Trial factors of safety	INCS	FOS(I), I = 1, INCS				
	6	1.0	1.5	2.0	2.3	2.4	2.5

Figure 6.13 Mesh and data for Program 6.1

```
     FOS        Largest displacements
 .1000E+01      .7580E-04
         2      Iterations
 .1500E+01      .7580E-04
         2
 .2000E+01      .7748E-04
        16
 .2300E+01      .8495E-04
        29
 .2400E+01      .9024E-04
       140
 .2500E+01      .1200E-03
       250
```

Figure 6.14 Results from Program 6.1

parameters are given as $\phi = 40°$, $c = 1\,\text{kN/m}^2$ and the dilation angle ψ is put equal to zero. The unit weight of the material is given as $\gamma = 20\,\text{kN/m}^3$. The 'structure data' and 'node freedom data' follow a familiar pattern in which nodes are counted in the y direction. Three one-dimensional arrays read in the coordinate information, and six trial factors of safety ranging from 1.0 to 2.5 are to be attempted.

The output in Figure 6.14 gives the factor of safety, the maxium displacement (BIG) at convergence and the number of iterations to achieve convergence. The results have also been plotted in Figure 6.15, and these indicate that the factor of safety of the slope is around 2.5. Bishop and Morgenstern (1960) produced charts for slope stability analysis using slip circle techniques, and these give a factor of safety of 2.505 for the slope considered in this example. The displacement vectors

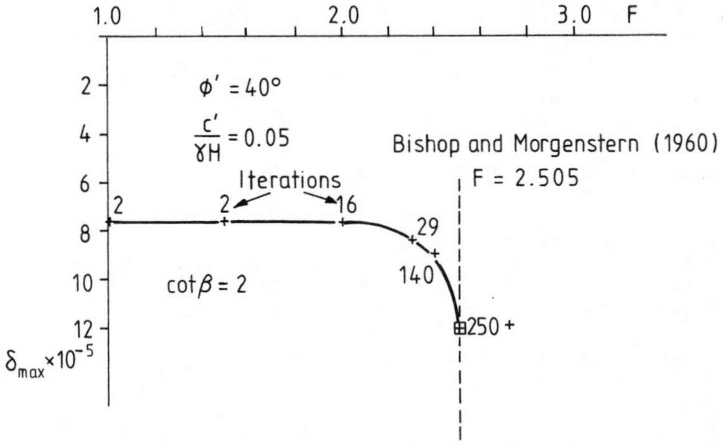

Figure 6.15 Plot of maximum displacement versus factor of safety

```
SLOPE FAILURE.

PHI   40.0                                   C    1.0

GAMA   20.0                                  F.O.S.   2.5
```

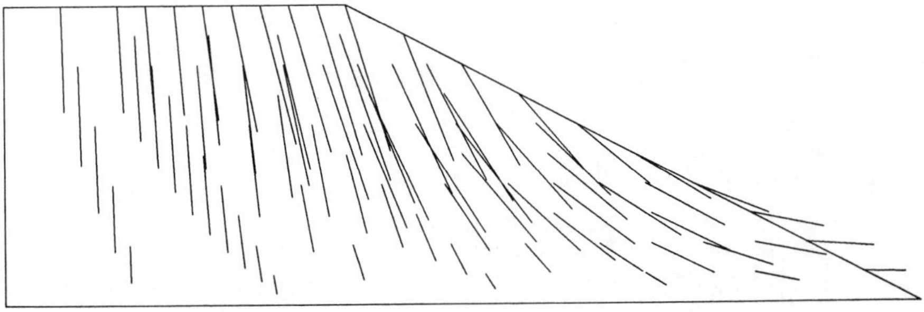

Figure 6.16 Displacement vectors at failure from Program 6.1

after 250 iterations with a factor of safety of 2.5 are given in Figure 6.16 to indicate the nature of the failure mechanism.

PROGRAM 6.2: PLANE STRAIN PASSIVE EARTH PRESSURE ANALYSIS—INITIAL STRESS METHOD USING THE MOHR–COULOMB CRITERION

```
C
C       PROGRAM 6.2 PLANE STRAIN OF AN ELASTIC-PLASTIC
C       (MOHR-COULOMB) SOLID USING 8-NODE QUADRILATERAL ELEMENTS
C       INITIAL STRESS METHOD
C
C
C       ALTER NEXT LINE TO CHANGE PROBLEM SIZE
C
      PARAMETER(IKB1=294,IKB2=48,ILOADS=294,INF=176,
     +          INX=20,INY=20,INO=10)
C
      REAL DEE(4,4),SAMP(4,2),COORD(8,2),JAC(2,2),JAC1(2,2),
     +DER(2,8),DERIV(2,8),BEE(4,16),DBEE(4,16),WIDTH(INX),DEPTH(INY),
     +BTDB(16,16),KM(16,16),ELD(16),EPS(4),SIGMA(4),BLOAD(16),
     +BT(16,4),FUN(8),KB(IKB1,IKB2),LOADS(ILOADS),ELOAD(16),
     +TOTD(ILOADS),BDYLDS(ILOADS),OLDIS(ILOADS),ELSO(4),
     +SX(INX,INY,4),SY(INX,INY,4),TXY(INX,INY,4),SZ(INX,INY,4),
     +STORKB(INO),STRESS(4),PL(4,4),GC(2)
      INTEGER NF(INF,2),G(16),NO(INO)
      DATA IDEE,IBEE,IDBEE,IH,IFLOW,IPL/6*4/,IDOF,IBTDB,IBT,IKM/4*16/
      DATA IJAC,IJAC1,NODOF,IT,IDER,IDERIV/6*2/,ICOORD,NOD/2*8/
      DATA ISAMP/4/
C
C       INPUT AND INITIALISATION
C
      READ(5,*)PHI,C,PSI,GAMA,EPKO,E,V,NXE,NYE,N,IW,NN,NR,NGP,ITS
      CALL READNF(NF,INF,NN,NODOF,NR)
      READ(5,*)(WIDTH(I),I=1,NXE+1)
      READ(5,*)(DEPTH(I),I=1,NYE+1)
      IWP1=IW+1
      CALL NULL(KB,IKB1,N,IWP1)
      CALL NULVEC(OLDIS,N)
      CALL NULVEC(TOTD,N)
      CALL NULVEC(BDYLDS,N)
      CALL FMDRAD(DEE,IDEE,E,V)
      CALL GAUSS(SAMP,ISAMP,NGP)
C
C       ELEMENT STIFFNESS INTEGRATION AND ASSEMBLY
C
      DO 10 IP=1,NXE
      DO 10 IQ=1,NYE
      CALL GEOV8Y(IP,IQ,NYE,WIDTH,DEPTH,COORD,ICOORD,G,NF,INF)
      CALL NULL(KM,IKM,IDOF,IDOF)
      IG=0
      DO 20 I=1,NGP
      DO 20 J=1,NGP
      IG=IG+1
      CALL FMQUAD(DER,IDER,FUN,SAMP,ISAMP,I,J)
      CALL GCOORD(FUN,COORD,ICOORD,NOD,IT,GC)
      SY(IP,IQ,IG)=GC(2)*GAMA
      SX(IP,IQ,IG)=GC(2)*GAMA*EPKO
      SZ(IP,IQ,IG)=GC(2)*GAMA*EPKO
      TXY(IP,IQ,IG)=0.
```

```
      CALL MATMUL(DER,IDER,COORD,ICOORD,JAC,IJAC,IT,NOD,IT)
      CALL TWOBY2(JAC,IJAC,JAC1,IJAC1,DET)
      CALL MATMUL(JAC1,IJAC1,DER,IDER,DERIV,IDERIV,IT,IT,NOD)
      CALL NULL(BEE,IBEE,IH,IDOF)
      CALL FORMB(BEE,IBEE,DERIV,IDERIV,NOD)
      CALL MATMUL(DEE,IDEE,BEE,IBEE,DBEE,IDBEE,IH,IH,IDOF)
      CALL MATRAN(BT,IBT,BEE,IBEE,IH,IDOF)
      CALL MATMUL(BT,IBT,DBEE,IDBEE,BTDB,IBTDB,IDOF,IH,IDOF)
      QUOT=DET*SAMP(I,2)*SAMP(J,2)
      CALL MSMULT(BTDB,IBTDB,QUOT,IDOF,IDOF)
   20 CALL MATADD(KM,IKM,BTDB,IBTDB,IDOF,IDOF)
   10 CALL FORMKB(KB,IKB1,KM,IKM,G,IW,IDOF)
C
C       READ PRESCRIBED FREEDOMS
C       AUGMENT AND REDUCE STIFFNESS MATRIX
C
      READ(5,*)NL,(NO(I),I=1,NL),PRESC,INCS
      DO 30 I=1,NL
      KB(NO(I),IWP1)=KB(NO(I),IWP1)+1.E20
   30 STORKB(I)=KB(NO(I),IWP1)
      CALL CHOLIN(KB,IKB1,N,IW)
C
C       DISPLACEMENT INCREMENT LOOP
C
      DO 40 IY=1,INCS
      PTOT=PRESC*IY
      ITERS=0
C
C       ITERATION LOOP
C
   50 ITERS=ITERS+1
      CALL NULVEC(LOADS,N)
      DO 60 I=1,NL
   60 LOADS(NO(I))=STORKB(I)*PRESC
      CALL VECADD(LOADS,BDYLDS,LOADS,N)
      CALL CHOBAC(KB,IKB1,LOADS,N,IW)
      CALL NULVEC(BDYLDS,N)
C
C       CHECK CONVERGENCE
C
      CALL CHECON(LOADS,OLDIS,N,0.001,ICON)
      IF(ITERS.EQ.1)ICON=0
C
C       INSPECT ALL GAUSS POINTS
C
      DO 70 IP=1,NXE
      DO 70 IQ=1,NYE
      CALL NULVEC(BLOAD,IDOF)
      CALL GEOV8Y(IP,IQ,NYE,WIDTH,DEPTH,COORD,ICOORD,G,NF,INF)
      DO 80 M=1,IDOF
      IF(G(M).EQ.0)ELD(M)=0.
   80 IF(G(M).NE.0)ELD(M)=LOADS(G(M))
      IG=0
      DO 90 I=1,NGP
      DO 90 J=1,NGP
      IG=IG+1
      CALL NULVEC(ELSO,IH)
      CALL FMQUAD(DER,IDER,FUN,SAMP,ISAMP,I,J)
      CALL MATMUL(DER,IDER,COORD,ICOORD,JAC,IJAC,IT,NOD,IT)
      CALL TWOBY2(JAC,IJAC,JAC1,IJAC1,DET)
      CALL MATMUL(JAC1,IJAC1,DER,IDER,DERIV,IDERIV,IT,IT,NOD)
      CALL NULL(BEE,IBEE,IH,IDOF)
```

```
      CALL FORMB(BEE,IBEE,DERIV,IDERIV,NOD)
      CALL MATRAN(BT,IBT,BEE,IBEE,IH,IDOF)
      CALL MVMULT(BEE,IBEE,ELD,IH,IDOF,EPS)
      CALL MVMULT(DEE,IDEE,EPS,IH,IH,SIGMA)
      STRESS(1)=SIGMA(1)+SX(IP,IQ,IG)
      STRESS(2)=SIGMA(2)+SY(IP,IQ,IG)
      STRESS(3)=SIGMA(3)+TXY(IP,IQ,IG)
      STRESS(4)=SIGMA(4)+SZ(IP,IQ,IG)
      CALL INVAR(STRESS,SIGM,DSBAR,THETA)
C
C      CHECK WHETHER YIELD IS VIOLATED
C
      CALL MOCOUF(PHI,C,SIGM,DSBAR,THETA,FNEW)
      IF(FNEW.LT.0.)GOTO 100
      STRESS(1)=SX(IP,IQ,IG)
      STRESS(2)=SY(IP,IQ,IG)
      STRESS(3)=TXY(IP,IQ,IG)
      STRESS(4)=SZ(IP,IQ,IG)
      CALL INVAR(STRESS,SIGM,DSBAR,THETA)
      CALL MOCOUF(PHI,C,SIGM,DSBAR,THETA,F)
      FAC=FNEW/(FNEW-F)
      STRESS(1)=SX(IP,IQ,IG)+(1.-FAC)*SIGMA(1)
      STRESS(2)=SY(IP,IQ,IG)+(1.-FAC)*SIGMA(2)
      STRESS(3)=TXY(IP,IQ,IG)+(1.-FAC)*SIGMA(3)
      STRESS(4)=SZ(IP,IQ,IG)+(1.-FAC)*SIGMA(4)
      CALL MOCOPL(PHI,PSI,E,V,STRESS,PL)
      CALL MSMULT(PL,IPL,FAC,IH,IH)
      CALL MVMULT(PL,IPL,EPS,IH,IH,ELSO)
      CALL MVMULT(BT,IBT,ELSO,IDOF,IH,ELOAD)
      QUOT=DET*SAMP(I,2)*SAMP(J,2)
      DO 110 K=1,IDOF
  110 BLOAD(K)=BLOAD(K)+ELOAD(K)*QUOT
  100 IF(ICON.NE.1.AND.ITERS.NE.ITS)GOTO 90
C
C      UPDATE GAUSS POINT STRESSES
C
      SX(IP,IQ,IG)=SX(IP,IQ,IG)+SIGMA(1)-ELSO(1)
      SY(IP,IQ,IG)=SY(IP,IQ,IG)+SIGMA(2)-ELSO(2)
      TXY(IP,IQ,IG)=TXY(IP,IQ,IG)+SIGMA(3)-ELSO(3)
      SZ(IP,IQ,IG)=SZ(IP,IQ,IG)+SIGMA(4)-ELSO(4)
   90 CONTINUE
C
C      COMPUTE TOTAL BODYLOADS VECTOR
C
      DO 120 M=1,IDOF
      IF(G(M).EQ.0)GOTO 120
      BDYLDS(G(M))=BDYLDS(G(M))+BLOAD(M)
  120 CONTINUE
   70 CONTINUE
      IF(ICON.NE.1.AND.ITERS.NE.ITS)GOTO 50
      CALL VECADD(TOTD,LOADS,TOTD,N)
      PAV=.5*((DEPTH(1)-DEPTH(2))*(SX(1,1,2)+SX(1,1,4))
     +        +(DEPTH(2)-DEPTH(3))*(SX(1,2,2)+SX(1,2,4))
     +        +(DEPTH(3)-DEPTH(4))*(SX(1,3,2)+SX(1,3,4))
     +        +(DEPTH(4)-DEPTH(5))*(SX(1,4,2)+SX(1,4,4)))
      WRITE(6,1000)PTOT,PAV,ITERS
      IF(ITERS.EQ.ITS)GOTO 130
   40 CONTINUE
  130 CONTINUE
 1000 FORMAT(2E12.4,I12)
      STOP
      END
```

The initial stress method of stress redistribution is demonstrated in a problem of passive earth pressure, in which a smooth wall is translated into a bed of sand. As in Program 6.0, a rectangular mesh of eight-noded elements is generated with nodes and freedoms counted in the y direction. An additional feature of this program which appears in the element integration and assembly section is the generation of initial self-weight stresses. The subroutine GCOORD calculates the coordinates of each Gauss point using the isoparametric property

$$x = \sum_{i=1}^{8} N_i x_i$$
$$y = \sum_{i=1}^{8} N_i y_i \tag{6.41}$$

The x and y coordinates that result are held in the one-dimensional array GC, in positions 1 and 2 respectively. Only the y coordinate is required in this case and the vertical stress σ_y is obtained after multiplication by the unit weight γ held in GAMA. The normal effective stresses σ_x and σ_z are obtained by multiplying σ_y by the 'at rest' earth pressure coefficient K_0 held in EPKO.

After the stiffness matrix formulation, the freedoms which are to receive prescribed displacements are read, followed by the magnitude of the displacement increment held in PRESC. In this program, INCS represents the number of constant displacement increments that are to be applied. The upper limit on iterations held in ITS does not need to be as large as it was in load-controlled problems. This is because convergence takes less iterations when using displacement control, especially as failure conditions are approached.

The 'stiff-spring' technique is used to implement the prescribed displacements, as was first demonstrated in Program 4.1 and described more fully in Section 3.6.

The program follows a familiar course until the calculation of the failure function. Initially, the failure function FNEW is obtained after adding the full elastic stress increment to those stresses existing previously. If FNEW is positive, indicating a yielding Gauss point, then the failure function F is obtained using just those stresses existing previously. The scaling parameter FAC is then calculated as described in equation (6.35). The plastic stress–strain matrix $\mathbf{D}^{\mathrm{P}}$ for a Mohr–Coulomb material is formed by the routine MOCOPL (if implementing the von Mises criterion, the subroutine VMPL should be substituted) using stresses that have been factored to ensure they lie on the failure surface. The resulting matrix PL is multiplied by the scaling parameter FAC and then by the total strain increment array EPS to yield the 'plastic' stress increment array ELSO. Integrals of the type described by equation (6.34) then follow and the array BDYLDS is accumulated from each element. It may be noted that in the algorithm presented here, the body-loads vector is completely reformed at each iteration. This is in contrast to the visco-plasticity algorithm presented in Programs 6.0 and 6.1 in which the body-loads vector was accumulated at each iteration.

At convergence, the stresses must be updated ready for the next displacement (load) increment. This involves adding to the stresses remaining from the previous increment, the one-dimensional array of total stress increments (SIGMA) minus the one-dimensional array of corrective 'plastic' stresses (ELSO).

The example problem shown in Figure 6.17 represents a sand with strength parameters $\phi = 30°$, $c = 0$ and $\psi = 0°$, subjected to prescribed displacements along the left face. The boundary condition is applied to the x components of displacement at the nine nodes adjacent to the hypothetical smooth, rigid wall shown hatched. The initial stresses in the ground are calculated assuming the unit weight $\gamma = 20\,\text{kN/m}^3$ and 'at rest' earth pressure coefficient $K_0 = 1$.

Following each displacement increment, and after numerical convergence, the resultant force, PAV, acting on the wall is computed by averaging the σ_x stresses at the eight Gauss points closest to the 'wall'.

The output shown in Figure 6.18 gives the wall displacement, PTOT, the resultant force and the number of iterations to convergence at each step, and these are

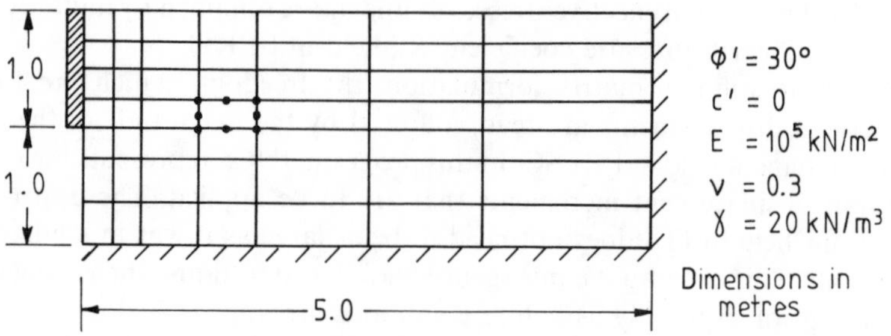

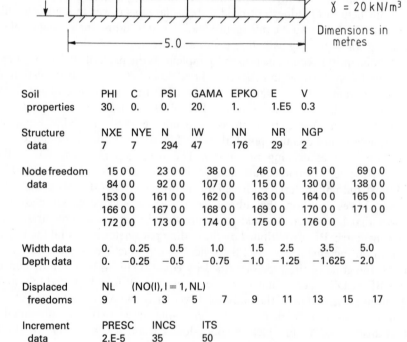

Soil properties	PHI 30.	C 0.	PSI 0.	GAMA 20.	EPKO 1.	E 1.E5	V 0.3
Structure data	NXE 7	NYE 7	N 294	IW 47	NN 176	NR 29	NGP 2

Node freedom data	15 0 0	23 0 0	38 0 0	46 0 0	61 0 0	69 0 0
	84 0 0	92 0 0	107 0 0	115 0 0	130 0 0	138 0 0
	153 0 0	161 0 0	162 0 0	163 0 0	164 0 0	165 0 0
	166 0 0	167 0 0	168 0 0	169 0 0	170 0 0	171 0 0
	172 0 0	173 0 0	174 0 0	175 0 0	176 0 0	

Width data	0.	0.25	0.5	1.0	1.5	2.5	3.5	5.0
Depth data	0.	−0.25	−0.5	−0.75	−1.0	−1.25	−1.625	−2.0

Displaced freedoms	NL	(NO(I), I = 1, NL)							
	9	1	3	5	7	9	11	13	15	17

Increment data	PRESC 2.E-5	INCS 35	ITS 50

Figure 6.17 Mesh and data for Program 6.2

Displacement	Force	Iterations
.2000E-04	-.1097E+02	2
.4000E-04	-.1194E+02	2
.6000E-04	-.1292E+02	4
.8000E-04	-.1385E+02	12
.1000E-03	-.1477E+02	10
.1200E-03	-.1568E+02	6
.1400E-03	-.1659E+02	8
.1600E-03	-.1750E+02	3
.1800E-03	-.1841E+02	7
.2000E-03	-.1932E+02	3
.2200E-03	-.2019E+02	15
.2400E-03	-.2106E+02	10
.2600E-03	-.2192E+02	6
.2800E-03	-.2277E+02	9
.3000E-03	-.2361E+02	9
.3200E-03	-.2441E+02	16
.3400E-03	-.2529E+02	15
.3600E-03	-.2616E+02	7
.3800E-03	-.2698E+02	15
.4000E-03	-.2770E+02	21
.4200E-03	-.2834E+02	19
.4400E-03	-.2891E+02	18
.4600E-03	-.2941E+02	16
.4800E-03	-.2989E+02	9
.5000E-03	-.3034E+02	14
.5200E-03	-.3058E+02	30
.5400E-03	-.3061E+02	38
.5600E-03	-.3063E+02	10
.5800E-03	-.3065E+02	9
.6000E-03	-.3065E+02	11
.6200E-03	-.3065E+02	12
.6400E-03	-.3066E+02	3
.6600E-03	-.3066E+02	3
.6800E-03	-.3067E+02	7
.7000E-03	-.3068E+02	3

Figure 6.18 Results from Program 6.2

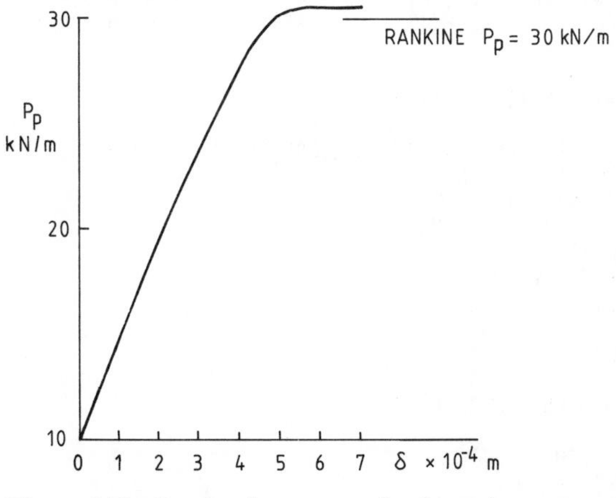

Figure 6.19 Passive force versus horizontal movement

plotted in Figure 6.19. The force is seen to build up a maximum value of around 31 kN/m. This is in close agreement with the closed form Rankine solution of 30 kN/m.

The initial stress algorithm presented in this program will tend to overestimate collapse loads, especially if the displacement (load) steps are made too big. Users are recommended to try one or two different increment sizes to test the sensitivity of the solutions. The problem is caused by incremental 'drift' of the stress state at individual Gauss points into illegal stress space in spite of apparent numerical convergence. Although not included in the present work, various strategies are available (e.g. Nayak and Zienkiewicz, 1972) for drift correction.

PROGRAM 6.3: AXISYMMETRIC 'UNDRAINED' ANALYSIS—VISCO-PLASTIC METHOD USING THE MOHR–COULOMB CRITERION

```
C
C      PROGRAM 6.3 AXISYMMETRIC STRAIN OF AN UNDRAINED
C      ELASTIC-PLASTIC (MOHR-COULOMB) SOLID USING
C      8-NODED QUADRILATERAL ELEMENTS,VISCOPLASTIC STRAIN METHOD
C
C
C      ALTER NEXT LINE TO CHANGE PROBLEM SIZE
C
      PARAMETER(IKV=3500,ILOADS=100,INF=100,
     +          IEVPT=500,INR=20,IND=20,INO=10)
C
      REAL DEE(4,4),SAMP(4,2),COORD(8,2),JAC(2,2),JAC1(2,2),
     +DER(2,8),DERIV(2,8),BEE(4,16),DBEE(4,16),RAD(INR),DEP(IND),
     +BTDB(16,16),KM(16,16),ELD(16),EPS(4),SIGMA(4),BLOAD(16),
     +BT(16,4),FUN(8),KV(IKV),LOADS(ILOADS),ELOAD(16),
     +TOTD(ILOADS),BDYLDS(ILOADS),EVPT(IEVPT),OLDIS(ILOADS),
     +SX(INR,IND,4),SY(INR,IND,4),TXY(INR,IND,4),SZ(INR,IND,4),
     +EX(INR,IND,4),EY(INR,IND,4),GXY(INR,IND,4),EZ(INR,IND,4),
     +STORKV(INO),ERATE(4),EVP(4),DEVP(4),M1(4,4),M2(4,4),M3(4,4),
     +FLOW(4,4),STRESS(4),PORE(INR,IND,4)
      INTEGER NF(INF,2),G(16),NO(INO)
      DATA IDEE,IBEE,IDBEE,IH,IFLOW/5*4/,IDOF,IBTDB,IBT,IKM/4*16/
      DATA IJAC,IJAC1,NODOF,IT,IDER,IDERIV/6*2/,ICOORD,NOD/2*8/
      DATA ISAMP/4/
C
C      INPUT AND INITIALISATION
C
      READ(5,*)PHI,C,PSI,E,V,BW,CONS,NRE,NDE,N,IW,NN,NR,NGP,ITS
      CALL READNF(NF,INF,NN,NODOF,NR)
      READ(5,*)(RAD(I),I=1,NRE+1)
      READ(5,*)(DEP(I),I=1,NDE+1)
      IR=N*(IW+1)
      CALL NULVEC(KV,IR)
      CALL NULVEC(OLDIS,N)
      CALL NULVEC(TOTD,N)
      CALL FMDRAD(DEE,IDEE,E,V)
C
C      ADD FLUID BULK MODULUS
C
      DO 10 I=1,IH
      DO 10 J=1,IH
      IF(I.EQ.3.OR.J.EQ.3)GOTO 10
```

```
      DEE(I,J)=DEE(I,J)+BW
   10 CONTINUE
      CALL GAUSS(SAMP,ISAMP,NGP)
      SNPH=SIN(PHI*4.*ATAN(1.)/180.)
      DT=4.*(1.+V)*(1.-2.*V)/(E*(1.-2.*V+SNPH+SNPH))
C
C       ELEMENT STIFFNESS INTEGRATION AND ASSEMBLY
C
      DO 20 IP=1,NRE
      DO 20 IQ=1,NDE
      CALL GEOV8Y(IP,IQ,NDE,RAD,DEP,COORD,ICOORD,G,NF,INF)
      CALL NULL(KM,IKM,IDOF,IDOF)
      IG=0
      DO 30 I=1,NGP
      DO 30 J=1,NGP
      IG=IG+1
      SX(IP,IQ,IG)=CONS
      SY(IP,IQ,IG)=CONS
      TXY(IP,IQ,IG)=0.
      SZ(IP,IQ,IG)=CONS
      EX(IP,IQ,IG)=0.
      EY(IP,IQ,IG)=0.
      GXY(IP,IQ,IG)=0.
      EZ(IP,IQ,IG)=0.
      CALL FMQUAD(DER,IDER,FUN,SAMP,ISAMP,I,J)
      CALL MATMUL(DER,IDER,COORD,ICOORD,JAC,IJAC,IT,NOD,IT)
      CALL TWOBY2(JAC,IJAC,JAC1,IJAC1,DET)
      CALL MATMUL(JAC1,IJAC1,DER,IDER,DERIV,IDERIV,IT,IT,NOD)
      CALL NULL(BEE,IBEE,IH,IDOF)
      CALL FMBRAD(BEE,IBEE,DERIV,IDERIV,FUN,COORD,ICOORD,SUM,NOD)
      CALL MATMUL(DEE,IDEE,BEE,IBEE,DBEE,IDBEE,IH,IH,IDOF)
      CALL MATRAN(BT,IBT,BEE,IBEE,IH,IDOF)
      CALL MATMUL(BT,IBT,DBEE,IDBEE,BTDB,IBTDB,IDOF,IH,IDOF)
      QUOT=SUM*DET*SAMP(I,2)*SAMP(J,2)
      CALL MSMULT(BTDB,IBTDB,QUOT,IDOF,IDOF)
   30 CALL MATADD(KM,IKM,BTDB,IBTDB,IDOF,IDOF)
   20 CALL FORMKV(KV,KM,IKM,G,N,IDOF)
C
C       READ PRESCRIBED DISPLACEMENTS
C       AUGMENT AND REDUCE STIFFNESS MATRIX
C
      READ(5,*)NL,(NO(I),I=1,NL),PRESC,INCS
      DO 40 I=1,NL
      KV(NO(I))=KV(NO(I))+1.E20
   40 STORKV(I)=KV(NO(I))
      CALL BANRED(KV,N,IW)
C
C       DISPLACEMENT INCREMENT LOOP
C
      CALL FMDRAD(DEE,IDEE,E,V)
      DO 50 IY=1,INCS
      PTOT=PRESC*IY
      ITERS=0
      CALL NULVEC(BDYLDS,N)
      CALL NULVEC(EVPT,NRE*NDE*IH*NGP*NGP)
C
C       ITERATION LOOP
C
   60 ITERS=ITERS+1
      CALL NULVEC(LOADS,N)
      DO 70 I=1,NL
   70 LOADS(NO(I))=STORKV(I)*PRESC
```

```
      CALL VECADD(LOADS,BDYLDS,LOADS,N)
      CALL BACSUB(KV,LOADS,N,IW)
C
C       CHECK CONVERGENCE
C
      CALL CHECON(LOADS,OLDIS,N,0.0001,ICON)
      IF(ITERS.EQ.1)ICON=0
C
C       INSPECT ALL GAUSS POINTS
C
      NM=0
      DO 80 IP=1,NRE
      DO 80 IQ=1,NDE
      NM=NM+1
      CALL NULVEC(BLOAD,IDOF)
      CALL GEOV8Y(IP,IQ,NDE,RAD,DEP,COORD,ICOORD,G,NF,INF)
      DO 90 M=1,IDOF
      IF(G(M).EQ.0)ELD(M)=0.
   90 IF(G(M).NE.0)ELD(M)=LOADS(G(M))
      IG=0
      DO 100 I=1,NGP
      DO 100 J=1,NGP
      IG=IG+1
      IN=NGP*NGP*IH*(NM-1)+IH*(IG-1)
      CALL FMQUAD(DER,IDER,FUN,SAMP,ISAMP,I,J)
      CALL MATMUL(DER,IDER,COORD,ICOORD,JAC,IJAC,IT,NOD,IT)
      CALL TWOBY2(JAC,IJAC,JAC1,IJAC1,DET)
      CALL MATMUL(JAC1,IJAC1,DER,IDER,DERIV,IDERIV,IT,IT,NOD)
      CALL NULL(BEE,IBEE,IH,IDOF)
      CALL FMBRAD(BEE,IBEE,DERIV,IDERIV,FUN,COORD,ICOORD,SUM,NOD)
      CALL MATRAN(BT,IBT,BEE,IBEE,IH,IDOF)
      CALL MVMULT(BEE,IBEE,ELD,IH,IDOF,EPS)
      DO 110 K=1,IH
  110 EPS(K)=EPS(K)-EVPT(IN+K)
      CALL MVMULT(DEE,IDEE,EPS,IH,IH,SIGMA)
      STRESS(1)=SIGMA(1)+SX(IP,IQ,IG)
      STRESS(2)=SIGMA(2)+SY(IP,IQ,IG)
      STRESS(3)=SIGMA(3)+TXY(IP,IQ,IG)
      STRESS(4)=SIGMA(4)+SZ(IP,IQ,IG)
      CALL INVAR(STRESS,SIGM,DSBAR,THETA)
C
C       CHECK WHETHER YIELD IS VIOLATED
C
      CALL MOCOUF(PHI,C,SIGM,DSBAR,THETA,F)
      IF(F.LT.0.)GOTO 120
      CALL MOCOUQ(PSI,DSBAR,THETA,DQ1,DQ2,DQ3)
      CALL FORMM(STRESS,M1,M2,M3)
      DO 130 L=1,IH
      DO 130 M=1,IH
  130 FLOW(L,M)=F*(M1(L,M)*DQ1+M2(L,M)*DQ2+M3(L,M)*DQ3)
      CALL MVMULT(FLOW,IFLOW,STRESS,IH,IH,ERATE)
      DO 140 K=1,IH
      EVP(K)=ERATE(K)*DT
  140 EVPT(IN+K)=EVPT(IN+K)+EVP(K)
      CALL MVMULT(DEE,IDEE,EVP,IH,IH,DEVP)
      CALL MVMULT(BT,IBT,DEVP,IDOF,IH,ELOAD)
      QUOT=SUM*DET*SAMP(I,2)*SAMP(J,2)
      DO 150 K=1,IDOF
  150 BLOAD(K)=BLOAD(K)+ELOAD(K)*QUOT
  120 IF(ICON.NE.1.AND.ITERS.NE.ITS)GOTO 100
C
C       UPDATE GAUSS POINT STRESSES AND STRAINS
```

```
C       CALCULATE PORE PRESSURES
C
      SX(IP,IQ,IG)=STRESS(1)
      SY(IP,IQ,IG)=STRESS(2)
      TXY(IP,IQ,IG)=STRESS(3)
      SZ(IP,IQ,IG)=STRESS(4)
      EX(IP,IQ,IG)=EX(IP,IQ,IG)+EPS(1)+EVPT(IN+1)
      EY(IP,IQ,IG)=EY(IP,IQ,IG)+EPS(2)+EVPT(IN+2)
      GXY(IP,IQ,IG)=GXY(IP,IQ,IG)+EPS(3)+EVPT(IN+3)
      EZ(IP,IQ,IG)=EZ(IP,IQ,IG)+EPS(4)+EVPT(IN+4)
      PORE(IP,IQ,IG)=(EX(IP,IQ,IG)+EY(IP,IQ,IG)+EZ(IP,IQ,IG))*BW
  100 CONTINUE
C
C       COMPUTE TOTAL BODYLOADS VECTOR
C
      DO 160 M=1,IDOF
      IF(G(M).EQ.0)GOTO 160
      BDYLDS(G(M))=BDYLDS(G(M))+BLOAD(M)
  160 CONTINUE
   80 CONTINUE
      IF(ICON.NE.1.AND.ITERS.NE.ITS)GOTO 60
      CALL VECADD(TOTD,LOADS,TOTD,N)
      WRITE(6,1000)PTOT
      WRITE(6,1000)SX(1,1,1),SY(1,1,1),SZ(1,1,1)
      WRITE(6,1000)DSBAR,PORE(1,1,1)
      WRITE(6,2000)ITERS
      IF(ITERS.EQ.ITS)GOTO 170
   50 CONTINUE
  170 CONTINUE
 1000 FORMAT(10E12.4)
 2000 FORMAT(10I12)
      STOP
      END
```

Little mention has been made so far of the role of the dilation angle ψ on the prediction of collapse loads in Mohr–Coulomb materials. The reason is that the dilation angle governs volumetric strains during plastic yield and will have little influence on collapse loads in 'unconfined' problems. The examples considered so far in this chapter have been relatively unconfined (e.g. slope stability, earth pressures) and in such cases it is recommended that the dilation angle is put equal to zero, as it has been found (e.g. Griffiths, 1982) that this usually requires slightly less computational effort.

'Undrained' soils are two-phase particulate materials in which the voids between the particles are full of water. In addition, the permeability of the material must be sufficiently low or the loads applied so quickly that pore water pressure that are generated have no time to dissipate during the time scale of the analysis.

In the case of undrained clays that have soft soil skeletons, the shear strength appears to be constant and given by an undrained 'cohesion' c_u and $\phi_u = 0$. In such materials, the von Mises or Tresca failure criterion can be successfully applied, as was demonstrated in Program 6.0.

In the case of saturated soils with hard skeletons such as dense quartz sand, shear stresses will tend to cause dilation which will be resisted by tensile water pressures in the voids of the soil. In turn, the effective stresses between particles will rise and, in a frictional material, the shear stresses necessary to cause failure will also rise. Thus, a dense sand, far from exhibiting a constant shear strength when sheared undrained, would have infinite strength provided the pore fluid could sustain infinite suction and the grains did not crush. In reality, a finite shear strength is recorded due to either gain crushing or pore fluid cavitation.

To perform analyses of this type it is necessary to separate stresses into pore water pressures (isotropic) and effective interparticle stresses (isotropic + shear). Such a treatment has already been described in Chapter 2 in terms of time dependent 'consolidation' properties of two-phase materials (Biot's poro-elastic theory) and programs to deal with this will be found in Chapter 9. However, the undrained problem pertaining at the beginning of the Biot process is so important in soil mechanics that it merits special treatment.

Naylor (1974) has described a method of separating the stresses into pore pressures and effective stresses. The method uses as its basis the concept of effective stress in matrix notation; thus

$$\boldsymbol{\sigma} = \boldsymbol{\sigma}' + \mathbf{u} \tag{6.42}$$

in which the one-dimensional array **u** contains no shear terms. The stress–strain relationships can be written as

$$\boldsymbol{\sigma}' = \mathbf{D}'\boldsymbol{\varepsilon} \tag{6.43}$$

and

$$\mathbf{u} = \mathbf{D}_{\mathrm{u}}\boldsymbol{\varepsilon} \tag{6.44}$$

which combine to give

$$\boldsymbol{\sigma} = \mathbf{D}\boldsymbol{\varepsilon} \tag{6.45}$$

where

$$\mathbf{D} = \mathbf{D}' + \mathbf{D}_{\mathrm{u}} \tag{6.46}$$

The matrix $\mathbf{D}'$ is the familiar elastic stress–strain matrix in terms of effective Young's modulus (E') and Poisson's ratio (ν'). The matrix $\mathbf{D}_{\mathrm{u}}$ contains the apparent bulk modulus of the fluid K_{e} in the following locations:

$$\mathbf{D}_{\mathrm{u}} = \begin{bmatrix} K_{\mathrm{e}} & K_{\mathrm{e}} & 0 & K_{\mathrm{e}} \\ & K_{\mathrm{e}} & 0 & K_{\mathrm{e}} \\ & & 0 & 0 \\ \text{Symmetrical} & & & K_{\mathrm{e}} \end{bmatrix} \tag{6.47}$$

assuming that the third column corresponds to the shear terms in a two-dimensional analysis.

To implement this method in the programs described in this chapter, it is necessary to form the global stiffness matrix using the total stress–strain matrix **D**, while effective stresses for use in the failure function are computed from total strains using the effective stress–strain matrix $\mathbf{D}'$. Pore pressures are simply computed from

$$u = K_{\mathrm{e}}(\varepsilon_r + \varepsilon_z + \varepsilon_\theta) \tag{6.48}$$

For relatively large values of K_e, the analysis is insensitive to the exact magnitude of K_e. For axisymmetric analyses, Griffiths (1985) defined the dimensionless grouping

$$\beta = \frac{(1-2\nu')K_e}{E'} \tag{6.49}$$

and showed that results were essentially constant for $\beta \geqslant 20$. The example shown in Figure 6.20 represents a single axisymmetric eight-node element subjected to vertical compressive displacement increments along its top face. In order to compute pore pressures during undrained loading, it is necessary to

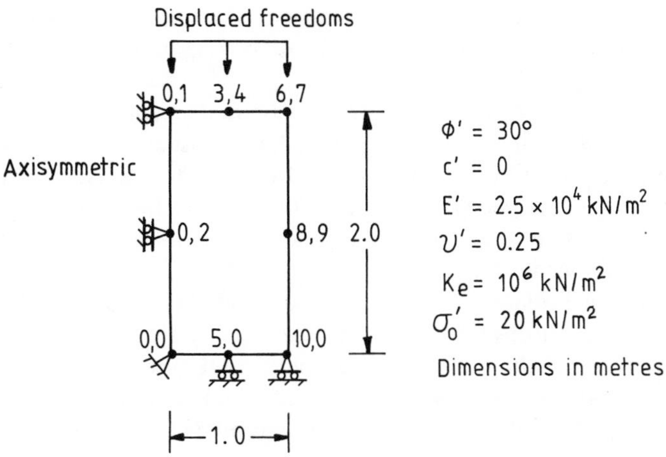

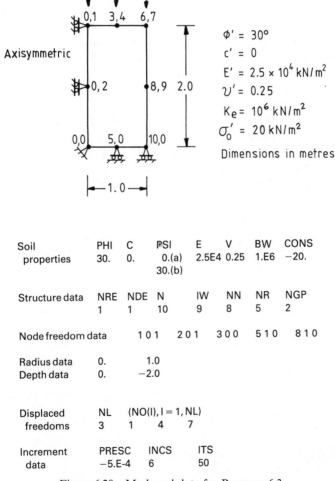

Soil properties	PHI	C	PSI	E	V	BW	CONS
	30.	0.	0.(a) 30.(b)	2.5E4	0.25	1.E6	−20.

Structure data	NRE	NDE	N	IW	NN	NR	NGP
	1	1	10	9	8	5	2

Node freedom data 1 0 1 2 0 1 3 0 0 5 1 0 8 1 0

Radius data	0.	1.0
Depth data	0.	−2.0

Displaced freedoms	NL	(NO(I), I = 1, NL)		
	3	1	4	7

Increment data	PRESC	INCS	ITS
	−5.E-4	6	50

Figure 6.20 Mesh and data for Program 6.3

update strains as well as stresses after each increment; hence the following additional three-dimensional arrays are declared:

ER, EZ, GRZ, ET	accumulates ε_r, ε_z, γ_{rz} and ε_θ at each Gauss point
PORE	pore pressure at each Gauss point

Additional input parameters are as follows:

BW	apparent pore fluid bulk modulus K_e
CONS	initial effective stress σ'_r, σ'_z, σ'_θ, $(\tau_{rz} = 0)$

The coordinates of the elements are read into one-dimensional arrays RAD and DEP following the notation established for axisymmetric problems by Program 5.6.

After the effective stress–strain matrix has been augmented by the fluid bulk modulus, the global stiffness matrix is formed in the usual way. It may be noted that in this listing the stiffness matrix is stored as a one-dimensional array (KV) in preparation for Gaussian direct methods of solution. Information about the magnitude of the prescribed displacement increments and the freedoms to which they are to be applied is read and the 'stiff-spring' technique duly implemented.

Just before the displacement increment loop begins, the routine FMDRAD is called to form the effective stress–strain matrix. The only other difference from earlier programs is that both stresses and strains are updated after each increment. The pore pressure is also computed from equation (6.48).

The data shown in Figure 6.20 is for an undrained sand with the following properties:

Effective soil properties	$\phi' = 30°$, $c' = 0$, $E' = 2.5 \times 10^4\,\text{kN/m}^2$, $\nu' = 0.25$
Bulk modulus	$K_e = 10^6\,\text{kN/m}^2$

The triaxial specimen has been consolidated under a cell pressure of $20\,\text{kN/m}^2$ before undrained loading commences.

The output of two analyses is presented in Figure 6.21. In analysis (a), $\psi = 0$ and in analysis (b), $\psi = 30°$. As expected, the inclusion of dilation has a considerable impact on the response in this 'confined' problem. The deviator stress and pore pressure versus axial strain have been plotted for both cases in Figure 6.22. Case(a), in which there is no plastic volume change ($\psi = 0$), reaches a peak deviator stress that is less than the drained failure load. This is due to the increase in pore pressure occurring in the elastic phase of compression. The closed form solution (Griffiths, 1985) for this case is seen to be in close agreement for $\beta = 20$. Case(b), on the other hand, shows no sign of failure due to the tendency for dilation. In this case, the pore pressures would continue to fall and the deviator

```
-.5000E-03      displacement
-.1755E+02   -.2502E+02   -.1755E+02  σr, σz, σθ
 .7475E+01   -.2451E+01   deviator stress and pore pressure
         2      iterations
-.1000E-02
-.1510E+02   -.3005E+02   -.1510E+02
 .1495E+02   -.4902E+01
         2
-.1500E-02
-.1265E+02   -.3507E+02   -.1265E+02
 .2243E+02   -.7353E+01
         2
-.2000E-02
-.1195E+02   -.3651E+02   -.1195E+02
 .2456E+02   -.8053E+01
         6
-.2500E-02
-.1197E+02   -.3646E+02   -.1197E+02
 .2448E+02   -.8027E+01
         7
-.3000E-02
-.1197E+02   -.3645E+02   -.1197E+02
 .2448E+02   -.8026E+01
         7
```

case (a) $\psi = 0°$

```
-.5000E-03   displacement
-.1755E+02   -.2502E+02   -.1755E+02  σr, σz, σθ
 .7475E+01   -.2451E+01   deviator stress and pore pressure
         2   iterations
-.1000E-02
-.1510E+02   -.3005E+02   -.1510E+02
 .1495E+02   -.4902E+01
         2
-.1500E-02
-.1265E+02   -.3507E+02   -.1265E+02
 .2243E+02   -.7353E+01
         2
-.2000E-02
-.1333E+02   -.4007E+02   -.1333E+02
 .2674E+02   -.5999E+01
         6
-.2500E-02
-.1498E+02   -.4505E+02   -.1498E+02
 .3007E+02   -.3466E+01
         6
-.3000E-02
-.1664E+02   -.5003E+02   -.1664E+02
 .3339E+02   -.9223E+00
         6
```

case (b) $\psi = 30°$

Figure 6.21 Results from Program 6.3

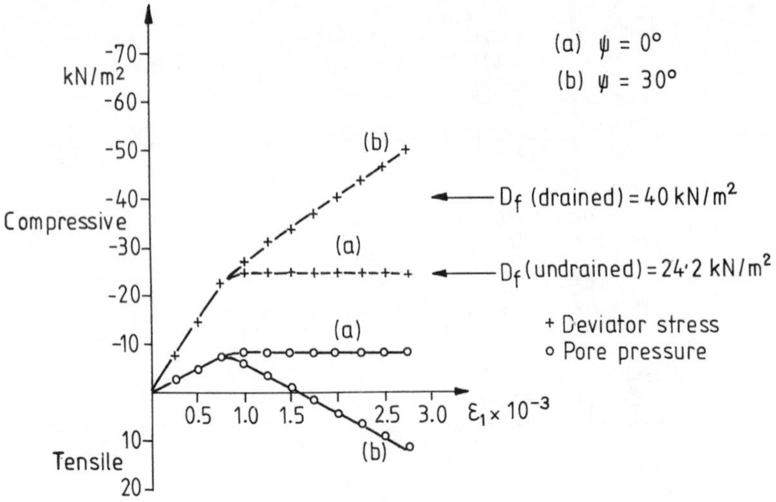

Figure 6.22 Deviator stress and pore pressure versus axial strain

stress continue to rise indefinitely unless some additional criterion was introduced. Analyses of this type are described in Griffiths (1980).

PROGRAM 6.4: THREE-DIMENSIONAL ELASTO-PLASTIC ANALYSIS—VISCO-PLASTIC METHOD USING THE MOHR–COULOMB CRITERION

```
C
C      PROGRAM 6.4 THREE-DIMENSIONAL ELASTO-PLASTIC
C      ANALYSIS USING 20-NODE BRICK ELEMENTS
C      MOHR-COULOMB'S FAILURE CRITERION
C
C
C      ALTER NEXT LINE TO CHANGE PROBLEM SIZE
C
      PARAMETER(IKV=1000,ILOADS=200,INF=100,IEL=10,
     +          IEVPT=500,INO=20)
C
      REAL DEE(6,6),SAMP(3,2),COORD(20,3),JAC(3,3),JAC1(3,3),
     +DER(3,20),DERIV(3,20),BEE(6,60),DBEE(6,60),BTDB(60,60),KM(60,60),
     +ELD(60),EPS(6),SIGMA(6),BT(60,6),FUN(20),KV(IKV),LOADS(ILOADS),
     +SX(IEL,27),SY(IEL,27),SZ(IEL,27),TXY(IEL,27),TYZ(IEL,27),
     +TZX(IEL,27),BLOAD(60),ELOAD(60),TOTD(ILOADS),OLDIS(ILOADS),
     +VAL(INO),ERATE(6),EVP(6),EVPT(IEVPT),BDYLDS(ILOADS),DEVP(6),
     +M1(6,6),M2(6,6),M3(6,6),FLOW(6,6),STRESS(6),STORKV(INO)
      INTEGER NO(INO),G(60),NF(INF,3),KDIAG(ILOADS)
      DATA IBTDB,IKM,IBT,IDOF/4*60/,IDEE,IBEE,IDBEE,IH,IFLOW/5*6/
      DATA ICOORD,NOD/2*20/,IJAC,IJAC1,IDER,IDERIV,NODOF,IT,ISAMP/7*3/
C
C       INPUT AND INITIALISATION
C
      READ(5,*)PHI,C,PSI,E,V,CONS,NXE,NYE,NZE,N,NN,NR,NGP,ITS,AA,BB,CC
      CALL READNF(NF,INF,NN,NODOF,NR)
```

```
      DO 10 I=1,N
   10 KDIAG(I)=0
      DO 20 IP=1,NXE
      DO 20 IQ=1,NYE
      DO 20 IS=1,NZE
      CALL GE203D(IP,IQ,IS,NXE,NZE,AA,BB,CC,COORD,ICOORD,G,NF,INF)
   20 CALL FKDIAG(KDIAG,G,IDOF)
      KDIAG(1)=1
      DO 30 I=2,N
   30 KDIAG(I)=KDIAG(I)+KDIAG(I-1)
      IR=KDIAG(N)
      CALL NULVEC(KV,IR)
      CALL NULVEC(TOTD,N)
      CALL NULVEC(OLDIS,N)
      CALL FORMD3(DEE,IDEE,E,V)
      CALL GAUSS(SAMP,ISAMP,NGP)
      SNPH=SIN(PHI*ACOS(-1.)/180.)
      DT=4.*(1.+V)*(1.-2.*V)/(E*(1.-2.*V+SNPH*SNPH))
C
C       ELEMENT STIFFNESS INTEGRATION AND ASSEMBLY
C
      NM=0
      DO 40 IP=1,NXE
      DO 40 IQ=1,NYE
      DO 40 IS=1,NZE
      NM=NM+1
      CALL GE203D(IP,IQ,IS,NXE,NZE,AA,BB,CC,COORD,ICOORD,G,NF,INF)
      CALL NULL(KM,IKM,IDOF,IDOF)
      IG=0
      DO 50 I=1,NGP
      DO 50 J=1,NGP
      DO 50 K=1,NGP
      IG=IG+1
      SX(NM,IG)=CONS
      SY(NM,IG)=CONS
      SZ(NM,IG)=CONS
      TXY(NM,IG)=0.
      TYZ(NM,IG)=0.
      TZX(NM,IG)=0.
      CALL FMQUA3(DER,IDER,FUN,SAMP,ISAMP,I,J,K)
      CALL MATMUL(DER,IDER,COORD,ICOORD,JAC,IJAC,IT,NOD,IT)
      CALL TREEX3(JAC,IJAC,JAC1,IJAC1,DET)
      CALL MATMUL(JAC1,IJAC1,DER,IDER,DERIV,IDERIV,IT,IT,NOD)
      CALL NULL(BEE,IBEE,IH,IDOF)
      CALL FORMB3(BEE,IBEE,DERIV,IDERIV,NOD)
      CALL MATMUL(DEE,IDEE,BEE,IBEE,DBEE,IDBEE,IH,IH,IDOF)
      CALL MATRAN(BT,IBT,BEE,IBEE,IH,IDOF)
      CALL MATMUL(BT,IBT,DBEE,IDBEE,BTDB,IBTDB,IDOF,IH,IDOF)
      QUOT=DET*SAMP(I,2)*SAMP(J,2)*SAMP(K,2)
      CALL MSMULT(BTDB,IBTDB,QUOT,IDOF,IDOF)
   50 CALL MATADD(KM,IKM,BTDB,IBTDB,IDOF,IDOF)
   40 CALL FSPARV(KV,KM,IKM,G,KDIAG,IDOF)
C
C       READ PRESCRIBED FREEDOMS
C       AUGMENT AND REDUCE STIFFNESS MATRIX
C
      READ(5,*)NL,(NO(I),I=1,NL),PRESC,INCS
      DO 60 I=1,NL
      KV(KDIAG(NO(I)))=KV(KDIAG(NO(I)))+1.E20
   60 STORKV(I)=KV(KDIAG(NO(I)))
      CALL SPARIN(KV,N,KDIAG)
C
```

```
C      DISPLACEMENT INCREMENT LOOP
C
      DO 70 IY=1,INCS
      PTOT=PRESC*IY
      ITERS=0
      CALL NULVEC(BDYLDS,N)
      CALL NULVEC(EVPT,NXE*NYE*NZE*IH*NGP*NGP*NGP)
C
C      ITERATION LOOP
C
   80 ITERS=ITERS+1
      CALL NULVEC(LOADS,N)
      DO 90 I=1,NL
   90 LOADS(NO(I))=STORKV(I)*PRESC
      CALL VECADD(LOADS,BDYLDS,LOADS,N)
      CALL SPABAC(KV,LOADS,N,KDIAG)
C
C      CHECK CONVERGENCE
C
      CALL CHECON(LOADS,OLDIS,N,.0001,ICON)
      IF(ITERS.EQ.1)ICON=0
      IF(ICON.EQ.1.OR.ITERS.EQ.ITS)CALL NULVEC(BDYLDS,N)
C
C      INSPECT ALL GAUSS POINTS
C
      NM=0
      DO 100 IP=1,NXE
      DO 100 IQ=1,NYE
      DO 100 IS=1,NZE
      NM=NM+1
      CALL NULVEC(BLOAD,IDOF)
      CALL GE203D(IP,IQ,IS,NXE,NZE,AA,BB,CC,COORD,ICOORD,G,NF,INF)
      DO 110 M=1,IDOF
      IF(G(M).EQ.0)ELD(M)=0.
  110 IF(G(M).NE.0)ELD(M)=LOADS(G(M))
      IG=0
      DO 120 I=1,NGP
      DO 120 J=1,NGP
      DO 120 K=1,NGP
      IG=IG+1
      IN=NGP*NGP*NGP*IH*(NM-1)+IH*(IG-1)
      CALL FMQUA3(DER,IDER,FUN,SAMP,ISAMP,I,J,K)
      CALL MATMUL(DER,IDER,COORD,ICOORD,JAC,IJAC,IT,NOD,IT)
      CALL TREEX3(JAC,IJAC,JAC1,IJAC1,DET)
      CALL MATMUL(JAC1,IJAC1,DER,IDER,DERIV,IDERIV,IT,IT,NOD)
      CALL NULL(BEE,IBEE,IH,IDOF)
      CALL FORMB3(BEE,IBEE,DERIV,IDERIV,NOD)
      CALL MATRAN(BT,IBT,BEE,IBEE,IH,IDOF)
      CALL MVMULT(BEE,IBEE,ELD,IH,IDOF,EPS)
      DO 130 L=1,IH
  130 EPS(L)=EPS(L)-EVPT(IN+L)
      CALL MVMULT(DEE,IDEE,EPS,IH,IH,SIGMA)
      STRESS(1)=SIGMA(1)+SX(NM,IG)
      STRESS(2)=SIGMA(2)+SY(NM,IG)
      STRESS(3)=SIGMA(3)+SZ(NM,IG)
      STRESS(4)=SIGMA(4)+TXY(NM,IG)
      STRESS(5)=SIGMA(5)+TYZ(NM,IG)
      STRESS(6)=SIGMA(6)+TZX(NM,IG)
      CALL INVAR3(STRESS,SIGM,DSBAR,THETA)
C
C      CHECK WHETHER YIELD IS VIOLATED
C
```

```
      CALL MOCOUF(PHI,C,SIGM,DSBAR,THETA,F)
      IF(ICON.EQ.1.OR.ITERS.EQ.ITS)GOTO 140
      IF(F.LT.0.)GOTO 150
      CALL MOCOUQ(PSI,DSBAR,THETA,DQ1,DQ2,DQ3)
      CALL FORMM3(STRESS,M1,M2,M3)
      DO 160 L=1,IH
      DO 160 M=1,IH
  160 FLOW(L,M)=F*(M1(L,M)*DQ1+M2(L,M)*DQ2+M3(L,M)*DQ3)
      CALL MVMULT(FLOW,IFLOW,STRESS,IH,IH,ERATE)
      DO 170 L=1,IH
      EVP(L)=ERATE(L)*DT
  170 EVPT(IN+L)=EVPT(IN+L)+EVP(L)
      CALL MVMULT(DEE,IDEE,EVP,IH,IH,DEVP)
      GOTO 180
  140 CALL VECCOP(STRESS,DEVP,IH)
  180 CALL MVMULT(BT,IBT,DEVP,IDOF,IH,ELOAD)
      QUOT=DET*SAMP(I,2)*SAMP(J,2)*SAMP(K,2)
      DO 190 L=1,IDOF
  190 BLOAD(L)=BLOAD(L)+ELOAD(L)*QUOT
  150 IF(ICON.NE.1.AND.ITERS.NE.ITS)GOTO 120
C
C      UPDATE GAUSS POINT STRESSES
C
      SX(NM,IG)=STRESS(1)
      SY(NM,IG)=STRESS(2)
      SZ(NM,IG)=STRESS(3)
      TXY(NM,IG)=STRESS(4)
      TYZ(NM,IG)=STRESS(5)
      TZX(NM,IG)=STRESS(6)
  120 CONTINUE
C
C      COMPUTE TOTAL BODYLOADS VECTOR
C
      DO 200 M=1,IDOF
      IF(G(M).EQ.0)GOTO 200
      BDYLDS(G(M))=BDYLDS(G(M))+BLOAD(M)
  200 CONTINUE
  100 CONTINUE
      IF(ICON.NE.1.AND.ITERS.NE.ITS)GOTO 80
      CALL VECADD(TOTD,LOADS,TOTD,N)
      WRITE(6,1000)PTOT
      WRITE(6,1000)SZ(1,1),SX(1,1),SY(1,1)
      WRITE(6,2000)ITERS
      IF(ITERS.EQ.ITS)GOTO 210
   70 CONTINUE
  210 CONTINUE
 1000 FORMAT(10E12.4)
 2000 FORMAT(10I12)
      STOP
      END
```

Although storage requirements increase considerably it is simple conceptually to extend any of the programs in this chapter to deal with three-dimensional strain conditions. The program described here is based on Program 5.10, and modified to account for elasto-plastic material behaviour.

The first change occurs in the PARAMETER statement which holds the new constant IEL which must be greater than or equal to the number of elements in the mesh. This constant appears in the array declarations (SX, SY, SZ, etc.) which now have only two subscripts, with the first counting elements and the second counting Gauss points up to a maximum of 27.

The program contains only two completely new subroutines, INVAR3 and FORMM3, which are the three-dimensional equivalents of INVAR and FORMM. These subroutines take in six components of stress instead of the four components required in the two-dimensional case.

As in Program 5.10, a skyline strategy is adopted for storage of the global stiffness matrix. As the length of the resulting one-dimensional array KV is not known in advance, it is suggested that, for large problems, a preliminary run of the program is made to obtain this length held in KDIAG(N). The constant IKV in the PARAMETER statement can then be initialised accordingly. After formation of the global stiffness matrix, the 'stiff-spring' method is implemented

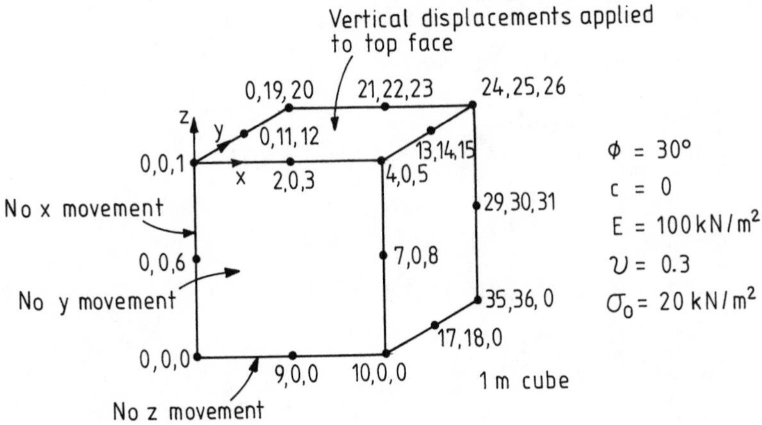

Soil properties	PHI	C	PSI	E	V	CONS		
	30.	0.	0.	100.	0.3	−20.		
Structure data	NXE	NYE	NZE	N	NN	NR	NGP	
	1	1	1	36	20	16	3	
Element data	AA	BB	CC					
	1.	1.	1.					
Node freedom data	1 0 0 1	2 1 0 1	3 1 0 1	4 0 0 1	5 1 0 1			
	6 0 0 0	7 1 0 0	8 1 0 0	9 0 1 1	11 0 1 0			
	12 1 1 0	13 0 1 1	16 0 1 1	18 0 1 0	19 1 1 0			
	20 1 1 0							
Displaced freedoms	NL	(NO(I), I = 1, NL)						
	8	1	3	5	12	15	20	23	26
Increment data	PRESC	INCS	ITS					
	−0.05	10	50					

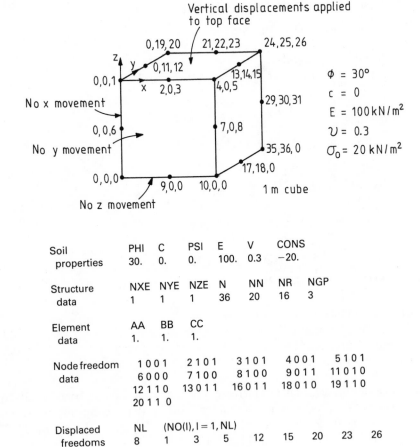

Figure 6.23 Mesh and data for Program 6.4

```
-.5000E-01   displacement
-.2500E+02   -.2000E+02   -.2000E+02 σx,σy,σz
         2   iterations
-.1000E+00
-.3000E+02   -.2000E+02   -.2000E+02
         2
-.1500E+00
-.3500E+02   -.2000E+02   -.2000E+02
         2
-.2000E+00
-.4000E+02   -.2000E+02   -.2000E+02
         2
-.2500E+00
-.4500E+02   -.2000E+02   -.2000E+02
         2
-.3000E+00
-.5000E+02   -.2000E+02   -.2000E+02
         2
-.3500E+00
-.5500E+02   -.2000E+02   -.2000E+02
         2
-.4000E+00
-.6000E+02   -.2000E+02   -.2000E+02
         2
-.4500E+00
-.6000E+02   -.2000E+02   -.2000E+02
        16
-.5000E+00
-.6000E+02   -.2000E+02   -.2000E+02
        16
```

Figure 6.24 Results from Program 6.4

and it is worth noting how the locations of the diagonal terms in KV are found using the one-dimensional array KDIAG.

The example shown in Figure 6.23 represents a cube of material subjected to prescribed displacements along its top face. The symmetry of the boundary conditions implies that the element represents one-eighth of a cube of side length 2 m. The material properties are those of a cohesionless material ($\phi = 30$, $c = 0$) and an initial isotropic stress state of 20 kN/m^2 is imposed. The output given in Figure 6.24 shows the axial displacement, the three normal stresses, σ_x, σ_y and σ_z (in this case equal to the principal stresses), and the number of iterations to reach convergence after each displacement increment.

As loads are applied axially and the cube is free to expand in the other orthogonal directions, two of the principal stresses remain constant and equal to 20 kN/m^2 throughout. As expected for a Mohr–Coulomb material with $\phi = 30$ and $c = 0$, the axial stress (σ_z or σ_1) builds up to a maximum of 60 kN/m^2 and remains constant at that value.

6.8 REFERENCES

Bishop, A. W., and Morgenstern, N. (1960) Stability coefficients for earth slopes. *Géotechnique*, **10**, No. 4, 129–50.

Cormeau, I. C. (1975) Numerical stability in quasi-static elasto-viscoplasticity. *Int. J. Num. Meth. Eng.*, **9**, No. 1, 109–27.

Duncan, J. M., and Chang, C. Y. (1970) Nonlinear analysis of stress and strain in soils. *J. Soil Mech. Found. Eng. Div., ASCE*, **96**, No. SM5, 1629–53.

Griffiths, D. V. (1980) Finite element analyses of walls, footings and slopes. Ph.D. Thesis, University of Manchester.

Griffiths, D. V. (1982) Computation of bearing capacity factors using finite elements. *Géotechnique*, **32**, No. 3, 195–202.

Griffiths, D. V. (1985) The effect of pore fluid compressibility on failure loads in elastic plastic soil. *Int. J. Num. Anal. Meth. Geomech.*, **9**, 253–9.

Griffiths, D. V., and Willson, S. M. (1986) An explicit form of the plastic matrix for Mohr–Coulomb materials. *Comm. App. Num. Meths.*, **2**.

Hill, R. (1950) *The Mathematical Theory of Plasticity*. Oxford University Press.

Nayak, G. C., and Zienkiewicz, O. C. (1972) Elasto/plastic stress analysis. A generalisation for various constitutive relationships including strain softening. *Int. J. Num. Meth. Eng.*, **5**, 113–35.

Naylor, D. J. (1974) Stresses in nearly incompressible materials by finite elements with application to the calculation of excess pore pressures. *Int. J. Num. Meth. Eng.*, **8**, 443–60.

Prevost, J.-H. (1977) Mathematical modelling of monotonic and cyclic undrained clay behaviour. *Int. J. Num. Anal. Meth. Geomech.*, **1**, 195–216.

Teng, C. K. (1981) The influence of geometric nonlinearity in geomechanics. M.Sc. Thesis, University of Manchester.

Yamada, Y., Yoshimura, N., and Sakurai, T. (1968) Plastic stress–strain matrix and its application for the solution of elastic plastic problems by the finite element method. *Int. J. Mech. Sci.*, **10**, 343–54.

Zienkiewicz, O. C., Valliappan, S., and King, I. P. (1969) Elasto-plastic solutions of engineering problems, 'initial-stress' finite element approach. *Int. J. Num. Meth. Eng.*, **1**, 75–100.

Zienkiewicz, O. C., and Cormeau, I. C. (1974) Viscoplasticity, plasticity and creep in elastic solids. A unified numerical solution approach. *Int. J. Num. Meth. Eng.*, **8**, 821–45.

Zienkiewicz. O. C., Valliappan, S., and King, I. P. (1969) Elasto-plastic solutions of associated viscoplasticity and plasticity in soil mechanics. *Geotechnique*, **25**, No. 4, 671–89.

Zienkiewicz, O. C. (1977) *The Finite Element Method*, 3rd edition. McGraw-Hill, New York.

CHAPTER 7
Steady State Flow

7.0 INTRODUCTION

The two programs presented in this chapter solve steady state problems governed by Laplace's equation (2.111). Typical examples of this type of problem include steady seepage through soils and steady heat flow through a conductor. Both programs are for two-dimensional planar conditions and use four-node quadrilateral elements. Each node has only one unknown or degree of freedom associated with it. The unknown would represent, in the case of seepage problems, the fluid potential and, in the case of conduction problems, the temperature.

Systems that are governed by Laplace's equation require boundary conditions to be prescribed at all points around a closed domain. These boundary conditions commonly take the form of fixed values of the potential or values of the first derivative of the potential normal to the boundary. The problem amounts to finding the values of the fluid potential at points within the closed domain.

Being 'elliptic' in character, solution of Laplace's equation quite closely resembles the solution of the equilibrium equations (2.54) in solid elasticity. Both methods ultimately require the solution of a set of linear simultaneous equations. The element 'stiffness' matrices are formed numerically, as described by equations (3.45) to (3.49), and assembled into a global 'stiffness' matrix which is symmetrical and banded. Taking the analogy with equilibrium analyses from Chapter 5 one step further, 'displacements' now become potentials and 'loads' now become net inflow/outflow.

Program 7.0 describes a conventional solution method for Laplace's equation over a fixed rectangular domain. Program 7.1 is rather more specialised and enables the mesh to deform iteratively until the free surface boundary conditions are satisfied.

PROGRAM 7.0: SOLUTION OF LAPLACE'S EQUATION OVER A PLANE AREA USING FOUR-NODE QUADRILATERALS

```
C
C      PROGRAM 7.0 SOLUTION OF LAPLACE'S EQUATION OVER A
C      PLANE AREA USING 4-NODE QUADRILATERALS
```

```
C
C      ALTER NEXT LINE TO CHANGE PROBLEM SIZE
C
      PARAMETER(IKV=1000,ILOADS=150,INF=50,INO=10)
C
      REAL JAC(2,2),JAC1(2,2),KAY(2,2),SAMP(3,2),DTKD(4,4),KP(4,4),
     +COORD(4,2),DER(2,4),DERIV(2,4),DERIVT(4,2),KDERIV(2,4),FUN(4),
     +VAL(INO),KVH(IKV),KV(IKV),LOADS(ILOADS),DISPS(ILOADS)
      INTEGER G(4),NO(INO),NF(INF,1)
      DATA IT,IJAC,IJAC1,IKAY,IDER,IDERIV,IKDERV/7*2/,ISAMP/3/
      DATA IDTKD,IKP,ICOORD,IDERVT,NOD/5*4/,NODOF/1/
C
C      INPUT AND INITIALISATION
C
      READ(5,*)NXE,NYE,N,IW,NN,NR,NGP,AA,BB,PERMX,PERMY
      CALL READNF(NF,INF,NN,NODOF,NR)
      IR=N*(IW+1)
      CALL NULVEC(KV,IR)
      CALL NULL(KAY,IKAY,IT,IT)
      KAY(1,1)=PERMX
      KAY(2,2)=PERMY
      CALL GAUSS(SAMP,ISAMP,NGP)
C
C      ELEMENT INTEGRATION AND ASSEMBLY
C
      DO 10 IP=1,NXE
      DO 10 IQ=1,NYE
      CALL GEO4X1(IP,IQ,NXE,AA,BB,COORD,ICOORD,G,NF,INF)
      CALL NULL(KP,IKP,NOD,NOD)
      DO 20 I=1,NGP
      DO 20 J=1,NGP
      CALL FORMLN(DER,IDER,FUN,SAMP,ISAMP,I,J)
      CALL MATMUL(DER,IDER,COORD,ICOORD,JAC,IJAC,IT,NOD,IT)
      CALL TWOBY2(JAC,IJAC,JAC1,IJAC1,DET)
      CALL MATMUL(JAC1,IJAC1,DER,IDER,DERIV,IDERIV,IT,IT,NOD)
      CALL MATMUL(KAY,IKAY,DERIV,IDERIV,KDERIV,IKDERV,IT,IT,NOD)
      CALL MATRAN(DERIVT,IDERVT,DERIV,IDERIV,IT,NOD)
      CALL MATMUL(DERIVT,IDERVT,KDERIV,IKDERV,DTKD,IDTKD,NOD,IT,NOD)
      QUOT=DET*SAMP(I,2)*SAMP(J,2)
      CALL MSMULT(DTKD,IDTKD,QUOT,NOD,NOD)
   20 CALL MATADD(KP,IKP,DTKD,IDTKD,NOD,NOD)
   10 CALL FORMKV(KV,KP,IKP,G,N,NOD)
      CALL VECCOP(KV,KVH,IR)
C
C      SPECIFY FIXED POTENTIALS AND REDUCE EQUATIONS
C
      READ(5,*)IFIX,(NO(I),VAL(I),I=1,IFIX)
      CALL NULVEC(LOADS,N)
      DO 30 I=1,IFIX
      KV(NO(I))=KV(NO(I))+1.E20
   30 LOADS(NO(I))=KV(NO(I))*VAL(I)
      CALL BANRED(KV,N,IW)
C
C      SOLVE EQUATIONS AND RETRIEVE FLOW RATE
C
      CALL BACSUB(KV,LOADS,N,IW)
      CALL PRINTV(LOADS,N)
      CALL LINMUL(KVH,LOADS,DISPS,N,IW)
      REACT=0.
      DO 40 I=1,IFIX
```

```
40 REACT=REACT+DISPS(NO(I))
   WRITE(6,'(E12.4)')REACT
   STOP
   END
```

Figure 7.1 shows a typical problem of steady seepage beneath an impermeable sheet pile wall. Laplace's equation governs the flow through the system, and the head loss across the flow domain has been normalised in terms of a fluid potential which falls by 100 units. Due to symmetry, exactly half the head will be lost by the time flow reaches the centreline and so, in this case, only the left half of the problem has been analysed. Figure 7.2 shows a square mesh of four-node elements through which steady seepage is occurring. The fluid potential is fixed at zero along the top surface and at 50 along the lower half of the right hand surface. The downstream zero potentials are obtained by 'restraining' the relevant freedoms so that they are never assembled into the global system of equations. The 'centreline' potentials of 50 units are prescribed using the 'stiff-spring' technique first described in Section 3.6. All other boundaries are impermeable and thus a 'natural' boundary condition ($(\partial\phi/\partial n)=0$) is required which is obtained automatically in the finite element formulation.

After declaration of arrays and input/initialisation, the program enters the element integration and assembly stage. The element 'stiffness' matrices require derivatives of the shape functions in global coordinates (DERIV) and these are formed by the familiar sequence of subroutine calls used in Program 5.3 and given previously by expressions (3.32). Finally, the sequence of operations and integrations described by equations (3.45) to (3.49) lead to formation of the element matrix KP. The element matrices are assembled into the global matrix KV of which a copy is held in KVH for use later when retrieving the flow rate.

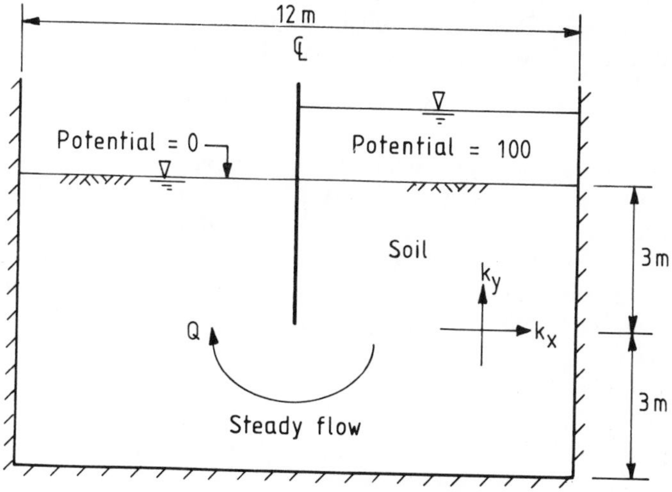

Figure 7.1 Flow under a sheet pile wall

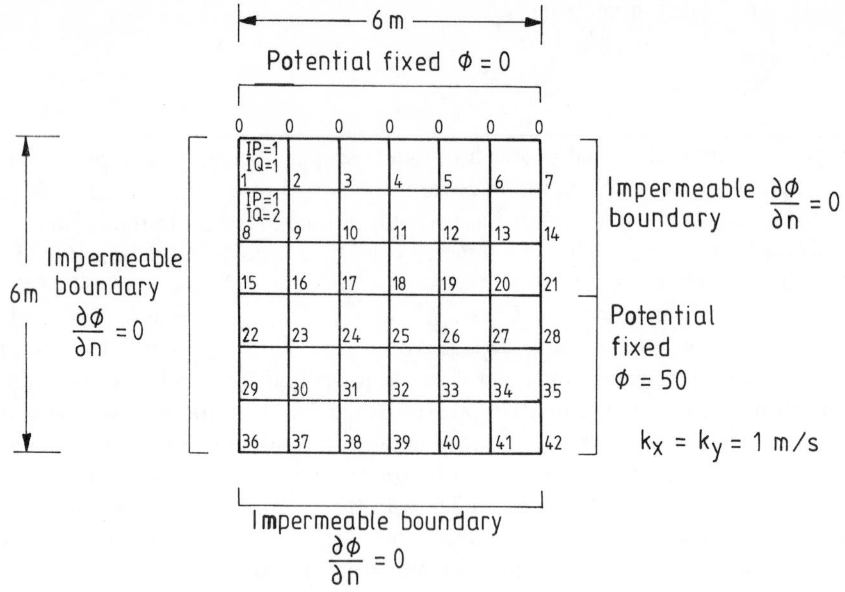

Mesh data	NXE	NYE	N	IW	NN	NR	NGP
	6	6	42	8	49	7	2

Element data	AA	BB	PERMX	PERMY
	1.	1.	1.	1.

Node freedom data: 1 0 2 0 3 0 4 0 5 0 6 0 7 0

Fixed freedom data	IFIX	(NO(I), VAL(I), I = 1, IFIX)							
	4	21	50.	28	50.	35	50.	42	50.

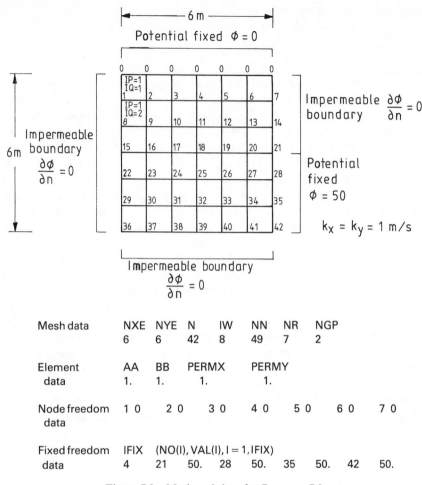

Figure 7.2 Mesh and data for Program 7.0

The program has much in common with Program 5.3 in which the use of the four-node quadrilateral was first demonstrated. Both programs assume rectangular elements of constant size with nodes and freedoms counted in the x direction. The geometry routines differ in that the one used in this program (GEO4X1) must take account of only one freedom per node in the formation of the steering vector G. Thus, the top left element (IP = 1, IQ = 1) in Figure 7.2 has a steering vector

1 0 0 2

and its neighbour (IP = 1, IQ = 2) has a steering vector

8 1 2 9

Returning to the main program, with the exception of simple integer counters, the meanings of the variable names with reference to the mesh in Figure 7.2 are given below:

NXE	number of elements in x direction (6)
NYE	number of elements in y direction (6)
N	number of degrees of freedom in mesh (42)
IW	half-bandwidth of mesh (8)
NN	number of nodes in mesh (49)
NR	number of restrained nodes in mesh (7)
NGP	number of integrating points per dimension (2)
AA	dimension of elements in x direction (1.0)
BB	dimension of elements in y direction (1.0)
PERMX	permeability in x direction (1.0)
PERMY	permeability in y direction (1.0)
IFIX	number of prescribed freedoms (4)
IDER IDERIV IJAC IJAC1 IKAY IKDERV ISAMP IDTKD IKP ICOORD IDERVT IKV ILOADS INF INO	working size of matrices, DER, DERIV, JAC, JAC1, KAY, KDERIV, SAMP, DTKD, KP, COORD, DERIVT, KV, KVH, LOADS, DISPS, NF, NO, VAL initialised using DATA and PARAMETER statements
IR	N∗(IW + 1)—working length of vectors KV and KVH
IT	dimensions of problem (2)
NOD	number of nodes per element (4)
NODOF	number of freedoms per node (1)
DET	determinant of element Jacobian matrix
QUOT	scaled weighting coefficient
REACT	sum of nodal inflow/outflow

Space is reserved for small fixed length arrays:

SAMP	quadrature abscissae and weights
JAC	Jacobian matrix
JAC1	inverse of Jacobian matrix
KAY	permeability matrix
DTKD	product $\text{DERIV}^T * \text{KAY} * \text{DERIV}$

KP	element 'stiffness' matrix
COORD	element nodal coordinates
DER	derivatives of shape functions in local coordinates
DERIV	derivatives of shape functions in global coordinates
DERIVT	transpose of DERIV
KDERIV	product KAY*DERIV
FUN	element shape functions in local coordinates
G	element steering vector

and then for variable length arrays which can be changed to alter the problem size:

KV	global 'stiffness' matrix (IKV $\geqslant$ IR)
KVH	copy of KV
LOADS	global potentials (ILOADS $\geqslant$ N)
DISPS	net inflow/outflow
NF	nodal freedom array (INF $\geqslant$ NN)
NO	prescribed freedoms (INO $\geqslant$ IFIX)
VAL	value to be fixed

Solution of the equilibrium equations by calls to BANRED and BACSUB gives the values of the nodal potentials held in LOADS. The flow rate through the system involves finding the nodal 'reactions' at those freedoms where the potential was fixed. This is achieved by multiplying the one-dimensional array LOADS by the global matrix held in KVH. This product is formed by the routine LINMUL which takes into account the storage ordering within KVH. The resulting array DISPS holds the nodal values of net inflow/outflow. The only non-zero components within this one-dimensional array correspond to the freedoms held in NO at which the potential was fixed. These values are summed and held in REACT.

The output from the program given in Figure 7.3 gives the non-zero nodal potentials (LOADS) and the flow rate through the system (REACT). As expected, the potentials all lie in the range 0 to 50 and gradually decrease towards the downstream boundary. The flow through the system is computed to be 48.57 $m^3/(s\,m)$ which is in good agreement with other methods of solution. For example, using the method of fragments (Griffiths, 1984) the flow rate for this problem is estimated to be around 47 $m^3/(s\,m)$.

```
.5716E+01   .5921E+01   .6553E+01   .7651E+01   .9225E+01   .1068E+02 )
.1136E+02   .1103E+02   .1143E+02   .1264E+02   .1479E+02   .1800E+02 |
.2269E+02   .2415E+02   .1558E+02   .1613E+02   .1783E+02   .2085E+02 |  nodal
.2554E+02   .3256E+02   .5000E+02   .1905E+02   .1972E+02   .2176E+02 >  potential
.2536E+02   .3084E+02   .3925E+02   .5000E+02   .2122E+02   .2196E+02 |
.2420E+02   .2807E+02   .3378E+02   .4130E+02   .5000E+02   .2196E+02 |
.2272E+02   .2502E+02   .2897E+02   .3464E+02   .4185E+02   .5000E+02 )
.4857E+02  Flow rate
```

Figure 7.3 Results from Program 7.0

PROGRAM 7.1: ANALYSIS OF PLANAR FREE SURFACE FLOW USING FOUR-NODE QUADRILATERALS

```
C
C      PROGRAM 7.1 SOLUTION OF LAPLACE'S EQUATION FOR
C      PLANE FREE-SURFACE FLOW USING 4-NODED QUADRILATERALS
C
C      ALTER NEXT LINE TO CHANGE PROBLEM SIZE
C
      PARAMETER(IKV=1000,ILOADS=150,INF=70,INO=12,INX=10)
C
      REAL JAC(2,2),JAC1(2,2),KAY(2,2),SAMP(3,2),DTKD(4,4),KP(4,4),
     +COORD(4,2),DER(2,4),DERIV(2,4),DERIVT(4,2),KDERIV(2,4),FUN(4),
     +VAL(INO),KVH(IKV),KV(IKV),LOADS(ILOADS),DISPS(ILOADS),
     +OLDPOT(ILOADS),WIDTH(INX),SURF(INX)
      INTEGER G(4),NO(INO),NF(INF,1)
      DATA IT,IJAC,IJAC1,IKAY,IDER,IDERIV,IKDERV/7*2/,ISAMP/3/
      DATA IDTKD,IKP,ICOORD,IDERVT,NOD/5*4/,NODOF/1/
C
C      INPUT AND INITIALISATION
C
      READ(5,*)NXE,NYE,N,IW,NN,NR,NGP,ITS,PERMX,PERMY
      READ(5,*)(WIDTH(I),I=1,NXE+1)
      READ(5,*)(SURF(I),I=1,NXE+1)
      CALL READNF(NF,INF,NN,NODOF,NR)
      IR=N*(IW+1)
      CALL NULVEC(OLDPOT,N)
      CALL NULL(KAY,IKAY,IT,IT)
      KAY(1,1)=PERMX
      KAY(2,2)=PERMY
      CALL GAUSS(SAMP,ISAMP,NGP)
C
C      ITERATE FOR POSITION OF FREESURFACE
C
      ITERS=0
   10 ITERS=ITERS+1
      CALL NULVEC(KV,IR)
C
C      ELEMENT INTEGRATION AND ASSEMBLY
C
      DO 20 IP=1,NXE
      DO 20 IQ=1,NYE
      CALL WELGEO(IP,IQ,NXE,NYE,WIDTH,SURF,COORD,ICOORD,G,NF,INF)
      CALL NULL(KP,IKP,NOD,NOD)
      DO 30 I=1,NGP
      DO 30 J=1,NGP
      CALL FORMLN(DER,IDER,FUN,SAMP,ISAMP,I,J)
      CALL MATMUL(DER,IDER,COORD,ICOORD,JAC,IJAC,IT,NOD,IT)
      CALL TWOBY2(JAC,IJAC,JAC1,IJAC1,DET)
      CALL MATMUL(JAC1,IJAC1,DER,IDER,DERIV,IDERIV,IT,IT,NOD)
      CALL MATMUL(KAY,IKAY,DERIV,IDERIV,KDERIV,IKDERV,IT,IT,NOD)
      CALL MATRAN(DERIVT,IDERVT,DERIV,IDERIV,IT,NOD)
      CALL MATMUL(DERIVT,IDERVT,KDERIV,IKDERV,DTKD,IDTKD,NOD,IT,NOD)
      QUOT=DET*SAMP(I,2)*SAMP(J,2)
      CALL MSMULT(DTKD,IDTKD,QUOT,NOD,NOD)
   30 CALL MATADD(KP,IKP,DTKD,IDTKD,NOD,NOD)
   20 CALL FORMKV(KV,KP,IKP,G,N,NOD)
      CALL VECCOP(KV,KVH,IR)
C
C      SPECIFY FIXED POTENTIALS AND REDUCE EQUATIONS
```

```
C
      IF(ITERS.EQ.1)READ(5,*)IFIX,(NO(I),I=1,IFIX)
      CALL NULVEC(LOADS,N)
      DO 40 I=1,IFIX
      KV(NO(I))=KV(NO(I))+1.E20
   40 LOADS(NO(I))=KV(NO(I))*SURF(NXE+1)
      DO 50 IQ=1,NYE-1
      J=IQ*(NXE+1)+1
   50 LOADS(J)=KV(J)*(NYE-IQ)*SURF(1)/NYE
      CALL BANRED(KV,N,IW)
C
C       SOLVE EQUATIONS
C
      CALL BACSUB(KV,LOADS,N,IW)
      CALL VECCOP(LOADS,SURF,NXE)
C
C       CHECK CONVERGENCE
C
      CALL CHECON(LOADS,OLDPOT,N,0.001,ICON)
      IF(ITERS.NE.ITS.AND.ICON.EQ.0)GOTO 10
      CALL LINMUL(KVH,LOADS,DISPS,N,IW)
      REACT=0.
      DO 60 I=1,NYE
   60 REACT=REACT+DISPS(I*(NXE+1))
      REACT=REACT+DISPS(N)
      CALL PRINTV(LOADS,N)
      WRITE(6,'(E12.4)')REACT
      WRITE(6,'(I10)')ITERS
      STOP
      END
```

In this program we consider a boundary condition frequently met in geomechanics in relation to flow of water through dams. Free surface problems involve an upper boundary that is not known *a priori* and so an iterative procedure is required to find it. This iteration can be done in various ways; for example, a fixed mesh can be used and nodes separated into 'active' and 'inactive' ones depending upon whether fluid exists at that point. An alternative strategy is used in the present program, whereby the mesh is deformed so that its upper surface ultimately coincides with the free surface; a summary of the boundary conditions required is given in Figure 7.4.

The analysis starts by assuming an initial position for the free surface. Solution of Laplace's equation gives values of the fluid potential, which will not in general equal the elevations of the upper surface. The coordinates of the nodes along the upper surface are therefore adjusted to equal the potential values that have just been calculated. In order to avoid the occurrence of distorted elements, the geometry routine WELGEO ensures that the y coordinates of nodes on any vertical line within the mesh are equally spaced. As the geometry of the finite element mesh has changed, it is necessary to form the element matrices again and assemble them into a new global matrix. Imposition of boundary conditions and solution of the system of equations leads to a new set of potentials and a further updating of the mesh coordinates. This process is repeated until the change in computed potential values from one iteration to the next is smaller than a tolerance value held in the parameter list of the routine CHECON.

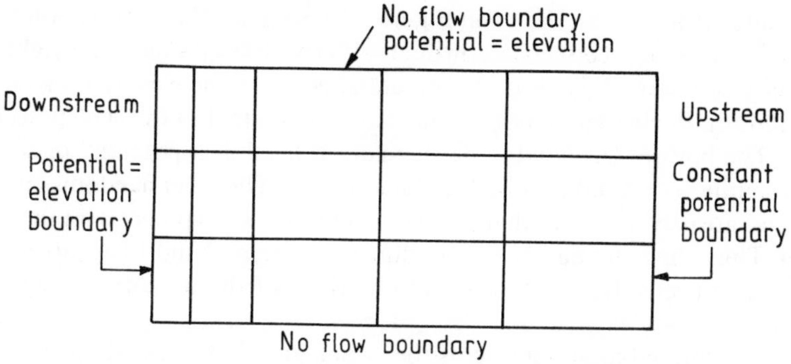

Figure 7.4 Boundary conditions for free surface flow

Referring to the program, the new integer variable ITS is read in and represents the maximum number of iterations that is permitted. Iterations are counted by the variable ITERS. The new arrays

WIDTH the x coordinates of the nodes
SURF the y coordinates of the upper mesh boundary
OLDPOT potentials at the previous iteration used to check convergence

are declared at the top of the main program, and the PARAMETER statement includes the integer $INX \geqslant NXE + 1$.

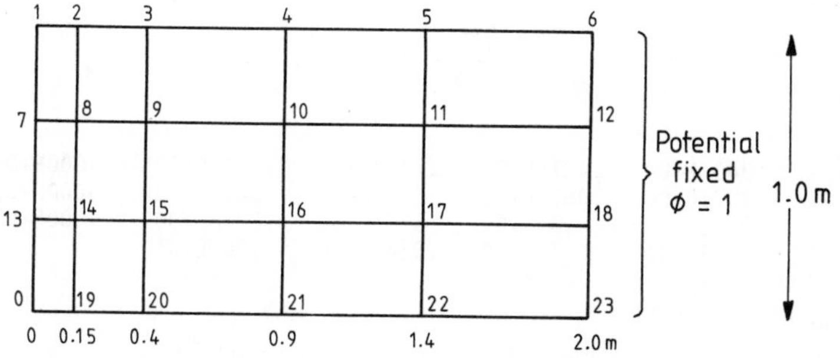

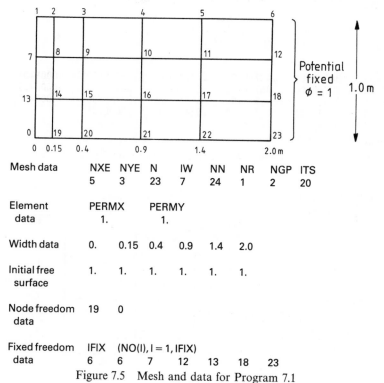

Mesh data	NXE	NYE	N	IW	NN	NR	NGP	ITS
	5	3	23	7	24	1	2	20
Element data	PERMX		PERMY					
	1.		1.					
Width data	0.	0.15	0.4	0.9	1.4	2.0		
Initial free surface	1.	1.	1.	1.	1.	1.		
Node freedom data	19	0						
Fixed freedom data	IFIX	(NO(I), I = 1, IFIX)						
	6	6	7	12	13	18	23	

Figure 7.5 Mesh and data for Program 7.1

The undeformed mesh shown in Figure 7.5 represents the starting point for an analysis of free surface seepage through a vertical faced dam. The right hand boundary of the mesh is fixed at a potential equal to the elevation of 1 m. The potentials on the left hand face are fixed at values equal to the elevations of the nodes. The bottom left hand node therefore is fixed at a potential of zero and consequently not assembled into the global system. The other freedoms on the left hand face are fixed at values that depend on the computed potential at the top left corner. Thus, the potentials corresponding to freedoms 7 and 13 are always fixed to equal, respectively, two-thirds and one-third of the potential computed at freedom 1.

The undeformed mesh represents the initial 'guess' as to the shape of the free surface, and these elevations of the top boundary are read into the one-dimensional array SURF. At each iteration, the computed values of the potentials along the top boundary are used as the new coordinates of this boundary for the next iteration. At convergence, the upper boundary of the mesh coincides with the free surface position.

The output for the present analysis is given in Figure 7.6 and gives the converged nodal potentials (LOADS), the flow rate through the system (REACT) and the number of iterations required (ITERS). The final shape of the deformed mesh is given in Figure 7.7 and the computed flow rate through the mesh is in perfect agreement with Dupuit's solution for this problem:

$$Q = \frac{kH^2}{2D} = 0.25\,\mathrm{m^3/(s\,m)} \tag{7.1}$$

```
 .1839E+00   .3407E+00   .4879E+00   .7012E+00   .8593E+00   .1000E+01 )
 .1232E+00   .2971E+00   .4573E+00   .6765E+00   .8403E+00   .1000E+01 } nodal
 .6159E-01   .2536E+00   .4337E+00   .6601E+00   .8286E+00   .1000E+01 ( potentials
 .2389E+00   .4236E+00   .6544E+00   .8246E+00   .1000E+01             )
.2500E+00 - Flow rate
    11    - No. of iterations
```

Figure 7.6 Results from Program 7.1

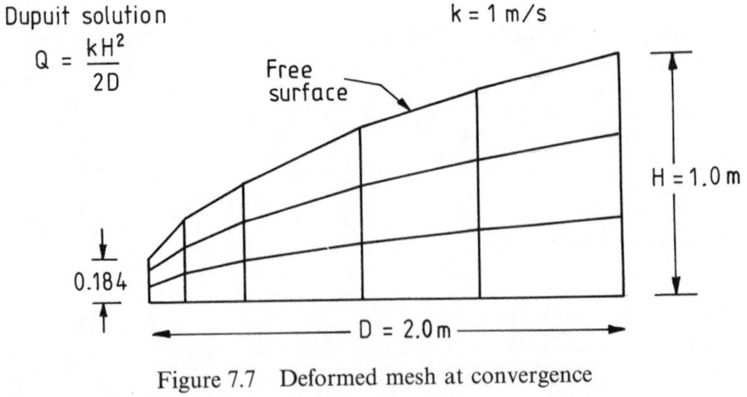

Figure 7.7 Deformed mesh at convergence

where

$$k = \text{permeability } (1.0\,\text{m/s})$$
$$H = \text{head loss } (1.0\,\text{m})$$
$$D = \text{width of dam } (2.0\,\text{m})$$

7.1 REFERENCE

Griffiths, D. V. (1984) Rationalised charts for the method of fragments applied to confined seepage. *Géotechnique*, **34**, No. 2, 229–38.

CHAPTER 8

Transient Problems: First Order (Uncoupled)

8.0 INTRODUCTION

In the previous chapter, programs for the solution of steady state potential flow problems were described. Typically, Laplace's equation (2.111) was discretised into an equilibrium equation (2.112) involving the solution of a set of simultaneous equations. For well-posed problems there are usually no associated numerical difficulties.

When a flow process is transient, or time dependent, the simplest extension of equation (2.111), or reduction of the Navier–Stokes equations, is provided by equations like (2.119). There is still a single dependent variable (for example potential), and so the analysis is 'uncoupled'. After discretisation in space, a typical element equation is given by equation (2.120). This is a set of first order ordinary differential equations, the solution of which is no longer a simple numerical task for large numbers of assembled elements.

Some of the many solution techniques available were described in Chapter 3. Possibly the simplest, and most robust, are the 'implicit' methods described by equation (3.94) and by the structure chart in Figure 3.18. These θ methods form the basis of Program 8.0.

PROGRAM 8.0: SOLUTION OF THE CONDUCTION EQUATION OVER A RECTANGULAR AREA

```
C
C      PROGRAM 8.0 SOLUTION OF THE CONDUCTION EQUATION
C      OVER A RECTANGULAR AREA USING 4-NODED QUADRILATERALS
C      IMPLICIT INTEGRATION IN TIME BY 'THETA' METHOD
C
C      ALTER NEXT LINE TO CHANGE PROBLEM SIZE
C
      PARAMETER(IBK=1000,ILOADS=150,INF=70)
C
      REAL JAC(2,2),JAC1(2,2),KAY(2,2),SAMP(3,2),FTF(4,4),DTKD(4,4),
     +COORD(4,2),DER(2,4),DERIV(2,4),DERIVT(4,2),KDERIV(2,4),FUN(4),
     +PM(4,4),KP(4,4),NEWLO(ILOADS),LOADS(ILOADS),BP(IBK),BK(IBK)
      INTEGER G(4),NF(INF,1)
```

```
      DATA IT,IJAC,IJAC1,IKAY,IDER,IDERIV,IKDERV/7*2/,ISAMP/3/
      DATA IDTKD,IKP,ICOORD,IDERVT,NOD,IFTF,IPM/7*4/,NODOF/1/
C
C      INPUT AND INITIALISATION
C
      READ(5,*)NXE,NYE,N,IW,NN,NR,NGP,AA,BB,PERMX,PERMY
      READ(5,*)DTIM,ISTEP,THETA,NPRI,NRES
      CALL READNF(NF,INF,NN,NODOF,NR)
      IR=N*(IW+1)
      CALL NULVEC(BK,IR)
      CALL NULVEC(BP,IR)
      CALL NULL(KAY,IKAY,IT,IT)
      KAY(1,1)=PERMX
      KAY(2,2)=PERMY
      CALL GAUSS(SAMP,ISAMP,NGP)
C
C      ELEMENT INTEGRATION AND ASSEMBLY
C
      DO 10 IP=1,NXE
      DO 10 IQ=1,NYE
      CALL GEO4X1(IP,IQ,NXE,AA,BB,COORD,ICOORD,G,NF,INF)
      CALL NULL(KP,IKP,NOD,NOD)
      CALL NULL(PM,IPM,NOD,NOD)
      DO 20 I=1,NGP
      DO 20 J=1,NGP
      CALL FORMLN(DER,IDER,FUN,SAMP,ISAMP,I,J)
      CALL MATMUL(DER,IDER,COORD,ICOORD,JAC,IJAC,IT,NOD,IT)
      CALL TWOBY2(JAC,IJAC,JAC1,IJAC1,DET)
      CALL MATMUL(JAC1,IJAC1,DER,IDER,DERIV,IDERIV,IT,IT,NOD)
      CALL MATMUL(KAY,IKAY,DERIV,IDERIV,KDERIV,IKDERV,IT,IT,NOD)
      CALL MATRAN(DERIVT,IDERVT,DERIV,IDERIV,IT,NOD)
      CALL MATMUL(DERIVT,IDERVT,KDERIV,IKDERV,DTKD,IDTKD,NOD,IT,NOD)
      QUOT=DET*SAMP(I,2)*SAMP(J,2)
      DO 25 K=1,NOD
      DO 25 L=1,NOD
      FTF(K,L)=FUN(K)*FUN(L)*QUOT
   25 DTKD(K,L)=DTKD(K,L)*QUOT*THETA*DTIM
      CALL MATADD(PM,IPM,FTF,IFTF,NOD,NOD)
   20 CALL MATADD(KP,IKP,DTKD,IDTKD,NOD,NOD)
      CALL FORMKV(BP,PM,IPM,G,N,NOD)
   10 CALL FORMKV(BK,KP,IKP,G,N,NOD)
C
C      REDUCTION OF LEFT HAND SIDE
C
      DO 40 I=1,IR
      BP(I)=BP(I)+BK(I)
   40 BK(I)=BP(I)-BK(I)/THETA
      CALL BANRED(BP,N,IW)
      READ(5,*)VALO
      DO 45 I=1,N
   45 LOADS(I)=VALO
C
C      TIME STEPPING RECURSION
C
      DO 50 J=1,ISTEP
      TIME=J*DTIM
      CALL LINMUL(BK,LOADS,NEWLO,N,IW)
      CALL BACSUB(BP,NEWLO,N,IW)
      DO 60 I=1,N
   60 LOADS(I)=NEWLO(I)
      IF(J/NPRI*NPRI.EQ.J)WRITE(6,'(2E12.4)')TIME,LOADS(NRES)
   50 CONTINUE
      STOP
      END
```

In the absence of sources or sinks, equation (3.94) reduces to

$$(\mathbf{PM} + \theta\,\Delta t\,\mathbf{KP})\mathbf{\Phi}_1 = [\mathbf{PM} - (1-\theta)\Delta t\,\mathbf{KP}]\mathbf{\Phi}_0 \tag{8.1}$$

or

$$(\mathbf{PM} + \theta\,\Delta t\,\mathbf{KP})\mathbf{\Phi}_1 = (\mathbf{PM} - \Delta t\,\mathbf{KP} + \theta\,\Delta t\,\mathbf{KP})\mathbf{\Phi}_0 \tag{8.2}$$

which is the form used in the present program. The element **KP** matrices are assembled, including the multiple $\theta\,\Delta t$ (THETA*DTIM in program terminology), into a global matrix **BK** and the **PM** matrices are assembled into the

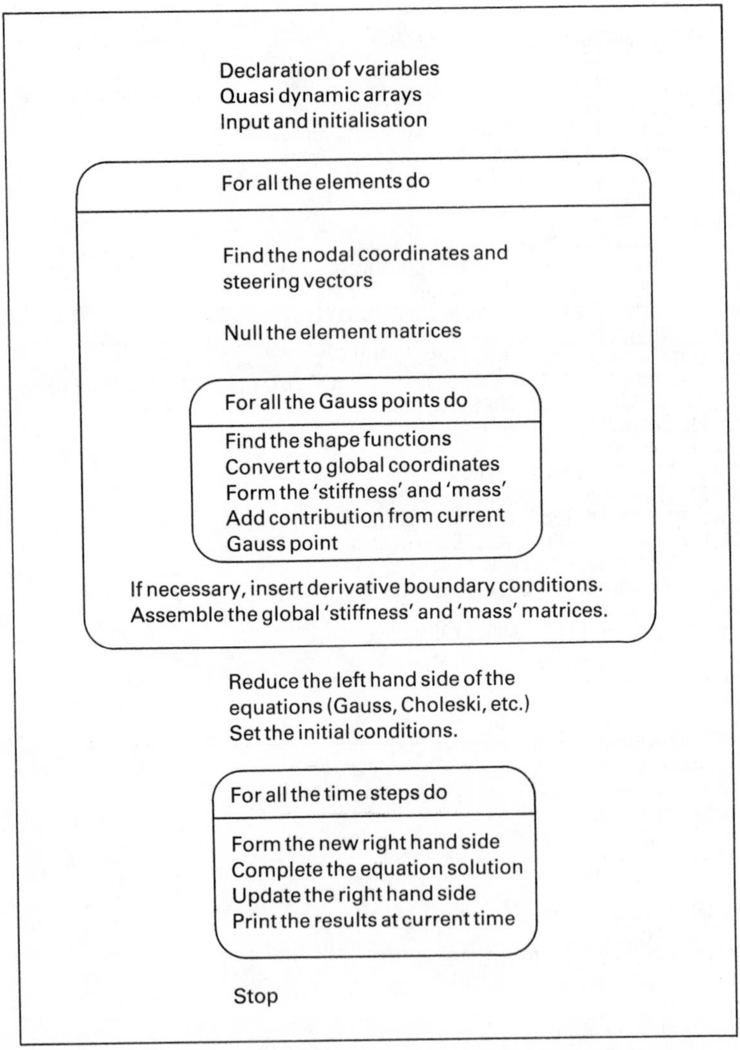

Figure 8.1 Structure chart for implicit analyses of transient problems

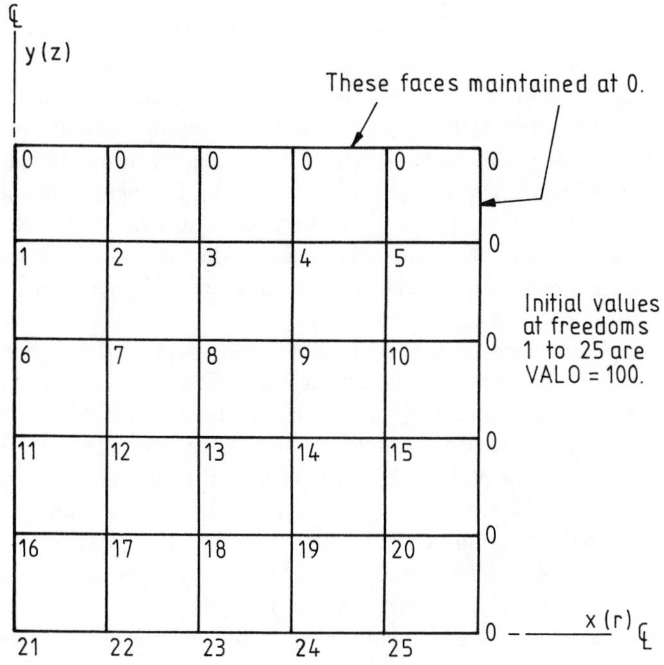

Mesh data	NXE	NYE	N	IW	NN	NR	NGP
	5	5	25	6	36	11	3

Element data	AA	BB	PERMX	PERMY
	0.2	0.2	1.0	1.0

Time integration	DTIM	ISTEP	THETA
data	0.01	150	0.5

Printing control	NPRI	NRES
data	10	21

Node freedom	1	0	2	0	3	0	4	0	5	0
data	6	0	12	0	18	0	24	0	30	0
	36	0								

Initial value	VALO
data	100.0

Note: Omit IW and THETA in Program 8.2.
Omit IW in Program 8.3.

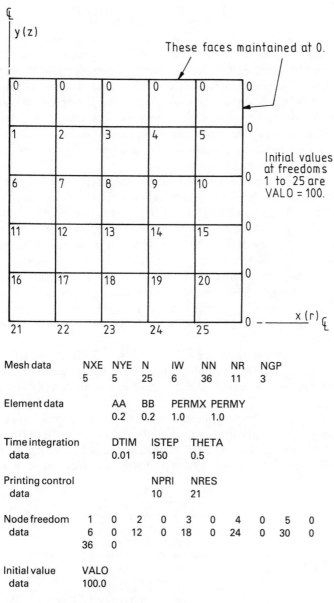

Figure 8.2 Mesh and data for Programs 8.0 to 8.3

global matrix **BP**. Thus, the global system to be solved is

$$(\mathbf{BP} + \mathbf{BK})\boldsymbol{\Phi}_1 = \left(\mathbf{BP} + \mathbf{BK} - \frac{1}{\theta}\mathbf{BK}\right)\boldsymbol{\Phi}_0 \tag{8.3}$$

with **BP** and **BK** being stored as vectors using the assembly routine FORMKV. For constant θ and Δt the left hand side of (8.3) is constant and so the strategy will be to form **BP** + **BK**, factorise the resulting matrix once only and then for each Δt to carry out the matrix-by-vector multiplication on the right hand side of (8.3) followed by a back-substitution. The process is described in detail by the structure chart in Figure 8.1. Note, however, that the matrix-by-vector multiplication on the right hand side could be done using element-by-element summation, avoiding storage of one large matrix.

The example chosen is shown in Figure 8.2 and could represent dissipation of excess porewater pressure from a rectangle of soil. Two boundaries have zero pressure while the remainder of the soil, represented by freedoms 1 to 25, has constant initial pressure VALO. The problem is to compute the pressures at freedoms 1 to 25 as time passes. The early part of the program, involving element integration and assembly, closely resembles Program 7.0. Since four-noded elements are used, with numbering in the x direction, the appropriate geometry subroutine is GEO4X1. The elements have coefficients of consolidation in the x and y directions, denoted by PERMX and PERMY (assumed constant for all elements). At the end of the element assembly **BP** and **BK**, as required by (8.3), have been stored.

In the section headed 'reduction of left hand side', **BP** is reset to **BP** + **BK**, and **BK** is reset to (new) **BP** $-$ $(1/\theta)$**BK**. The matrix **BP** is then factorised using BANRED and the non-zero initial pressures set to VALO (in this case 100.0 units).

The final section of the program consists of the time-stepping loop completed ISTEP times. The matrix-by-vector multiplication is carried out by LINMUL and back-substitution by BACSUB. The pressure at freedom 21 (NRES), the centre of the mesh, is output every NPRI (10 in this case) time steps, that is at time

T = TIME	Pressure at node 21
.1000E+00	.9009E+02
.2000E+00	.5845E+02
.3000E+00	.3580E+02
.4000E+00	.2178E+02
.5000E+00	.1324E+02
.6000E+00	.8051E+01
.7000E+00	.4895E+01
.8000E+00	.2976E+01
.9000E+00	.1809E+01
.1000E+01	.1100E+01
.1100E+01	.6687E+00
.1200E+01	.4065E+00
.1300E+01	.2472E+00
.1400E+01	.1503E+00
.1500E+01	.9135E-01

Figure 8.3 Results from Program 8.0

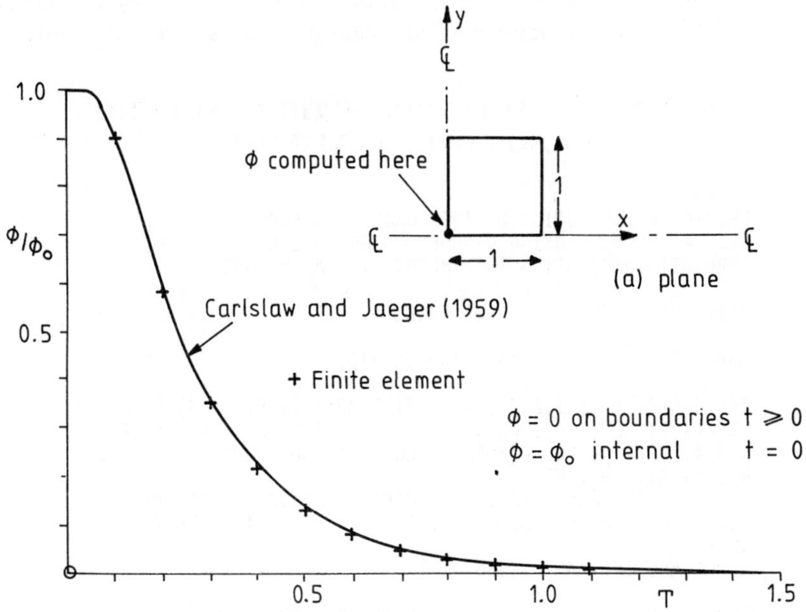

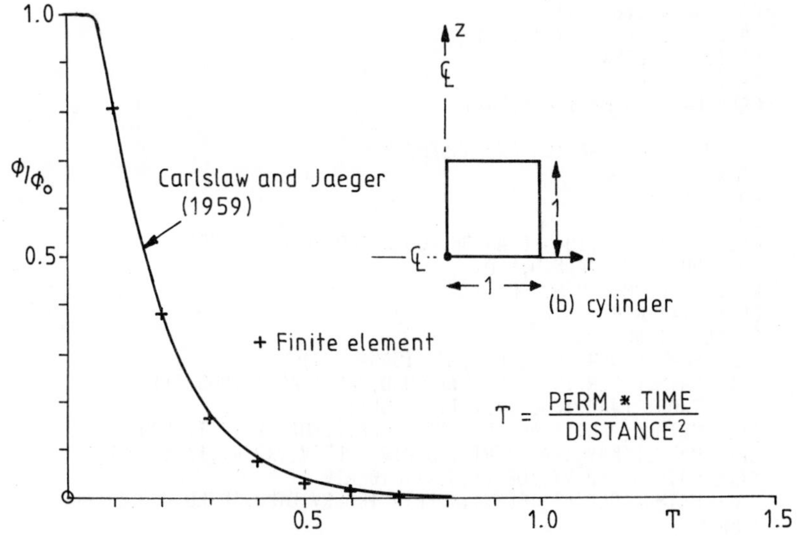

Figure 8.4 Comparisons of finite element results with series solutions

intervals of 0.1, and is listed in Figure 8.3. The results are plotted in Figure 8.4(a), where they are compared with series solution values obtained by Carslaw and Jaeger (1959). The crude finite element idealisation gives excellent results.

PROGRAM 8.1: SOLUTION OF THE CONDUCTION EQUATION OVER A CYLINDER

```
C
C      PROGRAM 8.1 SOLUTION OF THE CONDUCTION EQUATION
C      OVER A CYLINDER USING 4-NODED QUADRILATERALS
C      IMPLICIT INTEGRATION IN TIME BY 'THETA' METHOD
C
C      ALTER NEXT LINE TO CHANGE PROBLEM SIZE
C
      PARAMETER(IBK=1000,ILOADS=150,INF=70)
C
      REAL JAC(2,2),JAC1(2,2),KAY(2,2),SAMP(3,2),FTF(4,4),DTKD(4,4),
     +COORD(4,2),DER(2,4),DERIV(2,4),DERIVT(4,2),KDERIV(2,4),FUN(4),
     +PM(4,4),KP(4,4),NEWLO(ILOADS),LOADS(ILOADS),BP(IBK),BK(IBK)
      INTEGER G(4),NF(INF,1)
      DATA IT,IJAC,IJAC1,IKAY,IDER,IDERIV,IKDERV/7*2/,ISAMP/3/
      DATA IDTKD,IKP,ICOORD,IDERVT,NOD,IFTF,IPM/7*4/,NODOF/1/
C
C      INPUT AND INITIALISATION
C
      READ(5,*)NXE,NYE,N,IW,NN,NR,NGP,AA,BB,PERMX,PERMY
      READ(5,*)DTIM,ISTEP,THETA,NPRI,NRES
      CALL READNF(NF,INF,NN,NODOF,NR)
      IR=N*(IW+1)
      CALL NULVEC(BK,IR)
      CALL NULVEC(BP,IR)
      CALL NULL(KAY,IKAY,IT,IT)
      KAY(1,1)=PERMX
      KAY(2,2)=PERMY
      CALL GAUSS(SAMP,ISAMP,NGP)
C
C      ELEMENT INTEGRATION AND ASSEMBLY
C
      DO 10 IP=1,NXE
      DO 10 IQ=1,NYE
      CALL GEO4X1(IP,IQ,NXE,AA,BB,COORD,ICOORD,G,NF,INF)
      CALL NULL(KP,IKP,NOD,NOD)
      CALL NULL(PM,IPM,NOD,NOD)
      DO 20 I=1,NGP
      DO 20 J=1,NGP
      CALL FORMLN(DER,IDER,FUN,SAMP,ISAMP,I,J)
      CALL MATMUL(DER,IDER,COORD,ICOORD,JAC,IJAC,IT,NOD,IT)
      CALL TWOBY2(JAC,IJAC,JAC1,IJAC1,DET)
      CALL MATMUL(JAC1,IJAC1,DER,IDER,DERIV,IDERIV,IT,IT,NOD)
      CALL MATMUL(KAY,IKAY,DERIV,IDERIV,KDERIV,IKDERV,IT,IT,NOD)
      CALL MATRAN(DERIVT,IDERVT,DERIV,IDERIV,IT,NOD)
      CALL MATMUL(DERIVT,IDERVT,KDERIV,IKDERV,DTKD,IDTKD,NOD,IT,NOD)
      SUM=0.
      DO 30 K=1,NOD
   30 SUM=SUM+FUN(K)*COORD(K,1)
      QUOT=SUM*DET*SAMP(I,2)*SAMP(J,2)
      DO 40 K=1,NOD
      DO 40 L=1,NOD
```

```
      FTF(K,L)=FUN(K)*FUN(L)*QUOT
   40 DTKD(K,L)=DTKD(K,L)*QUOT*THETA*DTIM
      CALL MATADD(PM,IPM,FTF,IFTF,NOD,NOD)
   20 CALL MATADD(KP,IKP,DTKD,IDTKD,NOD,NOD)
      CALL FORMKV(BP,PM,IPM,G,N,NOD)
   10 CALL FORMKV(BK,KP,IKP,G,N,NOD)
C
C      REDUCTION OF LEFT HAND SIDE
C
      DO 50 I=1,IR
      BP(I)=BP(I)+BK(I)
   50 BK(I)=BP(I)-BK(I)/THETA
      CALL BANRED(BP,N,IW)
      READ(5,*)VALO
      DO 60 I=1,N
   60 LOADS(I)=VALO
C
C      TIME STEPPING RECURSION
C
      DO 70 J=1,ISTEP
      TIME=J*DTIM
      CALL LINMUL(BK,LOADS,NEWLO,N,IW)
      CALL BACSUB(BP,NEWLO,N,IW)
      CALL VECCOP(NEWLO,LOADS,N)
      IF(J/NPRI*NPRI.EQ.J)WRITE(6,'(2E12.4)')TIME,LOADS(NRES)
   70 CONTINUE
      STOP
      END
```

In the same way as a few modifications to Program 5.3 for plane elasticity led to Program 5.6 for axisymmetric conditions, Program 8.0 is readily modified to cylindrical coordinates. The coordinate x is associated with the radial coordinate r while y is associated with the axial coordinate z.

The mesh in Figure 8.2 now represents a symmetrical quarter of a cross-section through a right cylinder. In soil mechanics, the physical analogue would be a 'triaxial' specimen of soil draining from all its boundaries. However, the data are identical to those for the plane case. Integration in the circumferential direction is

T = TIME	Pressure at Node 21
.1000E+00	.8096E+02
.2000E+00	.3798E+02
.3000E+00	.1667E+02
.4000E+00	.7280E+01
.5000E+00	.3176E+01
.6000E+00	.1386E+01
.7000E+00	.6046E+00
.8000E+00	.2638E+00
.9000E+00	.1151E+00
.1000E+01	.5022E-01
.1100E+01	.2191E-01
.1200E+01	.9560E-02
.1300E+01	.4171E-02
.1400E+01	.1820E-02
.1500E+01	.7940E-03

Figure 8.5 Results from Program 8.1

carried out over 1 radian and the only additional code needed is to form SUM which is the radius of the current Gauss point (contained in the axisymmetric integrals, e.g. equation 3.42).

Program 8.1 differs from Program 8.0 only in the four lines from SUM = 0 onwards.

The results are listed in Figure 8.5 and compared with series solutions in Figure 8.4(b). The agreement is rather poorer than for the plane case due to the approximate integration as r tends to zero. However, the results are perfectly acceptable. Note that a 3×3 point Gaussian integration was used. It would have been sufficient in Program 8.0 to use 2×2.

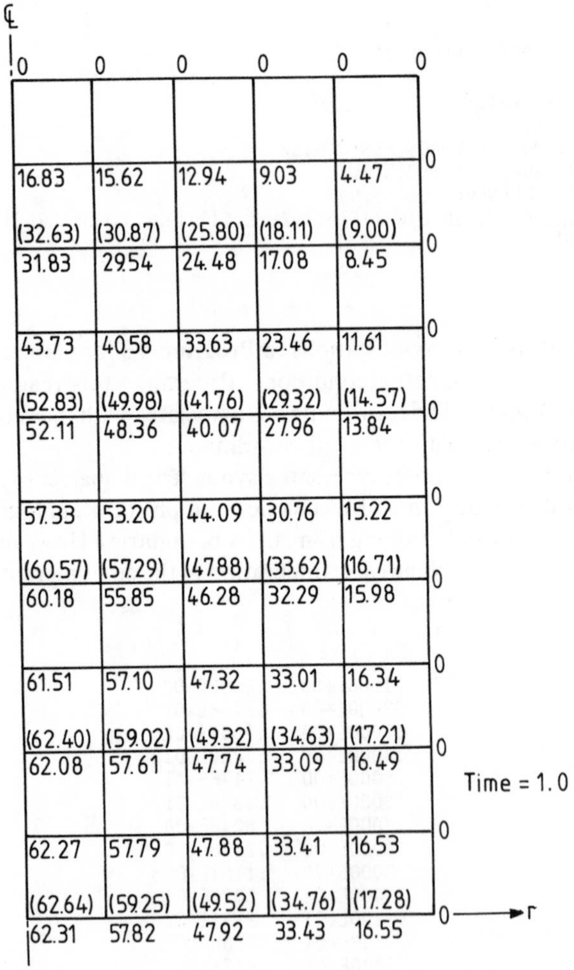

Figures in parentheses are series solution values (Carlslaw and Jaeger, 1959)

Figure 8.6 Results from triaxial consolidation analysis

To give an idea of the accuracy obtained for points other than the centre of a cylinder, Figure 8.6 shows results after 100 steps with DTIM = 0.01 for a 5 × 10 quadrant measuring 2.5 units radially and 5.0 units vertically. Otherwise, parameters are those of Figure 8.2.

PROGRAM 8.2: SOLUTION OF THE CONDUCTION EQUATION OVER A RECTANGLE USING AN EXPLICIT METHOD

```
C
C      PROGRAM 8.2 SOLUTION OF THE CONDUCTION EQUATION
C      OVER A RECTANGULAR AREA USING 4-NODED QUADRILATERALS
C      SIMPLE EXPLICIT INTEGRATION IN TIME USING
C      ELEMENT BY ELEMENT SUMMATION
C
C      ALTER NEXT LINE TO CHANGE PROBLEM SIZE
C
      PARAMETER(IELS=30,ILOADS=50,INF=70)
C
      REAL JAC(2,2),JAC1(2,2),KAY(2,2),SAMP(3,2),FTF(4,4),DTKD(4,4),
     +COORD(4,2),DER(2,4),DERIV(2,4),DERIVT(4,2),KDERIV(2,4),FUN(4),
     +PM(4,4),KP(4,4),NEWLO(ILOADS),LOADS(ILOADS),ELD(4),
     +STPM(4,4,IELS),GLOBMA(INF),MASS(4)
      INTEGER G(4),NF(INF,1)
      DATA IT,IJAC,IJAC1,IKAY,IDER,IDERIV,IKDERV/7*2/,ISAMP/3/
      DATA IDTKD,IKP,ICOORD,IDERVT,NOD,IFTF,IPM/7*4/,NODOF/1/
C
C      INPUT AND INITIALISATION
C
      READ(5,*)NXE,NYE,N,NN,NR,NGP,AA,BB,PERMX,PERMY
      READ(5,*)DTIM,ISTEP,NPRI,NRES
      CALL NULVEC(GLOBMA,NN)
      CALL NULL(KAY,IKAY,IT,IT)
      KAY(1,1)=PERMX
      KAY(2,2)=PERMY
      CALL GAUSS(SAMP,ISAMP,NGP)
      DO 2 I=1,NN
    2 NF(I,1)=I
C
C      ELEMENT INTEGRATION AND STORAGE
C
      NM=0
      DO 10 IP=1,NXE
      DO 10 IQ=1,NYE
      NM=NM+1
      CALL GEO4X1(IP,IQ,NXE,AA,BB,COORD,ICOORD,G,NF,INF)
      CALL NULL(KP,IKP,NOD,NOD)
      CALL NULL(PM,IPM,NOD,NOD)
      DO 20 I=1,NGP
      DO 20 J=1,NGP
      CALL FORMLN(DER,IDER,FUN,SAMP,ISAMP,I,J)
      CALL MATMUL(DER,IDER,COORD,ICOORD,JAC,IJAC,IT,NOD,IT)
      CALL TWOBY2(JAC,IJAC,JAC1,IJAC1,DET)
      CALL MATMUL(JAC1,IJAC1,DER,IDER,DERIV,IDERIV,IT,IT,NOD)
      CALL MATMUL(KAY,IKAY,DERIV,IDERIV,KDERIV,IKDERV,IT,IT,NOD)
      CALL MATRAN(DERIVT,IDERVT,DERIV,IDERIV,IT,NOD)
      CALL MATMUL(DERIVT,IDERVT,KDERIV,IKDERV,DTKD,IDTKD,NOD,IT,NOD)
      QUOT=DET*SAMP(I,2)*SAMP(J,2)
      DO 25 K=1,NOD
```

```
      DO 25 L=1,NOD
      FTF(K,L)=FUN(K)*FUN(L)*QUOT
   25 DTKD(K,L)=DTKD(K,L)*QUOT*DTIM
      CALL MATADD(PM,IPM,FTF,IFTF,NOD,NOD)
   20 CALL MATADD(KP,IKP,DTKD,IDTKD,NOD,NOD)
      DO 11 I=1,NOD
      QUOT=0.
      DO 12 J=1,NOD
   12 QUOT=QUOT+PM(I,J)
      MASS(I)=QUOT
   11 GLOBMA(G(I))=GLOBMA(G(I))+MASS(I)
      CALL NULL(PM,IPM,NOD,NOD)
      DO 1 I=1,NOD
    1 PM(I,I)=MASS(I)
      DO 15 I=1,NOD
      DO 15 J=1,NOD
   15 STPM(I,J,NM)=PM(I,J)-KP(I,J)
   10 CONTINUE
C
C      INITIAL CONDITIONS AND REARRANGE GLOBMA INTO M-1
C
      CALL READNF(NF,INF,NN,NODOF,NR)
      J=0
      DO 41 I=1,NN
      IF(NF(I,1).EQ.0)GO TO 41
      J=J+1
      GLOBMA(J)=1./GLOBMA(I)
   41 CONTINUE
      READ(5,*)VALO
      DO 45 I=1,N
   45 LOADS(I)=VALO
C
C      TIME STEPPING RECURSION
C
      DO 50 J=1,ISTEP
      CALL NULVEC(NEWLO,N)
      TIME=J*DTIM
      NM=0
      DO 100 IP=1,NXE
      DO 100 IQ=1,NYE
      NM=NM+1
      CALL GEO4X1(IP,IQ,NXE,AA,BB,COORD,ICOORD,G,NF,INF)
      DO 16 I=1,NOD
      DO 16 K=1,NOD
   16 PM(I,K)=STPM(I,K,NM)
      DO 34 I=1,NOD
      IF(G(I).EQ.0)ELD(I)=0.
   34 IF(G(I).NE.0)ELD(I)=LOADS(G(I))
      CALL MVMULT(PM,IPM,ELD,NOD,NOD,FUN)
      DO 35 I=1,NOD
   35 IF(G(I).NE.0)NEWLO(G(I))=NEWLO(G(I))+FUN(I)
  100 CONTINUE
      DO 60 I=1,N
   60 LOADS(I)=NEWLO(I)*GLOBMA(I)
      IF(J/NPRI*NPRI.EQ.J)WRITE(6,'(2E12.4)')TIME,LOADS(NRES)
   50 CONTINUE
      STOP
      END
```

The basis equation for simple explicit time-stepping was given by equation (3.97) as

$$\mathbf{PM}\boldsymbol{\Phi}_1 = (\mathbf{PM} - \Delta t\,\mathbf{KP})\boldsymbol{\Phi}_0 \tag{8.4}$$

If **PM** is lumped (diagonalised) this can be written as

$$\boldsymbol{\Phi}_1 = \mathbf{PM}^{-1}(\mathbf{PM} - \Delta t\,\mathbf{KP})\boldsymbol{\Phi}_0 \tag{8.5}$$

Since $\mathbf{PM} - \Delta t\,\mathbf{KP}$ are element matrices which when summed give global matrices, the summation can as well be done of the vectors $(\mathbf{PM} - \Delta t\,\mathbf{KP})\boldsymbol{\Phi}_0$ without there being any need to store a global matrix at all. This having been done, the global $\boldsymbol{\Phi}_1$ is computed by multiplying $\boldsymbol{\Phi}_0$ by the inverse of the global

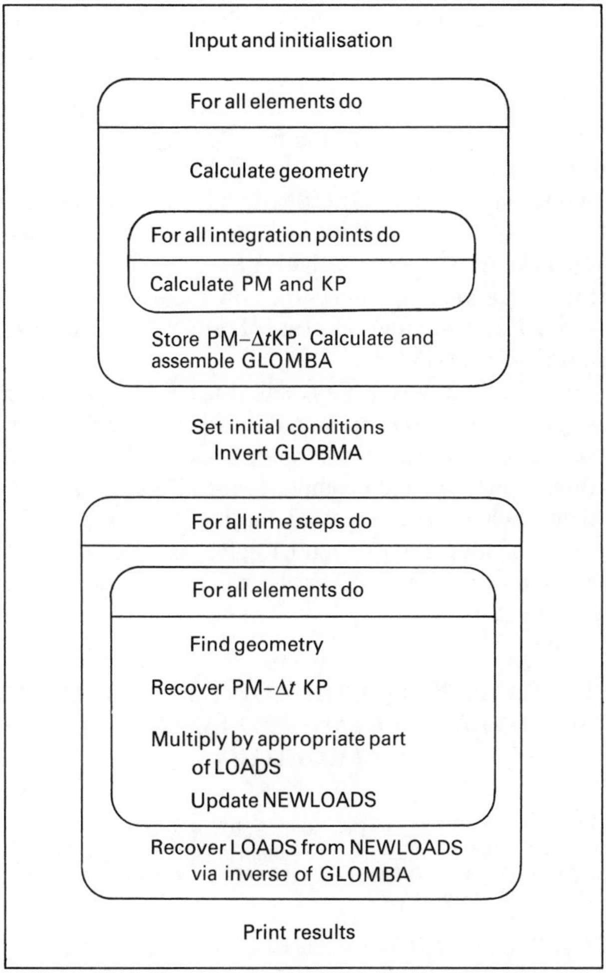

Figure 8.7 Structure chart for explicit time integration

T=Time	Pressure at Node 21
.1000E+00	.8972E+02
.2000E+00	.5881E+02
.3000E+00	.3624E+02
.4000E+00	.2215E+02
.5000E+00	.1353E+02
.6000E+00	.8258E+01
.7000E+00	.5042E+01
.8000E+00	.3078E+01
.9000E+00	.1879E+01
.1000E+01	.1147E+01
.1100E+01	.7005E+00
.1200E+01	.4277E+00
.1300E+01	.2611E+00
.1400E+01	.1594E+00
.1500E+01	.9733E-01

Figure 8.8 Results from Program 8.2

mass matrix $\sum \mathbf{PM}^{-1}$ (called GLOBMA in the program). The process is illustrated by the structure chart in Figure 8.7.

The program is readily derived from Program 8.0 and contains only three new arrays: the global mass vector GLOBMA, the element lumped mass vector MASS and a storage array for element matrices STPM. The PARAMETER statement now includes IELS ($\geqslant$ NXE $*$ NYE).

The element integration loop follows the now standard course but the element matrix $\mathbf{PM} - \Delta t\,\mathbf{KP}$ is stored for each element rather than assembled as would be the case for an implicit technique.

After the initial conditions have been established, GLOBMA is redefined as the inverse of the global mass vector required by equation (8.5). Then, in the time-stepping loop, element matrices are recovered from STPM and the multiplication and summation required by (8.5) completed for all elements.

After the elements have been inspected, the 'loads' are updated by multiplying by the inverse of the global mass vector to give the solution at the new time.

The problem shown on Figure 8.2 has been analysed again and the results are listed as Figure 8.8.

PROGRAM 8.3: SOLUTION OF THE CONDUCTION EQUATION OVER A RECTANGLE USING AN EBE PRODUCT ALGORITHM

```
C
C      PROGRAM 8.3 SOLUTION OF THE CONDUCTION EQUATION
C      OVER A RECTANGULAR AREA USING 4-NODED QUADRILATERALS
C      ELEMENT BY ELEMENT PRODUCT INTEGRATION IN TIME
C      TWO PASS ALGORITHM
C
C      ALTER NEXT LINE TO CHANGE PROBLEM SIZE
C
      PARAMETER(IEL=100,ILOADS=150,INF=70)
C
      REAL JAC(2,2),JAC1(2,2),KAY(2,2),SAMP(3,2),FTF(4,4),DTKD(4,4),
```

```
     +COORD(4,2),DER(2,4),DERIV(2,4),DERIVT(4,2),KDERIV(2,4),FUN(4),
     +PM(4,4),KP(4,4),LOADS(ILOADS),MASS(4),GLOBMA(ILOADS),STKP(4,4,IEL)
      INTEGER G(4),NF(INF,1)
      DATA IT,IJAC,IJAC1,IKAY,IDER,IDERIV,IKDERV/7*2/,ISAMP/3/
      DATA IDTKD,IKP,ICOORD,IDERVT,NOD,IFTF,IPM/7*4/,NODOF/1/
C
C       INPUT AND INITIALISATION
C
      READ(5,*)NXE,NYE,N,NN,NR,NGP,AA,BB,PERMX,PERMY
      READ(5,*)DTIM,ISTEP,THETA,NPRI,NRES
      CALL NULVEC(GLOBMA,NN)
      CALL NULL(KAY,IKAY,IT,IT)
      KAY(1,1)=PERMX
      KAY(2,2)=PERMY
      CALL GAUSS(SAMP,ISAMP,NGP)
      DO 5 I=1,NN
    5 NF(I,1)=I
C
C       ELEMENT INTEGRATION AND LUMPED MASS ASSEMBLY
C
      NM=0
      DO 10 IP=1,NXE
      DO 10 IQ=1,NYE
      NM=NM+1
      CALL GEO4X1(IP,IQ,NXE,AA,BB,COORD,ICOORD,G,NF,INF)
      CALL NULL(KP,IKP,NOD,NOD)
      CALL NULL(PM,IPM,NOD,NOD)
      DO 20 I=1,NGP
      DO 20 J=1,NGP
      CALL FORMLN(DER,IDER,FUN,SAMP,ISAMP,I,J)
      CALL MATMUL(DER,IDER,COORD,ICOORD,JAC,IJAC,IT,NOD,IT)
      CALL TWOBY2(JAC,IJAC,JAC1,IJAC1,DET)
      CALL MATMUL(JAC1,IJAC1,DER,IDER,DERIV,IDERIV,IT,IT,NOD)
      CALL MATMUL(KAY,IKAY,DERIV,IDERIV,KDERIV,IKDERV,IT,IT,NOD)
      CALL MATRAN(DERIVT,IDERVT,DERIV,IDERIV,IT,NOD)
      CALL MATMUL(DERIVT,IDERVT,KDERIV,IKDERV,DTKD,IDTKD,NOD,IT,NOD)
      QUOT=DET*SAMP(I,2)*SAMP(J,2)
      DO 25 K=1,NOD
      DO 25 L=1,NOD
      FTF(K,L)=FUN(K)*FUN(L)*QUOT
   25 DTKD(K,L)=DTKD(K,L)*QUOT
      CALL MATADD(PM,IPM,FTF,IFTF,NOD,NOD)
   20 CALL MATADD(KP,IKP,DTKD,IDTKD,NOD,NOD)
      DO 30 I=1,NOD
      DO 30 J=1,NOD
   30 STKP(I,J,NM)=KP(I,J)
      DO 31 I=1,NOD
      QUOT=0.
      DO 32 J=1,NOD
   32 QUOT=QUOT+PM(I,J)
   31 GLOBMA(G(I))=GLOBMA(G(I))+QUOT
   10 CONTINUE
C
C       CALCULATE AND STORE ELEMENT A AND B MATRICES
C
      NM=0
      DO 45 IP=1,NXE
      DO 45 IQ=1,NYE
      NM=NM+1
      CALL GEO4X1(IP,IQ,NXE,AA,BB,COORD,ICOORD,G,NF,INF)
      DO 50 I=1,NOD
      DO 50 J=1,NOD
```

```
      KP(I,J)=-STKP(I,J,NM)*(1.-THETA)*DTIM*.5
   50 FTF(I,J)=STKP(I,J,NM)*THETA*DTIM*.5
      DO 55 I=1,NOD
      FTF(I,I)=FTF(I,I)+GLOBMA(G(I))
   55 KP(I,I)=KP(I,I)+GLOBMA(G(I))
      CALL MATINV(FTF,IFTF,NOD)
      CALL MATMUL(FTF,IFTF,KP,IKP,PM,IPM,NOD,NOD,NOD)
      DO 60 I=1,NOD
      DO 60 J=1,NOD
   60 STKP(I,J,NM)=PM(I,J)
   45 CONTINUE
C
C       TAKE ACCOUNT OF INITIAL AND BOUNDARY CONDITIONS
C
      CALL READNF(NF,INF,NN,NODOF,NR)
      READ(5,*)VALO
      DO 62 I=1,N
   62 LOADS(I)=VALO
C
C       TIME STEPPING LOOP
C
      DO 70 J=1,ISTEP
      TIME=DTIM*J
      DO 80 IPASS=1,2
      IF(IPASS.EQ.1)THEN
      ILP=1
      IHP=NXE
      INCP=1
      ILQ=1
      IHQ=NYE
      NM=0
      ELSE
      ILP=NXE
      IHP=1
      INCP=-1
      ILQ=NYE
      IHQ=1
      NM=NXE*NYE+1
      END IF
      DO 85 IP=ILP,IHP,INCP
      DO 85 IQ=ILQ,IHQ,INCP
      NM=NM+INCP
      CALL GEO4X1(IP,IQ,NXE,AA,BB,COORD,ICOORD,G,NF,INF)
      DO 95 K=1,NOD
      DO 95 L=1,NOD
   95 FTF(K,L)=STKP(K,L,NM)
      DO 100 K=1,NOD
      IF(G(K).EQ.0)MASS(K)=0.
  100 IF(G(K).NE.0)MASS(K)=LOADS(G(K))
      CALL MVMULT(FTF,IFTF,MASS,NOD,NOD,FUN)
      DO 105 K=1,NOD
  105 IF(G(K).NE.0)LOADS(G(K))=FUN(K)
   85 CONTINUE
   80 CONTINUE
      IF(J/NPRI*NPRI.EQ.J)WRITE(6,'(2E12.4)')TIME,LOADS(NRES)
   70 CONTINUE
      STOP
      END
```

The motivation in using this algorithm is to preserve the storage economy achieved by the previous explicit technique while attaining the stability properties enjoyed by implicit methods typified by Program 8.0. The process is described in Chapter 3 by equation (3.101) and in structure by Figure 3.19. The program bears a strong resemblance to Program 8.2. The element integration loop is employed to store the element **KP** matrices in a storage array STKP. In addition, the element consistent mass matrices **PM** are diagonalised and the global mass vector, GLOBMA, assembled. A second loop over the elements is then made, headed 'calculate and store element **A** and **B** matrices'. These are the matrices given in Figure 3.19 by $[\mathbf{M} - (1-\theta)\Delta t\,\mathbf{KP}/2]$ and $[\mathbf{M} + \theta\,\Delta t\,\mathbf{KP}/2]$ respectively, and they are called KP and FTF respectively in the program. The algorithm calls for **B** (FTF) to be inverted, which is done using the library routine MATINV. Then **A** is formed as $\mathbf{B}^{-1}[\mathbf{M} - (1-\theta)\Delta t\,\mathbf{KP}/2]$, that is by multiplying FTF and KP. The result, called PM in the program, is stored as STKP.

Initial conditions can then be prescribed and the time-stepping loop entered. Within that loop, two passes, controlled by the counter IPASS, are made over the elements from first to last and back again. Half of the total $\Delta t\,\mathbf{KP}$ increment, operates on each pass, and this has been accounted for already in forming **A** and **B**. Some rather involved FORTRAN counting is necessary to pass forwards and backwards, but the essential coding recovers each element $\mathbf{B}^{-1}\mathbf{A}$ matrix from STKP and multiplies it by the appropriate part of the solution 'loads'. Note that in this product algorithm the solution is continually being updated so there is no need for any 'new loads' vector such as had to be employed in the explicit summation algorithm.

The process is fully detailed in Figure 3.19. Results for the problem described by Figure 8.2 are shown in Figure 8.9.

T=TIME	Pressure at Node 21
.1000E+00	.8907E+02
.2000E+00	.6102E+02
.3000E+00	.3890E+02
.4000E+00	.2447E+02
.5000E+00	.1536E+02
.6000E+00	.9635E+01
.7000E+00	.6044E+01
.8000E+00	.3791E+01
.9000E+00	.2378E+01
.1000E+01	.1492E+01
.1100E+01	.9357E+00
.1200E+01	.5870E+00
.1300E+01	.3682E+00
.1400E+01	.2309E+00
.1500E+01	.1449E+00

Figure 8.9 Results from Program 8.3

8.1 COMPARISON OF PROGRAMS 8.0, 8.2 AND 8.3

These three programs already described in this chapter can all be used to solve plane conduction or uncoupled consolidation problems. Comparison of Figures 8.3, 8.8 and 8.9 shows that for the chosen problem—at the time step used (that is 0.01)—all solutions are accurate, and indeed the explicit solution (Figure 8.8) is probably as accurate as any despite being the simplest and cheapest to obtain.

It must, however, be remembered that as the time step Δt is increased, the explicit algorithm will lead to unstable results (the stability limit for the chosen problem is about $\Delta t = 0.02$). At that time step, Program 8.0 with $\theta = 0.5$ will tend to produce oscillatory results, which can be damped, at the expense of average accuracy, by increasing θ towards 1.0. Typical behaviour of the implicit algorithm is illustrated in Figure 8.10.

Program 8.3, while retaining the storage economies of Program 8.2, allows the time step to be increased well beyond the explicit limit. For example, in the selected problem, reasonable results are still produced at $\Delta t = 10\Delta t_{crit}$. However, as Δt is increased still further, accuracy becomes poorer and Program 8.0 yields the best solutions for very large Δt.

It will be clear that algorithm choice in this area is not a simple one and depends on the nature of the problem (degree of non-linearity, etc.) and on the hardware employed.

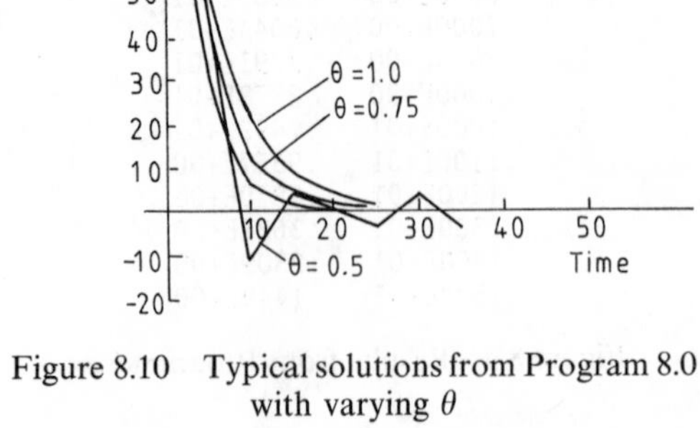

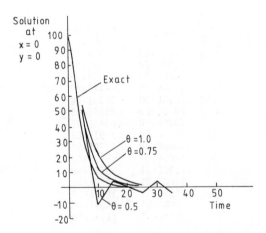

Figure 8.10 Typical solutions from Program 8.0 with varying θ

PROGRAM 8.4: DIFFUSION–CONVECTION PROBLEM OVER A RECTANGLE—TRANSFORMED ANALYSIS

```
C
C      PROGRAM 8.4  DIFFUSION CONVECTION EQUATION ON A RECTANGLE
C      USING 4-NODED QUADRILATERAL ELEMENTS. SELF-ADJOINT
C      TRANSFORMATION; IMPLICIT INTEGRATION
C
C      ALTER NEXT LINE TO CHANGE PROBLEM SIZE
C
      PARAMETER(IKB1=100,IKB2=10,INF=200)
C
      REAL FUN(4),COORD(4,2),DER(2,4),DERIV(2,4),JAC(2,2),
     +JAC1(2,2),SAMP(3,2),DTKD(4,4),KP(4,4),PM(4,4),FTF(4,4),
     +KB(IKB1,IKB2),PB(IKB1,IKB2),LOADS(IKB1),ANS(IKB1)
      INTEGER G(4),NF(INF,1)
      DATA ICOORD,IDERVT,IDTKD,IKP,IPM,IFTF,NOD/7*4/,ISAMP/3/
      DATA IDER,IDERIV,IKDERV,IJAC,IJAC1,IT/6*2/,NODOF/1/
C
C      INPUT AND INITIALISATION
C
      READ(5,*)NXE,NYE,N,IW,NN,NGP,AA,BB,PERMX,PERMY,UX,UY,
     +         DTIM,ISTEP,THETA
      IWP1=IW+1
      CALL NULL(KB,IKB1,N,IWP1)
      CALL NULL(PB,IKB1,N,IWP1)
      CALL NULVEC(LOADS,N)
      CALL GAUSS(SAMP,ISAMP,NGP)
      DO 10 I=1,NN
   10 NF(I,1)=I
C
C      ELEMENT INTEGRATION AND ASSEMBLY
C
      DO 20 IP=1,NXE
      DO 20 IQ=1,NYE
      CALL GEO4X1(IP,IQ,NXE,AA,BB,COORD,ICOORD,G,NF,INF)
      CALL NULL(KP,IKP,NOD,NOD)
      CALL NULL(PM,IPM,NOD,NOD)
      DO 30 I=1,NGP
      DO 30 J=1,NGP
      CALL FORMLN(DER,IDER,FUN,SAMP,ISAMP,I,J)
      CALL MATMUL(DER,IDER,COORD,ICOORD,JAC,IJAC,IT,NOD,IT)
      CALL TWOBY2(JAC,IJAC,JAC1,IJAC1,DET)
      CALL MATMUL(JAC1,IJAC1,DER,IDER,DERIV,IDERIV,IT,IT,NOD)
      QUOT=DET*SAMP(I,2)*SAMP(J,2)
      DO 40 K=1,NOD
      DO 40 L=1,NOD
      PART1=PERMX*DERIV(1,K)*DERIV(1,L)+PERMY*DERIV(2,K)*DERIV(2,L)
      PART2=(UX*UX/PERMX+UY*UY/PERMY)*FUN(K)*FUN(L)*.25
      DTKD(K,L)=QUOT*(PART1+PART2)
      FTF(K,L)=FUN(K)*FUN(L)*QUOT/(THETA*DTIM)
   40 CONTINUE
      CALL MATADD(KP,IKP,DTKD,IDTKD,NOD,NOD)
      CALL MATADD(PM,IPM,FTF,IFTF,NOD,NOD)
   30 CONTINUE
C
C      INSERT DERIVATIVE BOUNDARY CONDITIONS
C
      IF(IQ.NE.1)GOTO 50
      KP(2,2)=KP(2,2)+UY*AA/6.
```

```
      KP(2,3)=KP(2,3)+UY*AA/12.
      KP(3,2)=KP(3,2)+UY*AA/12.
      KP(3,3)=KP(3,3)+UY*AA/6.
   50 CONTINUE
      IF(IQ.NE.NYE)GOTO 60
      KP(1,1)=KP(1,1)+UY*AA/6.
      KP(1,4)=KP(1,4)+UY*AA/12.
      KP(4,1)=KP(4,1)+UY*AA/12.
      KP(4,4)=KP(4,4)+UY*AA/6.
   60 CONTINUE
      CALL FORMKB(KB,IKB1,KP,IKP,G,IW,NOD)
      CALL FORMKB(PB,IKB1,PM,IPM,G,IW,NOD)
   20 CONTINUE
C
C       REDUCTION OF LEFT HAND SIDE
C
      F1=UY*AA/(2.*THETA)
      F2=F1
      CALL MATADD(PB,IKB1,KB,IKB1,N,IWP1)
      DO 70 I=1,N
      DO 70 J=1,IWP1
   70 KB(I,J)=PB(I,J)-KB(I,J)/THETA
      CALL CHOLIN(PB,IKB1,N,IW)
C
C       TIME STEPPING RECURSION
C
      DO 80 J=1,ISTEP
      CALL BANMUL(KB,IKB1,LOADS,ANS,N,IW)
      ANS(N)=ANS(N)+F1
      ANS(N-1)=ANS(N-1)+F2
      CALL CHOBAC(PB,IKB1,ANS,N,IW)
      DO 90 I=1,N
   90 LOADS(I)=ANS(I)
      CALL PRINTV(LOADS,N)
   80 CONTINUE
      STOP
      END
```

When convection terms are retained in the simplified flow equations, (2.106) or (2.121) have to be solved. Again many techniques could be employed but, in this book, implicit algorithms based on equation (3.94) are used. Thus this program is an extension of Program 8.0.

When the transformation of equation (2.123) is employed, the equation to be solved becomes

$$c_x \frac{\partial^2 h}{\partial x^2} + c_y \frac{\partial^2 h}{\partial y^2} - \left(\frac{u^2}{4c_x} + \frac{v^2}{4c_y} \right) h = \frac{\partial h}{\partial t} \tag{8.6}$$

Thus the extra term involving h distinguishes the process from a simple diffusion one. However, reference to Table 2.1 shows that the semi-discretised 'stiffness' matrix for this problem will still be symmetrical, the h term involving an element matrix of the 'mass matrix' type $\iint N_i N_j \,dx\,dy$.

Comparison with Program 8.0 will show essentially the same array declarations and input parameters although rectangular array storage involving PARAMETER statements IKB1 $\geqslant$ N and IKB2 $\geqslant$ IW + 1 have been substituted. Extra variables are the velocities in the x and y directions, UX and UY respectively.

The problem chosen is the one-dimensional example shown in Figure 8.11. The dependent variable ϕ refers to concentration of sediment picked up by the flow from the base of the mesh, and distributed with time in the y direction. Thus, UX is zero and for numerical reasons PERMX is set to a small number, 1×10^{-6}, which is effectively zero. The equation to be solved is in effect

$$c_y \frac{\partial^2 h}{\partial y^2} - \frac{v^2}{4c_y} h = \frac{\partial h}{\partial t} \tag{8.7}$$

subject to the boundary conditions

$$\frac{\partial \phi}{\partial y} = \frac{v}{c_y} \tag{8.8}$$

$$= C_2 \text{ (constant)}$$

at $y = 0$ and

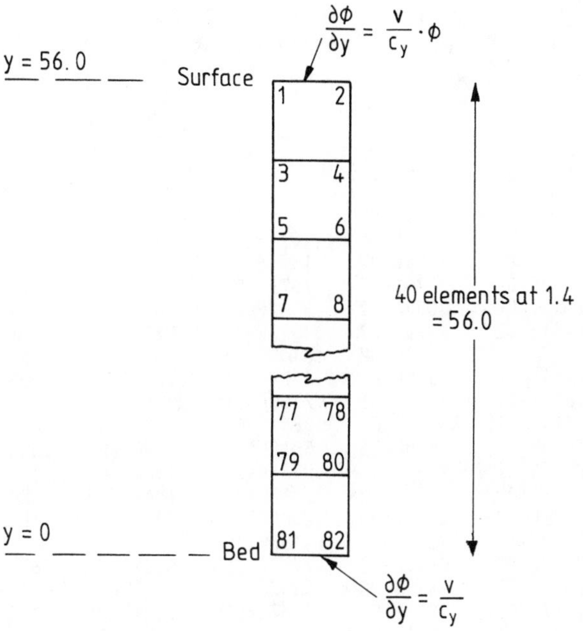

Mesh data	NXE	NYE	N	IW	NN	NGP
	1	40	82	3	82	2
Element data	AA	BB	PERMX	PERMY	UX	UY
	1.4	1.4	1.E-6	.49	.0	.0135
Time integration data	DTIM		ISTEP	THETA		
	300.0		2	0.5		

Figure 8.11 Mesh and data for Program 8.4

$$\frac{\partial \phi}{\partial y} = \frac{v}{c_y}\phi = C_1\phi \text{ where } C_1 \text{ is constant} \tag{8.9}$$

at $y = 56.0$.

After transformation, these conditions become

$$\frac{\partial h}{\partial y} = \frac{-v}{2c_y}h + \frac{v}{c_y} \tag{8.10}$$

and

$$\frac{\partial h}{\partial y} = +\frac{v}{2c_y}h \tag{8.11}$$

Boundary condition (8.11) is clearly of the type described in Section 3.6, equation (3.16). Therefore, at that boundary, the element matrix will have to be augmented by the matrix shown in equation (3.20). The multiple $C_1 c_y(x_k - x_j)/6$ in (3.20) is just $v(x_k - x_j)/12$ or UY * AA/12 in the program. This is carried out in the section of program headed 'insert derivative boundary conditions'.

The condition (8.10) contains a similar contribution, but in addition the term v/c_y is of the type described by equation (3.17). Thus, an addition must be made

Nodal concentrations

.1037E-02	.1037E-02	.1071E-02	.1071E-02	.1135E-02	.1135E-02	
.1229E-02	.1229E-02	.1356E-02	.1356E-02	.1521E-02	.1521E-02	
.1726E-02	.1726E-02	.1979E-02	.1979E-02	.2285E-02	.2285E-02	
.2653E-02	.2653E-02	.3094E-02	.3094E-02	.3618E-02	.3618E-02	
.4241E-02	.4241E-02	.4979E-02	.4979E-02	.5852E-02	.5852E-02	Timestep
.6884E-02	.6884E-02	.8103E-02	.8103E-02	.9542E-02	.9542E-02	1
.1124E-01	.1124E-01	.1324E-01	.1324E-01	.1561E-01	.1561E-01	
.1839E-01	.1839E-01	.2168E-01	.2168E-01	.2556E-01	.2556E-01	
.3013E-01	.3013E-01	.3551E-01	.3551E-01	.4187E-01	.4187E-01	
.4936E-01	.4936E-01	.5819E-01	.5819E-01	.6860E-01	.6860E-01	
.8087E-01	.8087E-01	.9534E-01	.9534E-01	.1124E+00	.1124E+00	
.1325E+00	.1325E+00	.1562E+00	.1562E+00	.1842E+00	.1842E+00	
.2171E+00	.2171E+00	.2560E+00	.2560E+00	.3018E+00	.3018E+00	
.3558E+00	.3558E+00	.4195E+00	.4195E+00			
.7562E-02	.7562E-02	.7783E-02	.7783E-02	.8157E-02	.8157E-02	
.8693E-02	.8693E-02	.9398E-02	.9398E-02	.1029E-01	.1029E-01	
.1137E-01	.1137E-01	.1267E-01	.1267E-01	.1421E-01	.1421E-01	
.1601E-01	.1601E-01	.1811E-01	.1811E-01	.2053E-01	.2053E-01	
.2331E-01	.2331E-01	.2649E-01	.2649E-01	.3013E-01	.3013E-01	Timestep
.3427E-01	.3427E-01	.3898E-01	.3898E-01	.4430E-01	.4430E-01	
.5032E-01	.5032E-01	.5709E-01	.5709E-01	.6471E-01	.6471E-01	2
.7324E-01	.7324E-01	.8277E-01	.8277E-01	.9338E-01	.9338E-01	
.1052E+00	.1052E+00	.1182E+00	.1182E+00	.1325E+00	.1325E+00	
.1481E+00	.1481E+00	.1651E+00	.1651E+00	.1835E+00	.1835E+00	
.2032E+00	.2032E+00	.2240E+00	.2240E+00	.2458E+00	.2458E+00	
.2682E+00	.2682E+00	.2908E+00	.2908E+00	.3128E+00	.3128E+00	
.3335E+00	.3335E+00	.3515E+00	.3515E+00	.3653E+00	.3653E+00	
.3727E+00	.3727E+00	.3712E+00	.3712E+00			

Figure 8.12 Results from Program 8.4

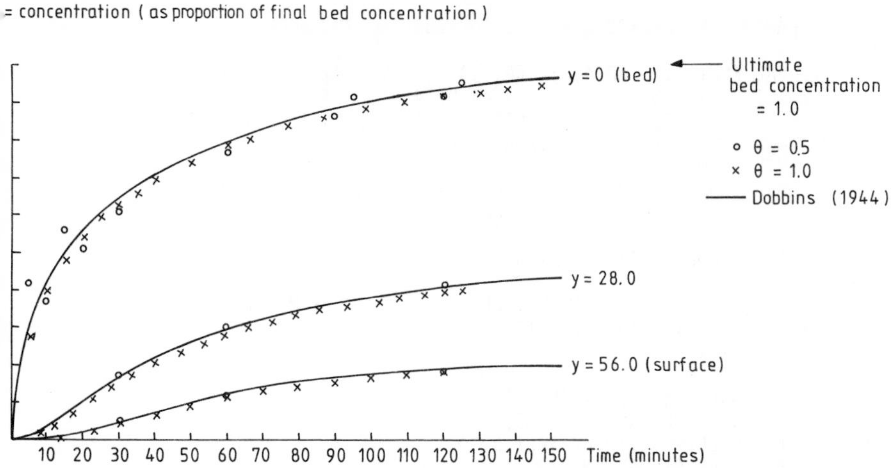

Figure 8.13 Graph of concentration versus time for Program 8.4

to the right hand side of the equations at such a boundary in accordance with equation (3.22). In this case the terms in equation (3.22) are $v(x_k - x_j)/2$ and are incorporated in the program immediately after the comment 'reduction of left hand side'.

This example shows that quite complicated coding would be necessary to permit very general boundary conditions to be specified in all problems. The authors prefer to write specialised code when necessary.

Comparing Program 8.4 with Program 8.0, it can be seen that the array storages KB and PB are used for global matrices in place of vectors BK and BP. In the present example there are no zero freedoms and so READNF is not required. In the element integration and assembly loop PART1 accumulates the diffusive part of the element 'stiffness' and PART2 the convective part.

After insertion of boundary conditions the (constant) global left hand side is factorised using CHOLIN. The time-stepping loop is as before. Output for two steps is listed as Figure 8.12, while Figure 8.13 shows how the finite element solution compares with an "analytical" one due to Dobbins (1944).

It should be remembered that solutions are in terms of the transformed variable h, and the true solution ϕ has to be recovered using (2.123).

PROGRAM 8.5: DIFFUSION–CONVECTION PROBLEM OVER A RECTANGLE—UNTRANSFORMED ANALYSIS

```
C
C      PROGRAM 8.5  DIFFUSION CONVECTION EQUATION ON A RECTANGLE
C      USING 4-NODED QUADRILATERAL ELEMENTS. UNTRANSFORMED
C      EQUATION BY GALERKIN'S METHOD; IMPLICIT INTEGRATION
C
C      ALTER NEXT LINE TO CHANGE PROBLEM SIZE
C
```

```
      PARAMETER(IKB1=250,IKB2=10,INF=250,INO=20,IWORK=5)
C
      REAL FUN(4),COORD(4,2),DER(2,4),DERIV(2,4),JAC(2,2),
     +JAC1(2,2),SAMP(3,2),DTKD(4,4),KP(4,4),PM(4,4),FTF(4,4),
     +KB(IKB1,IKB2),PB(IKB1,IKB2),LOADS(IKB1),ANS(IKB1),STORPB(INO),
     +WORK(IWORK,IKB1),COPY(IWORK,IKB1)
      INTEGER NO(INO),G(4),NF(INF,1)
      DATA ICOORD,IDERVT,IDTKD,IKP,IPM,IFTF,NOD/7*4/,ISAMP/3/
      DATA IDER,IDERIV,IKDERV,IJAC,IJAC1,IT/6*2/,NODOF/1/
C
C       INPUT AND INITIALISATION
C
      READ(5,*)NXE,NYE,N,IW,NN,NGP,AA,BB,PERMX,PERMY,UX,UY,
     +         DTIM,ISTEP,THETA
      IWP1=IW+1
      IBAND=2*IWP1-1
      CALL NULL(KB,IKB1,N,IBAND)
      CALL NULL(PB,IKB1,N,IBAND)
      CALL NULL(WORK,IWORK,IWP1,N)
      CALL NULVEC(LOADS,N)
      CALL GAUSS(SAMP,ISAMP,NGP)
      DO 10 I=1,NN
   10 NF(I,1)=I
C
C       ELEMENT INTEGRATION AND ASSEMBLY
C
      DO 20 IP=1,NXE
      DO 20 IQ=1,NYE
      CALL GEO4X1(IP,IQ,NXE,AA,BB,COORD,ICOORD,G,NF,INF)
      CALL NULL(KP,IKP,NOD,NOD)
      CALL NULL(PM,IPM,NOD,NOD)
      DO 30 I=1,NGP
      DO 30 J=1,NGP
      CALL FORMLN(DER,IDER,FUN,SAMP,ISAMP,I,J)
      CALL MATMUL(DER,IDER,COORD,ICOORD,JAC,IJAC,IT,NOD,IT)
      CALL TWOBY2(JAC,IJAC,JAC1,IJAC1,DET)
      CALL MATMUL(JAC1,IJAC1,DER,IDER,DERIV,IDERIV,IT,IT,NOD)
      QUOT=DET*SAMP(I,2)*SAMP(J,2)
      DO 40 K=1,NOD
      DO 40 L=1,NOD
      PART1=PERMX*DERIV(1,K)*DERIV(1,L)+PERMY*DERIV(2,K)*DERIV(2,L)
      PART2=UX*FUN(K)*DERIV(1,L)+UY*FUN(K)*DERIV(2,L)
      DTKD(K,L)=QUOT*(PART1-PART2)
      FTF(K,L)=FUN(K)*FUN(L)*QUOT/(THETA*DTIM)
   40 CONTINUE
      CALL MATADD(KP,IKP,DTKD,IDTKD,NOD,NOD)
      CALL MATADD(PM,IPM,FTF,IFTF,NOD,NOD)
   30 CONTINUE
      CALL FORMTB(KB,IKB1,KP,IKP,G,IW,NOD)
      CALL FORMTB(PB,IKB1,PM,IPM,G,IW,NOD)
   20 CONTINUE
C
C       SPECIFY FIXED NODAL VALUES
C
      CALL MATADD(PB,IKB1,KB,IKB1,N,IBAND)
      DO 50 I=1,N
      DO 50 J=1,IBAND
   50 KB(I,J)=PB(I,J)-KB(I,J)/THETA
      READ(5,*)IFIX,(NO(I),I=1,IFIX)
      DO 60 I=1,IFIX
      PB(NO(I),IWP1)=PB(NO(I),IWP1)+1.E20
```

```
   60 STORPB(I)=PB(NO(I),IWP1)
C
C       REDUCTION OF LEFT HAND SIDE
C
      CALL GAUSBA(PB,IKB1,WORK,IWORK,N,IW)
C
C       TIME STEPPING RECURSION
C
      DO 80 J=1,ISTEP
      CALL MATCOP(WORK,IWORK,COPY,IWORK,IWP1,N)
      CALL BANTML(KB,IKB1,LOADS,ANS,N,IW)
      DO 90 I=1,IFIX
      IF(J*DTIM.LE..2)ANS(NO(I))=STORPB(I)
      IF(J*DTIM.GT..2)ANS(NO(I))=0.
   90 CONTINUE
      CALL SOLVBA(PB,IKB1,COPY,IWORK,ANS,N,IW)
      DO 100 I=1,N
  100 LOADS(I)=ANS(I)
      CALL PRINTV(ANS,N)
   80 CONTINUE
      STOP
      END
```

In the previous example, difficulties would have arisen had the ratio of u to c_x been large, because the transformation involving $\exp(u/c_x)$ would not have been numerically feasible (Smith *et al.*, 1973). Under these circumstances (and even when $c_x = c_y = 0$) equation (2.121) can still be solved, but with the penalty that the element and system matrices become unsymmetrical, although the latter are still banded.

Program 8.5 will be used to solve a purely convective problem, that is with $c_x = c_y = 0$. The problem chosen is one-dimensional (see Figure 8.14) so the equation is effectively

$$-v\frac{\partial \phi}{\partial y} = \frac{\partial \phi}{\partial t} \tag{8.12}$$

in the region $0 \leqslant y \leqslant 2$, subject to the boundary conditions $\phi = 1$ at $y = 0$ for $0 \leqslant t \leqslant 0.2$, $\phi = 0$ at $y = 0$ for $t > 0.2$ and $\partial\phi/\partial y = 0$ at $y = 2$ for all t.

Comparing with Program 8.4, the usual geometry routine for four-noded elements numbered in the x direction, GEO4X1, is used. Arrays NO and STORPB are used to read in the numbers of fixed freedoms and to store information about them during the time-stepping process. The system KB matrix is now unsymmetrical and is formed as a total band using FORMTB. Although PB is symmetrical it too is stored as a full band to be compatible with KB.

The PART2 contribution to the KP matrix is the only one in the present example and it is unsymmetrical (skew-symmetric in fact).

The structure of the program is modelled on the previous one. There are no zero freedoms and the solution routines are GAUSBA and SOLVBA which use an extra array WORK as working space. Matrix-by-vector multiplication needs the subroutine BANTML (see Chapter 3). The PARAMETER statement now redefines IKB2 $\geqslant$ 2∗IWP1 − 1, IWORK $\geqslant$ IWP1 and INO $\geqslant$ IFIX.

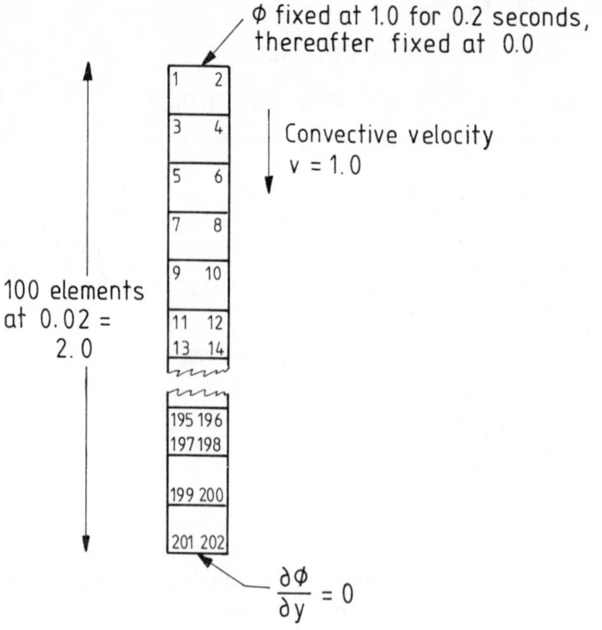

Mesh data		NXE	NYE	N	IW	NN	NGP
		1	100	202	3	202	2
Element data		AA	BB	PERMX	PERMY	UX	UY
		.02	.02	.0	.0	.0	1.0
Time integration data		DTIM	ISTEP	THETA			
		.04	1	.5			
Fixed freedom data	IFIX	(NO(I), I = 1, IFIX)					
	2	1	2				

Figure 8.14 Mesh and data for Program 8.5

In the section 'time-stepping recursion' it can be seen that the solution at nodes 1 and 2 is held at the value 1.0 for the first 0.2 seconds of convection.

For checking purposes, the first step of the solution is listed as Figure 8.15. The correct solution to the problem is described by a rectangular pulse moving with unit velocity in the y direction. Figure 8.16 shows the computed solution after one second. Spurious spatial oscillations are seen to have been introduced by the numerical solution. Measures to improve the solutions are beyond the scope of the present treatment, but see Smith (1976, 1979). Of course, in the present case improvements can be achieved by simply reducing Δy and Δt.

ϕ Values at first timestep

.1000E+01	.1000E+01	.3660E+00	.3660E+00	.1340E+00	.1340E+00
.4904E-01	.4904E-01	.1795E-01	.1795E-01	.6570E-02	.6570E-02
.2405E-02	.2405E-02	.8802E-03	.8802E-03	.3222E-03	.3222E-03
.1179E-03	.1179E-03	.4316E-04	.4316E-04	.1580E-04	.1580E-04
.5783E-05	.5783E-05	.2117E-05	.2117E-05	.7747E-06	.7747E-06
.2836E-06	.2836E-06	.1038E-06	.1038E-06	.3799E-07	.3799E-07
.1391E-07	.1391E-07	.5090E-08	.5090E-08	.1863E-08	.1863E-08
.6819E-09	.6819E-09	.2496E-09	.2496E-09	.9136E-10	.9136E-10
.3344E-10	.3344E-10	.1224E-10	.1224E-10	.4480E-11	.4480E-11
.1640E-11	.1640E-11	.6002E-12	.6002E-12	.2197E-12	.2197E-12
.8041E-13	.8041E-13	.2943E-13	.2943E-13	.1077E-13	.1077E-13
.3943E-14	.3943E-14	.1443E-14	.1443E-14	.5283E-15	.5283E-15
.1934E-15	.1934E-15	.7078E-16	.7078E-16	.2591E-16	.2591E-16
.9483E-17	.9483E-17	.3471E-17	.3471E-17	.1270E-17	.1270E-17
.4650E-18	.4650E-18	.1702E-18	.1702E-18	.6230E-19	.6230E-19
.2280E-19	.2280E-19	.8347E-20	.8347E-20	.3055E-20	.3055E-20
.1118E-20	.1118E-20	.4093E-21	.4093E-21	.1498E-21	.1498E-21
.5484E-22	.5484E-22	.2007E-22	.2007E-22	.7347E-23	.7347E-23
.2689E-23	.2689E-23	.9843E-24	.9843E-24	.3603E-24	.3603E-24
.1319E-24	.1319E-24	.4827E-25	.4827E-25	.1767E-25	.1767E-25
.6467E-26	.6467E-26	.2367E-26	.2367E-26	.8664E-27	.8664E-27
.3171E-27	.3171E-27	.1161E-27	.1161E-27	.4248E-28	.4248E-28
.1555E-28	.1555E-28	.5692E-29	.5692E-29	.2083E-29	.2083E-29
.7626E-30	.7626E-30	.2791E-30	.2791E-30	.1022E-30	.1022E-30
.3739E-31	.3739E-31	.1369E-31	.1369E-31	.5010E-32	.5010E-32
.1834E-32	.1834E-32	.6712E-33	.6712E-33	.2457E-33	.2457E-33
.8992E-34	.8992E-34	.3291E-34	.3291E-34	.1205E-34	.1205E-34
.4410E-35	.4410E-35	.1614E-35	.1614E-35	.5908E-36	.5908E-36
.2162E-36	.2162E-36	.7915E-37	.7915E-37	.2897E-37	.2897E-37
.1060E-37	.1060E-37	.3881E-38	.3881E-38	.1421E-38	.1421E-38
.5200E-39	.5200E-39	.1903E-39	.1903E-39	.6967E-40	.6967E-40
.2550E-40	.2550E-40	.9334E-41	.9334E-41	.3416E-41	.3416E-41
.1251E-41	.1251E-41	.4571E-42	.4571E-42	.1684E-42	.1684E-42
.6014E-43	.6014E-43	.2406E-43	.2406E-43		

Figure 8.15 Results from Program 8.5

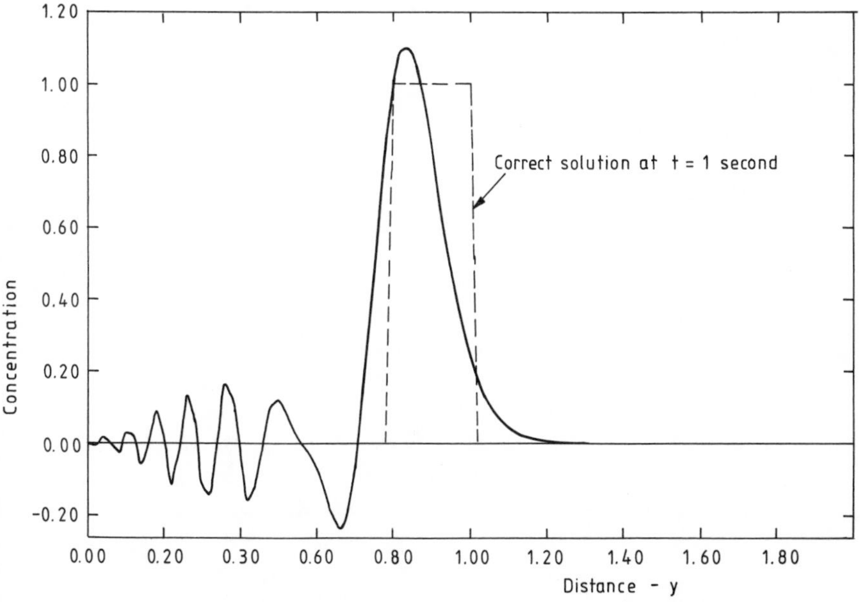

Figure 8.16 Solution from Program 8.5 after 1 second

8.2 REFERENCES

Carslaw, H. S., and Jaeger, J.C. (1959) *Conduction of Heat in Solids*, 2nd edition. Clarendon Press, Oxford.

Dobbins, W. E. (1944) Effect of turbulence on sedimentation. *Trans. Am. Soc. Civ. Eng.*, **109**, 629–656.

Smith, I. M. (1976) Integration in time of diffusion and diffusion-convection equations. In *Finite Elements in Water Resources*, eds W. G. Gay, G. Pinder and C. Brebbia, pp. 1.3–1.20. Pentech Press, Plymouth.

Smith, I. M. (1979) The diffusion-convection equation. In *Summary of Numerical Methods for Partial Differential Equations*, eds I. Gladwell and R. Wait, Chapter 11. Oxford University Press.

Smith, I. M., Farraday, R. V., and O'Connor, B. A. (1973) Rayleigh–Ritz and Galerkin finite elements for diffusion-convection problems. *Water Resources Research*, **9**, No. 3, 593–606.

CHAPTER 9
'Coupled' Problems

9.0 INTRODUCTION

In the previous chapter, flow problems were treated in terms of a single dependent variable, for example the 'potential' ϕ, and solutions involved only one degree of freedom per node in the finite element mesh. While this simplification may be adequate in many cases, it may sometimes be necessary to solve problems in which several degrees of freedom exist at the nodes of the mesh and the several dependent variables—for example velocities and pressures, or displacements and pressures—are 'coupled' in the differential equations. Strictly speaking, the equations of two- and three-dimensional elasticity involve coupling between the various components of displacement, but the term 'coupled problems' is really reserved for those in which variables of entirely different types are interdependent.

Both steady state and transient problems are considered in this chapter. As usual, the former involve the solution of sets of simultaneous equations, as in Chapters 4 to 7. For the problem chosen—solution of the Navier–Stokes equations—the simultaneous equations are, however, non-linear and involve unsymmetrical coefficient matrices.

The transient problems, although described by coupled systems of algebraic and differential equations, are cast as (linear) first order differential equations in the time variable, and solved by the techniques introduced in Chapter 8—specifically by implicit integration, as illustrated by Program 8.0.

The (steady state) Navier–Stokes equations were developed in Chapter 2, Section 2.17. The equilibrium equations to be solved are (2.104a), whose coefficients are themselves functions of the velocities $\mathbf{u}$ and $\mathbf{v}$ so that the equations are non-linear. Furthermore, the coefficient submatrices $\mathbf{C}$ are not, in general, symmetrical and reference to Section 3.8 shows that the subroutines GAUSBA and SOLVBA will be required to operate on the banded equation coefficients.

Section 3.13 illustrates how the element submatrices $\mathbf{C}$ are assembled and uses much of the program terminology already developed for uncoupled flow problems in Chapters 7 and 8.

PROGRAM 9.0: SOLUTION OF THE STEADY-STATE NAVIER–STOKES EQUATIONS OVER A RECTANGLE

```
C
C      PROGRAM 9.0 STEADY STATE NAVIER-STOKES
C      USING 8 NODE VELOCITY ELEMENTS
C      AND   4 NODE PRESSURE ELEMENTS
C
C      ALTER NEXT LINE TO CHANGE PROBLEM SIZE
C
      PARAMETER(IPB1=100,IPB2=50,IWORK=25,INF=100,INO=30,
     +          IWID=30,IDEP=30)
C
      REAL SAMP(3,2),COORD(8,2),DERIVT(8,2),UVEL(8),VVEL(8)
      REAL JAC(2,2),JAC1(2,2),KAY(2,2),DER(2,8),DERIV(2,8)
      REAL DTKD(8,8),KDERIV(2,8),KE(20,20),FUN(8),OLDLDS(IPB1)
      REAL FUNF(4),COORDF(4,2),DERF(2,4),DERIVF(2,4)
      REAL WIDTH(IWID),DEPTH(IDEP),VAL(INO),ROW(8),TEMP(8,8)
      REAL C11(8,8),C12(8,4),C21(4,8),C23(4,8),C32(8,4)
      REAL PB(IPB1,IPB2),WORK(IWORK,IPB1),LOADS(IPB1)
      INTEGER G(20),NO(INO),NF(INF,3)
      DATA IT,IJAC,IJAC1,IKAY,IDER,IDERIV,IKDERV,IDERF,IDERVF/9*2/
      DATA ISAMP,NODOF/2*3/,ICORDF,IC21,IC23,NODF/4*4/,IKE,ITOT/2*20/
      DATA ICOORD,IDERVT,IDTKD,ITEMP,IC11,IC12,IC32,NOD/8*8/
C
C      INPUT AND INITIALISATION
C
      READ(5,*)NXE,NYE,N,IW,NN,NR,NGP,VISC,RHO,ITS
      READ(5,*)(WIDTH(I),I=1,NXE+1)
      READ(5,*)(DEPTH(I),I=1,NYE+1)
      CALL READNF(NF,INF,NN,NODOF,NR)
      READ(5,*)IFIX,(NO(I),VAL(I),I=1,IFIX)
      IWP1=IW+1
      IBAND=2*IWP1-1
      CALL NULVEC(UVEL,NOD)
      CALL NULVEC(VVEL,NOD)
      CALL NULVEC(OLDLDS,N)
      CALL NULVEC(LOADS,N)
      CALL NULL(KAY,IKAY,IT,IT)
      KAY(1,1)=VISC/RHO
      KAY(2,2)=VISC/RHO
      CALL GAUSS(SAMP,ISAMP,NGP)
C
C      ITERATION LOOP
C
      ITERS=0
   10 ITERS=ITERS+1
      ICON=1
      CALL NULL(PB,IPB1,N,IBAND)
      CALL NULL(WORK,IWORK,IWP1,N)
      CALL NULL(KE,IKE,ITOT,ITOT)
C
C      ELEMENT MATRIX INTEGRATION AND ASSEMBLY
C
      DO 20 IP=1,NXE
      DO 20 IQ=1,NYE
      CALL GEVUPV(IP,IQ,NXE,COORD,ICOORD,COORDF,ICORDF,G,NF,INF,
     +            WIDTH,DEPTH)
      DO 30 M=1,NOD
      IF(G(M).EQ.0)UVEL(M)=0.
   30 IF(G(M).NE.0)UVEL(M)=(LOADS(G(M))+OLDLDS(G(M)))*.5
      DO 40 M=NOD+NODF+1,ITOT
```

```
      IF(G(M).EQ.0)VVEL(M-NOD-NODF)=0.0
   40 IF(G(M).NE.0)VVEL(M-NOD-NODF)=(LOADS(G(M))+ OLDLDS(G(M)))*.5
      CALL NULL(C11,IC11,NOD,NOD)
      CALL NULL(C12,IC12,NOD,NODF)
      CALL NULL(C21,IC21,NODF,NOD)
      CALL NULL(C23,IC23,NODF,NOD)
      CALL NULL(C32,IC32,NOD,NODF)
      DO 50 I=1,NGP
      DO 50 J=1,NGP
C
C       VELOCITY CONTRIBUTION
C
      CALL FMQUAD(DER,IDER,FUN,SAMP,ISAMP,I,J)
      UBAR=0.
      VBAR=0.
      DO 60 M=1,NOD
      UBAR=UBAR+FUN(M)*UVEL(M)
   60 VBAR=VBAR+FUN(M)*VVEL(M)
      IF(II.EQ.1)UBAR=1.0
      IF(II.EQ.1)VBAR=1.0
      CALL MATMUL(DER,IDER,COORD,ICOORD,JAC,IJAC,IT,NOD,IT)
      CALL TWOBY2(JAC,IJAC,JAC1,IJAC1,DET)
      CALL MATMUL(JAC1,IJAC1,DER,IDER,DERIV,IDERIV,IT,IT,NOD)
      CALL MATMUL(KAY,IKAY,DERIV,IDERIV,KDERIV,IKDERV,IT,IT,NOD)
      CALL MATRAN(DERIVT,IDERVT,DERIV,IDERIV,IT,NOD)
      CALL MATMUL(DERIVT,IDERVT,KDERIV,IKDERV,DTKD,IDTKD,NOD,IT,NOD)
      QUOT=DET*SAMP(I,2)*SAMP(J,2)
      DO 70 K=1,NOD
      DO 70 L=1,NOD
   70 DTKD(K,L)=DTKD(K,L)*QUOT
      CALL MATADD(C11,IC11,DTKD,IDTKD,NOD,NOD)
      DO 80 K=1,NOD
   80 ROW(K)=DERIV(1,K)
      PROD=QUOT*UBAR
      CALL VVMULT(FUN,ROW,TEMP,ITEMP,NOD,NOD)
      CALL MSMULT(TEMP,ITEMP,PROD,NOD,NOD)
      CALL MATADD(C11,IC11,TEMP,ITEMP,NOD,NOD)
      DO 90 K=1,NOD
   90 ROW(K)=DERIV(2,K)
      PROD=QUOT*VBAR
      CALL VVMULT(FUN,ROW,TEMP,ITEMP,NOD,NOD)
      CALL MSMULT(TEMP,ITEMP,PROD,NOD,NOD)
      CALL MATADD(C11,IC11,TEMP,ITEMP,NOD,NOD)
C
C       PRESSURE CONTRIBUTION
C
      CALL FORMLN(DERF,IDERF,FUNF,SAMP,ISAMP,I,J)
      CALL MATMUL(DERF,IDERF,COORDF,ICORDF,JAC,IJAC,IT,NODF,IT)
      CALL TWOBY2(JAC,IJAC,JAC1,IJAC1,DET)
      CALL MATMUL(JAC1,IJAC1,DERF,IDERF,DERIVF,IDERVF,IT,IT,NODF)
      QUOT=DET*SAMP(I,2)*SAMP(J,2)
      PROD=QUOT/RHO
      DO 100 K=1,NODF
  100 ROW(K)=DERIVF(1,K)
      CALL VVMULT(FUN,ROW,TEMP,ITEMP,NOD,NODF)
      CALL MSMULT(TEMP,ITEMP,PROD,NOD,NODF)
      CALL MATADD(C12,IC12,TEMP,ITEMP,NOD,NODF)
      DO 110 K=1,NODF
  110 ROW(K)=DERIVF(2,K)
      CALL VVMULT(FUN,ROW,TEMP,ITEMP,NOD,NODF)
      CALL MSMULT(TEMP,ITEMP,PROD,NOD,NODF)
      CALL MATADD(C32,IC32,TEMP,ITEMP,NOD,NODF)
```

```
      DO 120 K=1,NOD
  120 ROW(K)=DERIV(1,K)
      CALL VVMULT(FUNF,ROW,TEMP,ITEMP,NODF,NOD)
      CALL MSMULT(TEMP,ITEMP,QUOT,NODF,NOD)
      CALL MATADD(C21,IC21,TEMP,ITEMP,NODF,NOD)
      DO 130 K=1,NOD
  130 ROW(K)=DERIV(2,K)
      CALL VVMULT(FUNF,ROW,TEMP,ITEMP,NODF,NOD)
      CALL MSMULT(TEMP,ITEMP,QUOT,NODF,NOD)
      CALL MATADD(C23,IC23,TEMP,ITEMP,NODF,NOD)
   50 CONTINUE
      CALL FRMUPV(KE,IKE,C11,IC11,C12,IC12,C21,IC21,
     +            C23,IC23,C32,IC32,NOD,NODF,ITOT)
      CALL FORMTB(PB,IPB1,KE,IKE,G,IW,ITOT)
   20 CONTINUE
C
C       INSERT PRESCRIBED VALUES OF VELOCITY AND PRESSURE
C
      CALL NULVEC(LOADS,N)
      DO 140 I=1,IFIX
      PB(NO(I),IWP1)=PB(NO(I),IWP1)+1.E15
  140 LOADS(NO(I))=PB(NO(I),IWP1)*VAL(I)
C
C       SOLVE SIMULTANEOUS EQUATIONS
C
      CALL GAUSBA(PB,IPB1,WORK,IWORK,N,IW)
      CALL SOLVBA(PB,IPB1,WORK,IWORK,LOADS,N,IW)
      CALL CHECON(LOADS,OLDLDS,N,.001,ICON)
      IF(ICON.EQ.0)GOTO 10
      CALL PRINTV(LOADS,N)
      WRITE(6,1000)ITERS
 1000 FORMAT(1H ,I10)
      STOP
      END
```

The simple problem chosen to illustrate this program is shown in Figure 9.1. Flow is proceeding in the x direction between two parallel plates at $y = 0$ and $y = -3.0$. There is no flow in the y direction and since the nodal freedoms are in the order u, p, v, the third freedom is always zero. Note also that a dummy freedom has been inserted at mid-side nodes where there is no p variable. Thus, the second freedom at all mid-side nodes is zero.

The boundary conditions are that the top plate is moved with a u velocity of 3.0 relative to the bottom plate, which has a u velocity of zero. The fixed velocity of 3.0 at the top of the mesh is specified at freedoms 1, 3 and 4.

The pressure boundary conditions are a pressure of 1.0 on the left side ($x = 0$) and of -1.0 on the right side ($x = 1$). Thus, $\partial p/\partial x$ across the mesh is -2.0. The problem is to calculate the distribution of u with depth.

The meanings of the variables in the program (excluding simple counters) are as follows:

NXE	number of elements in x direction (1)
NYE	number of elements in y direction (3)
N	number of non-zero freedoms in mesh (23)
IW	half-bandwidth of equation coefficients (11)

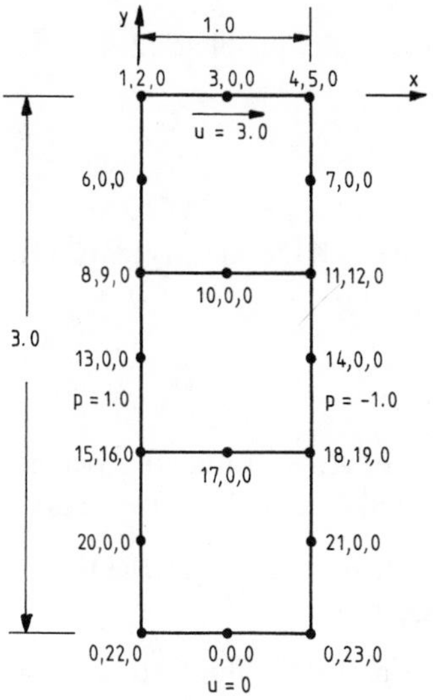

Mesh data

NXE	NYE	N	IW	NN	NR	NGP
1	3	23	11	18	18	2

Element data

VISC	RHO	ITS
1.	1.	5

Geometry data

WIDTH(I), I = 1, NXE + 1
0. 1.
DEPTH(I), I = 1, NYE + 1
0. −1. −2. −3.

Node freedom data

1	1	1	0	2	1	0	0	3	1	1	0
4	1	0	0	5	1	0	0	6	1	1	0
7	1	0	0	8	1	1	0	9	1	0	0
10	1	0	0	11	1	1	0	12	1	0	0
13	1	1	0	14	1	0	0	15	1	0	0
16	0	1	0	17	0	0	0	18	0	1	0

Fixed freedom data

IFIX (NO(I), VAL(I), I = 1, IFIX)
11

2	1.	9	1.	16	1.	22	1.
5	−1.	12	−1.	19	−1.	23	−1.
1	3.	3	3.	4	3.		

Figure 9.1 Mesh and data for Program 9.0

NN	total number of nodes in mesh (18)
NR	number of restrained nodes in mesh (18)
NGP	number of Gaussian integrating points in each direction (2)
VISC	molecular viscosity, μ (1.0)
RHO	density of the fluid, ρ (1.0)
ITS	maximum number of iterations allowed (5)
IFIX	number of freedoms with prescribed non-zero values (11)
IWPI	half-bandwidth plus 1
IBAND	total bandwidth of equation coefficients (2*IWP1 − 1)

IT	dimensions of the problem (2)
IJAC IJAC1 IKAY IDER IDERIV IKDERV IDERF IDERVF ISAMP	working size of arrays, JAC, JAC1, KAY, DER, DERIV, KDERIV, DERF, DERIVF, SAMP respectively
NODOF	number of degrees of freedom per node (3) (assumed constant, although mid-side nodes have really only 2 freedoms)
ICORDF IC21 IC23	working size of arrays COORDF, C21, C23 respectively
NODF	number of nodes for fluid pressure shape functions (4)
IKE	working size of array KE
ITOT	total number of degrees of freedom per element (20) (i.e. 16 velocity freedoms and 4 pressure freedoms)
ICOORD IDERVT IDTKD ITEMP IC11 IC12 IC32	working size of arrays, COORD, DERIVT, DTKD, TEMP, C11, C12, C32 respectively
NOD	number of nodes for velocity shape functions (8)

Space is reserved for the following small fixed length arrays:

SAMP	quadrature abscissae and weights
COORD	coordinates of 'velocity' nodes
UVEL	element nodal *u* velocity values
VVEL	element nodal *v* velocity values
JAC	Jacobian matrix
JAC1	inverse of Jacobian matrix
KAY	viscosity matrix
DER	derivatives of velocity shape functions in local coordinates
DERIV	derivatives of velocity shape functions in global coordinates
DTKD	product $DERIV^T*KAY*DERIV$
KDERIV	product KAY*DERIV
KE	element 'stiffness' matrix
FUN	velocity shape functions in local coordinates
FUNF	pressure shape functions in local coordinates
COORDF	coordinates of 'pressure' nodes
DERF	derivatives of pressure shape functions in local coordinates

DERIVF	derivatives of pressure shape functions in global coordinates
ROW	temporary storage of rows of DERIV, etc.
TEMP	temporary accumulation of C11, etc.
C11, C12, C21, C23, C32	element sub-matrices (see equation 2.104a)
G	element steering vector

The following arrays should have their dimensions adjusted via the PARAMETER statements to reflect the size of problem being analysed:

OLDLDS	solution at the previous iteration
WIDTH	x coordinates of the mesh lines
DEPTH	y coordinates of the mesh lines
VAL	values of fixed non-zero freedoms
PB	unsymmetrical band global 'stiffness' matrix
WORK	working space
LOADS	solution at the current iteration
NO	numbers of freedoms to be fixed at non-zero values
NF	node freedom array

Thus, the PARAMETER restrictions are:

IPB1 $\geqslant$ N
IPB2 $\geqslant$ IBAND
IWORK $\geqslant$ IWP1
INF $\geqslant$ NN
INO $\geqslant$ IFIX
IWID $\geqslant$ NXE + 1
IDEP $\geqslant$ NYE + 1

The structure of the program is described by the structure chart in Figure 9.2. After the data has been read in, various arrays must be initialised to zero. Note that the KAY matrix which, in the previous chapter, held coefficients of consolidation c_x etc., now holds viscosities in the form μ/ρ.

The iteration loop is then entered, controlled by the counter ITERS. System arrays PB and WORK must be nulled together with the element 'stiffness' matrix KE. Element matrix integration and assembly then proceeds as usual. The nodal coordinates and steering vector are formed by the geometry library routine GEVUPV. Nodal velocities used to form $\bar{u}$ and $\bar{v}$ in equation (2.104b) are taken to be the average of those computed in the last two iterations. Element submatrices C11, etc., are set to zero and the numerical integration loop entered. Average velocities $\bar{u}$ and $\bar{v}$ are recovered from UVEL and VVEL, except in the first iteration where the guess $\bar{u} = \bar{v} = 1.0$ is used. The submatrix C11 is formed as

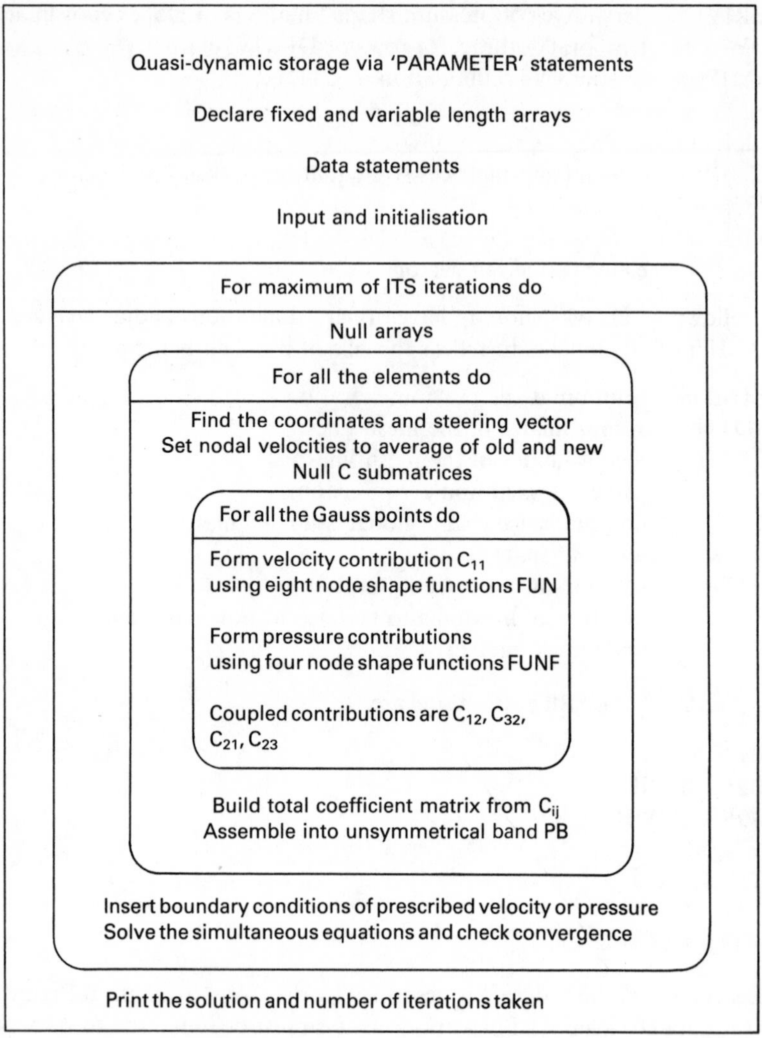

Quasi-dynamic storage via 'PARAMETER' statements
Declare fixed and variable length arrays
Data statements
Input and initialisation
For maximum of ITS iterations do
Null arrays
For all the elements do
Find the coordinates and steering vector
Set nodal velocities to average of old and new
Null C submatrices
For all the Gauss points do
Form velocity contribution C_{11} using eight node shape functions FUN
Form pressure contributions using four node shape functions FUNF
Coupled contributions are C_{12}, C_{32}, C_{21}, C_{23}
Build total coefficient matrix from C_{ij}
Assemble into unsymmetrical band PB
Insert boundary conditions of prescribed velocity or pressure
Solve the simultaneous equations and check convergence
Print the solution and number of iterations taken

Figure 9.2 Structure chart for Program 9.0

```
               Nodal velocities and pressures

 .3000E+01   .1000E+01   .3000E+01   .3000E+01  -.1000E+01   .3750E+01
 .3750E+01   .4000E+01   .1000E+01   .4000E+01   .4000E+01  -.1000E+01
 .3750E+01   .3750E+01   .3000E+01   .1000E+01   .3000E+01   .3000E+01
-.1000E+01   .1750E+01   .1750E+01   .1000E+01  -.1000E+01
         2   No. of iterations
```

Figure 9.3 Results from Program 9.0

required by equation (3.103). One component is $DERIV^T * KAY * DERIV$ which is formed as usual as DTKD. However, there are two extra components calculated by the two calls to the subroutine VVMULT.

Finally, submatrices C12, C32, C21 and C23 are computed as demanded by equation (3.104). The element 'stiffness' KE is built up by the subroutine FRMUPV and assembled into the global unsymmetrical band matrix by the assembly routine FORMTB.

It remains only to specify the fixed freedoms by the 'big spring' technique (see Section 3.6) and to complete the equation solution using GAUSBA and SOLVBA. The maximum number of iterations allowed is 5 but a convergence check of 0.1% is invoked by CHECON.

The results are listed as Figure 9.3 and illustrated in Figure 9.4. In fact, the solution converged in two iterations and in this simple case produces the exact solution to the problem.

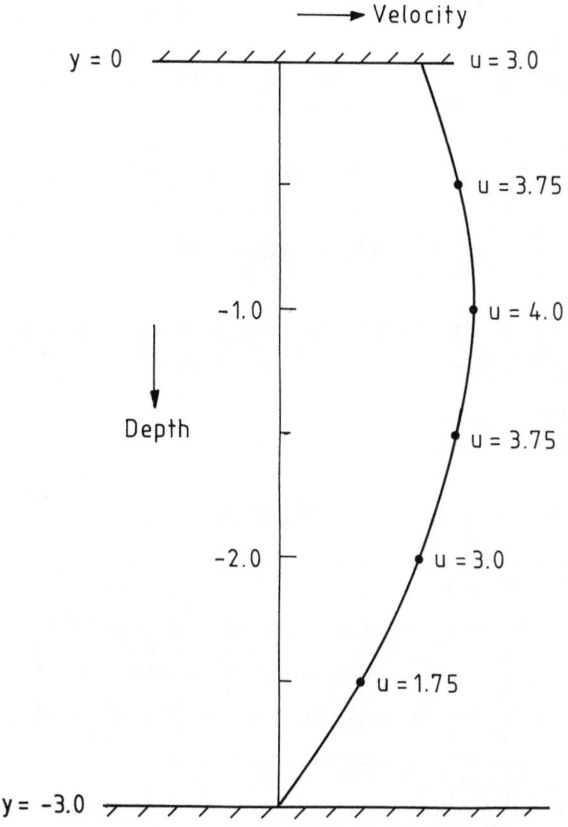

Figure 9.4 Distribution of u with depth from Program 9.0

9.1 ANALYSIS OF SOIL CONSOLIDATION USING BIOT'S THEORY

The analysis of the behaviour of porous elastic solids under load is in many ways analogous to the coupled flow analysis described in the previous section. The displacements of the soil skeleton take over the role of the velocities u and v, and the excess porewater pressure the role of the fluid pressure p.

The differential equations to be solved are (2.125), (2.126) and (2.129). Due to the coupling of fluid and solid phases there arises the complication that the applied 'total' stresses, σ, are divided between a portion carried by the soil skeleton, called the 'effective' stress σ', and a portion carried by the pore water, called in soil mechanics the 'pore pressure' and denoted in Chapter 2 by u_w to distinguish it from the mean total stress p.

After discretisation in space by finite elements, the coupled equations in u, v and u_w are given by (2.131). These can be seen to be partly algebraic equations and partly first order differential equations in time. Common techniques for discretisation in the time domain are shown by (3.112), in which θ has the same role as it had in Program 8.0, controlling the implicit integration.

Thus, equations (3.113) or (3.114) are in principle no different to (3.95) or (3.96) for uncoupled problems. Solutions will involve setting up the coupled 'stiffness' matrix on the left side of these equations, followed by an equation solution for every time step Δt to advance the solution from time 0 to time 1. For constant element properties and time step Δt, the equations need to be reduced only once, the remainder of the solution involving matrix-by-vector multiplication on the right hand side, followed by back-substitution. The process may be written:

$$\mathbf{KE}\delta_1 = \mathbf{KD}\delta_0 + \mathbf{F} \tag{9.1}$$

PROGRAM 9.1: BIOT CONSOLIDATION OF A RECTANGLE IN PLANE STRAIN (FOUR-NODE ELEMENTS)

```
C
C        PROGRAM 9.1 BIOT CONSOLIDATION OF AN ELASTIC SOLID IN PLANE
C        STRAIN USING 4-NODE QUADRILATERALS
C
C
C        ALTER NEXT LINE TO CHANGE PROBLEM SIZE
C
      PARAMETER(IBK=5000,IPB2=50,ILOADS=200,INF=100,IWID=30,
     +          IDEP=30)
C
      REAL DEE(3,3),SAMP(3,2),COORD(4,2),DERIVT(4,2),JAC(2,2)
      REAL JAC1(2,2),KAY(2,2),DER(2,4),DERIV(2,4),KDERIV(2,4)
      REAL BEE(3,8),DBEE(3,8),BT(8,3),BTDB(8,8),KM(8,8),ELD(8)
      REAL EPS(3),SIGMA(3),DTKD(4,4),KP(4,4),KE(12,12),KD(12,12)
      REAL FUN(4),C(8,4),VOLF(8,4),WIDTH(IWID),DEPTH(IDEP),BK(IBK),VOL(8
     +)
      REAL PB(ILOADS,IPB2),LOADS(ILOADS),ANS(ILOADS)
      INTEGER G(12),NF(INF,3)
      DATA IJAC,IJAC1,IKAY,IDER,IDERIV,IKDERV,IT/7*2/
      DATA IDEE,ISAMP,IBEE,IDBEE,NODOF,IH/6*3/
      DATA ICOORD,IDERVT,IDTKD,IKP,NOD/5*4/
```

```
      DATA IBT,IBTDB,IKM,IC,IVOLF,IDOF/6*8/,IKE,IKD,ITOT/3*12/
C
C      INPUT AND INITIALISATION
C
      READ(5,*)NXE,NYE,N,IW,NN,NR,NGP,PERMX,PERMY,E,V,DTIM,ISTEP,THETA
      READ(5,*)(WIDTH(I),I=1,NXE+1)
      READ(5,*)(DEPTH(I),I=1,NYE+1)
      CALL READNF(NF,INF,NN,NODOF,NR)
      IBAND=2*(IW+1)-1
      IR=N*(IW+1)
      CALL NULL(PB,ILOADS,N,IBAND)
      CALL NULVEC(LOADS,N)
      CALL NULVEC(BK,IR)
      CALL NULL(DEE,IDEE,IH,IH)
      CALL FMDEPS(DEE,IDEE,E,V)
      CALL NULL(KAY,IKAY,IT,IT)
      KAY(1,1)=PERMX
      KAY(2,2)=PERMY
      CALL GAUSS(SAMP,ISAMP,NGP)
C
C      ELEMENT MATRIX INTEGRATION AND ASSEMBLY
C
      DO 10 IP=1,NXE
      DO 10 IQ=1,NYE
      CALL GEV4X3(IP,IQ,NXE,COORD,ICOORD,G,NF,INF,WIDTH,DEPTH)
      CALL NULL(KM,IKM,IDOF,IDOF)
      CALL NULL(C,IC,IDOF,NOD)
      CALL NULL(KP,IKP,NOD,NOD)
      DO 20 I=1,NGP
      DO 20 J=1,NGP
      CALL FORMLN(DER,IDER,FUN,SAMP,ISAMP,I,J)
      CALL MATMUL(DER,IDER,COORD,ICOORD,JAC,IJAC,IT,NOD,IT)
      CALL TWOBY2(JAC,IJAC,JAC1,IJAC1,DET)
      CALL MATMUL(JAC1,IJAC1,DER,IDER,DERIV,IDERIV,IT,IT,NOD)
      CALL NULL(BEE,IBEE,IH,IDOF)
      CALL FORMB(BEE,IBEE,DERIV,IDERIV,NOD)
      CALL VOL2D(BEE,IBEE,VOL,NOD)
      CALL MATMUL(DEE,IDEE,BEE,IBEE,DBEE,IDBEE,IH,IH,IDOF)
      CALL MATRAN(BT,IBT,BEE,IBEE,IH,IDOF)
      CALL MATMUL(BT,IBT,DBEE,IDBEE,BTDB,IBTDB,IDOF,IH,IDOF)
      QUOT=DET*SAMP(I,2)*SAMP(J,2)
      CALL MSMULT(BTDB,IBTDB,QUOT,IDOF,IDOF)
      CALL MATADD(KM,IKM,BTDB,IBTDB,IDOF,IDOF)
      CALL MATMUL(KAY,IKAY,DERIV,IDERIV,KDERIV,IKDERV,IT,IT,NOD)
      CALL MATRAN(DERIVT,IDERVT,DERIV,IDERIV,IT,NOD)
      CALL MATMUL(DERIVT,IDERVT,KDERIV,IKDERV,DTKD,IDTKD,NOD,IT,NOD)
      PROD=QUOT*DTIM
      CALL MSMULT(DTKD,IDTKD,PROD,NOD,NOD)
      CALL MATADD(KP,IKP,DTKD,IDTKD,NOD,NOD)
      DO 30 K=1,IDOF
      DO 30 L=1,NOD
   30 VOLF(K,L)=VOL(K)*FUN(L)*QUOT
      CALL MATADD(C,IC,VOLF,IVOLF,IDOF,NOD)
   20 CONTINUE
      CALL FMKDKE(KM,IKM,KP,IKP,C,IC,KE,IKE,KD,IKD,IDOF,NOD,ITOT,THETA)
      CALL FORMKV(BK,KE,IKE,G,N,ITOT)
      CALL FORMTB(PB,ILOADS,KD,IKD,G,IW,ITOT)
   10 CONTINUE
C
C      REDUCE LEFT HAND SIDE
C
      CALL BANRED(BK,N,IW)
```

```
C
C       TIME STEPPING LOOP
C
      DO 40 NS=1,ISTEP
      CALL BANTML(PB,ILOADS,LOADS,ANS,N,IW)
C
C       RAMP LOADING
C
      X1=(.1*NS+.1*(THETA-1.))*.5
      X2=X1*2.
      IF(NS.GT.10)ANS(1)=ANS(1)-.5
      IF(NS.LE.10)ANS(1)=ANS(1)-X1
      IF(NS.GT.10)ANS(3)=ANS(3)-1.
      IF(NS.LE.10)ANS(3)=ANS(3)-X2
      IF(NS.GT.10)ANS(5)=ANS(5)-1.
      IF(NS.LE.10)ANS(5)=ANS(5)-X2
      IF(NS.GT.10)ANS(6)=ANS(6)-.5
      IF(NS.LE.10)ANS(6)=ANS(6)-X1
      CALL BACSUB(BK,ANS,N,IW)
      DO 50 I=1,N
   50 LOADS(I)=ANS(I)
      CALL PRINTV(ANS,N)
C
C       RECOVER ELEMENT EFFECTIVE STRESSES AT ELEMENT 'CENTRES'
C
      NGP=1
      CALL GAUSS(SAMP,ISAMP,NGP)
      DO 60 IP=1,NXE
      DO 60 IQ=1,NYE
      CALL GEV4X3(IP,IQ,NXE,COORD,ICOORD,G,NF,INF,WIDTH,DEPTH)
      DO 70 M=1,IDOF
      IF(G(M).EQ.0)ELD(M)=0.
   70 IF(G(M).NE.0)ELD(M)=ANS(G(M))
      DO 80 I=1,NGP
      DO 80 J=1,NGP
      CALL FORMLN(DER,IDER,FUN,SAMP,ISAMP,I,J)
      CALL MATMUL(DER,IDER,COORD,ICOORD,JAC,IJAC,IT,NOD,IT)
      CALL TWOBY2(JAC,IJAC,JAC1,IJAC1,DET)
      CALL MATMUL(JAC1,IJAC1,DER,IDER,DERIV,IDERIV,IT,IT,NOD)
      CALL NULL(BEE,IBEE,IH,IDOF)
      CALL FORMB(BEE,IBEE,DERIV,IDERIV,NOD)
      CALL MVMULT(BEE,IBEE,ELD,IH,IDOF,EPS)
      CALL MVMULT(DEE,IDEE,EPS,IH,IH,SIGMA)
      CALL PRINTV(SIGMA,IH)
   80 CONTINUE
   60 CONTINUE
   40 CONTINUE
      STOP
      END
```

The problem chosen is of a plane strain 'oedometer' specimen (Figure 9.5) subjected to a ramp loading at its surface as shown in Figure 9.6. There are roller boundaries at both sides of the mesh. The structure chart describing the process of the program is shown in Figure 9.7.

Since the same shape functions govern the variation of all the unknowns (u and v displacement and pore pressure u_w) the program is somewhat simpler than Program 9.0. However, the variables have the same significance, with the

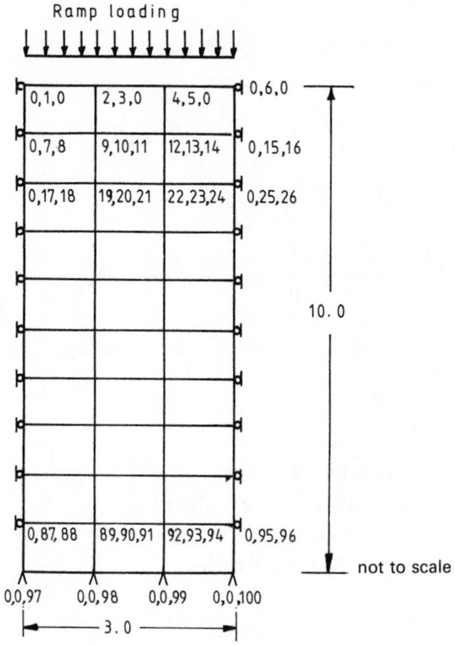

Mesh data	NXE	NYE	N	IW	NN	NR	NGP
	3	10	100	15	44	26	2

Element data	PERMX	PERMY	E	V
	1.0	1.0	1.0	0.0

Time integration data	DTIM	ISTEP	THETA
	1.0	1	.5

Geometry data	WIDTH(I), I = 1, NXE+1	0.0	1.0	2.0	3.0		
	DEPTH(I), I = 1, NYE+1	0.	−1.	−2.	−3.	−4.	−5.
		−6.	−7.	−8.	−9.	−10.	

Node freedom data

1	0	1	0	2	1	1	0	3	1	1	0	4	0	1	0
5	0	1	1	8	0	1	1	9	0	1	1	12	0	1	1
13	0	1	1	16	0	1	1	17	0	1	1	20	0	1	1
21	0	1	1	24	0	1	1	25	0	1	1	28	0	1	1
29	0	1	1	32	0	1	1	33	0	1	1	36	0	1	1
37	0	1	1	40	0	1	1								
41	0	0	1	42	0	0	1	43	0	0	1	44	0	0	1

Figure 9.5 Mesh and data for Program 9.1

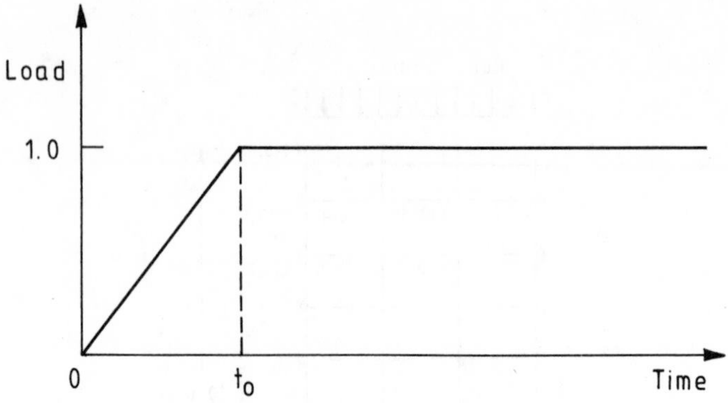

Figure 9.6 Ramp loading

following additions:

PERMX	permeability (D'Arcy) in the x direction/γ_w (see equation 2.129)
PERMY	permeability in the y direction/γ_w (see equation 2.129)
E	effective Young's modulus
V	effective Poisson's ratio
DTIM	time step
ISTEP	number of time steps
THETA	parameter in implicit time stepping process ($\frac{1}{2} \leqslant \theta \leqslant 1$)
IR	N*IWP1—storage required for symmetric global matrix
IDEE, IBEE, IDBEE	working sizes of arrays, DEE, BEE, DBEE
IH	number of components of stress and strain (3)
IKP, IBT, IBTDB, IKM, IC, IVOLF	working sizes of arrays, KP, BT, BTDB, KM C, VOLF respectively
IDOF	number of displacement degrees of freedom per element (8)
IKE, IKD	working sizes of arrays KE, KD

Arrays not encountered in Program 9.0 are:

DEE	effective stress–strain matrix (see equation 2.126)
BEE	strain-displacement matrix
DBEE	product of DEE and BEE
BT	transpose of BEE
BTDB	product of BT with DBEE
KM	element (solid) stiffness matrix

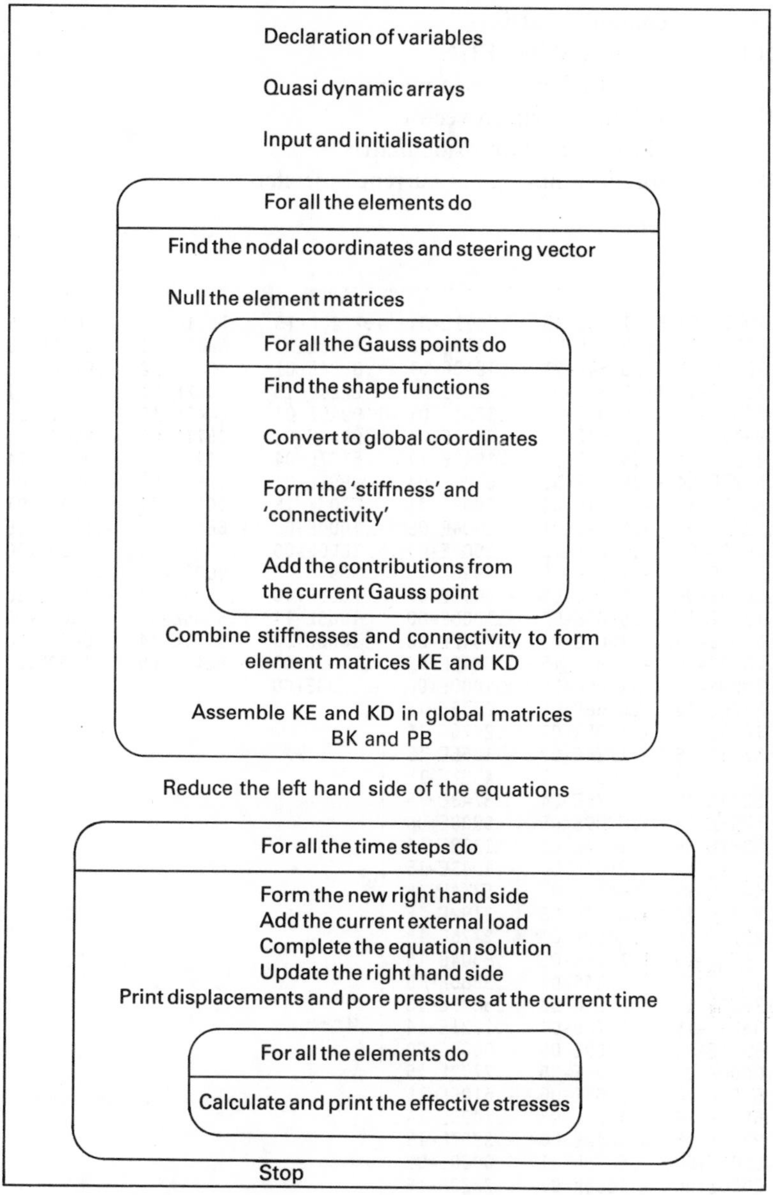

Figure 9.7 Structure chart for Programs 9.1 and 9.2

ELD	element nodal displacements
EPS	element strain vector
SIGMA	element effective stress vector
KP	element (fluid) stiffness matrix
KE, KD	element matrices (see equation 9.1)

C	coupling matrix
VOLF	product VOL∗FUN
BK	global left hand side matrix
VOL	volumetric strain vector
PB	global right hand side matrix
ANS	solution (answer) at current time step

```
                 Displacements and Porepressures

-.7071E-01  -.1665E-15  -.7071E-01  -.2220E-15  -.7071E-01  -.7071E-01
-.1213E-01  -.8284E-01   .1110E-15  -.1213E-01  -.8284E-01   .2220E-15
-.1213E-01  -.8284E-01  -.1213E-01  -.8284E-01  -.2082E-02  -.9706E-01
 .2220E-15  -.2082E-02  -.9706E-01  -.2220E-15  -.2082E-02  -.9706E-01
-.2082E-02  -.9706E-01  -.3571E-03  -.9949E-01   .4441E-15  -.3571E-03
-.9949E-01   .6661E-15  -.3571E-03  -.9949E-01  -.3571E-03  -.9949E-01
-.6127E-04  -.9991E-01   .1943E-14  -.6127E-04  -.9991E-01   .2331E-14
-.6127E-04  -.9991E-01  -.6127E-04  -.9991E-01  -.1051E-04  -.9999E-01
-.8882E-15  -.1051E-04  -.9999E-01  -.6106E-15  -.1051E-04  -.9999E-01
-.1051E-04  -.9999E-01  -.1804E-05  -.1000E+00  -.6661E-15  -.1804E-05
-.1000E+00  -.8327E-15  -.1804E-05  -.1000E+00  -.1804E-05  -.1000E+00
-.3095E-06  -.1000E+00   .4996E-15  -.3095E-06  -.1000E+00   .6661E-15
-.3095E-06  -.1000E+00  -.3095E-06  -.1000E+00  -.5305E-07  -.1000E+00
 .1943E-14  -.5305E-07  -.1000E+00   .1665E-14  -.5305E-07  -.1000E+00
-.5305E-07  -.1000E+00  -.8842E-08  -.1000E+00  -.1055E-14  -.8842E-08
-.1000E+00  -.8882E-15  -.8842E-08  -.1000E+00  -.8842E-08  -.1000E+00
-.1000E+00  -.1000E+00  -.1000E+00  -.1000E+00
-.2776E-16  -.5858E-01  -.2776E-16
 .1665E-15  -.1005E-01  -.2776E-16
 .3331E-15  -.1724E-02  -.1665E-15
 .1193E-14  -.2959E-03  -.4302E-15
 .5274E-15  -.5076E-04   .8743E-15
-.7772E-15  -.8709E-05   .0000E+00
-.8327E-16  -.1494E-05  -.1388E-16
 .1221E-14  -.2564E-06  -.1943E-15
 .4441E-15  -.4421E-07   .5274E-15
-.5274E-15  -.8842E-08  -.3192E-15
 .2776E-16  -.5858E-01   .2776E-15
-.1665E-15  -.1005E-01   .8604E-15
-.1110E-15  -.1724E-02   .3886E-15
 .3053E-15  -.2959E-03  -.6800E-15    Effective
 .3331E-15  -.5076E-04   .1221E-14    stresses
 .5551E-16  -.8709E-05   .0000E+00
 .0000E+00  -.1494E-05  -.7772E-15
-.5551E-16  -.2564E-06  -.6106E-15
-.5551E-16  -.4421E-07   .1499E-14
 .8327E-16  -.8842E-08  -.5967E-15
 .0000E+00  -.5858E-01   .0000E+00
 .0000E+00  -.1005E-01   .2220E-15
-.2220E-15  -.1724E-02   .1110E-15
-.1499E-14  -.2959E-03  -.1943E-15
-.8604E-15  -.5076E-04   .6245E-15
 .7216E-15  -.8709E-05   .5551E-16
 .8327E-16  -.1494E-05  -.4163E-16
-.1166E-14  -.2564E-06  -.1388E-15
-.3886E-15  -.4421E-07   .6384E-15
 .4441E-15  -.8842E-08  -.1110E-15
```

Figure 9.8 Results from Program 9.1

The PARAMETER size restrictions are:

IBK ⩾ N∗(IW + 1)
IPB2 ⩾ IBAND
ILOADS ⩾ N
INF ⩾ NN
IWID ⩾ NXE + 1
IDEP ⩾ NYE + 1

Turning to the program, the nodal coordinates and steering vector are provided by subroutine GEV4X3 and the subroutine FMKDKE builds up the KD and KE matrices required by equation (9.1). Then KE is assembled into the (symmetric) global matrix (actually vector) BK while KD is assembled into the unsymmetrical band matrix PB. When using (3.114) the presence of many zeros in KD means that element-by-element summation to avoid assembling KD would almost certainly be the better strategy.

The subroutine BANRED is used to factorise the symmetric left hand side global matrix BK summed from KE and the time-stepping loop is entered. The right hand side global matrix, PB, summed from KD is multiplied by the old solution LOADS using subroutine BANTML to give ANS, which is then modified

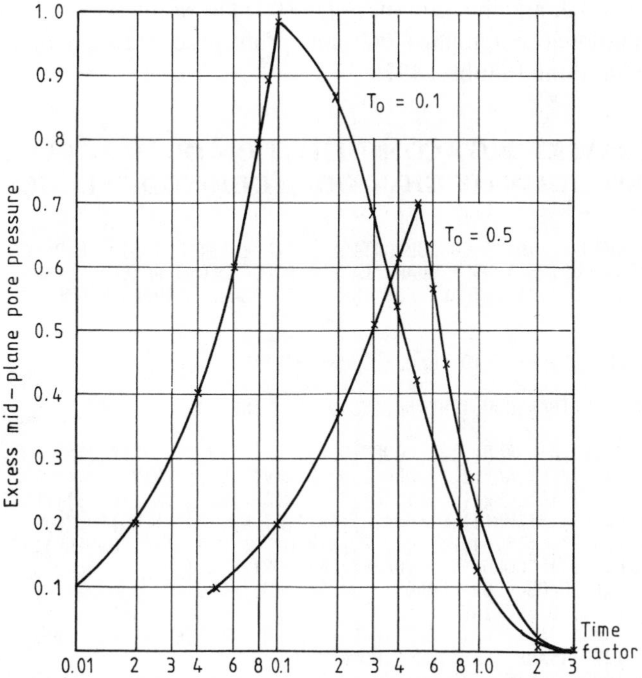

Figure 9.9 Mid-plane pore pressure response to ramp loading (Program 9.1)

for ramp loading. The subroutine BACSUB completes the solution and the element effective stresses are recovered at the element 'centre' ($\xi = \eta = 0$).

The results for the first time step are shown as Figure 9.8 and the mid-plane pore pressure (freedoms 97 to 100) is plotted against time in Figure 9.9 for two different rise-times, $t_o = 10$ and $t_o = 50$. The 'time factor' in Figure 9.9 is the dimensionless number

$$T = \frac{c_v t}{H^2}$$

where H is the 'drainage path distance' of 10.0 in the present instance. The coefficient of consolidation c_v is found from

$$c_v = \frac{k}{m_v \gamma_w} \tag{9.3}$$

where

$$m_v = \frac{(1 + v')(1 - 2v')}{E'(1 - v')} \tag{9.4}$$

and k is the 'permeability' (Chapter 7) with γ_w being the unit weight of water.

In the present example v' is 0 and $E' = 1.0$, so m_v is just 1.0. Similarly, k/γ_w is 1.0 so that $T = t/100$. Thus, for DT1M equal to 1.0, the solution at the first step is for $T = 0.01$ and the rise-time is $T_o = 0.1$. The results will be found to agree exactly with those of Schiffman (1960), and problems of practical importance can be solved (Smith and Hobbs, 1976).

PROGRAM 9.2: BIOT CONSOLIDATION OF A RECTANGLE IN PLANE STRAIN (EIGHT-NODE/FOUR-NODE ELEMENTS)

```
C
C         PROGRAM 9.2 BIOT CONSOLIDATION OF AN ELASTIC SOLID IN PLANE
C         STRAIN USING 4-NODE QUADRILATERALS FOR FLUID PHASE PRESSURE AND
C         8-NODE QUADRILATERALS FOR THE SOLID PHASE DISPLACEMENTS
C
C
C         ALTER NEXT LINE TO CHANGE PROBLEM SIZE
C
      PARAMETER(IBK=5000,IPB2=50,ILOADS=200,INF=100)
C
      REAL DEE(3,3),SAMP(3,2),COORD(8,2),DERIVT(8,2),JAC(2,2)
      REAL JAC1(2,2),KAY(2,2),DER(2,8),DERIV(2,8),KDERIV(2,8)
      REAL BEE(3,16),DBEE(3,16),BT(16,3),BTDB(16,16),KM(16,16),ELD(16)
      REAL EPS(3),SIGMA(3),DTKD(4,4),KP(4,4),KE(20,20),KD(20,20)
      REAL FUN(8),C(16,4),VOLF(16,4),WIDTH(30),DEPTH(30),BK(IBK),VOL(16)
      REAL FUNF(4),COORDF(4,2),DERF(2,4),DERIVF(2,4)
      REAL PB(ILOADS,IPB2),LOADS(ILOADS),ANS(ILOADS)
      INTEGER G(20),NF(INF,3)
      DATA IJAC,IJAC1,IKAY,IDER,IDERIV,IKDERV,IT,IDERF,IDERVF/9*2/
      DATA IDEE,ISAMP,IBEE,IDBEE,NODOF,IH/6*3/
      DATA ICORDF,IDTKD,IKP,NODF/4*4/
      DATA ICOORD,IDERVT,NOD/3*8/,IBT,IBTDB,IKM,IC,IVOLF,IDOF/6*16/
      DATA IKE,IKD,ITOT/3*20/
```

```
C
C      INPUT AND INITIALISATION
C
      READ(5,*)NXE,NYE,N,IW,NN,NR,NGP,PERMX,PERMY,E,V,DTIM,ISTEP,THETA
      READ(5,*)(WIDTH(I),I=1,NXE+1)
      READ(5,*)(DEPTH(I),I=1,NYE+1)
      CALL READNF(NF,INF,NN,NODOF,NR)
      IBAND=2*(IW+1)-1
      IR=N*(IW+1)
      CALL NULL(PB,ILOADS,N,IBAND)
      CALL NULVEC(LOADS,N)
      CALL NULVEC(BK,IR)
      CALL NULL(DEE,IDEE,IH,IH)
      CALL FMDEPS(DEE,IDEE,E,V)
      CALL NULL(KAY,IKAY,IT,IT)
      KAY(1,1)=PERMX
      KAY(2,2)=PERMY
      CALL GAUSS(SAMP,ISAMP,NGP)
C
C      ELEMENT MATRIX INTEGRATION AND ASSEMBLY
C
      DO 10 IP=1,NXE
      DO 10 IQ=1,NYE
      CALL GEVUVP(IP,IQ,NXE,COORD,ICOORD,COORDF,ICORDF,G,NF,INF,
     +            WIDTH,DEPTH)
      CALL NULL(KM,IKM,IDOF,IDOF)
      CALL NULL(C,IC,IDOF,NODF)
      CALL NULL(KP,IKP,NODF,NODF)
      DO 20 I=1,NGP
      DO 20 J=1,NGP
      CALL FMQUAD(DER,IDER,FUN,SAMP,ISAMP,I,J)
      CALL MATMUL(DER,IDER,COORD,ICOORD,JAC,IJAC,IT,NOD,IT)
      CALL TWOBY2(JAC,IJAC,JAC1,IJAC1,DET)
      CALL MATMUL(JAC1,IJAC1,DER,IDER,DERIV,IDERIV,IT,IT,NOD)
      CALL NULL(BEE,IBEE,IH,IDOF)
      CALL FORMB(BEE,IBEE,DERIV,IDERIV,NOD)
      CALL VOL2D(BEE,IBEE,VOL,NOD)
      CALL MATMUL(DEE,IDEE,BEE,IBEE,DBEE,IDBEE,IH,IH,IDOF)
      CALL MATRAN(BT,IBT,BEE,IBEE,IH,IDOF)
      CALL MATMUL(BT,IBT,DBEE,IDBEE,BTDB,IBTDB,IDOF,IH,IDOF)
      QUOT=DET*SAMP(I,2)*SAMP(J,2)
      CALL MSMULT(BTDB,IBTDB,QUOT,IDOF,IDOF)
      CALL MATADD(KM,IKM,BTDB,IBTDB,IDOF,IDOF)
C
C      FLUID CONTRIBUTION
C
      CALL FORMLN(DERF,IDERF,FUNF,SAMP,ISAMP,I,J)
      CALL MATMUL(DERF,IDERF,COORDF,ICORDF,JAC,IJAC,IT,NODF,IT)
      CALL TWOBY2(JAC,IJAC,JAC1,IJAC1,DET)
      CALL MATMUL(JAC1,IJAC1,DERF,IDERF,DERIVF,IDERVF,IT,IT,NODF)
      CALL MATMUL(KAY,IKAY,DERIVF,IDERVF,KDERIV,IKDERV,IT,IT,NODF)
      CALL MATRAN(DERIVT,IDERVT,DERIVF,IDERVF,IT,NODF)
      CALL MATMUL(DERIVT,IDERVT,KDERIV,IKDERV,DTKD,IDTKD,NODF,IT,NODF)
      QUOT=DET*SAMP(I,2)*SAMP(J,2)
      PROD=QUOT*DTIM
      CALL MSMULT(DTKD,IDTKD,PROD,NODF,NODF)
      CALL MATADD(KP,IKP,DTKD,IDTKD,NODF,NODF)
      DO 30 K=1,IDOF
      DO 30 L=1,NODF
   30 VOLF(K,L)=VOL(K)*FUNF(L)*QUOT
      CALL MATADD(C,IC,VOLF,IVOLF,IDOF,NODF)
   20 CONTINUE
```

```
      CALL FMKDKE(KM,IKM,KP,IKP,C,IC,KE,IKE,KD,IKD,IDOF,NODF,ITOT,THETA)
      CALL FORMKV(BK,KE,IKE,G,N,ITOT)
      CALL FORMTB(PB,ILOADS,KD,IKD,G,IW,ITOT)
   10 CONTINUE
C
C      REDUCE LEFT HAND SIDE
C
      CALL BANRED(BK,N,IW)
C
C      TIME STEPPING LOOP
C
      DO 40 NS=1,ISTEP
      CALL BANTML(PB,ILOADS,LOADS,ANS,N,IW)
C
C      RAMP LOADING
C
      X1=(.1*NS+.1*(THETA-1.))/6.
      X2=X1*4.
      IF(NS.GT.10)ANS(1)=ANS(1)-1./6.
      IF(NS.LE.10)ANS(1)=ANS(1)-X1
      IF(NS.GT.10)ANS(3)=ANS(3)-2./3.
      IF(NS.LE.10)ANS(3)=ANS(3)-X2
      IF(NS.GT.10)ANS(4)=ANS(4)-1./6.
      IF(NS.LE.10)ANS(4)=ANS(4)-X1
      CALL BACSUB(BK,ANS,N,IW)
      DO 50 I=1,N
   50 LOADS(I)=ANS(I)
      CALL PRINTV(ANS,N)
C
C      RECOVER ELEMENT EFFECTIVE STRESSES AT ELEMENT 'CENTRES'
C
      NGP=1
      CALL GAUSS(SAMP,ISAMP,NGP)
      DO 60 IP=1,NXE
      DO 60 IQ=1,NYE
      CALL GEVUVP(IP,IQ,NXE,COORD,ICOORD,COORDF,ICORDF,G,NF,INF,
     +            WIDTH,DEPTH)
      DO 70 M=1,IDOF
      IF(G(M).EQ.0)ELD(M)=0.
   70 IF(G(M).NE.0)ELD(M)=ANS(G(M))
      DO 80 I=1,NGP
      DO 80 J=1,NGP
      CALL FMQUAD(DER,IDER,FUN,SAMP,ISAMP,I,J)
      CALL MATMUL(DER,IDER,COORD,ICOORD,JAC,IJAC,IT,NOD,IT)
      CALL TWOBY2(JAC,IJAC,JAC1,IJAC1,DET)
      CALL MATMUL(JAC1,IJAC1,DER,IDER,DERIV,IDERIV,IT,IT,NOD)
      CALL NULL(BEE,IBEE,IH,IDOF)
      CALL FORMB(BEE,IBEE,DERIV,IDERIV,NOD)
      CALL MVMULT(BEE,IBEE,ELD,IH,IDOF,EPS)
      CALL MVMULT(DEE,IDEE,EPS,IH,IH,SIGMA)
      CALL PRINTV(SIGMA,IH)
   80 CONTINUE
   60 CONTINUE
   40 CONTINUE
      STOP
      END
```

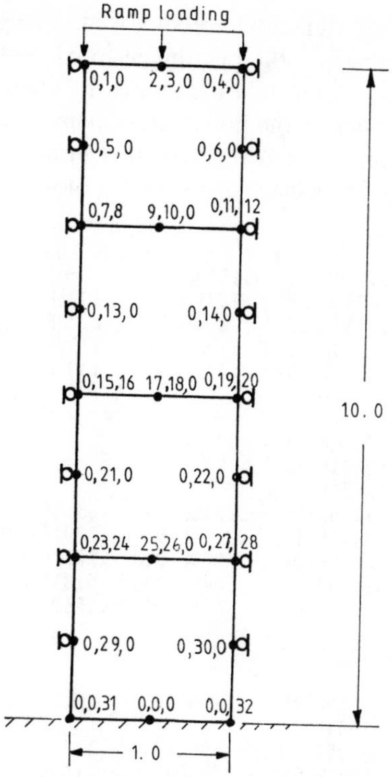

Mesh data	NXE	NYE	N	IW	NN	NR	NGP
	1	4	32	13	23	23	2

Element data	PERMX	PERMY	E	V
	1.0	1.0	1.0	.0

Time integration data	DTIM	ISTEP	THETA
	1.0	5	.5

Geometry data	WIDTH(I), I = 1, NXE+1	0.0	1.0			
	DEPTH(I), I = 1, NYE+1	0.0	−2.5	−5.0	−7.5	−10.0

Node freedom data	1	0	1	0	2	1	1	0	3	0	1	0
	4	0	1	0	5	0	1	0				
	6	0	1	1	7	1	1	0	8	0	1	1
	9	0	1	0	10	0	1	0				
	11	0	1	1	12	1	1	0	13	0	1	1
	14	0	1	0	15	0	1	0				
	16	0	1	1	17	1	1	0	18	0	1	1
	19	0	1	0	20	0	1	0				
	21	0	0	1	22	0	0	0	23	0	0	1

Figure 9.10 Mesh and data for Program 9.2

In this program, mixed shape functions are employed, as they were in Program 9.0.

The problem chosen is again the one-dimensional consolidation test (plane strain) and the mesh and input data are shown in Figure 9.10. The geometry subroutine GEVUVP provides the nodal coordinates and steering vector. The variables used have all been defined previously in relation to Programs 9.0 and 9.1.

Output from the program is listed as Figure 9.11 for five time steps. The mid-

```
-.1010E+00  -.2697E-15  -.1010E+00  -.1010E+00  -.1060E-01  -.1060E-01 )
 .1071E-01  -.1106E+00   .1943E-15   .1071E-01   .1071E-01  -.1106E+00 | Displacements
 .1124E-02   .1124E-02  -.1135E-02  -.9888E-01  -.2220E-15  -.1135E-02 > and
-.1135E-02  -.9888E-01  -.1193E-03  -.1193E-03   .1190E-03  -.1001E+00 | porepressures
-.3886E-15   .1190E-03   .1190E-03  -.1001E+00   .1397E-04   .1397E-04 |
-.9997E-01  -.9997E-01                                                  )
 .0000E+00  -.4470E-01   .2387E-15 )
 .0000E+00   .4739E-02  -.1395E-15 |
 .0000E+00  -.5018E-03   .3326E-16 > Effective stresses
 .0000E+00   .4762E-04  -.7771E-16 )                         Timestep  1
-.2516E+00  -.3469E-15  -.2516E+00  -.2516E+00  -.6312E-01  -.6312E-01
 .2204E-02  -.1970E+00   .4441E-15   .2204E-02   .2204E-02  -.1970E+00
 .4125E-02   .4125E-02   .2358E-02  -.2029E+00  -.7772E-15   .2358E-02
 .2358E-02  -.2029E+00  -.1645E-03  -.1645E-03  -.5131E-03  -.1994E+00
-.7772E-15  -.5131E-03  -.5131E-03  -.1994E+00  -.1559E-04  -.1559E-04
-.2002E+00  -.2002E+00
 .0000E+00  -.1015E+00   .1332E-15
 .0000E+00  -.6164E-04   .4607E-15
 .0000E+00   .1149E-02  -.6660E-15
 .0000E+00  -.2052E-03  -.1554E-15                                     2
-.4395E+00  -.8812E-15  -.4395E+00  -.4395E+00  -.1489E+00  -.1489E+00
-.2744E-01  -.2704E+00   .8882E-15  -.2744E-01  -.2744E-01  -.2704E+00
-.8689E-03  -.8689E-03   .4840E-02  -.3038E+00  -.1166E-14   .4840E-02
 .4840E-02  -.3038E+00   .1238E-02   .1238E-02  -.1181E-03  -.3002E+00
-.1332E-14  -.1181E-03  -.1181E-03  -.3002E+00  -.2053E-03  -.2053E-03
-.2997E+00  -.2997E+00
 .0000E+00  -.1648E+00   .1155E-14
 .0000E+00  -.1291E-01  -.3400E-16
 .0000E+00   .1983E-02   .3400E-16
 .0000E+00  -.4725E-04   .1778E-15                                     3
-.6585E+00  -.4961E-15  -.6585E+00  -.6585E+00  -.2633E+00  -.2633E+00
-.7775E-01  -.3354E+00   .6661E-15  -.7775E-01  -.7775E-01  -.3354E+00
-.1723E-01  -.1723E-01   .2781E-02  -.4002E+00  -.9992E-15   .2781E-02
 .2781E-02  -.4002E+00   .2237E-02   .2237E-02   .1080E-02  -.4012E+00
-.1443E-14   .1080E-02   .1080E-02  -.4012E+00   .7857E-04   .7857E-04
-.3997E+00  -.3997E+00
 .0000E+00  -.2323E+00  -.1332E-15
 .0000E+00  -.3221E-01   .7550E-15
 .0000E+00   .6802E-03  -.8007E-15
 .0000E+00   .4321E-03  -.1621E-14                                     4
-.9049E+00  -.1769E-14  -.9049E+00  -.9049E+00  -.4031E+00  -.4031E+00
-.1478E+00  -.3943E+00   .1554E-14  -.1478E+00  -.1478E+00  -.3943E+00
-.4632E-01  -.4632E-01  -.5897E-02  -.4921E+00  -.1776E-14  -.5897E-02
-.5897E-02  -.4921E+00   .1039E-02   .1039E-02   .2110E-02  -.5015E+00
-.2442E-14   .2110E-02   .2110E-02  -.5015E+00   .6515E-03   .6515E-03
-.5002E+00  -.5002E+00
 .0000E+00  -.3028E+00   .1465E-14                                     5
 .0000E+00  -.5678E-01   .6883E-15
 .0000E+00  -.3203E-02   .1346E-15
 .0000E+00   .8440E-03   .8431E-15
```

Figure 9.11 Results from Program 9.2

plane pore pressure is now given by freedoms 31 and 32 and can be seen to be following Figure 9.9. The only non-zero component of effective stress is σ'_y, which can be seen to have reached a value of -0.3028 at the centre of the top element at $T = 0.05$.

9.2 REFERENCES

Schiffman, R. L. (1960) Field applications of soil consolidation time-dependent loading and variable permeability. Highway Research Board, Bulletin 248, Washington, USA.

Smith, I. M., and Hobbs, R. (1976) Biot analysis of consolidation beneath embankment. *Géotechnique*, **26**, No. 1, 149–71.

CHAPTER 10
Eigenvalue Problems

10.0 INTRODUCTION

The ability to solve eigenvalue problems is important in many aspects of finite element work. For example, the number of zero eigenvalues of a 'stiffness' matrix (its rank deficiency) is an important guide to the suitability of that matrix. In this context, the problem to be solved is just

$$\mathbf{A}\mathbf{x} = \lambda\mathbf{x} \tag{10.1}$$

which is the eigenvalue problem in 'standard form'. More often, the eigenvalue equation will describe a physical situation such as free vibration of a solid or fluid. For example, equation (2.19) for a freely vibrating elastic solid was

$$\mathbf{KMa} = \omega^2\mathbf{MMa} \tag{10.2}$$

which, although no longer in standard form, can readily be converted to it. Section 3.11 describes the conversion process for the cases in which **MM** is 'lumped', that is diagonalised, and when it is not (the 'consistent' mass matrix).

The present chapter describes five programs for the determination of eigenvalues (and usually eigenvectors) of such elastic solids. Different algorithms, storage strategies and mass matrix assumptions are employed in the various cases. Since elastic solids are treated, the programs can be viewed as extensions of the programs described in Chapter 4 and 5. The same terminology is used.

PROGRAM 10.0: EIGENVALUES AND EIGENVECTORS OF A TWO-DIMENSIONAL FRAME OF BEAM-COLUMN ELEMENTS—CONSISTENT MASS

```
C
C      PROGRAM 10.0 EIGENVALUES AND EIGENVECTORS OF A 2-D FRAME
C      OF BEAM-COLUMN ELEMENTS,CONSISTENT MASS.
C
C
C      ALTER NEXT LINE TO CHANGE PROBLEM SIZE
C
      PARAMETER(IKB1=100,IKB2=20,IPROP=20,INF=30)
```

```
C
      REAL KB(IKB1,IKB2),MB(IKB1,IKB2),KM(6,6),MM(6,6),STOREC(IPROP,4)
      REAL BIGK(IKB1,IKB1),DIAG(IKB1),UDIAG(IKB1)
      INTEGER G(6),NF(INF,3)
      DATA IKM,IDOF/2*6/,NODOF/3/
C
C        INPUT SECTION
C
      READ(5,*)NXE,N,IW,NN,NR,NMODES
      CALL READNF(NF,INF,NN,NODOF,NR)
      IWP1=IW+1
      CALL NULL(KB,IKB1,N,IWP1)
      CALL NULL(MB,IKB1,N,IWP1)
      CALL NULL(BIGK,IKB1,N,N)
C
C        GLOBAL STIFFNESS AND MASS MATRIX ASSEMBLY
C
      DO 10 IP=1,NXE
      READ(5,*)EA,EI,RHO,AREA,(STOREC(IP,I),I=1,4)
      CALL BMCOL2(KM,EA,EI,IP,STOREC,IPROP)
      CALL BCMASS(MM,RHO,AREA,IP,STOREC,IPROP)
      CALL GSTRNG(IP,NODOF,NF,INF,G)
      CALL FORMKB(KB,IKB1,KM,IKM,G,IW,IDOF)
   10 CALL FORMKB(MB,IKB1,MM,IKM,G,IW,IDOF)
C
C       REDUCE TO STANDARD EIGENVALUE PROBLEM
C
      CALL CHOLIN(MB,IKB1,N,IW)
      CALL LBKBAN(MB,BIGK,KB,IKB1,IW,N)
      DO 20 I=1,N
      DO 20 J=I+1,N
   20 BIGK(I,J)=BIGK(J,I)
      CALL LBBT(MB,BIGK,IKB1,IW,N)
C
C       EXTRACT EIGENVALUES AND EIGENVECTORS
C
      CALL TRIDIA(N,1.E-293,BIGK,DIAG,UDIAG,BIGK,IKB1)
      IFAIL=1
      CALL EVECTS(N,1.E-293,DIAG,UDIAG,BIGK,IKB1,IFAIL)
      CALL PRINTV(DIAG,N)
      DO 30 J=1,NMODES
      DO 40 I=1,N
   40 UDIAG(I)=BIGK(I,J)
      CALL CHOBK2(MB,IKB1,UDIAG,N,IW)
   30 CALL PRINTV(UDIAG,N)
      STOP
      END
```

This program illustrates a free vibration analysis of a typical line structure and can be thought of as an extension of Programs 4.0 and 4.3. Only 'string' topologies are allowed and Figure 10.1 shows a simple cantilever beam made up from three line elements, with three degrees of freedom per node.

The simple variables used, with reference to Figure 10.1, are:

NXE	number of elements (3)
N	total number of non-zero degrees of freedom (9)
IW	half-bandwidth of coefficient matrices (5)
NN	total number of nodes (4)

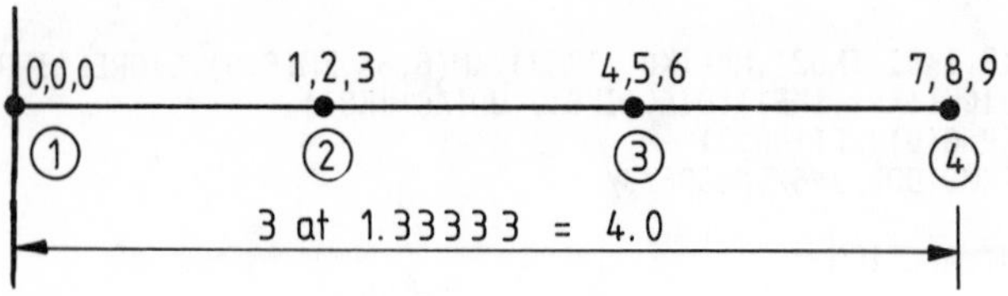

Structure data	NXE	N	IW	NN	NR			
	3	9	5	4	1			
Print control	NMODES							
	6							
Node freedom data	1	0	0	0				
Element data	EA	EI	RHO	AREA	X1	Y1	X2	Y2
	1.	.08333	1.	1.	0.	0.	1.33333	0.
	1.	.08333	1.	1.	1.33333	0.	2.66667	0.
	1.	.08333	1.	1.	2.66667	0.	4.	0.

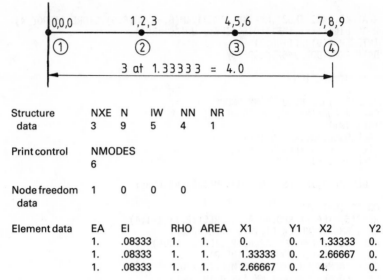

Figure 10.1 Mesh and data for Program 10.0

NR	number of restrained nodes (1)
NMODES	number of eigenvectors desired
IWP1	half-bandwidth plus 1
IKM	working size of array KM
IDOF	number of degrees of freedom per element (6)
NODOF	number of degrees of freedom per node (3)
EA	axial stiffness of element (1)
EI	flexural stiffness of element (0.08333)
RHO	mass per unit length of element (1)
AREA	cross-sectional area of element (1)

Fixed length arrays are:

KM	element stiffness matrix
MM	element mass matrix
G	steering vector

and adjustable length arrays are:

KB	global banded stiffness matrix
MB	global banded mass matrix
STOREC	element nodal coordinate matrix
BIGK	global standard eigenvalue matrix (not banded)
DIAG	eigenvalues
UDIAG	eigenvectors
NF	node freedom array

PARAMETER statement restrictions are:

IKB1 ⩾ N
IKB2 ⩾ IWP1
IPROP ⩾ NXE
INF ⩾ NN

After mesh data has been read in, including nodal freedom data requested by READNF, the global matrices are nulled. Then the global matrix assembly loop is entered, requiring axial stiffness, flexural stiffness, mass per unit length and cross-sectional area for each element, together with its nodal coordinates.

The subroutine BMCOL2 then constructs the appropriate element stiffness matrix (an amalgamation of 2.11 and 2.26) while BCMASS constructs the appropriate element consistent mass matrix (an amalgamation of 2.16 and 2.29).

The steering vector is built up by GSTRNG and symmetrical, banded global stiffness and mass (KB and MB respectively) constructed using FORMKB. The eigenvalue equation to be solved is now equation (3.82). As shown in (3.83) the next step is to factorise MB using CHOLIN, and then LBKBAN and LBBT produce the matrix BIGK described by (3.91). This matrix is in $N \times N$ form, and so the storage requirements are substantial for large N.

The eigenvalue solution routines are TRIDIA and EVECTS, after which the N eigenvalues are stored in array DIAG. The true eigenvectors must then be recovered (see equation 3.85) by calling the subroutine CHOBK2. This leaves the eigenvectors in array UDIAG, which is only done for the first NMODES eigenmodes. The higher frequency modes are unlikely to be of engineering significance.

The results from the program are listed as Figure 10.2. The square of the fundamental frequency, computed as ω_1^2, is 0.004025. This agrees with the exact result, $1.875^4 \times \mathrm{EI}/\rho AL^4$. Note from the eigenvectors, of which six are printed, that some modes, for example 1 and 3, represent flexural vibrations, whereas others, for example 2, represent longitudinal vibrations. The square of the fundamental axial frequency ω_2^2 is computed as 0.1578, which is in good agreement with the exact result, $\pi^2 \times E/4L^2\rho$.

```
 .4025E-02   .1578E+00   .1591E+00   .1270E+01   .1688E+01   .5554E+01
 .6441E+01   .2281E+02   .9068E+02
                                                Eigenvalues
-.3453E-13  -.1656E+00  -.2262E+00  -.6744E-13  -.5471E+00  -.3273E+00
-.9252E-13  -.1000E+01  -.3442E+00
                                                Eigenvector  1
 .3617E+00  -.8423E-11  -.6324E-11   .6265E+00  -.6101E-11   .1050E-10
 .7234E+00   .1415E-10   .1704E-10
                                                             2
-.5137E-11  -.5935E+00  -.4436E+00  -.8882E-11  -.4261E+00   .7444E+00
-.1023E-10   .1006E+01   .1204E+01
                                                             3
 .2844E-13  -.7476E+00   .4175E+00   .1106E-13   .6593E+00   .3187E+00
-.2504E-13  -.1005E+01  -.1997E+01
                                                             4
 .8660E+00   .1627E-13  -.3383E-13  -.2887E-05  -.2734E-13   .2429E-13
-.8660E+00   .6277E-13   .8537E-13
                                                             5
 .5751E+00  -.1357E-13   .1329E-12  -.9960E+00  -.5704E-15  -.1295E-12
 .1150E+01   .3230E-13   .1310E-12
                                                             6
```

Figure 10.2 Results from Program 10.0

PROGRAM 10.1: EIGENVALUES OF A RECTANGULAR SOLID IN PLANE STRAIN—FOUR-NODE QUADRILATERALS WITH LUMPED MASS

```
C
C      PROGRAM 10.1 EIGENVALUES OF A RECTANGULAR
C      SOLID IN PLANE STRAIN USING 4-NODE QUADRILATERALS
C      LUMPED MASS
C
C
C      ALTER NEXT LINE TO CHANGE PROBLEM SIZE
C
      PARAMETER(IKU1=200,IKU2=50,INF=100)
C
      REAL DEE(3,3),SAMP(3,2),COORD(4,2),FUN(4),JAC(2,2),JAC1(2,2)
      REAL DER(2,4),DERIV(2,4),BEE(3,8),DBEE(3,8),BTDB(8,8)
      REAL BT(8,3),KM(8,8),KU(IKU1,IKU2),EMM(8,8)
      REAL LOADS(IKU1),DIAG(IKU1),UDIAG(IKU1)
      INTEGER NF(INF,2),G(8)
      DATA IJAC,IJAC1,IDER,IDERIV,NODOF,IT/6*2/
      DATA IH,ISAMP,IDEE,IBEE,IDBEE/5*3/
      DATA ICOORD,NOD/2*4/,IBTDB,IKM,IBT,IEMM,IDOF/5*8/
C
C      INPUT AND INITIALISATION
C
      READ(5,*)NXE,NYE,N,IW,NN,NR,NGP,AA,BB,RHO,E,V
      CALL READNF(NF,INF,NN,NODOF,NR)
      IWP1=IW+1
      CALL NULL(EMM,IEMM,IDOF,IDOF)
      CALL NULVEC(DIAG,N)
      CALL NULL(KU,IKU1,N,IWP1)
      CALL FMDEPS(DEE,IDEE,E,V)
      CALL GAUSS(SAMP,ISAMP,NGP)
C
C          FORM ELEMENT LUMPED MASS MATRIX
C
      DO 30 I=1,IDOF
   30 EMM(I,I)=.25*AA*BB*RHO
C
C      ELEMENT STIFFNESS AND MASS INTEGRATION AND ASSEMBLY
C
      DO 10 IP=1,NXE
      DO 10 IQ=1,NYE
      CALL GEOM4Y(IP,IQ,NYE,AA,BB,COORD,ICOORD,G,NF,INF)
      CALL NULL(KM,IKM,IDOF,IDOF)
      DO 20 I=1,NGP
      DO 20 J=1,NGP
      CALL FORMLN(DER,IDER,FUN,SAMP,ISAMP,I,J)
      CALL MATMUL(DER,IDER,COORD,ICOORD,JAC,IJAC,IT,NOD,IT)
      CALL TWOBY2(JAC,IJAC,JAC1,IJAC1,DET)
      CALL MATMUL(JAC1,IJAC1,DER,IDER,DERIV,IDERIV,IT,IT,NOD)
      CALL NULL(BEE,IBEE,IH,IDOF)
      CALL FORMB(BEE,IBEE,DERIV,IDERIV,NOD)
      CALL MATMUL(DEE,IDEE,BEE,IBEE,DBEE,IDBEE,IH,IH,IDOF)
      CALL MATRAN(BT,IBT,BEE,IBEE,IH,IDOF)
      CALL MATMUL(BT,IBT,DBEE,IDBEE,BTDB,IBTDB,IDOF,IH,IDOF)
      QUOT=DET*SAMP(I,2)*SAMP(J,2)
      CALL MSMULT(BTDB,IBTDB,QUOT,IDOF,IDOF)
      CALL MATADD(KM,IKM,BTDB,IBTDB,IDOF,IDOF)
   20 CONTINUE
      CALL FORMKU(KU,IKU1,KM,IKM,G,IW,IDOF)
      CALL FMLUMP(DIAG,EMM,IEMM,G,IDOF)
```

```
   10 CONTINUE
      CALL PRINTV(DIAG,N)
C
C      REDUCE TO STANDARD EIGENVALUE PROBLEM
C
      DO 50 I=1,N
   50 DIAG(I)=1./SQRT(DIAG(I))
      DO 60 I=1,N
      IF(I.LE.N-IW)K=IWP1
      IF(I.GT.N-IW)K=N-I+1
      DO 60 J=1,K
   60 KU(I,J)=KU(I,J)*DIAG(I)*DIAG(I+J-1)
C
C      EXTRACT EIGENVALUES
C
      CALL BANDRD(N,IW,KU,IKU1,DIAG,UDIAG,LOADS)
      IFAIL=1
      CALL BISECT(N,1.E-293,DIAG,UDIAG,IFAIL)
      CALL PRINTV(DIAG,N)
      STOP
      END
```

This program is an extension of Program 5.3 and uses much of that code. For completeness, the significance of the variables is listed below. Simple variables read as data are:

NXE	number of elements in x direction (3)
NYE	number of elements in y direction (1)
N	total number of non-zero freedoms (12)
IW	half-bandwidth (7)
NN	total number of nodes in mesh (8)
NR	number of restrained nodes (2)
NGP	number of Gauss points in each direction (2)
AA	element size in x direction (1.33333)
BB	element size in y direction (1.0)
RHO	mass density per unit volume (1.0)
E	Young's modulus (1.0)
V	Poisson's ratio (0.3)

while those initialised by DATA statements are:

IJAC IJAC1 IDER IDERIV	working sizes of arrays JAC, JAC1, DER, DERIV1
NODOF	number of degrees of freedom per node (2)
IT	dimensions of problem (2)
IH	number of stress and strain components (3)
ISAMP IDEE IBEE IDBEE ICOORD	working sizes of arrays SAMP, DEE, BEE, DBEE, COORD

NOD	number of nodes per element (4)
IBTDB IKM IBT IEMM	working sizes of arrays BTDB, KM, BT, EMM
IDOF	number of degrees of freedom per element (8)

Fixed length arrays are:

DEE	stress–strain matrix
SAMP	Gaussian weights and abscissae
COORD	element coordinates
FUN	element shape functions
JAC	Jacobian matrix
JAC1	inverse of Jacobian matrix
DER	derivatives of shape functions in local coordinates
DERIV	derivatives of shape functions in global coordinates
BEE	strain-displacement matrix
DBEE	product of DEE*BEE
BTDB	product of BT*DEE*BEE
BT	transpose of BEE
KM	element stiffness matrix
EMM	element mass matrix
G	steering vector

while variable length arrays are:

KU	global stiffness matrix (upper triangle stored)
LOADS	working space vector
DIAG	lumped mass matrix (vector) overwritten by eigenvalues
UDIAG	working space vector
NF	node freedom array

PARAMETER restrictions are:

IKU1 ⩾ N
IKU2 ⩾ IWP1
INF ⩾ NN

An example problem is shown in Figure 10.3 and the structure of the program is shown in Figure 10.4. Nominally, this problem is the same as in Figure 10.1, representing an elastic solid cantilever 4.0 units long in the x direction with a cross-sectional area of 1.0 and flexural rigidity of $\frac{1}{12}$ or 0.08333. However, the solid in Figure 10.4 is truly two dimensional and has the additional property of Poisson's ratio, set in this example to 0.3. Since the stress–strain matrix DEE is formed by FMDEPS, the solid is assumed to be in a state of plane strain.

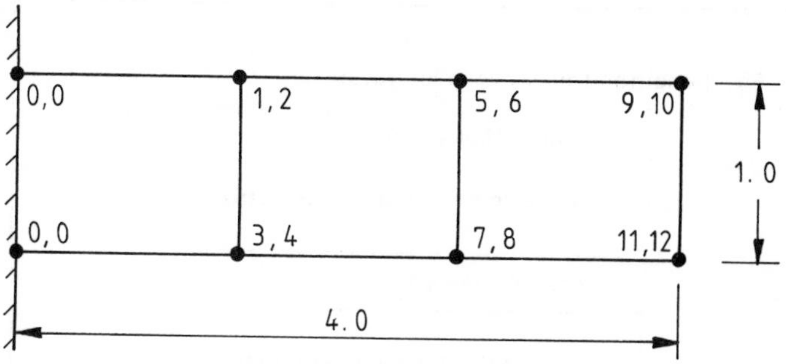

Structure data	NXE	NYE	N	IW	NN	NR	NGP
	3	1	12	7	8	2	2
Element data	AA		BB	RHO	E	V	
	1.33333		1.0	1.0	1.0	0.3	
Node freedom data	1	0	0	2	0	0	

Note: In Program 10.4, the additional print control integer NMODES(6 in this case) is read after V.

Figure 10.3 Mesh and data for Programs 10.1 and 10.4

The lumped mass matrix is readily formed with eight diagonal terms (equal to area × mass per unit area × 0.25), with one-quarter of the mass of the element 'lumped' at each corner and in each direction.

The stiffness and mass assembly loop is then entered for all elements, as was done for the stiffness in Chapter 5. In this case the global stiffness matrix is stored as an upper triangular band rectangle by subroutine FORMKU and the lumped mass matrix as a vector DIAG, by subroutine FMLUMP which is printed out. By factorising DIAG and altering the appropriate components of KU, the global band stiffness is modified into a banded version of BIGK, as described in Section 3.11.1, but still stored in KU.

The eigenvalues of this band matrix are then calculated using BANDRD and BISECT and printed, as listed in Figure 10.5.

It can be seen that the square of the fundamental frequency, printed as ω_1^2, is 0.006488, considerably higher than that calculated by Program 10.0 for a slender beam. Thus, the solid in plane strain, as represented by four-node elements, is a poor representation of a slender beam, at least in the flexural modes. The elements are too 'stiff'. However, the longitudinal modes for analyses 10.0 and 10.1 are much closer (and would agree more closely for Poisson's ratio of zero in Program 10.1).

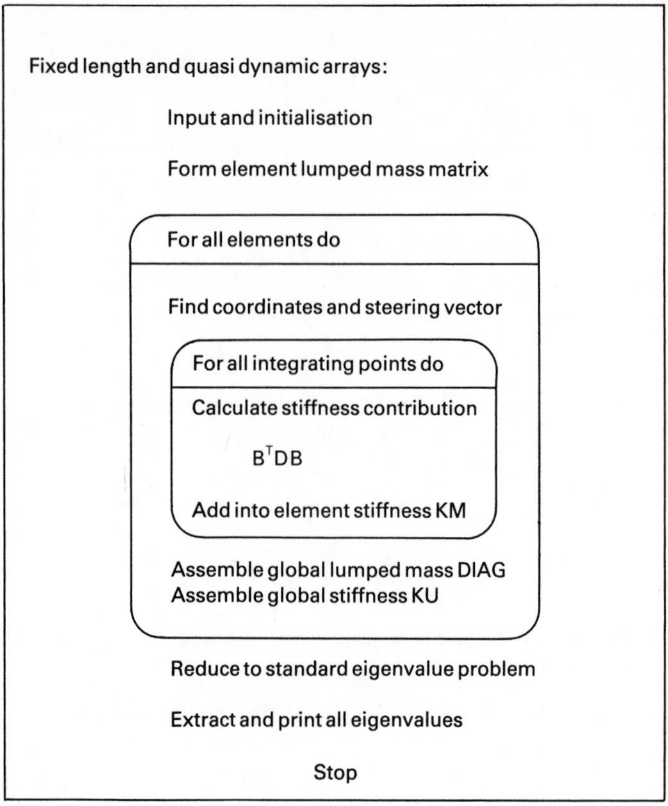

Figure 10.4 Structure chart for Program 10.1

.6667E+00	.6667E+00	.6667E+00	.6667E+00	.6667E+00	.6667E+00	Lumped mass matrix
.6667E+00	.6667E+00	.3333E+00	.3333E+00	.3333E+00	.3333E+00	
.6488E-02	.1253E+00	.1725E+00	.5294E+00	.1262E+01	.1543E+01	Eigenvalues
.1660E+01	.2025E+01	.2045E+01	.2639E+01	.4152E+01	.5474E+01	

Figure 10.5 Results from Program 10.1

PROGRAM 10.2: EIGENVALUES AND EIGENVECTORS OF A RECTANGULAR SOLID IN PLANE STRAIN—EIGHT-NODE QUADRILATERALS WITH LUMPED MASS

```
C
C      PROGRAM 10.2 EIGENVALUES AND EIGENVECTORS OF A RECTANGULAR
C      SOLID IN PLANE STRAIN USING 8-NODE QUADRILATERALS
C      LUMPED MASS
C
C
C      ALTER NEXT LINE TO CHANGE PROBLEM SIZE
C
      PARAMETER(IBIGK=103,INF=85)
```

```
C
      REAL DEE(3,3),SAMP(3,2),COORD(8,2),FUN(8),JAC(2,2),JAC1(2,2)
      REAL DER(2,8),DERIV(2,8),BEE(3,16),DBEE(3,16),BTDB(16,16)
      REAL BT(16,3),KM(16,16),BIGK(IBIGK,IBIGK),EMM(16,16)
      REAL LOADS(IBIGK),DIAG(IBIGK),UDIAG(IBIGK)
      INTEGER NF(INF,2),G(16)
      DATA IJAC,IJAC1,IDER,IDERIV,NODOF,IT/6*2/
      DATA IH,ISAMP,IDEE,IBEE,IDBEE/5*3/
      DATA ICOORD,NOD/2*8/,IBTDB,IKM,IBT,IEMM,IDOF/5*16/
C
C       INPUT AND INITIALISATION
C
      READ(5,*)NXE,NYE,N,NN,NR,NGP,AA,BB,RHO,E,V,NMODES
      CALL READNF(NF,INF,NN,NODOF,NR)
      IWP1=IW+1
      CALL NULL(BIGK,IBIGK,N,N)
      CALL NULL(DEE,IDEE,IH,IH)
      CALL FMDEPS(DEE,IDEE,E,V)
      CALL GAUSS(SAMP,ISAMP,NGP)
      CALL NULL(EMM,IEMM,IDOF,IDOF)
      CALL NULVEC(DIAG,N)
C
C       FORM ELEMENT LUMPED MASS MATRIX
C
      DO 30 I=1,IDOF
   30 EMM(I,I)=AA*BB*RHO*.2
      DO 31 I=1,13,4
   31 EMM(I,I)=EMM(3,3)*.25
      DO 32 I=2,14,4
   32 EMM(I,I)=EMM(3,3)*.25
C
C       ELEMENT STIFFNESS AND MASS INTEGRATION AND ASSEMBLY
C
      DO 10 IP=1,NXE
      DO 10 IQ=1,NYE
      CALL GEOM8Y(IP,IQ,NYE,AA,BB,COORD,ICOORD,G,NF,INF)
      CALL NULL(KM,IKM,IDOF,IDOF)
      DO 20 I=1,NGP
      DO 20 J=1,NGP
      CALL FMQUAD(DER,IDER,FUN,SAMP,ISAMP,I,J)
      CALL MATMUL(DER,IDER,COORD,ICOORD,JAC,IJAC,IT,NOD,IT)
      CALL TWOBY2(JAC,IJAC,JAC1,IJAC1,DET)
      CALL MATMUL(JAC1,IJAC1,DER,IDER,DERIV,IDERIV,IT,IT,NOD)
      CALL NULL(BEE,IBEE,IH,IDOF)
      CALL FORMB(BEE,IBEE,DERIV,IDERIV,NOD)
      CALL MATMUL(DEE,IDEE,BEE,IBEE,DBEE,IDBEE,IH,IH,IDOF)
      CALL MATRAN(BT,IBT,BEE,IBEE,IH,IDOF)
      CALL MATMUL(BT,IBT,DBEE,IDBEE,BTDB,IBTDB,IDOF,IH,IDOF)
      QUOT=DET*SAMP(I,2)*SAMP(J,2)
      CALL MSMULT(BTDB,IBTDB,QUOT,IDOF,IDOF)
      CALL MATADD(KM,IKM,BTDB,IBTDB,IDOF,IDOF)
   20 CONTINUE
      CALL FMBIGK(BIGK,IBIGK,KM,IKM,G,IDOF)
      CALL FMLUMP(DIAG,EMM,IEMM,G,IDOF)
   10 CONTINUE
      CALL PRINTV(DIAG,N)
C
C       REDUCE TO STANDARD EIGENVALUE PROBLEM
C
      DO 50 I=1,N
      DIAG(I)=1./SQRT(DIAG(I))
   50 LOADS(I)=DIAG(I)
```

```
      DO 60 I=1,N
      DO 60 J=1,N
   60 BIGK(I,J)=BIGK(I,J)*DIAG(I)*DIAG(J)
C
C      EXTRACT EIGENVALUES
C
      CALL TRIDIA(N,1.E-293,BIGK,DIAG,UDIAG,BIGK,IBIGK)
      IFAIL=1
      CALL EVECTS(N,1.E-293,DIAG,UDIAG,BIGK,IBIGK,IFAIL)
      CALL PRINTV(DIAG,N)
      DO 70 J=1,NMODES
      DO 80 I=1,N
   80 DIAG(I)=BIGK(I,J)*LOADS(I)
   70 CALL PRINTV(DIAG,N)
      STOP
      END
```

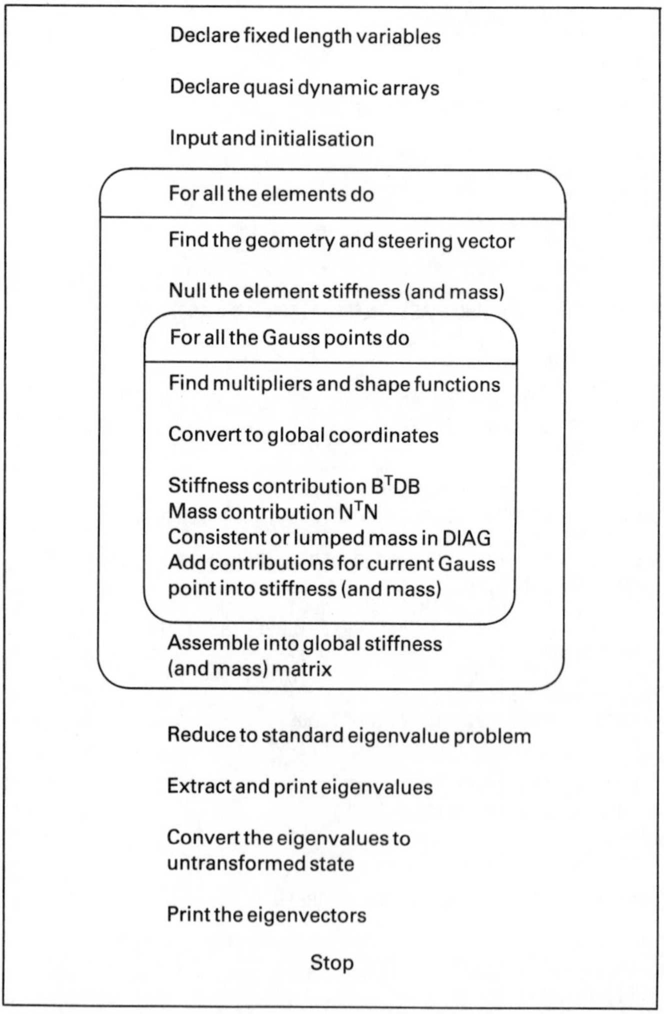

Declare fixed length variables
Declare quasi dynamic arrays
Input and initialisation
For all the elements do
Find the geometry and steering vector
Null the element stiffness (and mass)
For all the Gauss points do
Find multipliers and shape functions
Convert to global coordinates
Stiffness contribution B^TDB
Mass contribution N^TN
Consistent or lumped mass in DIAG
Add contributions for current Gauss point into stiffness (and mass)
Assemble into global stiffness (and mass) matrix
Reduce to standard eigenvalue problem
Extract and print eigenvalues
Convert the eigenvalues to untransformed state
Print the eigenvectors
Stop

Figure 10.6 Structure chart for Programs 10.2 and 10.3

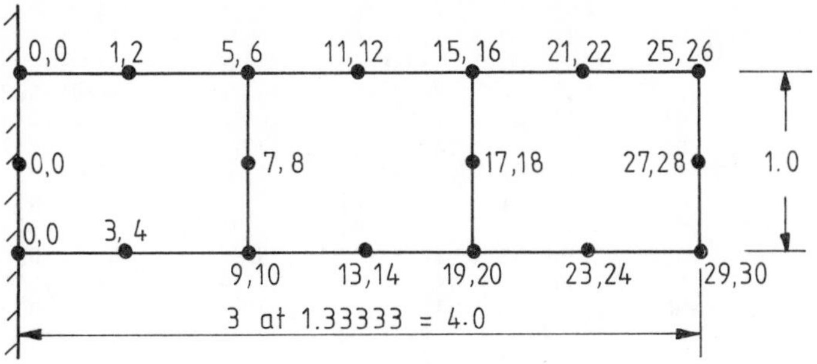

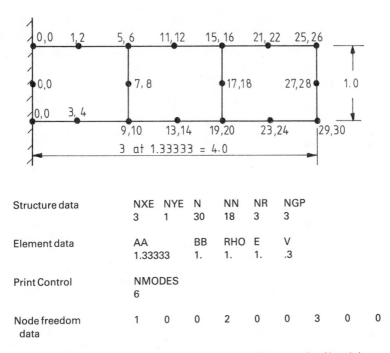

Structure data	NXE	NYE	N	NN	NR	NGP			
	3	1	30	18	3	3			
Element data	AA		BB	RHO	E	V			
	1.33333		1.	1.	1.	.3			
Print Control	NMODES								
	6								
Node freedom data	1	0	0	2	0	0	3	0	0

Note: In Program 10.3, the Semi-bandwidth IW = 15 is read after N and the integers NGPK and NGPM are read instead of NGP. Both have the value 3.

Figure 10.7 Mesh and data for Programs 10.2 and 10.3

A much better representation of flexural modes of 'beams' made up of solid elements is achieved by the use of eight-noded quadrilaterals. The present program uses these and a 'lumped' mass approximation. The process of mass lumping is not obvious for these elements. For example, the summation of rows of the consistent mass matrix leads to negative values at the corners. However, it can be shown (e.g. Smith, 1977) that a reasonable approximation is to lump the mass to the mid-point and corner nodes in the ratio 4:1 of the element area.

The structure chart for the program is listed in Figure 10.6 and the small example problem of a cantilever in Figure 10.7. Comparing with the previous program, the only new variables are NMODES, used to control the number of eigenmodes printed, and BIGK, which is a square ($N \times N$) global matrix replacing KU. Thus, the appropriate assembly routine is FMBIGK. The size of BIGK is set in the PARAMETER statement where IBIGK $\geqslant$ N.

Otherwise the program resembles the previous two closely. The output is shown in Figure 10.8 and consists of the diagonal global lumped mass matrix followed by the eigenvalues (as usual in the form ω^2) and the first six eigenvectors.

The square of the fundamental frequency is now $\omega_1^2 = 0.004114$, in much closer agreement with the value of 0.004025 produced by Program 10.0.

```
 .2667E+00  .2667E+00  .2667E+00  .2667E+00  .1333E+00  .1333E+00
 .5333E+00  .5333E+00  .1333E+00  .1333E+00  .2667E+00  .2667E+00   Lumped
 .2667E+00  .2667E+00  .1333E+00  .1333E+00  .5333E+00  .5333E+00   Mass
 .1333E+00  .1333E+00  .2667E+00  .2667E+00  .2667E+00  .2667E+00   Matrix
 .6667E-01  .6667E-01  .2667E+00  .2667E+00  .6667E-01  .6667E-01
 .4114E-02  .9594E-01  .1713E+00  .4599E+00  .1427E+01  .1812E+01
 .2674E+01  .3019E+01  .3231E+01  .3942E+01  .3964E+01  .4394E+01
 .5265E+01  .5435E+01  .6045E+01  .6858E+01  .7210E+01  .7509E+01   Eigenvalues
 .8503E+01  .8933E+01  .9704E+01  .9927E+01  .1033E+02  .1084E+02
 .1531E+02  .2256E+02  .2735E+02  .3866E+02  .3926E+02  .5617E+02
-.5290E-01  .4994E-01  .5290E-01  .4994E-01 -.9921E-01  .1659E+00
-.6518E-13  .1567E+00  .9921E-01  .1659E+00 -.1329E+00  .3292E+00   Eigenvector
 .1329E+00  .3292E+00 -.1526E+00  .5271E+00 -.1271E-12  .5259E+00
 .1526E+00  .5271E+00 -.1604E+00  .7413E+00  .1604E+00  .7413E+00   1
-.1632E+00  .9604E+00 -.1444E-12  .9600E+00  .1632E+00  .9604E+00
 .1287E+00 -.2497E+00 -.1287E+00 -.2497E+00  .9952E-01 -.5491E+00
 .1385E-12 -.5585E+00 -.9952E-01 -.5491E+00 -.6077E-01 -.6200E+00
 .6077E-01 -.6200E+00 -.2612E+00 -.4092E+00  .2853E-12 -.4447E+00   2
 .2612E+00 -.4092E+00 -.4104E+00  .9256E-01  .4104E+00  .9256E-01
-.4657E+00  .7244E+00  .3260E-12  .7243E+00  .4657E+00  .7244E+00
 .1726E+00 -.5985E-01  .1726E+00  .5985E-01  .3531E+00 -.4832E-01
 .3455E+00  .2015E-12  .3531E+00  .4832E-01  .4959E+00 -.4370E-01
 .4959E+00  .4370E-01  .6169E+00 -.3096E-01  .6099E+00  .3691E-12   3
 .6169E+00  .3096E-01  .6840E+00 -.1457E-01  .6840E+00  .1457E-01
 .7241E+00 -.1762E-01  .7057E+00 -.1559E-12  .7241E+00  .1762E-01
-.9707E-01  .4554E+00  .9707E-01  .4554E+00  .1600E+00  .6590E+00
 .1923E-13  .7062E+00 -.1600E+00  .6590E+00  .3277E+00  .8129E-01
-.3277E+00  .8129E-01  .8180E-01 -.4702E+00  .4875E-13 -.5438E+00   4
-.8180E-01 -.4702E+00 -.3334E+00 -.2316E+00  .3334E+00 -.2316E+00
-.5277E+00  .4872E+00  .2710E-13  .4736E+00  .5277E+00  .4872E+00
 .4403E+00 -.1462E+00  .4403E+00  .1462E+00  .7540E+00  .7608E-02
 .6861E+00 -.9012E-14  .7540E+00 -.7608E-02  .4866E+00  .1311E+00
 .4866E+00 -.1311E+00  .3186E-01  .1818E+00  .1547E-01 -.3332E-13   5
 .3186E-01 -.1818E+00 -.4657E+00  .1208E+00 -.4657E+00 -.1208E+00
-.8283E+00  .1702E+00 -.6682E+00  .8927E-13 -.8283E+00 -.1702E+00
-.2149E+00 -.6774E+00  .2149E+00 -.6774E+00 -.5991E+00 -.1042E+00
-.1086E-12 -.9155E-01  .5991E+00 -.1042E+00 -.1380E+00  .6831E+00
 .1380E+00  .6831E+00  .3104E+00  .5652E-02 -.8620E-14  .5742E-02   6
-.3104E+00  .5652E-02 -.2419E+00 -.6108E+00  .2419E+00 -.6108E+00
-.8178E+00  .2924E+00  .2853E-13  .2128E+00  .8178E+00  .2924E+00
```

Figure 10.8 Results from Program 10.2

PROGRAM 10.3: EIGENVALUES AND EIGENVECTORS OF A RECTANGULAR SOLID IN PLANE STRESS—EIGHT-NODE QUADRILATERALS WITH CONSISTENT MASS

```
C
C      PROGRAM 10.3 EIGENVALUES AND EIGENVECTORS OF A RECTANGULAR
C      SOLID IN PLANE STRESS USING 8-NODE QUADRILATERALS
C      CONSISTENT MASS
C
C
C      ALTER NEXT LINE TO CHANGE PROBLEM SIZE
C
      PARAMETER(IKB1=103,IKB2=16,INF=85)
C
      REAL DEE(3,3),SAMP(3,2),COORD(8,2),FUN(8),JAC(2,2),JAC1(2,2)
      REAL DER(2,8),DERIV(2,8),BEE(3,16),DBEE(3,16),BTDB(16,16)
      REAL BT(16,3),EMM(16,16),ECM(16,16),KM(16,16)
```

```
      REAL BIGK(IKB1,IKB1),DIAG(IKB1),UDIAG(IKB1)
      REAL TN(16,16),NT(16,2),KB(IKB1,IKB2),MB(IKB1,IKB2)
      INTEGER NF(INF,2),G(16)
      DATA IJAC,IJAC1,IDER,IDERIV,NODOF,IT/6*2/
      DATA IH,ISAMP,IDEE,IBEE,IDBEE/5*3/
      DATA ICOORD,NOD/2*8/,IBTDB,IKM,IBT,IEMM,IECM,IDOF,ITN,INT/8*16/
C
C       INPUT AND INITIALISATION
C
      READ(5,*)NXE,NYE,N,IW,NN,NR,NGPK,NGPM,AA,BB,RHO,E,V,NMODES
      CALL READNF(NF,INF,NN,NODOF,NR)
      IWP1=IW+1
      CALL NULL(KB,IKB1,N,IWP1)
      CALL NULL(MB,IKB1,N,IWP1)
      CALL NULL(BIGK,IKB1,N,N)
      CALL NULL(DEE,IDEE,IH,IH)
      CALL FMDSIG(DEE,IDEE,E,V)
C
C       ELEMENT STIFFNESS AND MASS INTEGRATION AND ASSEMBLY
C
      DO 10 IP=1,NXE
      DO 10 IQ=1,NYE
      CALL GEOM8Y(IP,IQ,NYE,AA,BB,COORD,ICOORD,G,NF,INF)
      CALL NULL(KM,IKM,IDOF,IDOF)
      CALL NULL(EMM,IEMM,IDOF,IDOF)
      CALL GAUSS(SAMP,ISAMP,NGPK)
      DO 20 I=1,NGPK
      DO 20 J=1,NGPK
      CALL FMQUAD(DER,IDER,FUN,SAMP,ISAMP,I,J)
      CALL MATMUL(DER,IDER,COORD,ICOORD,JAC,IJAC,IT,NOD,IT)
      CALL TWOBY2(JAC,IJAC,JAC1,IJAC1,DET)
      CALL MATMUL(JAC1,IJAC1,DER,IDER,DERIV,IDERIV,IT,IT,NOD)
      CALL NULL(BEE,IBEE,IH,IDOF)
      CALL FORMB(BEE,IBEE,DERIV,IDERIV,NOD)
      CALL MATMUL(DEE,IDEE,BEE,IBEE,DBEE,IDBEE,IH,IH,IDOF)
      CALL MATRAN(BT,IBT,BEE,IBEE,IH,IDOF)
      CALL MATMUL(BT,IBT,DBEE,IDBEE,BTDB,IBTDB,IDOF,IH,IDOF)
      QUOT=DET*SAMP(I,2)*SAMP(J,2)
      CALL MSMULT(BTDB,IBTDB,QUOT,IDOF,IDOF)
      CALL MATADD(KM,IKM,BTDB,IBTDB,IDOF,IDOF)
   20 CONTINUE
      CALL GAUSS(SAMP,ISAMP,NGPM)
      DO 30 I=1,NGPM
      DO 30 J=1,NGPM
      CALL FMQUAD(DER,IDER,FUN,SAMP,ISAMP,I,J)
      CALL MATMUL(DER,IDER,COORD,ICOORD,JAC,IJAC,IT,NOD,IT)
      CALL TWOBY2(JAC,IJAC,JAC1,IJAC1,DET)
      QUOT=DET*SAMP(I,2)*SAMP(J,2)*RHO
      CALL ECMAT(ECM,IECM,TN,ITN,NT,INT,FUN,NOD,NODOF)
      CALL MSMULT(ECM,IECM,QUOT,IDOF,IDOF)
      CALL MATADD(EMM,IEMM,ECM,IECM,IDOF,IDOF)
   30 CONTINUE
      CALL FORMKB(KB,IKB1,KM,IKM,G,IW,IDOF)
      CALL FORMKB(MB,IKB1,EMM,IEMM,G,IW,IDOF)
   10 CONTINUE
C
C       REDUCE TO STANDARD EIGENVALUE PROBLEM
C
      CALL CHOLIN(MB,IKB1,N,IW)
      CALL LBKBAN(MB,BIGK,KB,IKB1,IW,N)
      DO 40 I=1,N
      DO 40 J=I+1,N
```

```
   40 BIGK(I,J)=BIGK(J,I)
      CALL LBBT(MB,BIGK,IKB1,IW,N)
C
C      EXTRACT EIGENVALUES AND EIGENVECTORS
C
      CALL TRIDIA(N,1.E-293,BIGK,DIAG,UDIAG,BIGK,IKB1)
      IFAIL=1
      CALL EVECTS(N,1.E-293,DIAG,UDIAG,BIGK,IKB1,IFAIL)
      CALL PRINTV(DIAG,N)
      DO 50 J=1,NMODES
      DO 60 I=1,N
   60 UDIAG(I)=BIGK(I,J)
      CALL CHOBK2(MB,IKB1,UDIAG,N,IW)
   50 CALL PRINTV(UDIAG,N)
      STOP
      END
```

This program is a minor adaptation of the previous one. The only differences are that plane stress is assumed rather than plane strain, so the subroutine FMDSIG replaces FMDEPS. Second, the consistent mass matrix is used rather than the lumped approximation. The element mass matrices are formed by subroutine ECMAT where arrays TN and NT are working space. The data are essentially

.3850E-02	.1050E+00	.1557E+00	.6102E+00	.1402E+01	.1686E+01	
.3997E+01	.4188E+01	.5420E+01	.7671E+01	.8375E+01	.8847E+01	
.1020E+02	.1181E+02	.1300E+02	.1333E+02	.1440E+02	.1644E+02	Eigenvalues
.1737E+02	.2056E+02	.2375E+02	.2440E+02	.2493E+02	.2820E+02	
.3368E+02	.3749E+02	.3814E+02	.6621E+02	.6723E+02	.7016E+02	
-.5757E-01	.5207E-01	.5757E-01	.5207E-01	-.1046E+00	.1736E+00	
-.1281E-12	.1672E+00	.1046E+00	.1736E+00	-.1363E+00	.3430E+00	
.1363E+00	.3430E+00	-.1551E+00	.5447E+00	-.2289E-12	.5440E+00	Eigenvector 1
.1551E+00	.5447E+00	-.1618E+00	.7617E+00	.1618E+00	.7617E+00	
-.1634E+00	.9809E+00	-.2620E-12	.9808E+00	.1634E+00	.9809E+00	
.1485E+00	-.2786E+00	-.1485E+00	-.2786E+00	.1076E+00	-.5943E+00	
.1316E-12	-.6056E+00	-.1076E+00	-.5943E+00	-.9065E-01	-.6611E+00	
.9065E-01	-.6611E+00	-.3144E+00	-.3693E+00	.2259E-12	-.3956E+00	2
.3144E+00	-.3693E+00	-.4483E+00	.2028E+00	.4483E+00	.2028E+00	
-.4919E+00	.8684E+00	.2327E-12	.8713E+00	.4919E+00	.8684E+00	
.1777E+00	-.4165E-01	.1777E+00	.4165E-01	.3510E+00	-.3426E-01	
.3513E+00	.3545E-12	.3510E+00	.3426E-01	.4975E+00	-.3029E-01	
.4975E+00	.3029E-01	.6102E+00	-.2144E-01	.6142E+00	.2678E-12	3
.6102E+00	.2144E-01	.6817E+00	-.1094E-01	.6817E+00	.1094E-01	
.7061E+00	-.1533E-03	.7102E+00	.1696E-12	.7061E+00	.1533E-03	
-.8644E-01	.6049E+00	.8644E-01	.6049E+00	.2450E+00	.6499E+00	
-.6848E-13	.6906E+00	-.2450E+00	.6499E+00	.3473E+00	-.3912E-01	
-.3473E+00	-.3912E-01	.8177E-02	-.5491E+00	-.6716E-13	-.6161E+00	4
-.8177E-02	-.5491E+00	-.4899E+00	-.1448E+00	.4899E+00	-.1448E+00	
-.7280E+00	.8313E+00	-.4804E-13	.8163E+00	.7280E+00	.8313E+00	
.4726E+00	-.9905E-01	.4726E+00	.9905E-01	.6828E+00	.6879E-02	
.7248E+00	.6650E-13	.6828E+00	-.6879E-02	.4953E+00	.9324E-01	
.4953E+00	-.9324E-01	.6888E-02	.1513E+00	.8212E-02	-.7432E-14	5
.6888E-02	-.1513E+00	-.4841E+00	.9598E-01	-.4841E+00	-.9598E-01	
-.6801E+00	.7790E-02	-.7262E+00	-.7583E-13	-.6801E+00	-.7790E-02	
.1550E+00	.6773E+00	-.1550E+00	.6773E+00	.4907E+00	-.1536E-01	
-.1064E-12	-.3094E-01	-.4907E+00	-.1536E-01	.7069E-02	-.6051E+00	
-.7069E-02	-.6051E+00	-.2583E+00	.2107E+00	.3111E-13	.2249E+00	6
.2583E+00	.2107E+00	.4633E+00	.4204E+00	-.4633E+00	.4204E+00	
.1050E+01	-.8463E+00	.7707E-13	-.7488E+00	-.1050E+01	-.8463E+00	

Figure 10.9 Results from Program 10.3

those given in Figure 10.7 for a cantilever problem, with the slight sophistication that different orders of integration are allowed for the stiffness and mass matrices. This is controlled by NGPK and NGPM instead of the usual constant NGP, but in the present example NGPK = NGPM = 3.

Otherwise the program follows the familiar course, the global mass matrix being stored as MB. The reduction to standard form follows the same course as in Program 10.0 and the results are printed as Figure 10.9 in the form of the N eigenvalues followed by the first six eigenvectors.

The square of the fundamental frequency of the plane stress idealisation of the cantilever, $\omega_1^2 = 0.003850$ is somewhat lower than that of a slender beam. The same program run for the plane strain case gives $\omega_1^2 = 0.004356$.

PROGRAM 10.4: EIGENVALUES AND EIGENVECTORS OF A RECTANGULAR SOLID IN PLANE STRAIN—FOUR-NODE QUADRILATERALS WITH CONSISTENT MASS

```
C
C      PROGRAM 10.4 EIGENVALUES AND EIGENVECTORS OF A
C      RECTANGULAR SOLID IN PLANE STRAIN USING 4-NODE
C      QUADRILATERAL ELEMENTS : CONSISTENT MASS,LANCZOS METHOD
C
C      ALTER NEXT LINE TO CHANGE PROBLEM SIZE
C
      PARAMETER(ILOADS=200,IKB2=20,LALFA=500,LEIG=20,LX=80,LY=200,
     +          LZ=500,INF=110)
C
      REAL DEE(3,3),SAMP(3,2),COORD(4,2),FUN(4),JAC(2,2),JAC1(2,2)
      REAL DER(2,4),DERIV(2,4),BEE(3,8),DBEE(3,8),BTDB(8,8),KM(8,8)
      REAL BT(8,3),EMM(8,8),ECM(8,8),TN(8,8),NT(8,2)
      REAL UA(ILOADS),VA(ILOADS),EIG(LEIG),X(LX),DEL(LX),UDIAG(ILOADS)
      REAL ALFA(LALFA),BETA(LALFA),W1(ILOADS),Y(LY,LEIG),Z(LZ,LEIG)
      REAL KB(ILOADS,IKB2),MB(ILOADS,IKB2),DIAG(ILOADS)
      INTEGER G(8),NF(INF,2),NU(LX),JEIG(2,LEIG)
      DATA NODOF,IT,IJAC,IJAC1,IDER,IDERIV/6*2/,ICOORD,NOD/2*4/
      DATA IDEE,IBEE,IDBEE,IH,ISAMP/5*3/
      DATA IBTDB,IKM,IBT,IEMM,IECM,ITN,INT,IDOF/8*8/
C
C       INPUT AND INITIALISATION
C
      DATA EL/0./,ER/20./,ACC/1.E-6/,LP/6/,ITAPE/1/,IFLAG/-1/
      READ(5,*)NXE,NYE,N,IW,NN,NR,NGP,AA,BB,RHO,E,V,NMODES
      CALL READNF(NF,INF,NN,NODOF,NR)
      IWP1=IW+1
      CALL NULL(KB,ILOADS,N,IWP1)
      CALL NULL(MB,ILOADS,N,IWP1)
      CALL NULL(DEE,IDEE,IH,IH)
      CALL FMDEPS(DEE,IDEE,E,V)
      CALL GAUSS(SAMP,ISAMP,NGP)
C
C       ELEMENT STIFFNESS AND MASS INTEGRATION AND ASSEMBLY
C
      DO 3 IP=1,NXE
      DO 3 IQ=1,NYE
      CALL GEOM4Y(IP,IQ,NYE,AA,BB,COORD,ICOORD,G,NF,INF)
      CALL NULL(KM,IKM,IDOF,IDOF)
      CALL NULL(EMM,IEMM,IDOF,IDOF)
      DO 4 I=1,NGP
```

```
      DO 4 J=1,NGP
      CALL FORMLN(DER,IDER,FUN,SAMP,ISAMP,I,J)
      CALL MATMUL(DER,IDER,COORD,ICOORD,JAC,IJAC,IT,NOD,IT)
      CALL TWOBY2(JAC,IJAC,JAC1,IJAC1,DET)
      CALL MATMUL(JAC1,IJAC1,DER,IDER,DERIV,IDERIV,IT,IT,NOD)
      CALL NULL(BEE,IBEE,IH,IDOF)
      CALL FORMB(BEE,IBEE,DERIV,IDERIV,NOD)
      CALL MATMUL(DEE,IDEE,BEE,IBEE,DBEE,IDBEE,IH,IH,IDOF)
      CALL MATRAN(BT,IBT,BEE,IBEE,IH,IDOF)
      CALL MATMUL(BT,IBT,DBEE,IDBEE,BTDB,IBTDB,IDOF,IH,IDOF)
      CALL ECMAT(ECM,IECM,TN,ITN,NT,INT,FUN,NOD,NODOF)
      QUOT=DET*SAMP(I,2)*SAMP(J,2)
      DO 5 K=1,IDOF
      DO 5 L=1,IDOF
      ECM(K,L)=ECM(K,L)*QUOT*RHO
    5 BTDB(K,L)=BTDB(K,L)*QUOT
      CALL MATADD(KM,IKM,BTDB,IBTDB,IDOF,IDOF)
      CALL MATADD(EMM,IEMM,ECM,IECM,IDOF,IDOF)
    4 CONTINUE
      CALL FORMKB(KB,ILOADS,KM,IKM,G,IW,IDOF)
      CALL FORMKB(MB,ILOADS,EMM,IEMM,G,IW,IDOF)
    3 CONTINUE
C
C      FIND EIGENVALUES
C
      CALL CHOLIN(MB,ILOADS,N,IW)
      DO 30 ITER=1,LALFA
      CALL LANCZ1(N,EL,ER,ACC,LEIG,LX,LALFA,LP,ITAPE,
     +            IFLAG,UA,VA,EIG,JEIG,NEIG,X,DEL,NU,
     +            ALFA,BETA)
      IF(IFLAG.EQ.0)GOTO 40
      IF(IFLAG.GT.1)GOTO 70
C
C      IFLAG=1.   FORM U+AV
C
      CALL VECCOP(VA,UDIAG,N)
      CALL CHOBK2(MB,ILOADS,UDIAG,N,IW)
      CALL BANMUL(KB,ILOADS,UDIAG,DIAG,N,IW)
      CALL CHOBK1(MB,ILOADS,DIAG,N,IW)
      CALL VECADD(UA,DIAG,UA,N)
   30 CONTINUE
      GOTO 70
C
C      WRITE OUT SPECTRUM FOUND
C
   40 WRITE(6,100)ITER,EL,ER
      CALL PRINTV(EIG,NEIG)
C
C      CALCULATE EIGENVECTORS
C
      NEXQT=1
      IF(NEXQT.EQ.0)GOTO 130
      IF(NEIG.GT.10)NEIG=10
      CALL LANCZ2(N,LALFA,LP,ITAPE,EIG,JEIG,NEIG,ALFA,BETA,
     +            LY,LZ,JFLAG,Y,W1,Z)
      IF(JFLAG.NE.0)GOTO 80
C
C      JFLAG=0.  CALCULATE EIGENVECTORS
C
      DO 60 I=1,NMODES
      DO 50 J=1,N
   50 UDIAG(J)=Y(J,I)
      CALL CHOBK2(MB,ILOADS,UDIAG,N,IW)
```

```
      CALL PRINTV(UDIAG,N)
   60 CONTINUE
      GOTO 130
C
C      LANCZ1 IS SIGNALLING FAILURE
C
   70 WRITE(6,110)IFLAG
      GOTO 130
C
C      LANCZ2 IS SIGNALLING FAILURE
C
   80 WRITE(6,120)JFLAG
  130 CONTINUE
  100 FORMAT(1H ,I10,2E12.4)
  110 FORMAT(26X,'LANCZ1 HAS FAILED. IFLAG=',I2)
  120 FORMAT(26X,'LANCZ2 HAS FAILED. JFLAG=',I2)
      STOP
      END
```

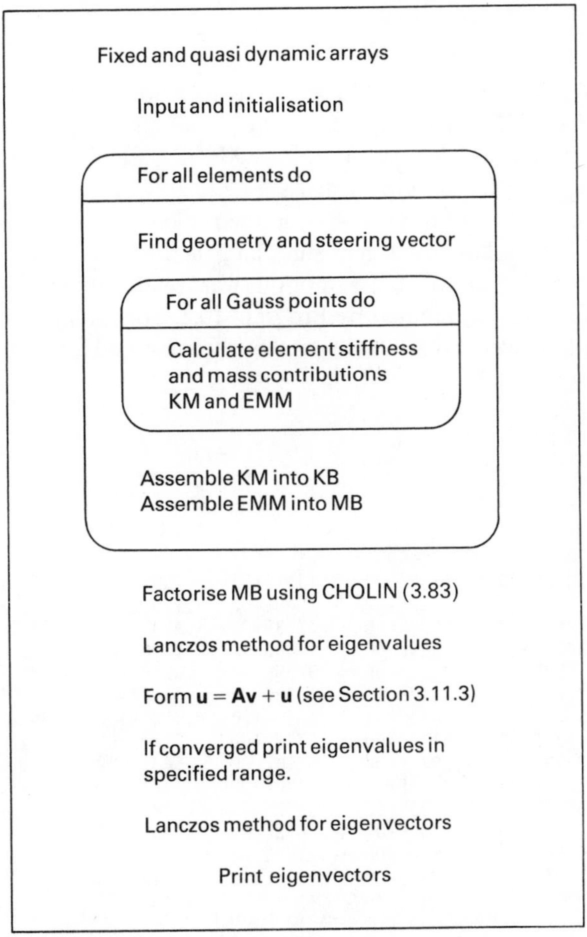

Figure 10.10 Structure chart for Program 10.4

In Programs 10.0, 10.2 and 10.3, Householder transformation has been used to solve the eigenvalue problem. Although reliable and robust, this method is time consuming for large problems (especially when $N \times N$ storage is used, as was done in these previous programs). For larger problems, iterative methods are more attractive, for example the Lanczos method (Parlett and Reid, 1981). The process is described in Chapter 3, Section 3.11.3, and is used in the present program, whose structure chart is given as Figure 10.10. The problem and data are essentially described by Figure 10.3 with the additional input parameter NMODES giving the number of eigenmodes required.

Turning to the program, relevant new variables are:

EL	lower limit of eigenvalue spectrum (0.0)
ER	upper limit of eigenvalue spectrum (20.0)
ACC	tolerance for iteration (1×10^{-6})
LP	output device (6)
ITAPE	scratch tape number (1)
IFLAG	failure option (−1)

The principal new vectors are **u** and **v**, as described in the structure chart, denoted by UA and VA in the program. The remaining arrays are used for work space. See also Smith (1984); subroutines LANCZ1 and LANCZ2 are used.

The output from the program is listed as Figure 10.11. It consists of the number of Lanczos iterations (20 in this case) and the range of the eigenvalue spectrum (0.0 to 20.0). Then follow the eigenvalues and the first six eigenmodes.

Since the consistent mass assumption was made, the eigenvalues differ somewhat from those computed by Program 10.1. For example, $\omega_1^2 = 0.007264$ compared with the value of 0.006488 previously computed for a similar problem with lumped mass.

Iterations	Range: Low	Range: High				
20	.0000E+00	.2000E+02				
.7264E-02	.1815E+00	.2165E+00	.1351E+01	.1981E+01	.5291E+01	Eigenvalues
.6406E+01	.7593E+01	.1219E+02	.1646E+02	.1765E+02	.1859E+02	Eigenvalues
-.1020E+00	.1753E+00	.1020E+00	.1753E+00	-.1490E+00	.5423E+00	Eigenvector 1
.1490E+00	.5423E+00	-.1600E+00	.9680E+00	.1600E+00	.9680E+00	Eigenvector 1
.3461E+00	-.6739E-01	.3461E+00	.6739E-01	.6277E+00	-.2783E-01	2
.6277E+00	.2783E-01	.7347E+00	-.1278E-01	.7347E+00	.1278E-01	2
.8434E-01	-.6955E+00	-.8434E-01	-.6955E+00	-.3085E+00	-.4420E+00	3
.3085E+00	-.4420E+00	-.5170E+00	.8778E+00	.5170E+00	.8778E+00	3
.2032E+00	.8229E+00	-.2032E+00	.8229E+00	-.5841E-02	-.8680E+00	4
.5841E-02	-.8680E+00	-.7429E+00	.7149E+00	.7429E+00	.7149E+00	4
-.8303E+00	.5292E-01	-.8303E+00	-.5292E-01	-.6686E-01	-.2033E+00	5
-.6686E-01	.2033E+00	.9060E+00	-.1673E+00	.9060E+00	.1673E+00	5
.9334E+00	-.5358E-01	-.9334E+00	-.5358E-01	.1025E+01	-.1182E+00	6
-.1025E+01	-.1182E+00	.7945E+00	.4230E+00	-.7945E+00	.4230E+00	6

Figure 10.11 Results from Program 10.4

10.1 REFERENCES

Parlett, B. N., and Reid, J. K. (1981) Tracking the progress of the Lanczos algorithm for large symmetric eigenproblems. *IMA Journal of Numerical Analysis*, **1**, 135–55.

Smith, I. M. (1977) Transient phenomena of offshore foundations. In *Numerical Methods in Offshore Engineering*, Chapter 14. John Wiley, Chichester.

Smith, I. M. (1984) Adaptability of truly modular software. *Engineering Computations*, **1**, No. 1, 25–35.

CHAPTER 11
Forced Vibrations

11.0 INTRODUCTION

In the previous chapter, programs were described which enable the calculation of the intrinsic dynamic properties of systems—their undamped natural frequencies and mode shapes. The next stage in a dynamic analysis is usually the calculation of the response of the system to an imposed time dependent disturbance. This chapter describes six programs which enable such calculations to be made.

The type of equations to be solved was derived early in the book (see, for example, 2.13). After semi-discretisation in space using finite elements, the resulting matrix equations are typified by (2.17), a set of second order ordinary differential equations in the time variable. On inclusion of damping, the relevant equations become (3.117) and Section 3.15 describes the principles behind the various solution procedures used below. For example, the first program described uses the real modal superposition method described in some detail in Section 3.15.1.

PROGRAM 11.0: FORCED VIBRATION OF A RECTANGULAR SOLID IN PLANE STRAIN USING EIGHT-NODE QUADRILATERALS, LUMPED MASS, MODAL SUPERPOSITION METHOD

```
C
C      PROGRAM 11.0 FORCED VIBRATION OF A RECTANGULAR SOLID IN
C      PLANE STRAIN USING 8-NODE QUADRILATERALS,LUMPED MASS,
C      MODAL SUPERPOSITION
C
C
C      ALTER NEXT LINE TO CHANGE PROBLEM SIZE
C
      PARAMETER(IBIGK=103,INF=85,IMOD=30)
C
      REAL DEE(3,3),SAMP(3,2),COORD(8,2),FUN(8),JAC(2,2),JAC1(2,2),
     +DER(2,8),DERIV(2,8),BEE(3,16),DBEE(3,16),BTDB(16,16),
     +BT(16,3),KM(16,16),BIGK(IBIGK,IBIGK),XMOD(IMOD),
     +LOADS(IBIGK),DIAG(IBIGK),UDIAG(IBIGK),EMM(16,16)
      INTEGER NF(INF,2),G(16)
```

```
      DATA IJAC,IJAC1,IDER,IDERIV,NODOF,IT/6*2/
      DATA IH,ISAMP,IDEE,IBEE,IDBEE/5*3/
      DATA ICOORD,NOD/2*8/,IBTDB,IKM,IBT,IEMM,IDOF/5*16/
C
C      INPUT AND INITIALISATION
C
      READ(5,*)NXE,NYE,N,NN,NR,NGP,AA,BB,RHO,E,V,DR,NMODES,
     +ISTEP,OMEGA,NPRI
      CALL READNF(NF,INF,NN,NODOF,NR)
      PI=ACOS(-1.)
      PERIOD=2.*PI/OMEGA
      DTIM=PERIOD/20.
      CALL NULL(BIGK,IBIGK,N,N)
      CALL NULL(DEE,IDEE,IH,IH)
      CALL FMDEPS(DEE,IDEE,E,V)
      CALL GAUSS(SAMP,ISAMP,NGP)
      CALL NULL(EMM,IEMM,IDOF,IDOF)
      CALL NULVEC(DIAG,N)
C
C      FORM LUMPED MASS MATRIX
C
      DO 30 I=1,IDOF
   30 EMM(I,I)=AA*BB*RHO*.2
      DO 40 I=1,13,4
   40 EMM(I,I)=EMM(3,3)*.25
      DO 50 I=2,14,4
   50 EMM(I,I)=EMM(3,3)*.25
C
C      ELEMENT STIFFNESS AND MASS INTEGRATION AND ASSEMBLY
C
      DO 10 IP=1,NXE
      DO 10 IQ=1,NYE
      CALL GEOM8Y(IP,IQ,NYE,AA,BB,COORD,ICOORD,G,NF,INF)
      CALL NULL(KM,IKM,IDOF,IDOF)
      DO 20 I=1,NGP
      DO 20 J=1,NGP
      CALL FMQUAD(DER,IDER,FUN,SAMP,ISAMP,I,J)
      CALL MATMUL(DER,IDER,COORD,ICOORD,JAC,IJAC,IT,NOD,IT)
      CALL TWOBY2(JAC,IJAC,JAC1,IJAC1,DET)
      CALL MATMUL(JAC1,IJAC1,DER,IDER,DERIV,IDERIV,IT,IT,NOD)
      CALL NULL(BEE,IBEE,IH,IDOF)
      CALL FORMB(BEE,IBEE,DERIV,IDERIV,NOD)
      CALL MATMUL(DEE,IDEE,BEE,IBEE,DBEE,IDBEE,IH,IH,IDOF)
      CALL MATRAN(BT,IBT,BEE,IBEE,IH,IDOF)
      CALL MATMUL(BT,IBT,DBEE,IDBEE,BTDB,IBTDB,IDOF,IH,IDOF)
      QUOT=DET*SAMP(I,2)*SAMP(J,2)
      CALL MSMULT(BTDB,IBTDB,QUOT,IDOF,IDOF)
   20 CALL MATADD(KM,IKM,BTDB,IBTDB,IDOF,IDOF)
      CALL FMLUMP(DIAG,EMM,IEMM,G,IDOF)
   10 CALL FMBIGK(BIGK,IBIGK,KM,IKM,G,IDOF)
C
C      REDUCE TO STANDARD EIGENVALUE PROBLEM
C
      DO 60 I=1,N
      DIAG(I)=1./SQRT(DIAG(I))
   60 LOADS(I)=DIAG(I)
      DO 70 I=1,N
      DO 70 J=1,N
   70 BIGK(I,J)=BIGK(I,J)*DIAG(I)*DIAG(J)
```

```
C
C       EXTRACT EIGENVALUES
C
      CALL TRIDIA(N,1.E-293,BIGK,DIAG,UDIAG,BIGK,IBIGK)
      IFAIL=1
      CALL EVECTS(N,1.E-293,DIAG,UDIAG,BIGK,IBIGK,IFAIL)
C
C       TIME STEPPING LOOP
C
      TIM=0.
      DO 80 J=1,ISTEP
      TIM=TIM+DTIM
      DO 90 M=1,NMODES
      DO 100 I=1,N
  100 UDIAG(I)=BIGK(I,M)*LOADS(I)
      F=UDIAG(N)
      X1=DIAG(M)-OMEGA*OMEGA
      X2=X1*X1+4.*OMEGA*OMEGA*DR*DR*DIAG(M)
      X3=F*X1/X2
      X4=F*2.*OMEGA*DR*SQRT(DIAG(M))/X2
      XMOD(M)=X3*COS(OMEGA*TIM)+X4*SIN(OMEGA*TIM)
   90 CONTINUE
C
C       SUPERIMPOSE THE MODES
C
      DO 110 I=1,N
      SUM=0.
      DO 120 M=1,NMODES
  120 SUM=SUM+BIGK(I,M)*LOADS(I)*XMOD(M)
  110 UDIAG(I)=SUM
      IF(J/NPRI*NPRI.EQ.J)WRITE(6,1000)TIM,COS(OMEGA*TIM),UDIAG(N)
   80 CONTINUE
 1000 FORMAT(3E12.4)
      END
```

Since the basis of this method is a synthesis of the undamped natural modes of the vibrating system, it follows very naturally from the programs of the previous chapter. Indeed, this program can be built up, with minor extensions, from Program 10.2.

The simple illustrative problem chosen for the first four programs in this chapter is shown in Figure 11.1. The same cantilever 'beam' as was analysed in the previous chapter is subjected to a harmonic vertical force $\cos \omega t$ at node 18 (degree of freedom 30). The damping ratio γ (see 3.132), called DR in the program, is 0.05 or 5% applied to all modes of the system.

The forcing frequency OMEGA was chosen to be 0.3. By reference to the results of Program 10.2 (see Figure 10.8), this is close to the second natural frequency $\omega_2 = \sqrt{0.09594}$ of the undamped system, and so at this frequency the influence of damping should be significant.

Since the program is based so closely on those of the previous chapter (see the listing of variables for Program 10.1), only additional variables are listed here.

New simple variables are:

DR	damping ratio γ (0.05)
NMODES	number of modes retained in the analysis (6)
ISTEP	number of time steps (20)

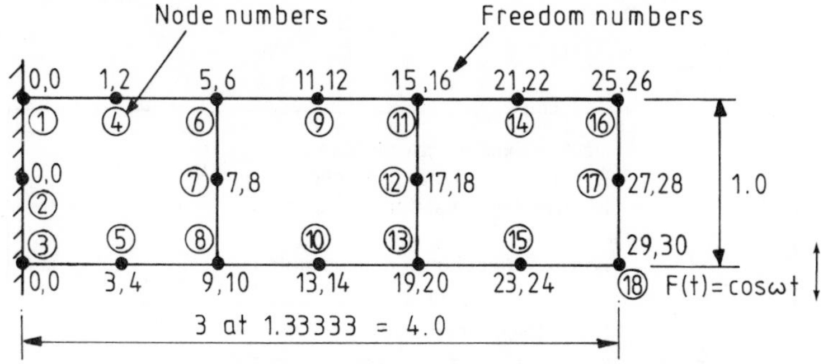

Structure data	NXE	NYE	N	NN	NR	NGP			
	3	1	30	18	3	3			
Element data	AA		BB	RHO	E	V	DR		
	1.33333		1.	1.	1.	.3	.05		
Time integration data	NMODES	ISTEP		OMEGA		NPRI			
	6		20	.3		1			
Node freedom data	1	0	0	2	0	0	3	0	0

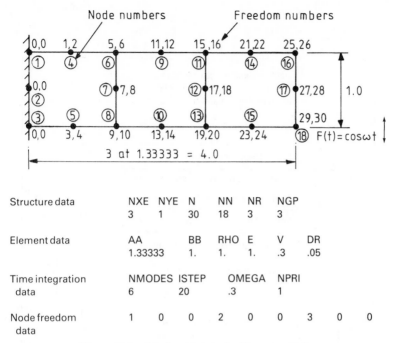

Figure 11.1 Mesh and data for Program 11.0

OMEGA	forcing frequency ω (0.3)
NPRI	print control—print every NPRI(1) steps
PERIOD	forcing period ($2\pi/\omega$)
DTIM	time-step (1/20 of forcing period)
TIM	total elapsed time

and a new array is:

XMOD normal coordinates **ap** (see 3.124, 3.125)

The size of XMOD is given in the PARAMETER statement where IMOD $\geqslant$ NMODES.

The structure of the program is illustrated in Figure 11.2. Up to the end of the section headed 'extract eigenvalues and eigenvectors' it is copied directly from Program 10.2.

When the time-stepping loop is entered, it must be remembered that the eigenvectors computed are those of the transformed problem and the true eigenvectors are first recovered as UDIAG. The eigenvalues, in the form ω_i^2, are contained in DIAG.

Turning to equation (3.136), P is called F in the program, the ω_i^2 are in DIAG and θ is the forcing frequency OMEGA. Therefore ap_i are readily computed and stored in XMOD.

In the final section of the program, the modes are superimposed as required by equation (3.125).

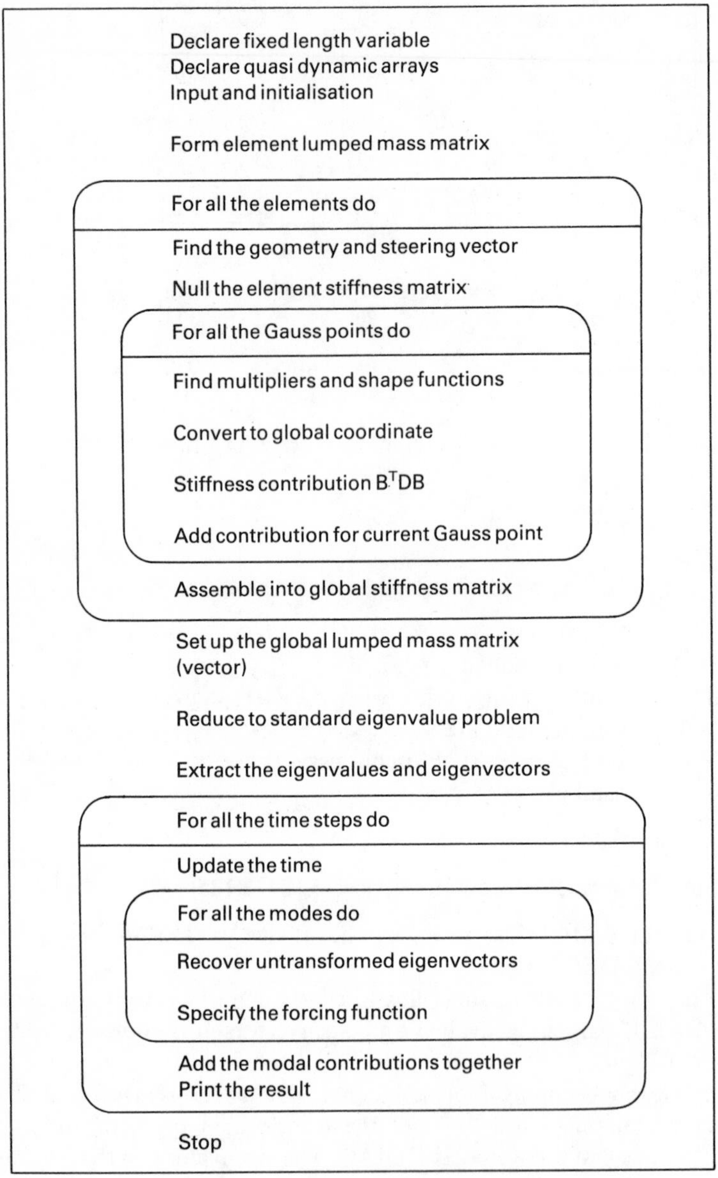
Declare fixed length variable
Declare quasi dynamic arrays
Input and initialisation
Form element lumped mass matrix
For all the elements do
Find the geometry and steering vector
Null the element stiffness matrix
For all the Gauss points do
Find multipliers and shape functions
Convert to global coordinate
Stiffness contribution B^TDB
Add contribution for current Gauss point
Assemble into global stiffness matrix
Set up the global lumped mass matrix (vector)
Reduce to standard eigenvalue problem
Extract the eigenvalues and eigenvectors
For all the time steps do
Update the time
For all the modes do
Recover untransformed eigenvectors
Specify the forcing function
Add the modal contributions together
Print the result
Stop

Figure 11.2 Structure chart for Program 11.0

.1047E+01	.9511E+00	.2732E+02
.2094E+01	.8090E+00	.3636E+02
.3142E+01	.5878E+00	.4184E+02
.4189E+01	.3090E+00	.4322E+02
.5236E+01	.5390E-14	.4037E+02
.6283E+01	-.3090E+00	.3357E+02
.7330E+01	-.5878E+00	.2348E+02
.8378E+01	-.8090E+00	.1110E+02
.9425E+01	-.9511E+00	-.2373E+01
.1047E+02	-.1000E+01	-.1561E+02
.1152E+02	-.9511E+00	-.2732E+02
.1257E+02	-.8090E+00	-.3636E+02
.1361E+02	-.5878E+00	-.4184E+02
.1466E+02	-.3090E+00	-.4322E+02
.1571E+02	-.4459E-13	-.4037E+02
.1676E+02	.3090E+00	-.3357E+02
.1780E+02	.5878E+00	-.2348E+02
.1885E+02	.8090E+00	-.1110E+02
.1990E+02	.9511E+00	.2373E+01
.2094E+02	.1000E+01	.1561E+02
Time t	cos ωt	Tip displacement

Figure 11.3 Results from Program 11.0

The results are listed as Figure 11.3. They consist of the elapsed time, the value of the force and the value of the displacement under the force every NPRI (in this case 1) steps. It can be seen that the amplitude of the displacement is 43.22 at a phase shift of the order of 90°.

PROGRAM 11.1: FORCED VIBRATION OF A RECTANGULAR SOLID IN PLANE STRAIN USING EIGHT-NODE QUADRILATERALS—LUMPED OR CONSISTENT MASS, IMPLICIT INTEGRATION BY THETA METHOD

```
C
C      PROGRAM 11.1 FORCED VIBRATION OF A RECTANGULAR SOLID IN
C      PLANE STRAIN USING 8-NODE QUADRILATERALS,
C      LUMPED OR CONSISTENT MASS,DIRECT INTEGRATION IN TIME USING
C      THE THETA-METHOD
C
C      ALTER NEXT LINE TO CHANGE PROBLEM SIZE
C
      PARAMETER(IKV=1000,ILOADS=103,INF=85)
C
      REAL DEE(3,3),SAMP(3,2),COORD(8,2),FUN(8),JAC(2,2),JAC1(2,2),
     +DER(2,8),DERIV(2,8),BEE(3,16),DBEE(3,16),BTDB(16,16),
     +BT(16,3),KM(16,16),EMM(16,16),ECM(16,16),TN(16,16),NT(16,2),
     +KV(IKV),MM(IKV),F1(IKV),LOADS(ILOADS),
     +X0(ILOADS),D1X0(ILOADS),D2X0(ILOADS),
     +X1(ILOADS),D1X1(ILOADS),D2X1(ILOADS)
      INTEGER NF(INF,2),G(16)
      DATA IJAC,IJAC1,IDER,IDERIV,NODOF,IT/6*2/
      DATA IH,ISAMP,IDEE,IBEE,IDBEE/5*3/
      DATA ICOORD,NOD/2*8/,IBTDB,IKM,IBT,IEMM,IECM,IDOF,ITN,INT/8*16/
```

```
C
C      INPUT AND INITIALISATION
C
      READ(5,*)NXE,NYE,N,IW,NN,NR,NGP,AA,BB,ITYPE,RHO,E,V,ALPHA,BETA,
     +          ISTEP,NPRI,THETA,OMEGA
      CALL READNF(NF,INF,NN,NODOF,NR)
      IR=N*(IW+1)
      PI=ACOS(-1.)
      PERIOD=2.*PI/OMEGA
      DTIM=PERIOD/20.
      CALL NULVEC(KV,IR)
      CALL NULVEC(MM,IR)
      CALL NULL(DEE,IDEE,IH,IH)
      CALL FMDEPS(DEE,IDEE,E,V)
      CALL GAUSS(SAMP,ISAMP,NGP)
C
C      ELEMENT STIFFNESS AND MASS INTEGRATION AND ASSEMBLY
C
      DO 10 IP=1,NXE
      DO 10 IQ=1,NYE
      AREA=0.
      CALL GEOM8Y(IP,IQ,NYE,AA,BB,COORD,ICOORD,G,NF,INF)
      CALL NULL(KM,IKM,IDOF,IDOF)
      CALL NULL(EMM,IEMM,IDOF,IDOF)
      DO 20 I=1,NGP
      DO 20 J=1,NGP
      CALL FMQUAD(DER,IDER,FUN,SAMP,ISAMP,I,J)
      CALL MATMUL(DER,IDER,COORD,ICOORD,JAC,IJAC,IT,NOD,IT)
      CALL TWOBY2(JAC,IJAC,JAC1,IJAC1,DET)
      CALL MATMUL(JAC1,IJAC1,DER,IDER,DERIV,IDERIV,IT,IT,NOD)
      CALL NULL(BEE,IBEE,IH,IDOF)
      CALL FORMB(BEE,IBEE,DERIV,IDERIV,NOD)
      CALL MATMUL(DEE,IDEE,BEE,IBEE,DBEE,IDBEE,IH,IH,IDOF)
      CALL MATRAN(BT,IBT,BEE,IBEE,IH,IDOF)
      CALL MATMUL(BT,IBT,DBEE,IDBEE,BTDB,IBTDB,IDOF,IH,IDOF)
      QUOT=DET*SAMP(I,2)*SAMP(J,2)
      AREA=AREA+QUOT
      IF(ITYPE.NE.1)THEN
      CALL ECMAT(ECM,IECM,TN,ITN,NT,INT,FUN,NOD,NODOF)
      PROD=QUOT*RHO
      CALL MSMULT(ECM,IECM,PROD,IDOF,IDOF)
      CALL MATADD(EMM,IEMM,ECM,IECM,IDOF,IDOF)
      END IF
      CALL MSMULT(BTDB,IBTDB,QUOT,IDOF,IDOF)
   20 CALL MATADD(KM,IKM,BTDB,IBTDB,IDOF,IDOF)
      IF(ITYPE.EQ.1)THEN
      DO 30 I=1,IDOF
   30 EMM(I,I)=AREA*RHO*.2
      DO 31 I=1,13,4
   31 EMM(I,I)=EMM(3,3)*.25
      DO 32 I=2,14,4
   32 EMM(I,I)=EMM(3,3)*.25
      END IF
      CALL FORMKV(KV,KM,IKM,G,N,IDOF)
   10 CALL FORMKV(MM,EMM,IEMM,G,N,IDOF)
C
C      REDUCTION OF LEFT HAND SIDE
C
      CALL NULVEC(X0,N)
      CALL NULVEC(D1X0,N)
```

```
      CALL NULVEC(D2X0,N)
      C1=(1.-THETA)*DTIM
      C2=BETA-C1
      C3=ALPHA+1./(THETA*DTIM)
      C4=BETA+THETA*DTIM
      DO 40 I=1,IR
   40 F1(I)=C3*MM(I)+C4*KV(I)
      CALL BANRED(F1,N,IW)
C
C      TIME STEPPING LOOP
C
      TIM=0.
      DO 50 J=1,ISTEP
      TIM=TIM+DTIM
      CALL NULVEC(LOADS,N)
      DO 60 I=1,N
   60 X1(I)=C3*X0(I)+D1X0(I)/THETA
      LOADS(N)=THETA*DTIM*COS(OMEGA*TIM)+C1*COS(OMEGA*(TIM-DTIM))
      CALL LINMUL(MM,X1,D1X1,N,IW)
      CALL VECADD(D1X1,LOADS,D1X1,N)
      DO 70 I=1,N
   70 LOADS(I)=C2*X0(I)
      CALL LINMUL(KV,LOADS,X1,N,IW)
      CALL VECADD(X1,D1X1,X1,N)
      CALL BACSUB(F1,X1,N,IW)
      DO 80 I=1,N
      D1X1(I)=(X1(I)-X0(I))/(THETA*DTIM)-D1X0(I)*(1.-THETA)/THETA
   80 D2X1(I)=(D1X1(I)-D1X0(I))/(THETA*DTIM)-D2X0(I)*(1.-THETA)/THETA
      IF(J/NPRI*NPRI.EQ.J)WRITE(6,1000)TIM,COS(OMEGA*TIM),X1(N)
      CALL VECCOP(X1,X0,N)
      CALL VECCOP(D1X1,D1X0,N)
      CALL VECCOP(D2X1,D2X0,N)
   50 CONTINUE
 1000 FORMAT(5E12.4)
      STOP
      END
```

In this program, the problem previously analysed by modal superposition is solved again by a direct integration procedure. The specific method is the same implicit technique as was used for first order problems in Program 8.0, where it is often called the 'Crank–Nicolson' approach. In second order problems it is also known as the 'Newmark' $\beta = \frac{1}{4}$ method.

The formulation was described in Section 3.15.2.1 where it was shown that in principle the algorithm is the same as that used in Program 8.0. To step from one time instant to the next, a set of simultaneous equations has to be solved. Since the differential equations are often linearised, this is not as great a numerical task as might be supposed because the equation coefficients are constant and need be reduced or factorised only once before the time-stepping procedure commences (see equation 3.139). Velocities and accelerations are computed by ancillary equations (3.140) and (3.141).

The problem layout and data are almost the same as in Figure 11.1 and are reproduced as Figure 11.4. The structure of the program is contained in Figure 11.5. Turning to the coding, the new variables are the simple variables:

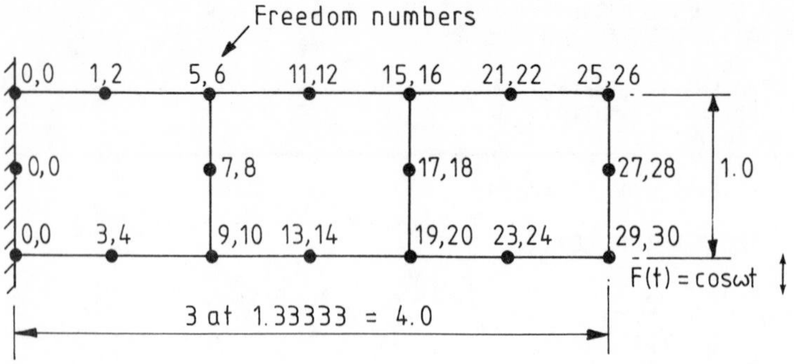

Structure data	NXE	NYE	N	IW	NN	NR	NGP
	3	1	30	15	18	3	3

Element data	AA	BB	ITYPE	RHO	E	V	ALPHA	BETA
	1.33333	1.	1	1.	1.	.3	.005	.272

Time integration data	ISTEP	NPRI	THETA	OMEGA
	20	1	.5 (Program 11.1) 1.4 (Program 11.2)	.3

Node freedom data	1	0	0	2	0	0	3	0	0

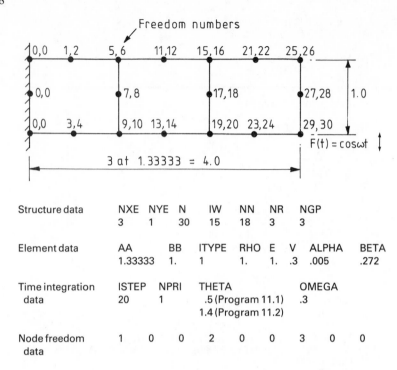

Figure 11.4 Mesh and data for Programs 11.1 and 11.2

ITYPE	switch—equals 1 for lumped mass, 0 for consistent
ALPHA, BETA	Rayleigh damping parameters α and β (see 3.127)
AREA	element area

and the arrays:

F1	$(\alpha + 1/\theta/\Delta t)\mathbf{M} + (\beta + \theta\,\Delta t)\mathbf{K}$ (see 3.139)
X0, D1X0, D2X0	displacement, velocity, acceleration at time 0
X1, D1X1, D2X1	displacement, velocity, acceleration at time 1

Up to the section headed 'element stiffness and mass integration and assembly' the program's task is the familiar one of generating the global stiffness and mass matrices, in this case stored as the vectors KV and MM respectively. The matrix arising on the left hand side of (3.139) is then created, called F1 and factorised using BANRED.

In the time-stepping loop, the matrix-by-vector multiplications and vector additions specified on the right hand side of (3.139) are carried out and equation solution is completed by BACSUB. It then remains only to compute the new velocities and accelerations using (3.140) and (3.141).

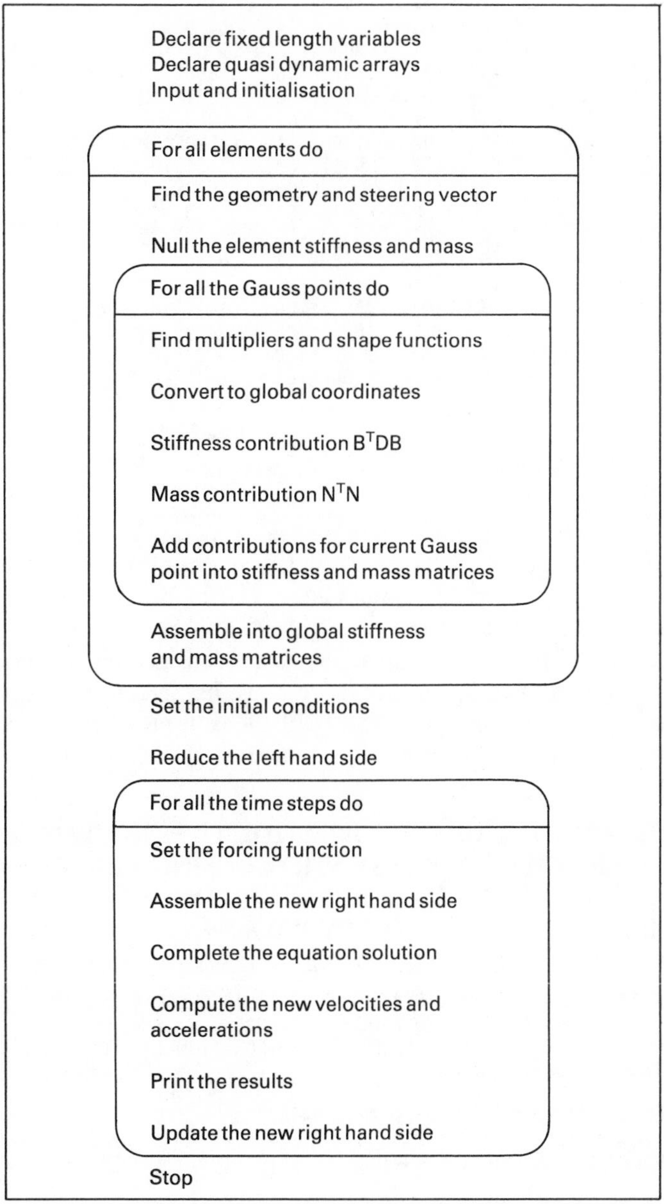
Declare fixed length variables
Declare quasi dynamic arrays
Input and initialisation
For all elements do
Find the geometry and steering vector
Null the element stiffness and mass
For all the Gauss points do
Find multipliers and shape functions
Convert to global coordinates
Stiffness contribution B^TDB
Mass contribution N^TN
Add contributions for current Gauss point into stiffness and mass matrices
Assemble into global stiffness and mass matrices
Set the initial conditions
Reduce the left hand side
For all the time steps do
Set the forcing function
Assemble the new right hand side
Complete the equation solution
Compute the new velocities and accelerations
Print the results
Update the new right hand side
Stop

Figure 11.5 Structure chart for Programs 11.1, 11.2 and 11.5

Time t	cos ωt	Tip displacement
.1047E+01	.9511E+00	.2291E+01
.2094E+01	.8090E+00	.5473E+01
.3142E+01	.5878E+00	.7771E+01
.4189E+01	.3090E+00	.1108E+02
.5236E+01	.5390E-14	.1470E+02
.6283E+01	-.3090E+00	.1735E+02
.7330E+01	-.5878E+00	.1866E+02
.8378E+01	-.8090E+00	.1882E+02
.9425E+01	-.9511E+00	.1768E+02
.1047E+02	-.1000E+01	.1520E+02
.1152E+02	-.9511E+00	.1154E+02
.1257E+02	-.8090E+00	.7069E+01
.1361E+02	-.5878E+00	.2196E+01
.1466E+02	-.3090E+00	-.2633E+01
.1571E+02	-.4459E-13	-.6880E+01
.1676E+02	.3090E+00	-.9971E+01
.1780E+02	.5878E+00	-.1147E+02
.1885E+02	.8090E+00	-.1116E+02
.1990E+02	.9511E+00	-.9027E+01
.2094E+02	.1000E+01	-.5315E+01

Figure 11.6 Results from Program 11.1

In order to compare results with those from Program 11.0, the lumped mass option has been chosen (ITYPE = 1) and, using (3.132), damping constants $\alpha = 0.005$ and $\beta = 0.272$ yield γ values close to 0.05 for the first three natural frequencies ω. The results are listed as Figure 11.6, where only the first complete cycle of forcing is shown. The displacements are considerably disturbed by the initial condition 'transients', but once the response has settled down, say after 200 time steps, the average amplitude is 45.59 with the same phase shift of about 90° as was found by the modal superposition technique.

PROGRAM 11.2: FORCED VIBRATION OF A RECTANGULAR ELASTIC SOLID IN PLANE STRAIN USING EIGHT-NODE QUADRILATERALS—LUMPED OR CONSISTENT MASS, IMPLICIT INTEGRATION BY WILSON THETA METHOD

```
C
C       PROGRAM 11.2 FORCED VIBRATION OF A RECTANGULAR SOLID IN
C       PLANE STRAIN USING 8-NODE QUADRILATERALS,
C       LUMPED OR CONSISTENT MASS,DIRECT INTEGRATION IN TIME USING
C       WILSON'S THETA METHOD
C
C       ALTER NEXT LINE TO CHANGE PROBLEM SIZE
C
      PARAMETER(IKV=1000,ILOADS=103,INF=85)
C
      REAL DEE(3,3),SAMP(3,2),COORD(8,2),FUN(8),JAC(2,2),JAC1(2,2),
     +DER(2,8),DERIV(2,8),BEE(3,16),DBEE(3,16),BTDB(16,16),
     +BT(16,3),KM(16,16),EMM(16,16),ECM(16,16),TN(16,16),NT(16,2),
     +KV(IKV),MM(IKV),F1(IKV),LOADS(ILOADS),
     +X0(ILOADS),D1X0(ILOADS),D2X0(ILOADS),
     +X1(ILOADS),D1X1(ILOADS),D2X1(ILOADS)
```

```
      INTEGER NF(INF,2),G(16)
      DATA IJAC,IJAC1,IDER,IDERIV,NODOF,IT/6*2/
      DATA IH,ISAMP,IDEE,IBEE,IDBEE/5*3/
      DATA ICOORD,NOD/2*8/,IBTDB,IKM,IBT,IEMM,IECM,IDOF,ITN,INT/8*16/
C
C      INPUT AND INITIALISATION
C
      READ(5,*)NXE,NYE,N,IW,NN,NR,NGP,AA,BB,ITYPE,RHO,E,V,ALPHA,BETA,
     +         ISTEP,NPRI,THETA,OMEGA
      CALL READNF(NF,INF,NN,NODOF,NR)
      IR=N*(IW+1)
      PI=ACOS(-1.)
      PERIOD=2.*PI/OMEGA
      DTIM=PERIOD/20.
      CALL NULVEC(KV,IR)
      CALL NULVEC(MM,IR)
      CALL NULL(DEE,IDEE,IH,IH)
      CALL FMDEPS(DEE,IDEE,E,V)
      CALL GAUSS(SAMP,ISAMP,NGP)
C
C      ELEMENT STIFFNESS AND MASS INTEGRATION AND ASSEMBLY
C
      DO 10 IP=1,NXE
      DO 10 IQ=1,NYE
      AREA=0.
      CALL GEOM8Y(IP,IQ,NYE,AA,BB,COORD,ICOORD,G,NF,INF)
      CALL NULL(KM,IKM,IDOF,IDOF)
      CALL NULL(EMM,IEMM,IDOF,IDOF)
      DO 20 I=1,NGP
      DO 20 J=1,NGP
      CALL FMQUAD(DER,IDER,FUN,SAMP,ISAMP,I,J)
      CALL MATMUL(DER,IDER,COORD,ICOORD,JAC,IJAC,IT,NOD,IT)
      CALL TWOBY2(JAC,IJAC,JAC1,IJAC1,DET)
      CALL MATMUL(JAC1,IJAC1,DER,IDER,DERIV,IDERIV,IT,IT,NOD)
      CALL NULL(BEE,IBEE,IH,IDOF)
      CALL FORMB(BEE,IBEE,DERIV,IDERIV,NOD)
      CALL MATMUL(DEE,IDEE,BEE,IBEE,DBEE,IDBEE,IH,IH,IDOF)
      CALL MATRAN(BT,IBT,BEE,IBEE,IH,IDOF)
      CALL MATMUL(BT,IBT,DBEE,IDBEE,BTDB,IBTDB,IDOF,IH,IDOF)
      QUOT=DET*SAMP(I,2)*SAMP(J,2)
      AREA=AREA+QUOT
      IF(ITYPE.NE.1)THEN
      CALL ECMAT(ECM,IECM,TN,ITN,NT,INT,FUN,NOD,NODOF)
      PROD=QUOT*RHO
      CALL MSMULT(ECM,IECM,PROD,IDOF,IDOF)
      CALL MATADD(EMM,IEMM,ECM,IECM,IDOF,IDOF)
      END IF
      CALL MSMULT(BTDB,IBTDB,QUOT,IDOF,IDOF)
   20 CALL MATADD(KM,IKM,BTDB,IBTDB,IDOF,IDOF)
      IF(ITYPE.EQ.1)THEN
      DO 30 I=1,IDOF
   30 EMM(I,I)=AREA*RHO*.2
      DO 31 I=1,13,4
   31 EMM(I,I)=EMM(3,3)*.25
      DO 32 I=2,14,4
   32 EMM(I,I)=EMM(3,3)*.25
      END IF
      CALL FORMKV(KV,KM,IKM,G,N,IDOF)
   10 CALL FORMKV(MM,EMM,IEMM,G,N,IDOF)
```

```
C
C      REDUCTION OF LEFT HAND SIDE
C
      CALL NULVEC(X0,N)
      CALL NULVEC(D1X0,N)
      CALL NULVEC(D2X0,N)
      C1=6./(THETA*DTIM)**2
      C2=6./(THETA*DTIM)
      C3=DTIM**2/6.
      C4=2.
      C5=3.*ALPHA/(THETA*DTIM)
      C6=3.*BETA/(THETA*DTIM)
      C7=.5*ALPHA*THETA*DTIM
      C8=.5*BETA*THETA*DTIM
      DO 40 I=1,IR
   40 F1(I)=(C1+C5)*MM(I)+(1.+C6)*KV(I)
      CALL BANRED(F1,N,IW)
C
C      TIME STEPPING LOOP
C
      TIM=0.
      DO 50 J=1,ISTEP
      TIM=TIM+DTIM
      CALL NULVEC(LOADS,N)
      DO 60 I=1,N
   60 X1(I)=(C1+C5)*X0(I)+(C2+2.*ALPHA)*D1X0(I)+(2.+C7)*D2X0(I)
      LOADS(N)=THETA*COS(OMEGA*TIM)+(1.-THETA)*COS(OMEGA*(TIM-DTIM))
      CALL LINMUL(MM,X1,D1X1,N,IW)
      CALL VECADD(D1X1,LOADS,D1X1,N)
      DO 70 I=1,N
   70 LOADS(I)=C6*X0(I)+2.*BETA*D1X0(I)+C8*D2X0(I)
      CALL LINMUL(KV,LOADS,X1,N,IW)
      CALL VECADD(X1,D1X1,X1,N)
      CALL BACSUB(F1,X1,N,IW)
      DO 80 I=1,N
      D2X1(I)=(X1(I)-X0(I))*C1-D1X0(I)*C2-D2X0(I)*C4
      D2X1(I)=D2X0(I)+(D2X1(I)-D2X0(I))/THETA
      D1X1(I)=D1X0(I)+.5*DTIM*(D2X1(I)+D2X0(I))
   80 X1(I)=X0(I)+DTIM*D1X0(I)+2.*C3*D2X0(I)+C3*D2X1(I)
      IF(J/NPRI*NPRI.EQ.J)WRITE(6,1000)TIM,COS(OMEGA*TIM),X1(N)
      CALL VECCOP(X1,X0,N)
      CALL VECCOP(D1X1,D1X0,N)
      CALL VECCOP(D2X1,D2X0,N)
   50 CONTINUE
 1000 FORMAT(5E12.4)
      STOP
      END
```

This algorithm is described in Section 3.15.2.2. The essential step is that shown in (3.142), which is of exactly the same form as (3.139), and so this program can be expected to resemble the previous one very closely. The structure chart of Figure 11.5 is again appropriate and the problem layout and data are again those of Figure 11.4. No new variables are involved, but the parameter θ has a stability limit of about 1.4 compared with 0.5 in the previous algorithm. All that need be said is that the F1 matrix is now constructed as demanded by (3.142). The remaining steps of (3.143) to (3.146) are carried out within the section headed 'time stepping loop'.

Time t	$\cos\omega t$	Tip displacement
.1047E+01	.9511E+00	.4557E+00
.2094E+01	.8090E+00	.2699E+01
.3142E+01	.5878E+00	.5166E+01
.4189E+01	.3090E+00	.7263E+01
.5236E+01	.5390E-14	.9409E+01
.6283E+01	-.3090E+00	.1138E+02
.7330E+01	-.5878E+00	.1244E+02
.8378E+01	-.8090E+00	.1213E+02
.9425E+01	-.9511E+00	.1041E+02
.1047E+02	-.1000E+01	.7466E+01
.1152E+02	-.9511E+00	.3476E+01
.1257E+02	-.8090E+00	-.1297E+01
.1361E+02	-.5878E+00	-.6472E+01
.1466E+02	-.3090E+00	-.1158E+02
.1571E+02	-.4459E-13	-.1612E+02
.1676E+02	.3090E+00	-.1962E+02
.1780E+02	.5878E+00	-.2171E+02
.1885E+02	.8090E+00	-.2211E+02
.1990E+02	.9511E+00	-.2076E+02
.2094E+02	.1000E+01	-.1780E+02

Figure 11.7 Results from Program 11.2

The results for one cycle are listed as Figure 11.7. Again these reflect mainly the influence of the start-up conditions and after 200 or so time steps the amplitude of vibration has settled down to 45.13, although with a slightly greater phase shift than was computed by the previous two methods.

PROGRAM 11.3: FORCED VIBRATION OF A RECTANGULAR SOLID IN PLANE STRAIN USING EIGHT-NODE QUADRILATERALS—LUMPED MASS, COMPLEX RESPONSE METHOD

```
C
C      PROGRAM 11.3 FORCED VIBRATION OF A RECTANGULAR SOLID IN
C      PLANE STRAIN USING 8-NODE QUADRILATERALS,
C      LUMPED MASS,COMPLEX RESPONSE METHOD
C
C      ALTER NEXT LINE TO CHANGE PROBLEM SIZE
C
      PARAMETER(IKC=1000,ILOADS=103,INF=85)
C
      REAL DEE(3,3),SAMP(3,2),COORD(8,2),FUN(8),JAC(2,2),JAC1(2,2),
     +DER(2,8),DERIV(2,8),BEE(3,16),DBEE(3,16),BTDB(16,16),
     +BT(16,3),KM(16,16),CM(16,16),EMM(16,16)
      COMPLEX KC(IKC),LOADS(ILOADS)
      INTEGER NF(INF,2),G(16)
      DATA IJAC,IJAC1,IDER,IDERIV,NODOF,IT/6*2/
      DATA IH,ISAMP,IDEE,IBEE,IDBEE/5*3/
      DATA ICOORD,NOD/2*8/,IBTDB,IKM,IBT,IEMM,IDOF/5*16/
C
C      INPUT AND INITIALISATION
C
      READ(5,*)NXE,NYE,N,IW,NN,NR,NGP,AA,BB,RHO,E,V,DR,OMEGA,
     +ISTEP,NPRI
```

```
      CALL READNF(NF,INF,NN,NODOF,NR)
      IR=N*(IW+1)
      PI=ACOS(-1.)
      PERIOD=2.*PI/OMEGA
      DTIM=PERIOD/20.
      CALL FMDEPS(DEE,IDEE,E,V)
      CALL GAUSS(SAMP,ISAMP,NGP)
C
C       SOLVE FOR SINGLE INPUT LOADING HARMONIC
C
      CALL NULL(EMM,IEMM,IDOF,IDOF)
      DO 10 I=1,IR
   10 KC(I)=(0.,0.)
      DO 20 I=1,N
   20 LOADS(I)=(0.,0.)
C
C       FORM ELEMENT LUMPED MASS MATRIX INCLUDING FREQUENCY
C
      DO 30 I=1,IDOF
   30 EMM(I,I)=AA*BB*RHO*.2*OMEGA**2
      DO 31 I=1,13,4
   31 EMM(I,I)=EMM(3,3)*.25
      DO 32 I=2,14,4
   32 EMM(I,I)=EMM(3,3)*.25
C
C       ELEMENT STIFFNESS INTEGRATION AND ASSEMBLY
C
      DO 40 IP=1,NXE
      DO 40 IQ=1,NYE
      CALL GEOM8Y(IP,IQ,NYE,AA,BB,COORD,ICOORD,G,NF,INF)
      CALL NULL(KM,IKM,IDOF,IDOF)
      DO 50 I=1,NGP
      DO 50 J=1,NGP
      CALL FMQUAD(DER,IDER,FUN,SAMP,ISAMP,I,J)
      CALL MATMUL(DER,IDER,COORD,ICOORD,JAC,IJAC,IT,NOD,IT)
      CALL TWOBY2(JAC,IJAC,JAC1,IJAC1,DET)
      CALL MATMUL(JAC1,IJAC1,DER,IDER,DERIV,IDERIV,IT,IT,NOD)
      CALL NULL(BEE,IBEE,IH,IDOF)
      CALL FORMB(BEE,IBEE,DERIV,IDERIV,NOD)
      CALL MATMUL(DEE,IDEE,BEE,IBEE,DBEE,IDBEE,IH,IH,IDOF)
      CALL MATRAN(BT,IBT,BEE,IBEE,IH,IDOF)
      CALL MATMUL(BT,IBT,DBEE,IDBEE,BTDB,IBTDB,IDOF,IH,IDOF)
      QUOT=DET*SAMP(I,2)*SAMP(J,2)*(1.-2.*DR*DR)
      CALL MSMULT(BTDB,IBTDB,QUOT,IDOF,IDOF)
   50 CALL MATADD(KM,IKM,BTDB,IBTDB,IDOF,IDOF)
C
C       COMPLEX MATRIX ASSEMBLY
C
      SNP=2.*DR*SQRT(1.-DR*DR)/(1.-2.*DR*DR)
      DO 60 I=1,IDOF
      DO 60 J=1,IDOF
      CM(I,J)=KM(I,J)*SNP
   60 KM(I,J)=KM(I,J)-EMM(I,J)
   40 CALL FORMKC(KC,KM,IKM,CM,IKM,G,N,IDOF)
C
C       COMPLEX EQUATION SOLUTION
C
      LOADS(N)=(1.,0.)
      CALL COMRED(KC,N,IW)
      CALL COMBAC(KC,LOADS,N,IW)
```

```
      A=REAL(LOADS(N))
      B=AIMAG(LOADS(N))
      TIM=0.
      DO 70 J=1,ISTEP
      TIM=TIM+DTIM
      IF(J/NPRI*NPRI.EQ.J)
     +WRITE(6,1000)TIM,A*COS(OMEGA*TIM)-B*SIN(OMEGA*TIM)
   70 CONTINUE
 1000 FORMAT(3E12.4)
      STOP
      END
```

This forced vibration analysis uses the 'complex response' or complex mode superposition method described in Section 3.15.4. The program terminology is essentially as before, but, as shown by (3.150) and (3.151), the equations to be solved involve complex numbers. Thus the appropriate element assembly routine is FORMKC and the appropriate equation solution routines COMRED and COMBAC.

The layout and data again follow Figure 11.1 and are reproduced as Figure 11.8. The lumped mass approximation is used and the mass matrix is built up containing the multiple ω^2 as required by (3.151). The stiffness matrix is built up incorporating the multiple $1-2\gamma^2$ (see 3.150) so that the complex part is obtained by multiplying the real part by $\mathrm{SNP}=2\gamma\sqrt{(1-\gamma^2)}/(1-2\gamma^2)$.

Therefore, the only unfamiliar variables are SNP and the complex part of the

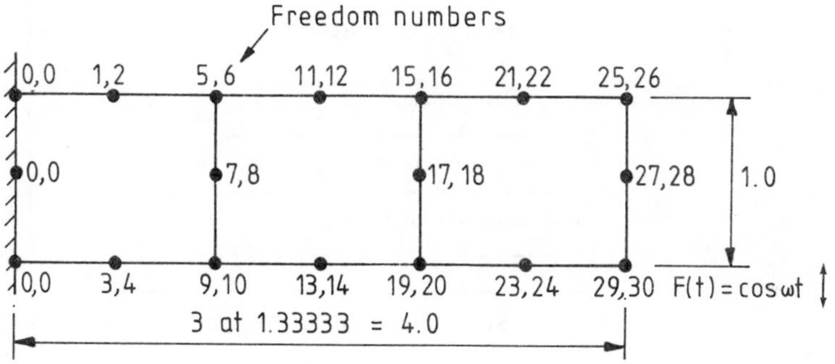

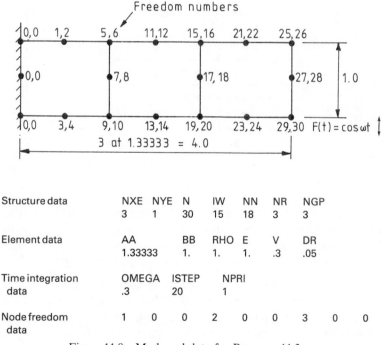

Structure data	NXE	NYE	N	IW	NN	NR	NGP		
	3	1	30	15	18	3	3		
Element data	AA		BB	RHO	E	V	DR		
	1.33333		1.	1.	1.	.3	.05		
Time integration data	OMEGA		ISTEP		NPRI				
	.3		20		1				
Node freedom data	1	0	0	2	0	0	3	0	0

Figure 11.8 Mesh and data for Program 11.3

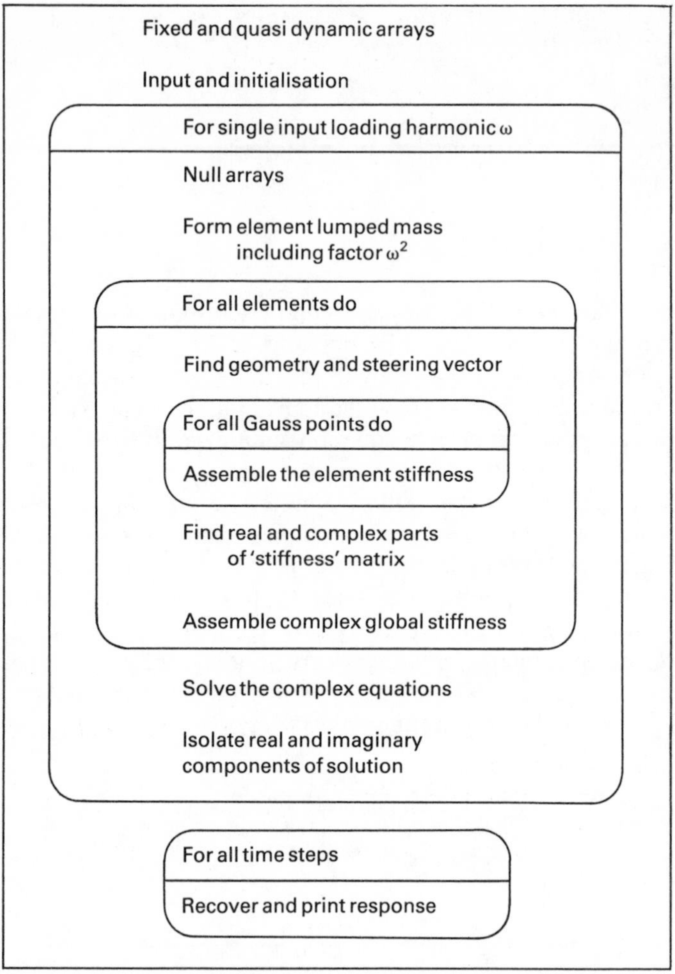

Figure 11.9 Structure chart for Program 11.3

'stiffness' matrix called CM. The global complex stiffness KC has adjustable dimension IKC $\geqslant$ IR. The structure of the program is described by Figure 11.9.

The results are shown in Figure 11.10 and show virtually the same amplitude (44.26 as against 43.22) and phase shift as computed by Program 11.0.

Thus, Programs 11.0 to 11.3 deliver essentially the same results. As was the case for first order time dependent problems described in Chapter 8, algorithm choice is a rather complicated matter and no simple rule can be laid down. In the final two programs of this chapter, algorithms involving 'explicit' integration are described.

```
.1047E+01    .2721E+02
.2094E+01    .3668E+02
.3142E+01    .4255E+02
.4189E+01    .4426E+02
.5236E+01    .4164E+02
.6283E+01    .3494E+02
.7330E+01    .2482E+02
.8378E+01    .1227E+02
.9425E+01   -.1480E+01
.1047E+02   -.1509E+02
.1152E+02   -.2721E+02
.1257E+02   -.3668E+02
.1361E+02   -.4255E+02
.1466E+02   -.4426E+02
.1571E+02   -.4164E+02
.1676E+02   -.3494E+02
.1780E+02   -.2482E+02
.1885E+02   -.1227E+02
.1990E+02    .1480E+01
.2094E+02    .1509E+02
  Time t        Tip
             displacement
```

Figure 11.10 Results from Program 11.3

PROGRAM 11.4: FORCED VIBRATION OF A RECTANGULAR ELASTO-PLASTIC SOLID IN PLANE STRAIN USING EIGHT-NODE QUADRILATERALS—LUMPED MASS, EXPLICIT INTEGRATION

```
C
C      PROGRAM 11.4 FORCED VIBRATION OF A RECTANGULAR
C      ELASTO-PLASTIC SOLID IN PLANE STRAIN USING 8-NODE
C      QUADRILATERAL ELEMENTS:LUMPED MASS:EXPLICIT INTEGRATION
C
C      ALTER NEXT LINE TO CHANGE PROBLEM SIZE
C
      PARAMETER(ILOADS=1000,INX=20,INY=20,INO=20,INF=50)
C
      REAL DEE(4,4),PL(4,4),SAMP(3,2),COORD(8,2),JAC(2,2),JAC1(2,2),
     +DER(2,8),DERIV(2,8),BEE(4,16),ELD(16),FUN(8),EMM(16),STRESS(4),
     +EPS(4),SIGMA(4),BT(16,4),ELOAD(16),BLOAD(16),VAL(INO),
     +X1(ILOADS),D1X1(ILOADS),D2X1(ILOADS),MM(ILOADS),BDYLDS(ILOADS),
     +SX(INX,INY,4),SY(INX,INY,4),TXY(INX,INY,4),SZ(INX,INY,4),
     +EX(INX,INY,4),EY(INX,INY,4),GXY(INX,INY,4),EZ(INX,INY,4)
      INTEGER G(16),NO(INO),NF(INF,2)
      DATA IDEE,IBEE,IH/3*4/,IJAC,IJAC1,IDER,IDERIV,NODOF,IT/6*2/
      DATA ICOORD,NOD/2*8/,IBT,IDOF/2*16/,ISAMP/3/
C
C      INPUT AND INITIALISATION
C
      READ(5,*)NXE,NYE,N,NN,NR,NGP,AA,BB,RHO,E,V,SBARY,PLOAD,
     +         DTIM,ISTEP,NPRI
      CALL READNF(NF,INF,NN,NODOF,NR)
      READ(5,*)NL,(NO(I),VAL(I),I=1,NL)
      IGTOT=NGP*NGP
      CALL NULL3(SX,INX,INY,NXE,NYE,IGTOT)
      CALL NULL3(SY,INX,INY,NXE,NYE,IGTOT)
```

```
      CALL NULL3(TXY,INX,INY,NXE,NYE,IGTOT)
      CALL NULL3(SZ,INX,INY,NXE,NYE,IGTOT)
      CALL NULL3(EX,INX,INY,NXE,NYE,IGTOT)
      CALL NULL3(EY,INX,INY,NXE,NYE,IGTOT)
      CALL NULL3(GXY,INX,INY,NXE,NYE,IGTOT)
      CALL NULL3(EZ,INX,INY,NXE,NYE,IGTOT)
      CALL NULVEC(X1,N)
      CALL NULVEC(D1X1,N)
      CALL NULVEC(D2X1,N)
      CALL NULVEC(MM,N)
      CALL GAUSS(SAMP,ISAMP,NGP)
C
C       EXPLICIT INTEGRATION LOOP
C
      TIM=0.
      WRITE(6,1000)TIM,X1(N),D1X1(N),D2X1(N)
      DO 10 JJ=1,ISTEP
      TIM=TIM+DTIM
C
C       APPLIED LOAD
C
      DO 20 I=1,N
   20 X1(I)=X1(I)+(D1X1(I)+D2X1(I)*DTIM*.5)*DTIM
      CALL NULVEC(BDYLDS,N)
C
C       FORM ELEMENT STRAIN-DISPLACEMENT RELATIONSHIPS
C
      DO 30 IP=1,NXE
      DO 30 IQ=1,NYE
      AREA=.0
      CALL NULVEC(BLOAD,IDOF)
      CALL GEOM8X(IP,IQ,NXE,AA,BB,COORD,ICOORD,G,NF,INF)
      DO 40 M=1,IDOF
      IF(G(M).EQ.0)ELD(M)=0.0
   40 IF(G(M).NE.0)ELD(M)=X1(G(M))
      IG=0
      DO 50 I=1,NGP
      DO 50 J=1,NGP
      IG=IG+1
      CALL NULL(DEE,IDEE,IH,IH)
      CALL FMDRAD(DEE,IDEE,E,V)
      CALL FMQUAD(DER,IDER,FUN,SAMP,ISAMP,I,J)
      CALL MATMUL(DER,IDER,COORD,ICOORD,JAC,IJAC,IT,NOD,IT)
      CALL TWOBY2(JAC,IJAC,JAC1,IJAC1,DET)
      CALL MATMUL(JAC1,IJAC1,DER,IDER,DERIV,IDERIV,IT,IT,NOD)
      CALL NULL(BEE,IBEE,IH,IDOF)
      CALL FORMB(BEE,IBEE,DERIV,IDERIV,NOD)
      QUOT=DET*SAMP(I,2)*SAMP(J,2)
      AREA=AREA+QUOT*RHO
      CALL MVMULT(BEE,IBEE,ELD,IH,IDOF,EPS)
      EPS(1)=EPS(1)-EX(IP,IQ,IG)
      EPS(2)=EPS(2)-EY(IP,IQ,IG)
      EPS(3)=EPS(3)-GXY(IP,IQ,IG)
      EPS(4)=EPS(4)-EZ(IP,IQ,IG)
      CALL MVMULT(DEE,IDEE,EPS,IH,IH,SIGMA)
      STRESS(1)=SIGMA(1)+SX(IP,IQ,IG)
      STRESS(2)=SIGMA(2)+SY(IP,IQ,IG)
      STRESS(3)=SIGMA(3)+TXY(IP,IQ,IG)
      STRESS(4)=SIGMA(4)+SZ(IP,IQ,IG)
      CALL INVAR(STRESS,SIGM,DSBAR,THETA)
      FNEW=DSBAR-SBARY
```

```
C
C       CHECK WHETHER YIELD IS VIOLATED
C
      IF(FNEW.LT.0.)GOTO 70
      STRESS(1)=SX(IP,IQ,IG)
      STRESS(2)=SY(IP,IQ,IG)
      STRESS(3)=TXY(IP,IQ,IG)
      STRESS(4)=SZ(IP,IQ,IG)
      CALL INVAR(STRESS,SIGM,SBAR,THETA)
      F=SBAR-SBARY
      FAC=FNEW/(FNEW-F)
      STRESS(1)=SX(IP,IQ,IG)+(1.-FAC)*SIGMA(1)
      STRESS(2)=SY(IP,IQ,IG)+(1.-FAC)*SIGMA(2)
      STRESS(3)=TXY(IP,IQ,IG)+(1.-FAC)*SIGMA(3)
      STRESS(4)=SZ(IP,IQ,IG)+(1.-FAC)*SIGMA(4)
      CALL VMPL(E,V,STRESS,PL)
      DO 60 K=1,IH
      DO 60 L=1,IH
   60 DEE(K,L)=DEE(K,L)-FAC*PL(K,L)
   70 CALL MVMULT(DEE,IDEE,EPS,IH,IH,SIGMA)
      SIGMA(1)=SIGMA(1)+SX(IP,IQ,IG)
      SIGMA(2)=SIGMA(2)+SY(IP,IQ,IG)
      SIGMA(3)=SIGMA(3)+TXY(IP,IQ,IG)
      SIGMA(4)=SIGMA(4)+SZ(IP,IQ,IG)
      CALL MATRAN(BT,IBT,BEE,IBEE,IH,IDOF)
      CALL MVMULT(BT,IBT,SIGMA,IDOF,IH,ELOAD)
      DO 80 K=1,IDOF
   80 BLOAD(K)=BLOAD(K)+ELOAD(K)*QUOT
C
C       UPDATE GAUSS POINT STRESSES AND STRAINS
C
      SX(IP,IQ,IG)=SIGMA(1)
      SY(IP,IQ,IG)=SIGMA(2)
      TXY(IP,IQ,IG)=SIGMA(3)
      SZ(IP,IQ,IG)=SIGMA(4)
      EX(IP,IQ,IG)=EX(IP,IQ,IG)+EPS(1)
      EY(IP,IQ,IG)=EY(IP,IQ,IG)+EPS(2)
      GXY(IP,IQ,IG)=GXY(IP,IQ,IG)+EPS(3)
      EZ(IP,IQ,IG)=EZ(IP,IQ,IG)+EPS(4)
   50 CONTINUE
      DO 90 M=1,IDOF
      IF(G(M).EQ.0)GO TO 90
      BDYLDS(G(M))=BDYLDS(G(M))-BLOAD(M)
   90 CONTINUE
      IF(JJ.NE.1)GO TO 30
C
C       FORM LUMPED MASS MATRIX
C
      DO 100 I=1,IDOF
  100 EMM(I)=.2*AREA
      DO 110 I=1,13,4
  110 EMM(I)=.05*AREA
      DO 120 I=2,14,4
  120 EMM(I)=.05*AREA
      DO 130 I=1,IDOF
  130 IF(G(I).NE.0)MM(G(I))=MM(G(I))+EMM(I)
   30 CONTINUE
      DO 140 I=1,NL
  140 BDYLDS(NO(I))=BDYLDS(NO(I))+VAL(I)*PLOAD
      DO 150 I=1,N
      BDYLDS(I)=BDYLDS(I)/MM(I)
```

```
      D1X1(I)=D1X1(I)+(D2X1(I)+BDYLDS(I))*.5*DTIM
  150 D2X1(I)=BDYLDS(I)
      IF(JJ.EQ.JJ/NPRI*NPRI)WRITE(6,1000)TIM,X1(N),D1X1(N),D2X1(N)
   10 CONTINUE
 1000 FORMAT(4E12.4)
      STOP
      END
```

In the same way as was done for first order problems in Program 8.2, θ can be set to zero in second order recurrence formulae such as (3.139). Then the only matrix remaining on the left hand side of the equation is **MM**; if this is lumped (diagonalised), the new solution $\mathbf{r}_1$ can be computed without solving simultaneous equations at all. Further, the right hand side products can again be completed using element-by-element summation and so no global matrices are involved. This procedure is particularly attractive in non-linear problems where the stiffness **KM** is a function of, for example, strain. In the present program, non-linearity is introduced in the form of elasto-plasticity, which was described in Chapter 6. The nomenclature used is therefore drawn from the earlier programs in this chapter and from those in Chapter 6, particularly Program 6.0 which dealt with von Mises solids.

The structure chart for the program is shown in Figure 11.11 and the problem layout and data in Figure 11.12. All the variables used, and their significance, are listed below:

Simple variables:

NXE	number of elements in x direction (1)
NYE	number of elements in y direction (6)
N	total number of non-zero freedoms in mesh (50)
NN	total number of nodes in mesh (33)
NR	number of restrained nodes (15)
NGP	number of 'Gauss points' in each direction (2)
AA	size of elements in x direction (1.)
BB	size of elements in y direction (2.5)
RHO	mass of elements per unit volume (0.733×10^{-3})
E	Young's modulus (3.0×10^{7})
SBARY	von Mises' yield stress (50,000)
PLOAD	load multiplier (180.0)
DTIM	time step (1.0×10^{-6})
ISTEP	number of time steps (300)
NPRI	printing control (every NPRI results printed) (50)
NL	number of loaded freedoms (13)
IGTOT	total number of Gauss points (4)
TIM	accumulated time
AREA	element area (includes multiple ρ)
DSBAR	deviatoric stress invariant $\bar{\sigma}$
FNEW	new value of yield function

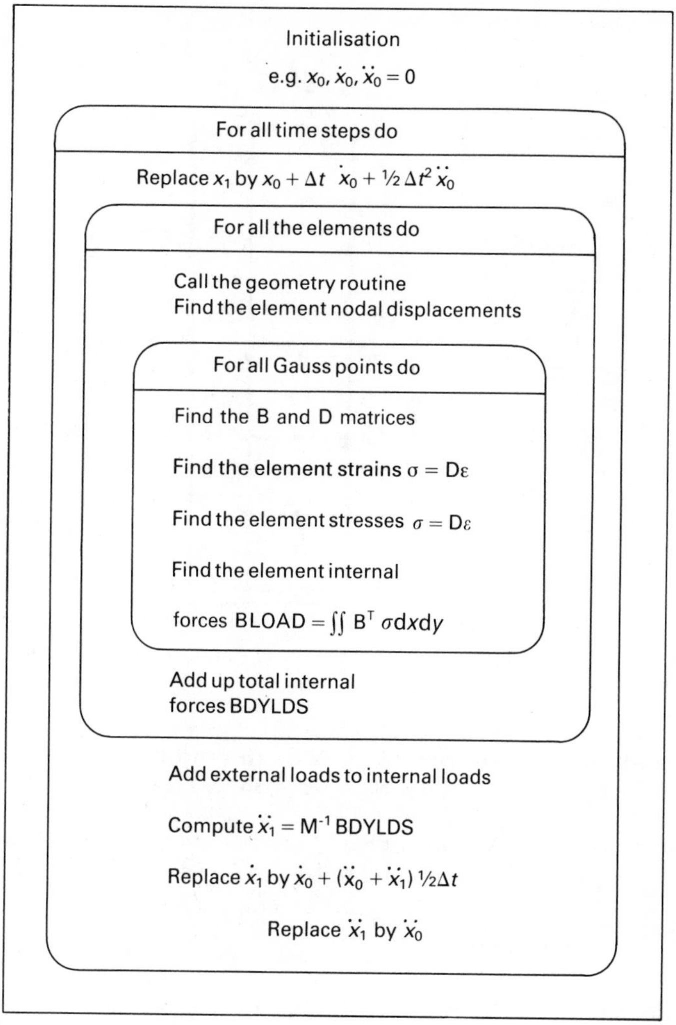

Figure 11.11 Structure chart for Program 11.4

THETA	lode angle (see Chapter 6)
F	yield function
SBAR	deviatoric stress invariant $\bar{\sigma}$
FAC	factor (see Figure 6.7)

Data constants:

IDEE, IBEE	sizes of arrays DEE and BEE
IH	components of stress and strain vectors (4)

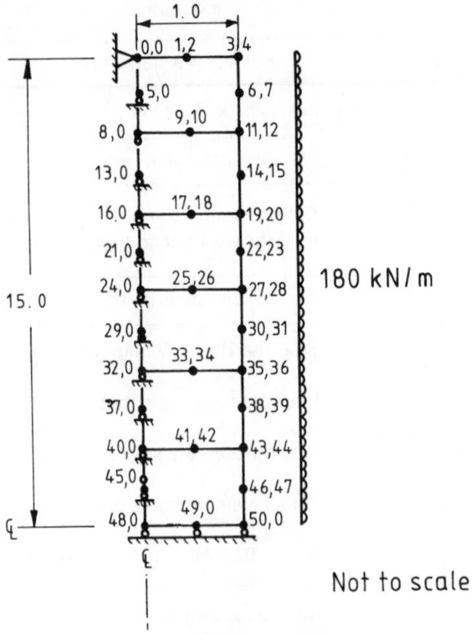

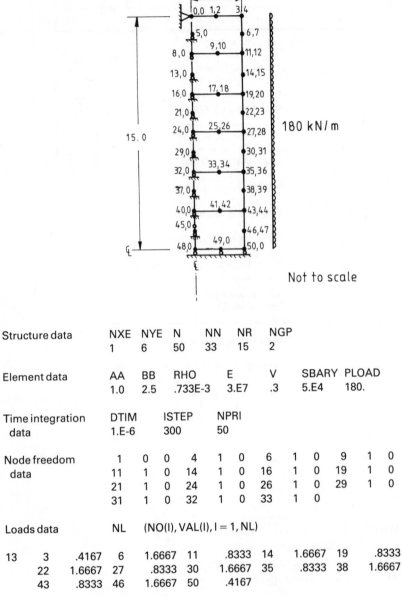

Structure data	NXE	NYE	N	NN	NR	NGP
	1	6	50	33	15	2

Element data	AA	BB	RHO	E	V	SBARY	PLOAD
	1.0	2.5	.733E-3	3.E7	.3	5.E4	180.

Time integration data	DTIM	ISTEP	NPRI
	1.E-6	300	50

Node freedom data												
	1	0	0	4	1	0	6	1	0	9	1	0
	11	1	0	14	1	0	16	1	0	19	1	0
	21	1	0	24	1	0	26	1	0	29	1	0
	31	1	0	32	1	0	33	1	0			

Loads data NL (NO(I), VAL(I), I = 1, NL)

13	3	.4167	6	1.6667	11	.8333	14	1.6667	19	.8333
	22	1.6667	27	.8333	30	1.6667	35	.8333	38	1.6667
	43	.8333	46	1.6667	50	.4167				

Figure 11.12 Mesh and data for Program 11.4

IJAC, IJAC1, IDER, IDER1V	sizes of arrays JAC, JAC1, DER, DER1V
NODOF	number of degrees of freedom per node (2)
IT	dimensions of problem (2)
ICOORD	size of array COORD
NOD	number of nodes per element (8)
IBT	size of array BT
IDOF	number of degrees of freedom per element (16)
ISAMP	size of array SAMP

Fixed size arrays:

DEE	stress–strain matrix
PL	plasticity matrix
SAMP	Gaussian abscissae and weights
COORD	element coordinates
JAC	Jacobian matrix
JAC1	inverse of Jacobian matrix
DER	derivatives of shape functions (local)
DER1V	derivatives of shape functions (global)
BEE	strain-displacement matrix
ELD	element nodal displacements
FUN	element shape functions
EMM	element mass matrix
STRESS	vector of current Gauss point stresses
EPS	vector of Gauss point strains
SIGMA	vector of elastic Gauss point element stresses
BT	transpose of BEE
BLOAD	element 'internal' nodal forces
G	steering vector

Variable size arrays:

VAL	values of nodal freedom loads
X1, D1X1, D2X1	displacement, velocity, acceleration
MM	global mass matrix
BDYLDS	global 'body-loads' vector (see Chapter 6)
SX, SY, TXY, SZ	accumulated Gauss point stresses
EX, EY, GXY, EZ	accumulated Gauss point strains
NF	node freedom array

PARAMETER statement restrictions are:

ILOADS $\geqslant$ N
INX $\geqslant$ NXE

INY ⩾ NYE
INO ⩾ NL
INF ⩾ NN

Turning to the program code, after input and initialisation, three dimensional arrays are nulled by subroutine NULL3 and von Mises plastic stress–strain matrix PL is formed by subroutine VMPL. The remainder of the program is a large explicit integration time-stepping loop. The displacement X1 is updated and then, scanning all elements and Gauss points, new strains can be computed. The constitutive relation then determines the appropriate level of stress and hence the **D** matrix which should operate (whether the yield has been violated or not). The difference between the true stresses and the elastic ones is redistributed as 'body-loads' BDYLDS, whence the new accelerations D2X1 can be found and

Time	Displacement	Velocity	Acceleration
.0000E+00	.0000E+00	.0000E+00	.0000E+00
.5000E-04	.2995E-03	.1199E+02	.2301E+06
.1000E-03	.1214E-02	.2381E+02	.4808E+06
.1500E-03	.2684E-02	.3354E+02	.4893E+06
.2000E-03	.4867E-02	.5286E+02	-.8480E+06
.2500E-03	.8084E-02	.7435E+02	.6151E+06
.3000E-03	.1231E-01	.9414E+02	.6978E+04

(Displacement, Velocity, Acceleration: at beam centreline)

Figure 11.13 Results from Program 11.4

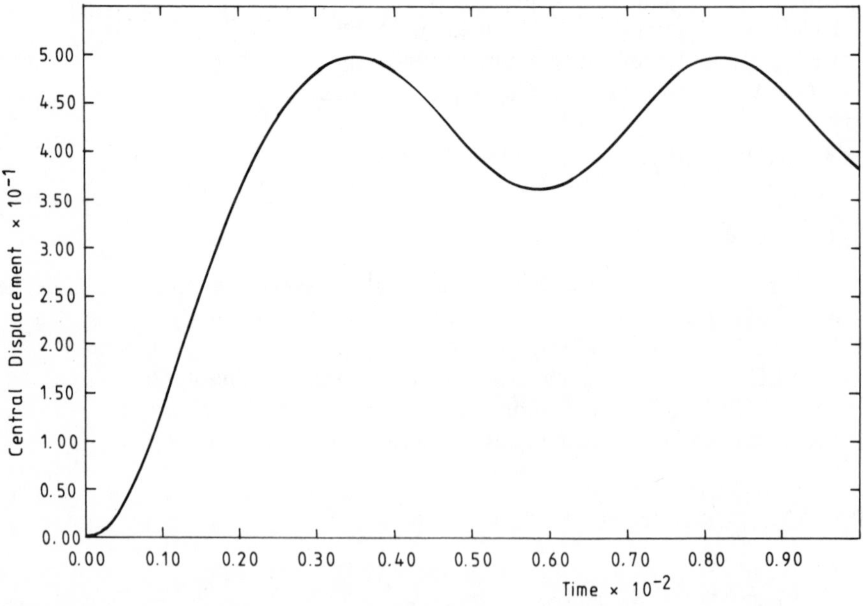

Figure 11.14 Plot of results from Program 11.4

integrated to find the new velocities, D1X1. Then the next cycle of displacements can be updated.

The results from the program are printed in Figure 11.13 in the form of elapsed time, displacement, velocity and acceleration at the upper surface centreline of the beam. Figure 11.14 gives the centreline displacement of the beam as a function of time computed over the first 10,000 time steps. The development of permanent, plastic deformation is clearly demonstrated.

PROGRAM 11.5: FORCED VIBRATION OF AN ELASTIC SOLID IN PLANE STRAIN USING FOUR-NODE QUADRILATERALS—LUMPED OR CONSISTENT MASS, MIXED EXPLICIT/IMPLICIT INTEGRATION

```
C
C      PROGRAM 11.5 FORCED VIBRATION OF A RECTANGULAR
C      SOLID IN PLANE STRAIN USING 4-NODE QUADRILATERALS
C      IMPLICIT AND EXPLICIT INTEGRATION
C      LUMPED OR CONSISTENT MASS
C
C      ALTER NEXT LINE TO CHANGE PROBLEM SIZE
C
      PARAMETER(IKV=1000,ILOADS=100,INF=100,INX=25,INY=25)
C
      REAL DEE(3,3),SAMP(3,2),COORD(4,2),JAC(2,2),JAC1(2,2),BTDB(8,8),
     +DER(2,4),DERIV(2,4),BEE(3,8),DBEE(3,8),KM(8,8),EMM(8,8),
     +ECM(8,8),FUN(4),BT(8,3),TN(8,8),NT(8,2),KV(IKV),MM(IKV),
     +LOADS(ILOADS),X0(ILOADS),D1X0(ILOADS),D2X0(ILOADS),
     +X1(ILOADS),D1X1(ILOADS),D2X1(ILOADS)
      INTEGER NF(INF,2),KDIAG(ILOADS),TYPE(INX,INY),G(8)
      DATA IBTDB,IKM,IEMM,IECM,IBT,ITN,INT,IDOF/8*8/,ICOORD,NOD/2*4/
      DATA ISAMP,IDEE,IBEE,IDBEE,IH/5*3/
      DATA IJAC,IJAC1,IDER,IDERIV,NODOF,IT/6*2/
C
C      INPUT AND INITIALISATION
C
      READ(5,*)NXE,NYE,N,IW,NN,NR,NGP,RHO,
     +         ISTEP,DTIM,AA,BB,V,E,GAMMA,BETA
      READ(5,*)((TYPE(I,J),J=1,NYE),I=1,NXE)
      CALL READNF(NF,INF,NN,NODOF,NR)
      IR=N*(IW+1)
      C1=1./DTIM/DTIM/BETA
      C2=GAMMA/DTIM/BETA
      CALL NULVEC(MM,IR)
      CALL NULL(DEE,IDEE,IH,IH)
      CALL FMDEPS(DEE,IDEE,E,V)
      CALL GAUSS(SAMP,ISAMP,NGP)
C
C      VARIABLE BANDWIDTH STORE
C
      DO 10 I=1,N
   10 KDIAG(I)=0
      DO 20 IP=1,NXE
      DO 20 IQ=1,NYE
      CALL GEOM4Y(IP,IQ,NYE,AA,BB,COORD,ICOORD,G,NF,INF)
      IF(TYPE(IP,IQ).NE.1)CALL FKDIAG(KDIAG,G,IDOF)
   20 CONTINUE
```

```
      DO 30 I=1,N
   30 IF(KDIAG(I).EQ.0)KDIAG(I)=1
      KDIAG(1)=1
      DO 40 I=2,N
   40 KDIAG(I)=KDIAG(I)+KDIAG(I-1)
      IR=KDIAG(N)
      CALL NULVEC(KV,IR)
C
C       ELEMENT STIFFNESS AND MASS INTEGRATION AND ASSEMBLY
C
      DO 50 IP=1,NXE
      DO 50 IQ=1,NYE
      AREA=0.
      CALL GEOM4Y(IP,IQ,NYE,AA,BB,COORD,ICOORD,G,NF,INF)
      CALL NULL(KM,IKM,IDOF,IDOF)
      CALL NULL(EMM,IEMM,IDOF,IDOF)
      IF(TYPE(IP,IQ).EQ.1)THEN
      DO 60 I=1,IDOF
   60 EMM(I,I)=1.
      END IF
      DO 70 I=1,NGP
      DO 70 J=1,NGP
      CALL FORMLN(DER,IDER,FUN,SAMP,ISAMP,I,J)
      CALL MATMUL(DER,IDER,COORD,ICOORD,JAC,IJAC,IT,NOD,IT)
      CALL TWOBY2(JAC,IJAC,JAC1,IJAC1,DET)
      CALL MATMUL(JAC1,IJAC1,DER,IDER,DERIV,IDERIV,IT,IT,NOD)
      CALL NULL(BEE,IBEE,IH,IDOF)
      CALL FORMB(BEE,IBEE,DERIV,IDERIV,NOD)
      CALL MATMUL(DEE,IDEE,BEE,IBEE,DBEE,IDBEE,IH,IH,IDOF)
      CALL MATRAN(BT,IBT,BEE,IBEE,IH,IDOF)
      CALL MATMUL(BT,IBT,DBEE,IDBEE,BTDB,IBTDB,IDOF,IH,IDOF)
      IF(TYPE(IP,IQ).NE.1)
     +CALL ECMAT(ECM,IECM,TN,ITN,NT,INT,FUN,NOD,NODOF)
      QUOT=DET*SAMP(I,2)*SAMP(J,2)
      AREA=AREA+QUOT
      DO 80 K=1,IDOF
      DO 80 L=1,IDOF
      BTDB(K,L)=BTDB(K,L)*QUOT
      IF(TYPE(IP,IQ).NE.1)ECM(K,L)=ECM(K,L)*QUOT*RHO*C1
   80 CONTINUE
      CALL MATADD(KM,IKM,BTDB,IBTDB,IDOF,IDOF)
      IF(TYPE(IP,IQ).NE.1)CALL MATADD(EMM,IEMM,ECM,IECM,IDOF,IDOF)
   70 CONTINUE
      AREA=AREA/NOD*RHO
      IF(TYPE(IP,IQ).EQ.1)THEN
      DO 90 K=1,IDOF
   90 EMM(K,K)=EMM(K,K)*AREA*C1
      CALL FDIAGV(KV,EMM,IEMM,G,KDIAG,IDOF)
      DO 100 I=1,IDOF
      DO 100 J=1,IDOF
  100 KM(I,J)=-KM(I,J)
      CALL FORMKV(MM,KM,IKM,G,N,IDOF)
      CALL FORMKV(MM,EMM,IEMM,G,N,IDOF)
      ELSE
      CALL FSPARV(KV,KM,IKM,G,KDIAG,IDOF)
      CALL FSPARV(KV,EMM,IEMM,G,KDIAG,IDOF)
      CALL FORMKV(MM,EMM,IEMM,G,N,IDOF)
      END IF
   50 CONTINUE
```

```
C
C      TIME INTEGRATION BY SIMPLE PREDICTOR-CORRECTOR
C
      TIM=0.
      DO 110 JR=1,N
      X0(JR)=0.
      D1X0(JR)=1.
  110 D2X0(JR)=0.
      CALL SPARIN(KV,N,KDIAG)
      DO 120 J=1,ISTEP
      TIM=TIM+DTIM
      DO 130 JR=1,N
      LOADS(JR)=.0
  130 D1X1(JR)=X0(JR)+D1X0(JR)*DTIM+D2X0(JR)*.5*DTIM*DTIM*
     +(1.-2.*BETA)
      CALL LINMUL(MM,D1X1,X1,N,IW)
      CALL VECADD(X1,LOADS,X1,N)
      CALL SPABAC(KV,X1,N,KDIAG)
      DO 140 JR=1,N
      D2X1(JR)=(X1(JR)-D1X1(JR))/DTIM/DTIM/BETA
  140 D1X1(JR)=D1X0(JR)+D2X0(JR)*DTIM*(1.-GAMMA)+D2X1(JR)*DTIM*
     +GAMMA
      WRITE(6,1000)TIM,X1(N),D1X1(N),D2X1(N)
      CALL VECCOP(X1,X0,N)
      CALL VECCOP(D1X1,D1X0,N)
      CALL VECCOP(D2X1,D2X0,N)
  120 CONTINUE
 1000 FORMAT(F8.5,3E12.4)
      STOP
      END
```

The final program in this chapter combines the methods of implicit and explicit integration, already introduced in Programs 11.1 and 11.4, in a single program. The idea is (e.g. Key, 1980) that a mesh may contain only a few elements which have a very small explicit stability limit and are therefore best integrated implicitly. The remainder of the mesh can be successfully integrated explicitly at reasonable time steps.

The recurrence relations (3.139) to (3.141) are cast in a slightly different form:

$$\left(\frac{1}{\Delta t^{2}\beta}\mathbf{MM}+\mathbf{KM}\right)\mathbf{r}_1=\mathbf{F}_1+\frac{1}{\Delta t^2\beta}\mathbf{MM}\bar{\mathbf{r}}_1 \tag{11.1}$$

for implicit elements and

$$\left(\frac{1}{\Delta t^2\beta}\mathbf{MM}\right)\mathbf{r}_1=\mathbf{F}_1+\left(\frac{1}{\Delta t^2\beta}\mathbf{MM}-\mathbf{KM}\right)\bar{\mathbf{r}}_1 \tag{11.2}$$

for explicit elements where

$$\bar{\mathbf{r}}_1=\mathbf{r}_0+\Delta t\,\dot{\mathbf{r}}_0+\tfrac{1}{2}\Delta t^2(1-2\beta)\ddot{\mathbf{r}}_0. \tag{11.3}$$

Accelerations and velocities are obtained from:

$$\ddot{\mathbf{r}}_1=(\mathbf{r}_1-\bar{\mathbf{r}}_1)/(\Delta t^2\beta) \tag{11.4}$$

and

$$\dot{\mathbf{r}}_1=\dot{\mathbf{r}}_0+\Delta t(1-\gamma)\ddot{\mathbf{r}}_0+\Delta t\gamma\ddot{\mathbf{r}}_1 \tag{11.5}$$

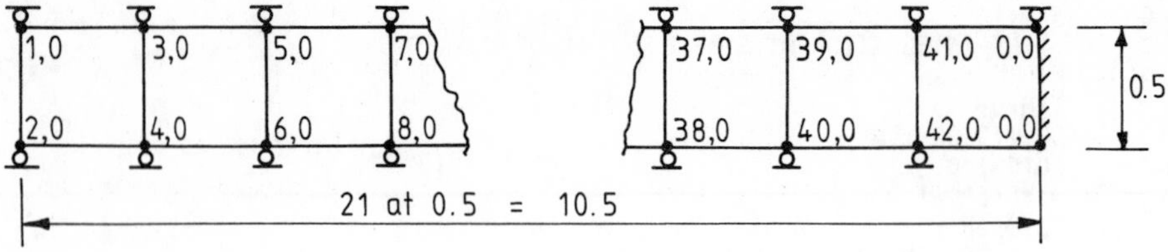

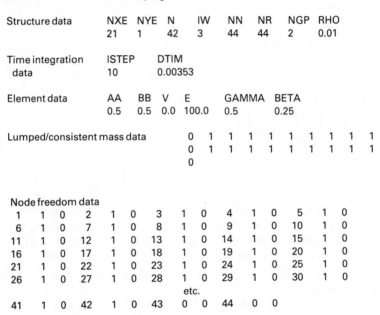

Initial condition $u = 1.0$ at all freedoms at $t = 0$.
No material damping.

Structure data	NXE	NYE	N	IW	NN	NR	NGP	RHO
	21	1	42	3	44	44	2	0.01

Time integration data	ISTEP	DTIM
	10	0.00353

Element data	AA	BB	V	E	GAMMA	BETA
	0.5	0.5	0.0	100.0	0.5	0.25

Lumped/consistent mass data

0	1	1	1	1	1	1	1	1	1
0	1	1	1	1	1	1	1	1	1
0									

Node freedom data

1	1	0	2	1	0	3	1	0	4	1	0	5	1	0
6	1	0	7	1	0	8	1	0	9	1	0	10	1	0
11	1	0	12	1	0	13	1	0	14	1	0	15	1	0
16	1	0	17	1	0	18	1	0	19	1	0	20	1	0
21	1	0	22	1	0	23	1	0	24	1	0	25	1	0
26	1	0	27	1	0	28	1	0	29	1	0	30	1	0
							etc.							
41	1	0	42	1	0	43	0	0	44	0	0			

Figure 11.15 Mesh and data for Program 11.5

The time integration parameters are conventionally called $\beta = \frac{1}{4}$ and $\gamma = \frac{1}{2}$ corresponding to $\theta = \frac{1}{2}$ in Program 8.1.

When an explicit element is not coupled to an implicit one, the half-bandwidth of the assembled equation coefficient matrix will only be 1, whereas the full half-bandwidth will apply for implicit elements.

This is clearly a case where variable bandwidth storage is essential. The assembly routine for 'skyline' storage FSPARV was introduced in Program 5.10 and is used again here (see Smith, 1984).

The problem chosen is illustrated in Figure 11.15. An elastic rod is constrained to vibrate in the horizontal direction only, and fixed at the right hand end. Initial conditions are that a uniform unit velocity at $t = 0$ is applied to all freedoms in the mesh. The appropriate structure chart is Figure 11.2 for implicit integration.

The only new parameters are GAMMA and BETA, used in the time integration (not to be confused with damping parameters), and constants $C1 = 1/\beta\Delta t^2$ and $C2 = \gamma/\Delta t\beta$. No damping is considered.

The integer array TYPE is used to distinguish between lumped and consistent mass. In the present example, elements 1, 11 and 21 have consistent mass, the others lumped. Also, the lumped elements are explicitly integrated while the consistent ones are implicitly integrated.

Early in the program, the variable bandwidth store is set up, using the knowledge that attached explicit elements have a half-bandwidth of 1.

Then the stiffness and mass are integrated as usual. When TYPE is 1 (explicit element) the element lumped mass is assembled into KV by subroutine FDIAGV

Time	Displacement (At node 42)	Velocity (At node 42)	Acceleration (At node 42)
.00353	.3071E-02	.7398E+00	-.1474E+03
.00706	.4783E-02	.2301E+00	-.1414E+03
.01059	.5072E-02	-.6625E-01	-.2649E+02
.01412	.4883E-02	-.4069E-01	.4097E+02
.01765	.4906E-02	.5353E-01	.1240E+02
.02118	.5055E-02	.3102E-01	-.2515E+02
.02471	.5054E-02	-.3168E-01	-.1037E+02
.02824	.4961E-02	-.2123E-01	.1629E+02
.03177	.4962E-02	.2199E-01	.8200E+01
.03530	.5029E-02	.1601E-01	-.1159E+02

Figure 11.16 Results from Program 11.5

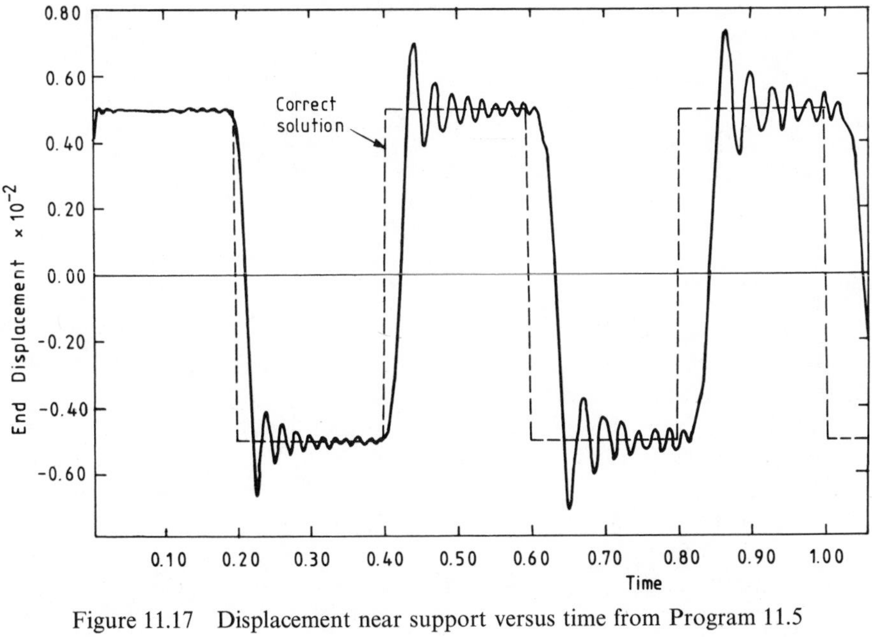

Figure 11.17 Displacement near support versus time from Program 11.5

while the element MM − KM (see equation 11.2) is assembled into MM by subroutine FORMKV. Conversely, when the element is an implicit one, MM + KM are assembled into KV by subroutine FSPARV while the consistent element mass is assembled into MM again by subroutine FORMKV.

The initial conditions are then set, with the starting velocity, D1XO, equal to 1.0. The global matrix factorisation (sparse matrix) is done by SPARIN and the time-stepping loop is entered. Equations (11.1) and (11.2) require the usual matrix-by-vector multiplication on the right hand side (by LINMUL in this case; if the right hand side matrix is sparse, the appropriate routine is LINMLS). Equation solution is completed by SPABAC and it remains only to update velocities, accelerations, etc., for the next timestep from (11.4) and (11.5).

The results are listed in Figure 11.16 for the first ten steps. The displacements close to the support (freedom 42) are compared graphically with the exact solution in Figure 11.17. Despite some spurious oscillations the response is reasonably modelled.

11.1 REFERENCES

Key, S. W. (1980) Transient response by time integration: a review of implicit and explicit operators. In *Advance in Structural Dynamics*, ed. J. Donea, pp. 71–95. Applied Science, London.

Smith, I. M. (1984) Adaptability of truly modular software. *Engineering Computations*, **1**, No. 1, 25–35.

CHAPTER 12
Analysis of Piles and Pile Groups

12.0 INTRODUCTION

The concepts developed in this book can be applied to a very wide variety of problem areas. As an illustration, this chapter deals with foundation engineering, and in particular with problems of deep foundations supported on piles. Analyses of these problems are particularly vital to the offshore industry where critical production facilities are so supported.

The first step in any foundation evaluation is an assessment of feasibility. For this, the capacity of a single pile to sustain a design load must be computed, together with its displacements. This must be done for the two conditions in which the pile is loaded axially and laterally.

Then, since foundations almost always consist of many piles installed close together, an assessment must be made of 'group' effects, that is the extent to which the behaviour of the composite system differs from that of a single pile, due to interaction effects. Again, both ultimate capacity and displacement at working load are required.

Finally, the piles have somehow to be installed in the ground. Particularly in offshore operations this is nearly always done by a process of 'driving', that is by applying an impact to the top of the pile using a hammer, thus causing the tip of the pile to penetrate into the underlying soil.

The four programs in this chapter address all of these different problems in turn. They make use of features of programs from earlier chapters. For example, the laterally loaded pile problem was addressed in Program 4.2. However, in practice the behaviour of soil is so non-linear that the concepts of elasto-plasticity described in Chapter 6 have to be incorporated in the analyses as well. The impact driving problem is one involving forced vibrations of the non-linear pile/soil system and so involves the procedures developed in Chapter 11.

PROGRAM 12.0: *t–z* ANALYSIS OF AXIALLY LOADED PILES USING TWO-NODED LINE ELEMENTS

```
C
C       PROGRAM 12.0 T-Z ANALYSIS OF AXIALLY LOADED PILES
C       USING 2-NODE LINE ELEMENTS
```

```
C
C      ALTER NEXT LINE TO CHANGE PROBLEM SIZE
C
      PARAMETER(IKV=200,ILOADS=100,INO=10,INF=100,IQINC=50)
C
      REAL KV(IKV),LOADS(ILOADS),DISPS(ILOADS),BDYLDS(ILOADS),
     +OLDIS(ILOADS),RU(INF),QU(INF),F(INF),DF(INF),SX(INF),
     +QINC(IQINC),ELD(2),KM(2,2)
      INTEGER G(2),NF(INF,1),NO(INO)
      DATA IKM,IDOF/2*2/,IW,NODOF/2*1/
C
C      INPUT AND INITIALISATION
C
      READ(5,*)NXE,N,NN,NR,CSA,E,ELL,ITS
      CALL READNF(NF,INF,NN,NODOF,NR)
      READ(5,*)(RU(I),I=1,NXE+1)
      READ(5,*)(QU(I),I=1,NXE+1)
      READ(5,*)NL,(NO(I),I=1,NL)
      READ(5,*)INCS,(QINC(I),I=1,INCS)
      IR=N*(IW+1)
      CALL NULVEC(BDYLDS,N)
      CALL NULVEC(OLDIS,N)
      CALL NULVEC(DISPS,N)
      CALL NULVEC(SX,NXE)
      CALL NULVEC(F,NN)
      CALL AXIKM(KM,CSA,E,ELL)
C
C      LOAD INCREMENT LOOP
C
      DO 10 L=1,INCS
      CALL NULVEC(KV,IR)
C
C      ASSEMBLE GLOBAL STIFFNESS MATRIX
C
      DO 20 IP=1,NXE
      CALL GSTRNG(IP,NODOF,NF,INF,G)
   20 CALL FORMKV(KV,KM,IKM,G,N,IDOF)
C
C      ADD T-Z SPRINGS
C
      DO 30 I=1,NN
      IF(QU(I).LT.1.E-6)GOTO 30
      IF(ABS(DISPS(I)).LE..75*QU(I))KV(I)=KV(I)+RU(I)/QU(I)
      IF(ABS(DISPS(I)).GT..75*QU(I))KV(I)=KV(I)+.1*RU(I)/QU(I)
   30 CONTINUE
C
C      REDUCE EQUATIONS
C
      CALL BANRED(KV,N,IW)
C
C      ITERATION LOOP
C
      ITERS=0
   40 ITERS=ITERS+1
      CALL NULVEC(LOADS,N)
      DO 50 I=1,NL
   50 LOADS(NO(I))=QINC(L)
      CALL VECADD(LOADS,BDYLDS,LOADS,N)
      CALL NULVEC(BDYLDS,N)
C
C      SOLVE EQUATIONS
C
```

```
      CALL BACSUB(KV,LOADS,N,IW)
C
C      CHECK CONVERGENCE
C
      IF(ITS.EQ.1)GOTO 60
      CALL CHECON(LOADS,OLDIS,N,0.00001,ICON)
      IF(ITERS.EQ.1)ICON=0
C
C      REDISTRIBUTE EXCESS SPRING FORCES
C
      DO 70 I=1,NN
      IF(QU(I).LT.1.E-6)GOTO 70
      DF(I)=LOADS(I)*RU(I)/QU(I)
      IF(ABS(DISPS(I)).GT..75*QU(I))DF(I)=.1*DF(I)
      IF(ABS(F(I)+DF(I)).GT.RU(I))THEN
      BDYLDS(I)=F(I)+DF(I)+RU(I)
      DF(I)=-RU(I)-F(I)
      END IF
   70 CONTINUE
      IF(ICON.EQ.0.AND.ITERS.NE.ITS)GOTO 40
C
C      COMPUTE ELEMENT STRESSES AND STRAINS
C
   60 DO 80 IP=1,NXE
      CALL GSTRNG(IP,NODOF,NF,INF,G)
      DO 90 J=1,2
      IF(G(J).EQ.0)ELD(J)=0.
   90 IF(G(J).NE.0)ELD(J)=LOADS(G(J))
      EPS=(ELD(2)-ELD(1))/ELL
      SIGMA=E*EPS
      SX(IP)=SX(IP)+SIGMA
   80 CONTINUE
      CALL VECADD(DISPS,LOADS,DISPS,N)
C
C      CHECK SUM OF SPRING FORCES
C
      SUM=0.
      DO 100 I=1,NN
      F(I)=F(I)+DF(I)
  100 SUM=SUM+F(I)
      WRITE(6,1000)SUM,DISPS(1),ITERS
      IF(ITERS.EQ.ITS)GOTO 110
   10 CONTINUE
  110 CONTINUE
 1000 FORMAT(2E12.4,I10)
      STOP
      END
```

The problem of pile/soil interaction is of course really three dimensional, or at least axisymmetric, and can be analysed as such using finite elements (e.g. Chow and Smith, 1982). However, useful results can be obtained from a simpler idealisation, as shown in Figure 12.1. This representation allows the pile to be analysed as a line, and so the simple rod elements introduced in equation (2.11) are appropriate. They are characterised only by length, cross-sectional area and axial stiffness.

The ground is represented by discrete 'springs' which have load-displacement characteristics typified by Figure 12.2; these springs are simply attached to the

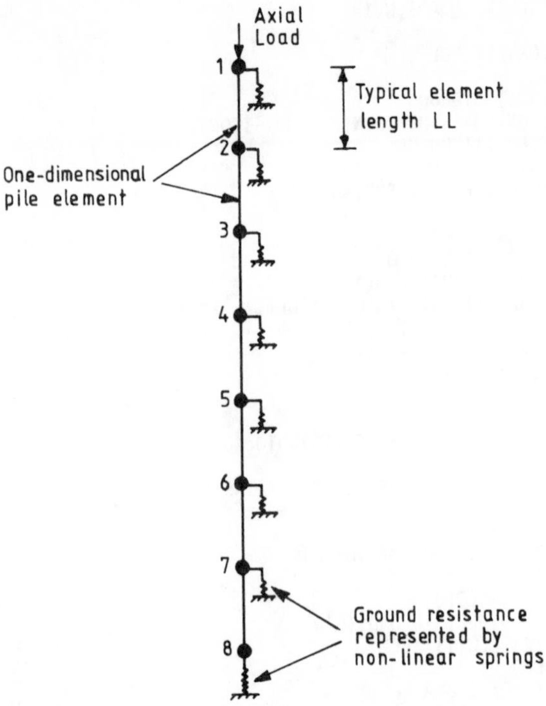

Figure 12.1 Representation of pile and ground

nodes of the finite elements, as shown in Figure 12.1. The load in a spring is traditionally termed 't' and the displacement 'z', so the load-displacement curves are called 't–z' curves. They can be quite general relationships, but in the drivability analyses described later, they are approximated by the bilinear form shown in Figure 12.2(a). The ultimate load in the spring is termed R_u and the displacement at which it is reached is termed Q_u, the so-called 'quake'. When the displacement exceeds Q_u, the fictitious extra load, CB or HG in Figure 12.2(b), is redistributed to other springs by the 'initial stress' technique discussed in Chapter 6. Thus, Figure 12.2(b) is analogous to Figure 6.7.

In the present program, the slightly more general t–z relationship shown in Figure 12.2(c) is allowed. When the displacement exceeds $0.75Q_u$, the stiffness is reduced to $0.1R_u/Q_u$ until the limiting force R_u is attained.

The variables in the program have the usual significance and are listed below. The example is just like Figure 12.1 but the pile consists of 19 elements and 20 nodes. The program data are shown in Figure 12.3 and the structure chart in Figure 12.4.

Simple variables:

NXE	number of elements (19)
N	number of non-zero freedoms (20)

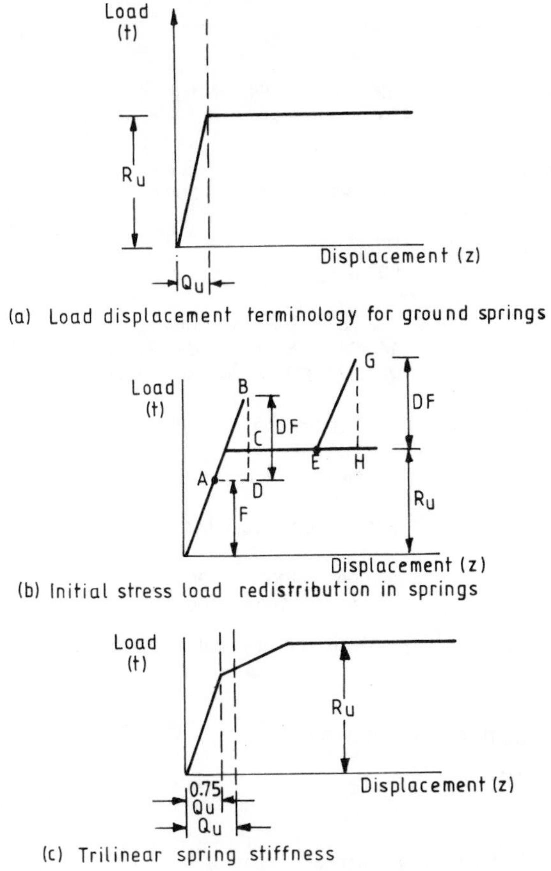

Figure 12.2 Treatment of ground behaviour

NN	number of nodes (20)
NR	number of restrained nodes (0)
CSA	cross-sectional area of elements (4129.67)
E	modulus of elements (2110.1)
ELL	length of elements (200.0)
ITS	maximum number of iterations allowed (50)
ITERS	iteration counter
IKM	working size of array KM
IDOF	number of degrees of freedom per element (2)
IW	half-bandwidth (1)
NODOF	number of degrees of freedom per node (1)

Fixed length arrays:

ELD	element nodal displacements

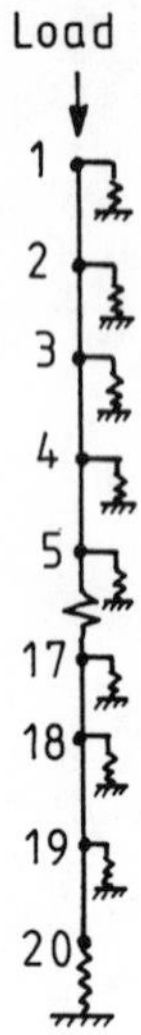

Structure data	NXE	N	NN	NR		
	19	20	20	0		
Element data	CSA	E	ELL			
	4129.67	2110.1	200.0			
Iteration data	ITS					
	50					
R_u data	42.73	136.74	(9 times)			
	140.68	168.30	199.85	231.40	262.96	
	294.52	326.07	341.84	512.78	1444.0	
Q_u data	.2	(19 times)	2.0			
LOADS data	NL, NO(I), I = 1, NL					
	1	1				
	INCS, QINC(I), I = 1, INCS.					
	11					
	−1000.	−1000.	−500.	−500.	−500.	−500. −300.
	−300.	−200.	−100.	−100.		

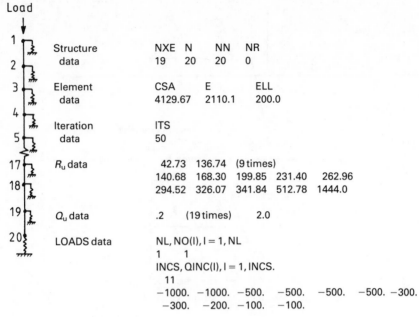

Figure 12.3 Mesh and data for Program 12.0

KM	element stiffness matrix
G	steering vector

Variable length arrays:

KV	global stiffness matrix (vector)
LOADS	current load/displacement increment
DISPS	updated displacements
BDYLDS	'Body-loads'
OLDIS	previous iteration displacements
RU	ultimate spring forces R_u
QU	elastic limit displacements Q_u
F	force in springs } see Figure 12.2
DF	change in force in springs } see Figure 12.2
SX	stress in pile
QINC	load increments
NO	numbers of loaded freedoms
NF	node freedom array

PARAMETER statement limitations:

IKV ⩾ N∗(IW + 1)
ILOADS ⩾ N

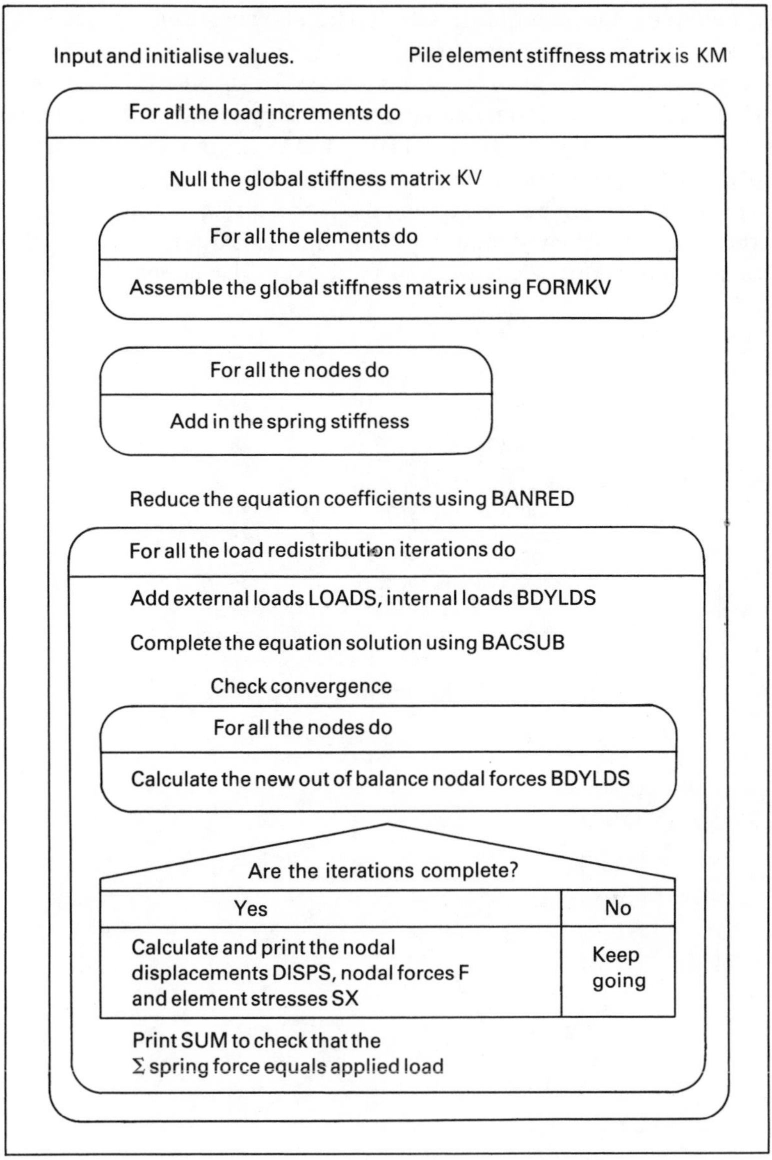

Figure 12.4 Structure chart for Program 12.0

INO $\geqslant$ NL
INF $\geqslant$ NN
IQINC $\geqslant$ INCS

After input and initialisation, the rod element stiffness matrix KM is formed by the subroutine AXIKM. Then the load increment loop is entered and the steering vector found by the subroutine GSTRNG, allowing element assembly by FORMKV. The spring stiffnesses are added to the nodes as R_u/Q_u, or $0.1R_u/Q_u$ as the case may be, and the global equations reduced by BANRED. The excess load redistribution loop is then commenced, controlled by counter ITERS. The nodal loads are the sum of the external loads, LOADS, and the internal loads, BDYLDS, allowing equation solution completion by BACSUB.

Convergence can then be checked and the excess spring forces redistributed. If convergence is not achieved, another iteration is called for; otherwise element stresses and strains are calculated and the current displacement increment is added to DISPS.

-.1000E+04	-.1848E+00	2
-.2000E+04	-.4337E+00	13
-.2500E+04	-.6092E+00	8
-.3000E+04	-.8114E+00	8
-.3500E+04	-.1039E+01	9
-.4000E+04	-.1505E+01	40
-.4300E+04	-.1989E+01	30
-.4600E+04	-.2536E+01	29
-.4800E+04	-.2900E+01	32
-.4900E+04	-.3082E+01	34
-.4986E+04	-.4318E+01	50
Axial load	Displacement under load	Iterations

Figure 12.5 Results from Program 12.0

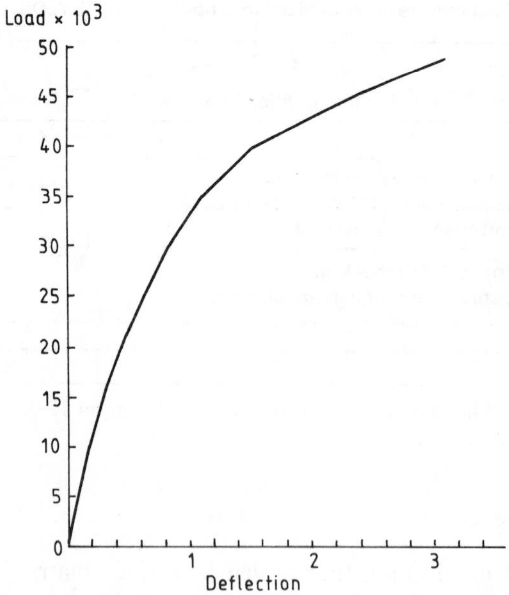

Figure 12.6 Load displacement for axially loaded pile

The output, listed as Figure 12.5, consists of the sum of the forces in the springs, which should of course equal the applied load, followed by the displacement of the top of the pile and the number of iterations to convergence.

The total ground resistance is 5196.0 and it can be seen that in the last computed increment the load reaches 4986.0 without convergence being achieved. In such a 'load control' algorithm, convergence to within the specified tolerance will be increasingly difficult to achieve as the ultimate load is approached.

The results are shown graphically as Figure 12.6 up to the last increment at which convergence was achieved. The iteration tolerance was a fine one of 0.00001, or a thousandth of 1%. If there had been no spring at any node, a very small value of Q_u (less than 10^{-6}) would have been specified and a nominal R_u, allowing that node to be skipped in the assembly process.

PROGRAM 12.1: *p–y* ANALYSIS OF LATERALLY LOADED PILES USING TWO-NODED BEAM ELEMENTS

```
C
C      PROGRAM 12.1 P-Y ANALYSIS OF LATERALLY LOADED PILES
C      USING 2-NODE BEAM ELEMENTS
C
C      ALTER NEXT LINE TO CHANGE PROBLEM SIZE
C
      PARAMETER(IKV=500,ILOADS=200,INO=10,ISEG=12,INF=100)
C
      REAL KV(IKV),LOADS(ILOADS),BDYLDS(ILOADS),DISPS(ILOADS),
     +OLDIS(ILOADS),RU(INF,ISEG),QU(INF,ISEG),F(INF),DF(INF),
     +MOM(INF),VAL(INO),STORE(INO),KM(4,4),KP(4,4),ACTION(4),ELD(4)
      INTEGER G(4),NF(INF,2),NO(INO)
      DATA IKM,IDOF/2*4/,NODOF/2/,IW/3/
C
C      INPUT AND INITIALISATION
C
      READ(5,*)NXE,N,NN,NR,PA,NP,EI,ELL,INCS,ITS
      CALL READNF(NF,INF,NN,NODOF,NR)
      READ(5,*)((RU(I,J),J=1,NP+1),I=1,NN)
      READ(5,*)((QU(I,J),J=1,NP+1),I=1,NN)
      READ(5,*)NL,(NO(I),VAL(I),I=1,NL)
      IR=N*(IW+1)
      CALL NULVEC(BDYLDS,N)
      CALL NULVEC(OLDIS,N)
      CALL NULVEC(DISPS,N)
      CALL NULVEC(F,NN)
      CALL NULVEC(MOM,NN)
      CALL BEAMKM(KM,EI,ELL)
      CALL BEAMKP(KP,ELL)
C
C      LOAD INCREMENT LOOP
C
      DO 10 L=1,INCS
      CALL NULVEC(KV,IR)
C
C      MODIFY ELEMENT STIFFNESS FOR AXIAL LOADING
C
      DO 20 I=1,IDOF
      DO 20 J=1,IDOF
```

```
   20 KP(I,J)=KM(I,J)-L*PA*KP(I,J)
C
C      ASSEMBLE GLOBAL STIFFNESS MATRIX
C
      DO 30 IP=1,NXE
      CALL GSTRNG(IP,NODOF,NF,INF,G)
   30 CALL FORMKV(KV,KP,IKM,G,N,IDOF)
C
C      ADD P-Y SPRINGS
C
      DO 40 I=1,NN
      II=2*I-1
   40 KV(II)=KV(II)+RU(I,2)/QU(I,2)
C
C      ADD 'BIG SPRINGS' FOR PRESCRIBED DISPLACEMENTS
C      AND REDUCE EQUATIONS
C
      DO 50 I=1,NL
      KV(NO(I))=KV(NO(I))+1.E20
   50 STORE(I)=KV(NO(I))
      CALL BANRED(KV,N,IW)
C
C      ITERATION LOOP
C
      ITERS=0
   60 ITERS=ITERS+1
      CALL NULVEC(LOADS,N)
      DO 70 I=1,NL
   70 LOADS(NO(I))=STORE(I)*VAL(I)
      CALL VECADD(LOADS,BDYLDS,LOADS,N)
      CALL NULVEC(BDYLDS,N)
C
C      SOLVE EQUATIONS
C
      CALL BACSUB(KV,LOADS,N,IW)
C
C      CHECK CONVERGENCE
C
      CALL CHECON(LOADS,OLDIS,N,.00001,ICON)
      IF(ITERS.EQ.1)ICON=0
C
C      REDISTRIBUTE EXCESS SPRING FORCES
C
      DO 80 I=1,NN
      II=2*I-1
      DF(I)=-LOADS(II)*RU(I,2)/QU(I,2)
      J=1
      DO 90 IP=2,NP+1
   90 IF(ABS(DISPS(II)+LOADS(II)).GT.QU(I,IP))J=J+1
      IF(J.EQ.1.AND.DISPS(II)*LOADS(II).GT.0.)GOTO 100
      FORCE=(ABS(DISPS(II)+LOADS(II))-QU(I,J))*(RU(I,J+1)-RU(I,J))
     +/(QU(I,J+1)-QU(I,J))+RU(I,J)
      IF(DISPS(II).LT.0.)BDYLDS(II)=-F(I)-DF(I)+FORCE
      IF(DISPS(II).GT.0.)BDYLDS(II)=-F(I)-DF(I)-FORCE
  100 DF(I)=BDYLDS(II)+DF(I)
   80 CONTINUE
      IF(ICON.EQ.0.AND.ITERS.NE.ITS)GOTO 60
C
C      COMPUTE ELEMENT MOMENTS AND SHEARS
C
      DO 110 IP=1,NXE
      CALL GSTRNG(IP,NODOF,NF,INF,G)
```

```
      DO 120 M=1,IDOF
      IF(G(M).EQ.0)ELD(M)=.0
  120 IF(G(M).NE.0)ELD(M)=LOADS(G(M))
      CALL MVMULT(KP,IKM,ELD,IDOF,IDOF,ACTION)
      MOM(IP)=MOM(IP)+ACTION(2)
      IF(IP.EQ.NXE)MOM(IP+1)=MOM(IP+1)+ACTION(4)
  110 CONTINUE
      CALL VECADD(DISPS,LOADS,DISPS,N)
      SUM=0.0
      DO 130 I=1,NN
      F(I)=F(I)+DF(I)
  130 SUM=SUM+F(I)
      WRITE(6,1000)SUM,DISPS(1),ITERS
      IF(ITERS.EQ.ITS)GOTO 140
   10 CONTINUE
  140 CONTINUE
 1000 FORMAT(2E12.4,I10)
      STOP
      END
```

In this type of analysis the pile is again a line such as in Figure 12.1, but the load is applied horizontally and resisted by ground springs oriented horizontally. The pile distributes the load by bending and the appropriate elements are flexural ones with stiffnesses typified by equations (2.26b). At each node a transverse displacement and a rotation are permitted. In order to allow for the effect of an axial force on the bending stiffness, the element stiffness matrices in the program are actually given by equation (2.33); that is equations (2.26b) are modified by (2.32). The matrix KP from (2.32) is formed by subroutine BEAMKP.

The springs now have load-displacement characteristics known as *p–y* curves directly analogous to the axial *t–z* ones. In this program, quite general 'curves' are allowed, by specifying any number of linear segments as shown in Figure 12.7, which happens to have six (that is NP) segments.

The data for the pile treated in the analysis are given in Figure 12.8 and the structure of the program by Figure 12.9. In the program coding, the nomenclature follows that of the previous program with the additions:

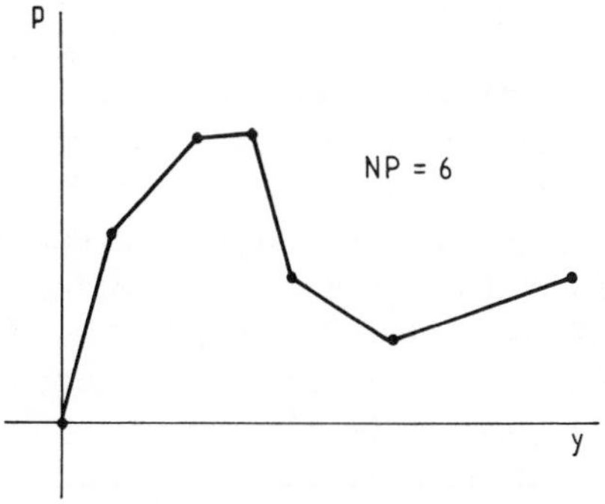

Figure 12.7 Typical *p–y* curve

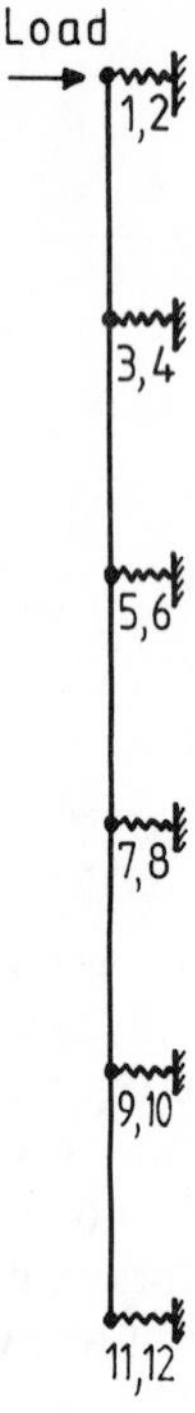

Structure data	NXE	N	NN	NR	PA	NP
	5	12	6	0	.0	7

Element data	EI	ELL
	1.795E11	12.0

Iteration data	INCS	ITS
	10	15

R_u data

.0	.0	.0	.0	.0	.0	.0	.0
.0	46.8	93.6	139.2	187.2	234.0	280.8	280.8
.0	730.8	1461.6	2199.6	2313.6	2164.8	858.0	858.0
.0	2008.8	4058.4	4546.8	4858.8	4545.8	1478.4	1478.4
.0	3566.4	6522.0	13084.8	13981.2	13084.8	2278.8	2278.8
.0	9358.8	16548.0	33192.0	35472.0	33198.0	3532.8	3532.8

Q_u data

.0	1.0	2.0	3.0	4.0	5.0	6.0	25.2
.0	.162	.325	.487	.650	.812	.975	25.2
.0	.025	.051	.076	.109	.145	.434	25.2
.0	.035	.070	.091	.136	.181	.543	25.2
.0	.014	.029	.086	.129	.172	.516	25.2
.0	.016	.031	.093	.139	.186	.558	25.2

Displacement increment data	NL	(NO(I), VAL(I), I = 1, NL)	
	1	1	0.1

Figure 12.8 Mesh and data for Program 12.1

PA	axial force in the pile (0.0)
NP	number of segments on p–y 'curves' (excluding the origin) (7)
EI	flexural stiffness of the pile (1.795×10^{11})
VAL	values of displacement increments
MOM	bending moments and shear forces in pile
STORE	storage for 'big spring' freedoms
KP	stiffness modification for axial load
ACTION	increments of bending moment and shear force

The PARAMETER section includes the new constant ISEG $\geqslant$ NP + 1

The arrays RU and QU have essentially the same significance as in the previous program but now hold the coordinates of the points on the p–y 'curves', including the origin.

The structure of the program follows the previous one with the exception that displacement is prescribed at the top of the pile rather than load. The total force is found by summing the forces in all the lateral springs as before, and the results are listed as Figure 12.10. For each increment, the output gives the total force, followed by the displacement of the pile top and the number of iterations to convergence. Note that the iteration counts are much less than in the previous program and, in general, it is more efficient to prescribe displacements rather than loads if this is possible (Smith, 1979). The results are plotted graphically as Figure 12.11.

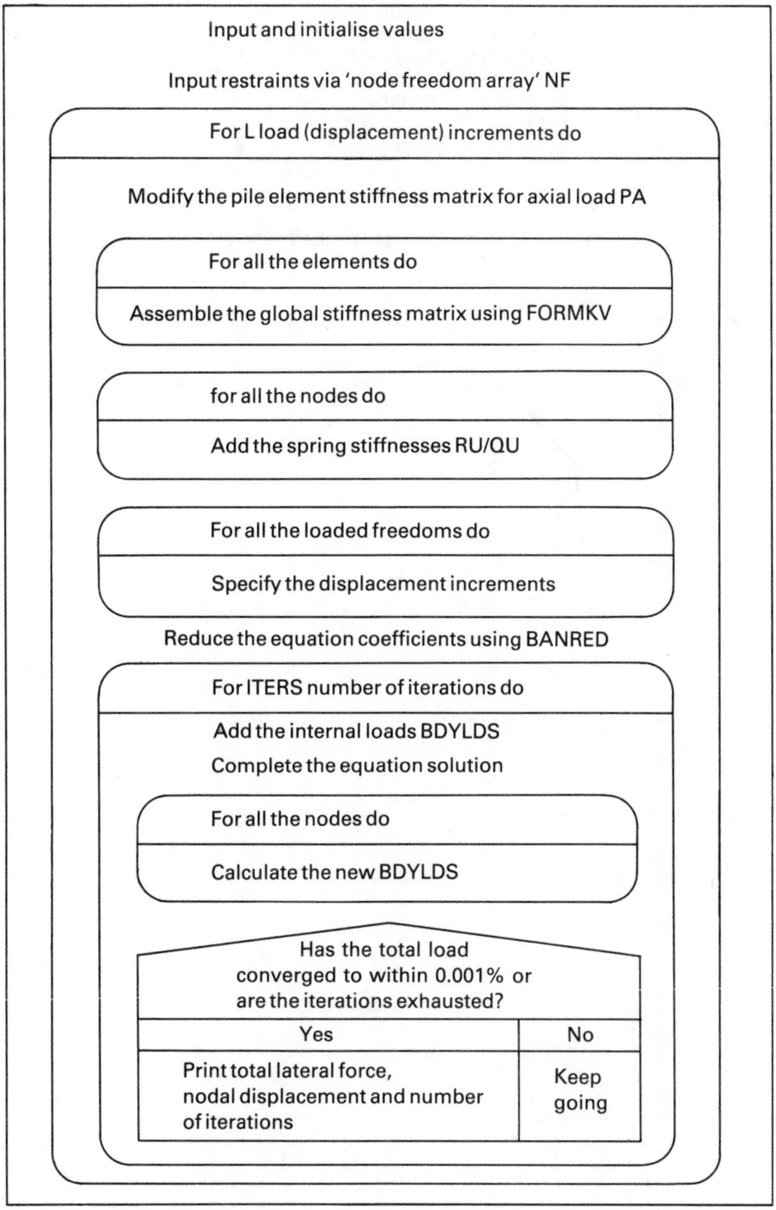

Figure 12.9 Structure chart for Program 12.1

-.2510E+04	.1000E+00	2
-.4320E+04	.2000E+00	5
-.4828E+04	.3000E+00	9
-.5132E+04	.4000E+00	7
-.5289E+04	.5000E+00	7
-.5197E+04	.6000E+00	10
-.5000E+04	.7000E+00	9
-.4809E+04	.8000E+00	9
-.4602E+04	.9000E+00	9
-.4372E+04	.1000E+01	8
Lateral load	Displacement under load	Iterations

Figure 12.10. Results from Program 12.1

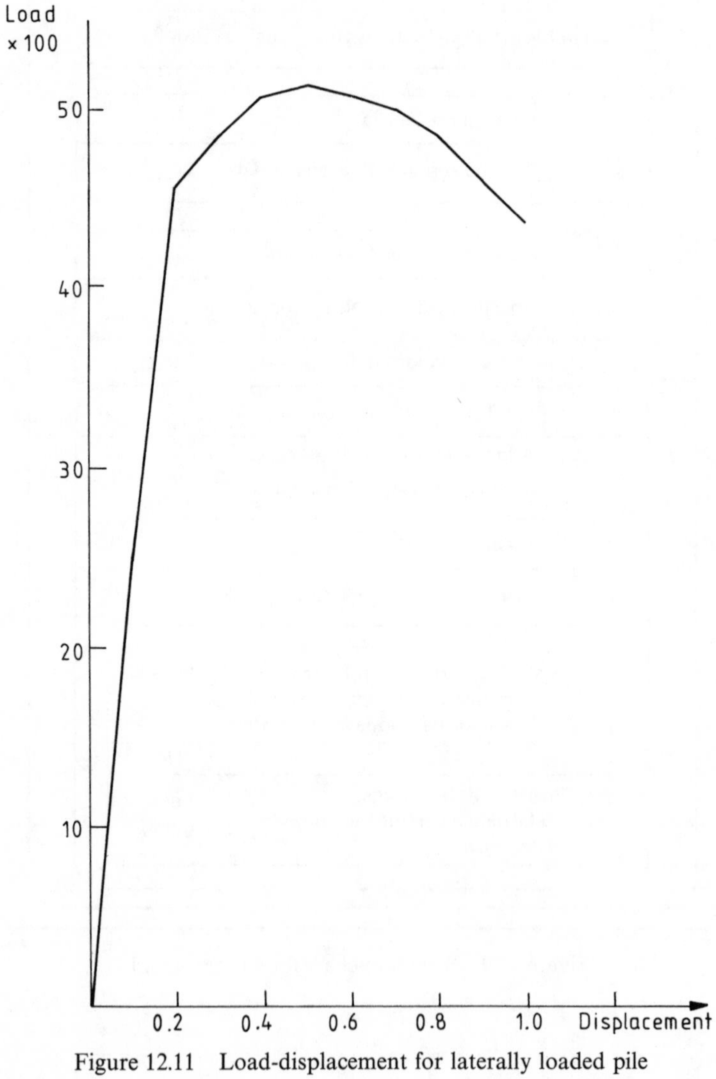

Figure 12.11 Load-displacement for laterally loaded pile

PROGRAM 12.2: ANALYSIS OF VERTICALLY LOADED PILE GROUPS USING TWO-NODED LINE ELEMENTS—*t–z* SPRINGS FOR THE SOIL AND MINDLIN'S EQUATION FOR THE INTERACTIONS†

```
C
C      PROGRAM 12.2 ANALYSIS OF PILE GROUPS USING 2-NODE LINE ELEMENTS
C      FOR THE PILES, T-Z SPRINGS FOR THE SOIL AND MINDLIN'S EQUATION
C      FOR THE INTERACTIONS
C
C      ALTER NEXT LINE TO CHANGE PROBLEM SIZE
C
      PARAMETER(IKV=4950,ILOADS=300,INO=25,INX=31)
C
      REAL KV(IKV),SM(IKV),LOADS(ILOADS),FORCE(ILOADS),KM(2,2),VAL(INO),
     +COORD(INO,2),ZCOORD(INX),FMAX(INX),FSOIL(INO,INX),SUM(INO),KS(2,2)
      INTEGER G(2),GP(INX),GQ(INX),NO(INO),NA(ILOADS)
      DATA IKM,IKS,IDOF/3*2/,RF/0.9/
C
C      INPUT AND INITIALISATION
C
      READ(5,*)NXE,E,PILEN,RO,GSURF,GRATE,VSOIL,RHO,NPILE,INCS,IGROUP
      READ(5,*)((COORD(I,J),J=1,2),I=1,NPILE)
      READ(5,*)(FMAX(I),I=1,NXE+1)
      READ(5,*)NL,(NO(I),VAL(I),I=1,NL)
      NN=NXE+1
      N=NN*NPILE
      ELL=PILEN/FLOAT(NXE)
      PI=ACOS(-1.)
      CSA=PI*RO*RO
      RM=2.5*RHO*(1.-VSOIL)*PILEN
      DO 10 I=1,NN
   10 ZCOORD(I)=FLOAT(I-1)*ELL
      CALL NULL(FSOIL,INO,NPILE,NN)
      FACT2=3.-4.*VSOIL
      FACT3=(1.-VSOIL)**2
C
C      LOCATION OF DIAGONAL ELEMENTS OF STIFFNESS MATRIX
C
      NSTO=0
      DO 20 J=1,N
      NSTO=NSTO+J
   20 NA(J)=NSTO
      IR=NA(N)
C
C      LOAD/DISPLACEMENTS APPLIED IN INCREMENTS
C
      DO 30 II=1,INCS
      WRITE(6,2000)II
      CALL NULVEC(KV,IR)
      CALL NULVEC(SM,IR)
      IF(IGROUP.EQ.0)GOTO 40
C
C      FORM SOIL FLEXIBILITY MATRIX USING MINDLIN'S EQUATION
C
      DO 50 IP=1,NPILE
      CALL GEOM(IP,NN,NPILE,GP)
      IF(IP+1.GT.NPILE)GOTO 50
      DO 60 IQ=IP+1,NPILE
      CALL GEOM(IQ,NN,NPILE,GQ)
```

† This program was contributed by Y. K. Chow, National University of Singapore; see Chow (1986).

```
      RR=SQRT((COORD(IP,1)-COORD(IQ,1))**2 +
     +        (COORD(IP,2)-COORD(IQ,2))**2)
      DO 70 I=1,NN
      IF(FSOIL(IP,I)/FMAX(I).GT..9999)GOTO 70
      ZZ=ZCOORD(I)
      GSOILI=GSURF+GRATE*ZZ
      IF(I.EQ.1)GSOILI=GSURF+GRATE*.25*ELL
      IF(I.EQ.NXE)GSOILI=GSURF+GRATE*(ZZ+.25*ELL)
      DO 80 J=1,NN
      IF(FSOIL(IQ,J)/FMAX(J).GT..9999)GOTO 80
      CC=ZCOORD(J)
      GSOILJ=GSURF+GRATE*CC
      IF(J.EQ.1)GSOILJ=GSURF+GRATE*.25*ELL
      IF(J.EQ.NXE)GSOILJ=GSURF+GRATE*(CC+.25*ELL)
      GSOIL=.5*(GSOILI+GSOILJ)
      FACT1=1./(16.*PI*GSOIL*(1.-VSOIL))
      ZMC2=(ZZ-CC)*(ZZ-CC)
      ZPC2=(ZZ+CC)*(ZZ+CC)
      R1=SQRT(RR*RR+ZMC2)
      R2=SQRT(RR*RR+ZPC2)
C
C       SOIL DISPLACEMENT AT IP DUE TO UNIT LOAD AT IQ
C       OBTAINED USING MINDLIN'S EQUATION
C
      KOUNT=MAX0(NA(GQ(J))-GQ(J)+GP(I),NA(GP(I))-GP(I)+GQ(J))
      KV(KOUNT)=FACT1*(FACT2/R1+(8.*FACT3-FACT2)/R2+ZMC2/R1**3
     +        +(FACT2*ZPC2-2.*CC*ZZ)/R2**3+6.*ZZ*CC*ZPC2/R2**5)
   80 CONTINUE
   70 CONTINUE
   60 CONTINUE
   50 CONTINUE
C
C       ADD IN FLEXIBILITY CONTRIBUTIONS FROM DISCRETE SOIL SPRINGS
C
   40 DO 90 IP=1,NPILE
      CALL GEOM(IP,NN,NPILE,GP)
      GSOIL=GSURF+GRATE*.25*ELL
      CALL FORMXI(FSOIL(IP,1),FMAX(1),RF,RM,RO,XI)
      FSHAFT=XI/(PI*GSOIL*ELL)
      IF(FSOIL(IP,1)/FMAX(1).GT.0.9999)FSHAFT=1.E12
      KV(NA(GP(1)))=KV(NA(GP(1)))+FSHAFT
      GSOIL=GSURF+GRATE*(ZCOORD(NXE)+.25*ELL)
      CALL FORMXI(FSOIL(IP,NXE),FMAX(NXE),RF,RM,RO,XI)
      FSHAFT=XI/(PI*GSOIL*3.*ELL)
      IF(FSOIL(IP,NXE)/FMAX(NXE).GT..9999)FSHAFT=1.E12
      KV(NA(GP(NXE)))=KV(NA(GP(NXE)))+FSHAFT
      DO 100 I=2,NN-2
      GSOIL=GSURF+GRATE*ZCOORD(I)
      CALL FORMXI(FSOIL(IP,I),FMAX(I),RF,RM,RO,XI)
      FSHAFT=XI/(2.*PI*GSOIL*ELL)
      IF(FSOIL(IP,I)/FMAX(I).GT..9999)FSHAFT=1.E12
  100 KV(NA(GP(I)))=KV(NA(GP(I)))+FSHAFT
C
C       ADD TIP FLEXIBILITY
C
      GSOIL=GSURF+GRATE*ZCOORD(NN)
      FTIP=(1.-VSOIL)/((4.*GSOIL*RO)*(1.-RF*FSOIL(IP,NN)/FMAX(NN))**2)
      IF(FSOIL(IP,NN)/FMAX(NN).GT..9999)FTIP=1.E12
   90 KV(NA(GP(NN)))=KV(NA(GP(NN)))+FTIP
C
C       INVERT SOIL FLEXIBILITY MATRIX TO GIVE STIFFNESS MATRIX
C
```

```
      CALL SPARIN(KV,N,NA)
      DO 110 I=1,N
      CALL NULVEC(LOADS,N)
      LOADS(I)=1.
      CALL SPABAC(KV,LOADS,N,NA)
      DO 120 J=1,I
  120 SM(NA(I)-I+J)=LOADS(J)
  110 CONTINUE
      CALL VECCOP(SM,KV,IR)
C
C       FORM AND ASSEMBLE STIFFNESS MATRIX
C
      CALL AXIKM(KM,CSA,E,ELL)
      DO 130 IP=1,NPILE
      CALL GEOM(IP,NN,NPILE,GP)
      DO 140 I=1,NXE
      G(1)=GP(I)
      G(2)=GP(I+1)
  140 CALL FSPARV(SM,KM,IKM,G,NA,IDOF)
  130 CONTINUE
C
C       PRESCRIBE DISPLACEMENTS : BIG SPRING TECHNIQUE
C
      CALL NULVEC(LOADS,N)
      DO 150 I=1,NL
      IX=NA(NO(I))
      SM(IX)=SM(IX)+1.E20
  150 LOADS(NO(I))=VAL(I)*SM(IX)
C
C       SOLVE SM TO GIVE PILE DISPLACEMENTS
C
      CALL SPARIN(SM,N,NA)
      CALL SPABAC(SM,LOADS,N,NA)
C
C       RECOVER SOIL SPRING FORCES
C
      CALL LINMLS(KV,LOADS,FORCE,N,NA)
      TOTSUM=0.
      DO 160 IP=1,NPILE
      CALL GEOM(IP,NN,NPILE,GP)
      SUM(IP)=0.
      DO 170 I=1,NN
      FSOIL(IP,I)=FSOIL(IP,I)+FORCE(GP(I))
      IF(FSOIL(IP,I).GT.FMAX(I))FSOIL(IP,I)=FMAX(I)
  170 SUM(IP)=SUM(IP)+FSOIL(IP,I)
  160 TOTSUM=TOTSUM+SUM(IP)
      WRITE(6,1000)TOTSUM,VAL(1)*II
      PAV=TOTSUM/FLOAT(NPILE)
      DO 180 IP=1,NPILE
  180 WRITE(6,2000)IP,SUM(IP)/PAV
   30 CONTINUE
 1000 FORMAT(2E12.4)
 2000 FORMAT(I10,E12.4)
      STOP
      END
```

The analysis of groups of piles closely spaced enough to interact with one another must attempt to model the (non-linear) behaviour of the individual piles, as was done in Program 12.0, and the interaction between them. An obvious simplification used in the present program is that interaction effects are described by the

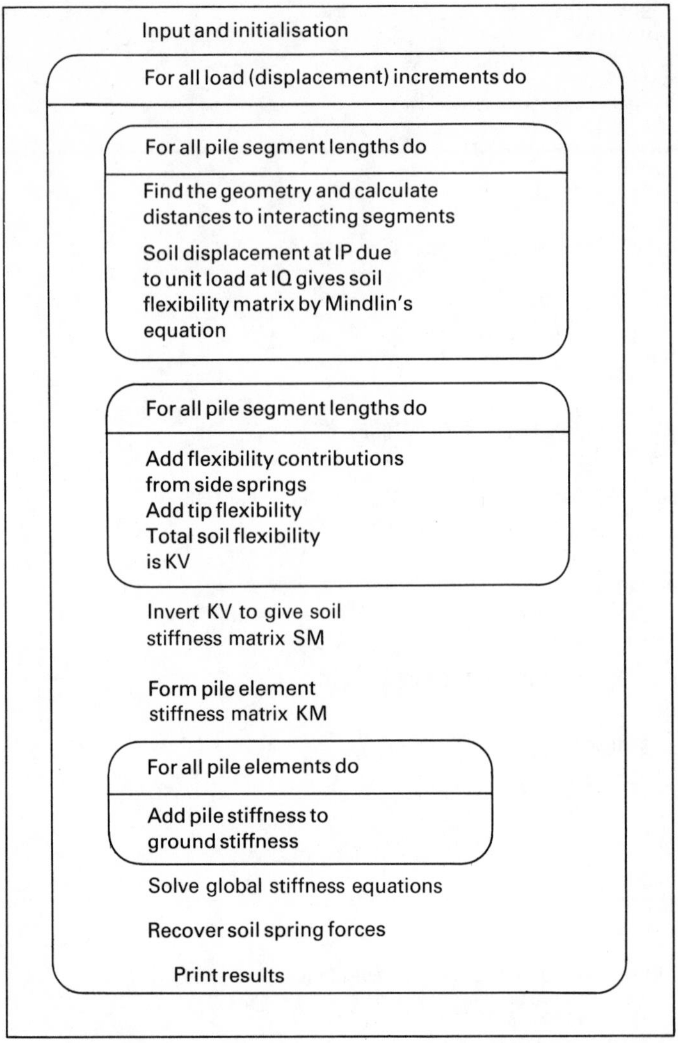

Figure 12.12 Structure chart for Program 12.2

theory of linear elasticity and thus the well-known Mindlin equation (e.g. Poulos and Davis, 1980) can be used to construct a foundation stiffness matrix to which the pile stiffnesses are added. The analysis can be done for general loading conditions but the present program deals with the special case of axially loaded groups. The elastic 'soil' is assumed to be infinitely deep (linear elastic half-space) and, by assuming an average, variable, soil stiffness over each pile element, non-homogeneous foundations can be approximated.

The structure of the program is shown in Figure 12.12 and the variable names are listed below. The particular case analysed is the 3 × 3 group of piles illustrated in Figure 12.13, together with input data.

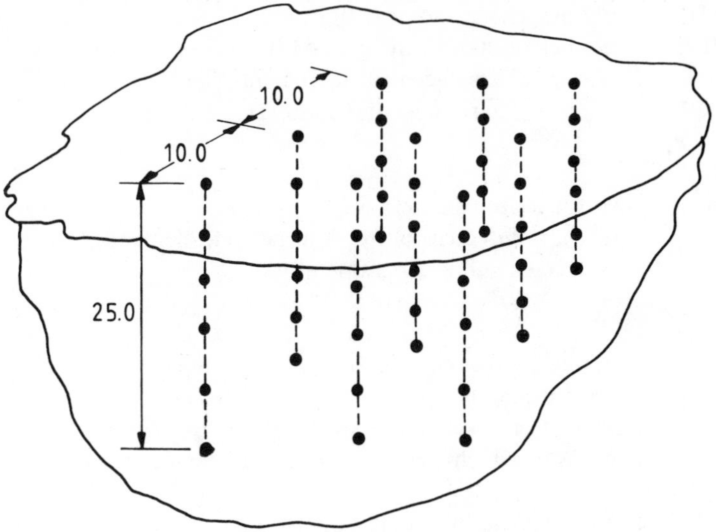

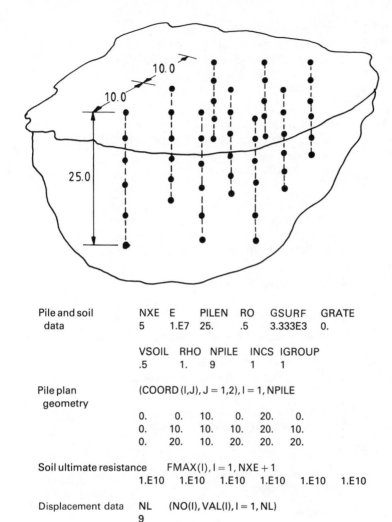

Pile and soil data	NXE	E	PILEN	RO	GSURF	GRATE				
	5	1.E7	25.	.5	3.333E3	0.				
	VSOIL	RHO	NPILE	INCS	IGROUP					
	.5	1.	9	1	1					
Pile plan geometry	(COORD (I,J), J = 1,2), I = 1, NPILE									
	0.	0.	10.	0.	20.	0.				
	0.	10.	10.	10.	20.	10.				
	0.	20.	10.	20.	20.	20.				
Soil ultimate resistance	FMAX(I), I = 1, NXE + 1									
	1.E10	1.E10	1.E10	1.E10	1.E10	1.E10				
Displacement data	NL	(NO(I), VAL(I), I = 1, NL)								
	9									
	1	1.	2	1.	3	1.	4	1.		
	5	1.	6	1.	7	1.	8	1.	9	1.

Figure 12.13 Mesh and data for Program 12.2

Simple variables:

NXE	number of elements per pile (5)
E	modulus of piles (1×10^7)
PILEN	length of each pile (25.0)
RO	radius of each pile (0.5)
GSURF	soil shear modulus at surface (3.333×10^3)
GRATE	rate of increase of G with depth (0.0)
VSOIL	Poisson ratio of soil (0.5)

RHO	inhomogeneity factor (1.0)
NPILE	number of piles in the group (9)
INCS	number of displacement increments (1)
IGROUP	switch—if zero, no interaction (1)

Data constants:

IKM	working size of arrays KM
IDOF	number of degrees of freedom per pile element (2)
RF	hyperbolic stiffness parameter (0.9)

Derived variables:

N	number of degrees of freedom in problem
ELL	element length
CSA	element cross-sectional area
RM	radius at which shear stress in soil becomes zero
FACT1, FACT2, FACT3	terms in Mindlin's equation
RR	distance to interaction node
GSOILI, GSOILJ	soil stiffnesses at top and bottom of segment
GSOIL	average soil shear stiffness
XI	tangent flexibility (Chow, 1986, equation 8)
FSHAFT	shaft flexibility
FTIP	tip flexibility
PAV	average load on a pile in a group

Fixed length arrays:

KM	pile element stiffness matrix
G	steering vector

Variable length arrays:

KV	global soil flexibility matrix
SM	global stiffness matrix of piles and soil
LOADS	loads on system to achieve prescribed displacements
FORCE	loads in soil springs
VAL	values of displacements prescribed
COORD	plan coordinates of piles
ZCOORD	depth coordinates of pile nodes
FMAX	maximum force in *tz* springs
FSOIL	current force in *tz* springs
SUM	load in each pile
GP, GQ	steering vectors

NO	freedoms at which displacements are prescribed
NA	diagonal elements of (sparse) global matrices

PARAMETER statement restrictions:

IKV $\geqslant$ IR (profile storage)
ILOADS $\geqslant$ N
INO $\geqslant$ NL
INX $\geqslant$ NN

Turning to the program coding, two global matrices are used: KV to hold the global soil flexibility matrix and SM to hold the global pile plus soil stiffness matrix. Profile storage is used and so the appropriate assembly routine is FSPARV and the solution routines are SPARIN and SPABAC.

After input and initialisation, the rest of the program is enclosed by a large loop in which the displacements are added in increments. The present analysis is linear, achieved by setting a high value of FMAX, and so the loop is only completed once. Analyses with no interaction at all can be achieved by setting IGROUP to zero. Nodal numbers for the elements are provided by subroutine GEOM.

The soil flexibility matrix is obtained from Mindlin's equation while a small subroutine FORMXI delivers the tangent stiffnesses of the soil springs. The tip flexibility is obtained from the solution for a rigid circular punch on an elastic half-space.

When KV has been found it is 'inverted' by Gaussian elimination and then the pile stiffnesses can be added, using FSPARV to find the global stiffness SM. These equations are then solved and the forces in the springs recovered. By adding up all the spring forces for each pile, the load in the piles is known, called SUM.

The output from the program is given in Figure 12.14. The increment number is printed (for a single increment in this linear case) followed by the total load on the group, TOTSUM and the value of the displacement accumulated (1.0 in this case). Then for all piles, the ratio of pile load to the average, if all piles were carrying the same load, is printed. In this example, the corner piles 1, 3, 7, 9 carry nearly 20% more than the average, while the central pile, number 5, carries less than 60% of the average. These figures are in excellent agreement with values

```
          1    Increment Number
   .4261E+06     .1000E+01  Total load  Displacement
          1    .1198E+01
          2    .9064E+00
          3    .1198E+01
          4    .9064E+00
          5    .5833E+00
          6    .9064E+00
          7    .1198E+01
          8    .9064E+00
          9    .1198E+01
       Pile    Proportion of
     number    total load carried
```

Figure 12.14 Results from Program 12.2

quoted by Poulos and Davis (1980, Table 6.4a) for the case where $E_{pile}/E_{soil} = 10^3$ and $L/d = 25$.

PROGRAM 12.3: PILE DRIVABILITY BY WAVE EQUATION—TWO-NODE LINE ELEMENTS—IMPLICIT INTEGRATION IN TIME

```
C
C      PROGRAM 12.3 PILE DRIVABILITY BY THE WAVE EQUATION
C      2-NODE LINE ELEMENTS, NEWMARK IMPLICIT INTEGRATION
C
C      ALTER NEXT LINE TO CHANGE PROBLEM SIZE
C
      PARAMETER(IKV=1000,ILOADS=200,INF=101)
C
      REAL E(INF),RHO(INF),SX(INF),ELL(INF),CSA(INF),KM(2,2),EMM(2,2),
     +RU(INF),QU(INF),JJ(INF),F(INF),DF(INF),KV(IKV),F1(IKV),MM(IKV),
     +LOADS(ILOADS),X0(ILOADS),D1X0(ILOADS),D2X0(ILOADS),ELD(2),
     +BDYLDS(ILOADS),X1(ILOADS),D1X1(ILOADS),D2X1(ILOADS)
      INTEGER NF(INF,1),G(2)
      DATA IKM,IEMM,IDOF/3*2/,NODOF,IW/2*1/
C
C      INPUT AND INITIALISATION
C
      READ(5,*)NXE,N,NN,NR,ALPHA,BETA,ITS,DTIM,ISTEP,THETA,ITYPE,
     +         RAMVEL,STATLD
      IR=N*(IW+1)
      CALL NULVEC(LOADS,N)
      CALL NULVEC(BDYLDS,N)
      CALL NULVEC(SX,NXE)
      CALL NULVEC(F,NN)
      CALL NULVEC(DF,NN)
      CALL NULVEC(KV,IR)
      CALL NULVEC(MM,IR)
      READ(5,*)(E(I),I=1,NXE)
      READ(5,*)(RHO(I),I=1,NXE)
      READ(5,*)(CSA(I),I=1,NXE)
      READ(5,*)(ELL(I),I=1,NXE)
      READ(5,*)(RU(I),I=1,NN)
      READ(5,*)(QU(I),I=1,NN)
      READ(5,*)(JJ(I),I=1,NN)
      CALL READNF(NF,INF,NN,NODOF,NR)
C
C      ELEMENT MASS AND STIFFNESS ASSEMBLY
C
      DO 10 IP=1,NXE
      CALL GSTRNG(IP,NODOF,NF,INF,G)
      CALL AXIKM(KM,CSA(IP),E(IP),ELL(IP))
      IF(ITYPE.EQ.1)GO TO 20
      EMM(1,1)=RHO(IP)*CSA(IP)*ELL(IP)/3.
      EMM(2,2)=EMM(1,1)
      EMM(2,1)=EMM(1,1)*.5
      EMM(1,2)=EMM(2,1)
      GOTO 30
   20 EMM(1,1)=RHO(IP)*CSA(IP)*ELL(IP)*.5
      EMM(2,2)=EMM(1,1)
      EMM(2,1)=.0
      EMM(1,2)=EMM(2,1)
   30 CALL FORMKV(KV,KM,IKM,G,N,IDOF)
   10 CALL FORMKV(MM,EMM,IEMM,G,N,IDOF)
      CALL VECCOP(KV,F1,IR)
```

```
C
C       PRELIMINARY STATIC SOLUTION
C
      F1(N)=F1(N)+1.E6
      CALL BANRED(F1,N,IW)
      LOADS(1)=STATLD
      CALL BACSUB(F1,LOADS,N,IW)
      CALL PRINTV(LOADS,N)
      CALL NULVEC(LOADS,N)
C
C       INITIAL CONDITIONS
C
      CALL NULVEC(X0,N)
      CALL NULVEC(D1X0,N)
      CALL NULVEC(D2X0,N)
      CALL NULVEC(D1X1,N)
      D1X0(2)=RAMVEL
      D1X0(1)=D1X0(2)
      C1=(1.-THETA)*DTIM
      C2=BETA-C1
      C3=ALPHA+1./(THETA*DTIM)
      C4=BETA+THETA*DTIM
C
C       NEWMARK THETA METHOD FOR TIME STEPPING
C
      TIM=0.0
      DO 40 J=1,ISTEP
      TIM=TIM+DTIM
      DO 50 I=1,IR
   50 F1(I)=C3*MM(I)+C4*KV(I)
      DO 60 I=1,NN
      IF(QU(I).LT.1.E-6)GOTO 60
      F1(I)=F1(I)+C4*RU(I)*(1.+JJ(I)*D1X1(I))/QU(I)
   60 CONTINUE
      CALL BANRED(F1,N,IW)
C
C       REDISTRIBUTE EXCESS SPRING FORCES
C
      DO 70 K=1,ITS
      DO 80 I=1,N
      LOADS(I)=DTIM*BDYLDS(I)
      BDYLDS(I)=0.0
   80 X1(I)=C3*X0(I)+D1X0(I)/THETA
      CALL LINMUL(MM,X1,D1X1,N,IW)
      CALL VECADD(D1X1,LOADS,D1X1,N)
      DO 90 I=1,N
   90 LOADS(I)=C2*X0(I)
      CALL LINMUL(KV,LOADS,X1,N,IW)
      CALL VECADD(X1,D1X1,X1,N)
      CALL BACSUB(F1,X1,N,IW)
      DO 100 I=1,N
      D1X1(I)=(X1(I)-X0(I))/(THETA*DTIM)-D1X0(I)*(1.-THETA)/THETA
  100 D2X1(I)=(D1X1(I)-D1X0(I))/(THETA*DTIM)-D2X0(I)*(1.-THETA)/THETA
      DO 110 I=1,NN
      IF(QU(I).LT.1.E-6)GO TO 110
      W1=1.+JJ(I)*D1X1(I)
      DF(I)=-(X1(I)-X0(I))*W1*RU(I)/QU(I)
      IF(ABS(F(I)+DF(I)).LT.W1*RU(I))GO TO 120
      IF(DF(I).GT..0)BDYLDS(I)=-F(I)-DF(I)+W1*RU(I)
      IF(DF(I).LE..0)BDYLDS(I)=-F(I)-DF(I)-W1*RU(I)
  120 DF(I)=-(X1(I)-X0(I))*W1*RU(I)/QU(I)+BDYLDS(I)
  110 CONTINUE
```

```
   70 CONTINUE
      DO 130 IP=1,NXE
      CALL GSTRNG(IP,NODOF,NF,INF,G)
      DO 140 M=1,2
      IF(G(M).EQ.0)ELD(M)=.0
  140 IF(G(M).NE.0)ELD(M)=X1(G(M))
      EPS=(ELD(2)-ELD(1))/ELL(IP)
      SIGMA=E(IP)*EPS
  130 SX(IP)=SIGMA
      CALL VECADD(F,DF,F,NN)
      WRITE(6,1000)TIM
      CALL PRINTV(X1,N)
      CALL PRINTV(D1X1,N)
      CALL PRINTV(F,NN)
      CALL PRINTV(SX,NXE)
C
C      UPDATE DISPLACEMENTS,VELOCITIES,ACCELERATIONS
C
      CALL VECCOP(X1,X0,N)
      CALL VECCOP(D1X1,D1X0,N)
      CALL VECCOP(D2X1,D2X0,N)
   40 CONTINUE
 1000 FORMAT(F8.5)
      STOP
      END
```

This type of analysis was pioneered by E. A. L. Smith (1960) who used a finite difference spatial semi-discretisation and an explicit integrator in the time domain. Although more insight into the real process of pile driving can be obtained by treating pile and ground as axisymmetric solids (Smith and Chow, 1982), one-dimensional spatial idealisations are still the norm in practice. In the present program, line finite elements are used and implicit integration in time. Such a dynamic process has already been described in Program 11.1, and so this analysis is a dynamic version for a pile such as that described in Program 12.0. A similar kind of 'spring' idealisation is used for the ground, i.e. with no coupling between 'springs', but a viscous component of resistance is added as shown in Figure 12.15. The empirical parameter J controls the damping assumption.

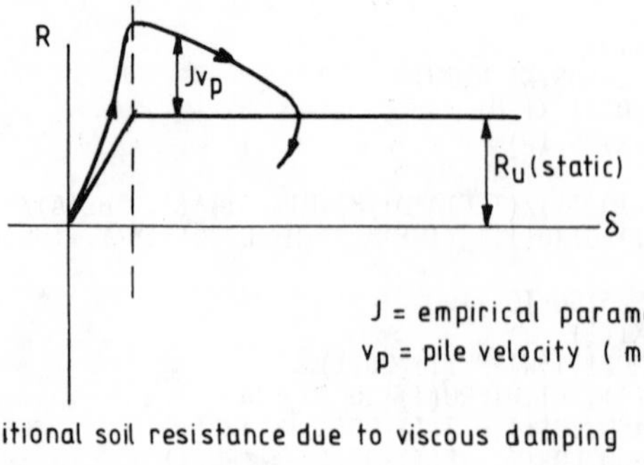

Figure 12.15 Soil spring behaviour for velocity dependent damping

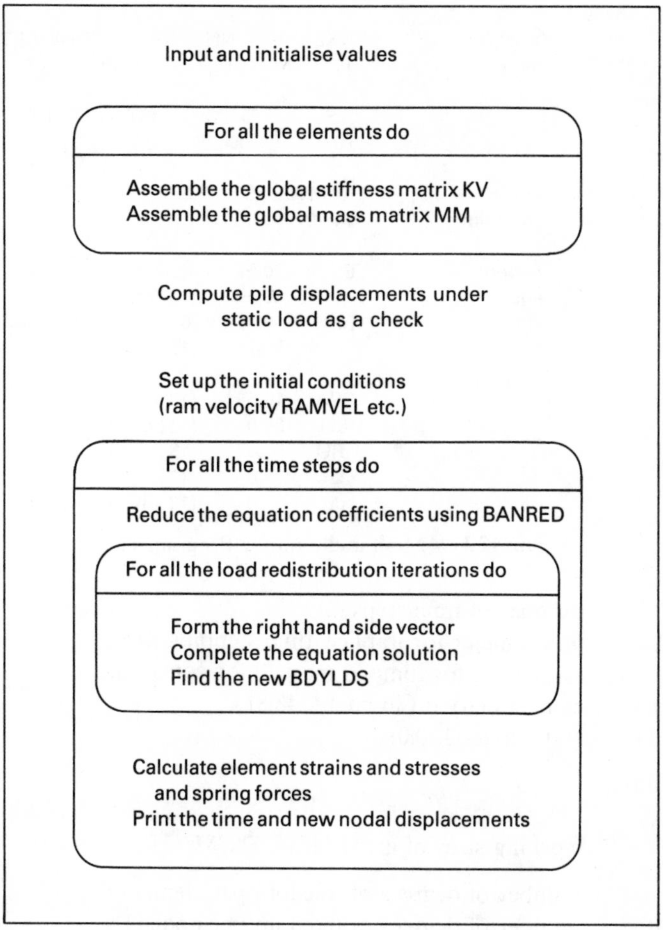

Figure 12.16 Structure chart for Program 12.3

The structure of the program is shown in Figure 12.16 and the test problem with input data in Figure 12.17. The variables have the following significance:

Input simple variables:

NXE	number of line elements in the pile plus hammer and cushion (13)
N	number of degrees of freedom in mesh (14)
NN	number of nodes in mesh (14)
NR	number of restrained nodes in mesh (0)
ALPHA, BETA	Rayleigh damping parameters for pile (both 0.0)
ITS	Maximum number of load redistribution iterations (10)
DTIM	time step (0.00025)

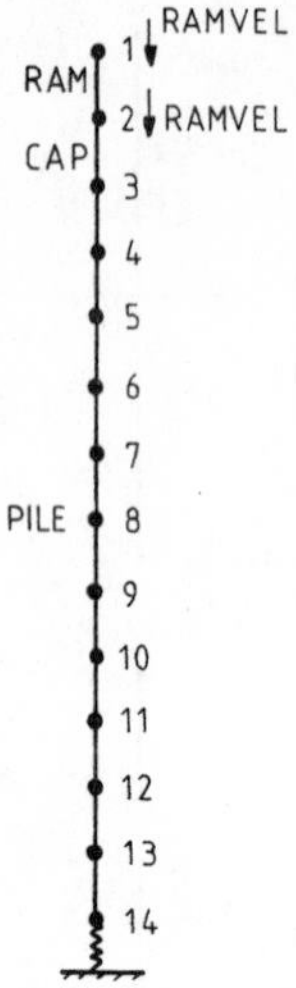

Structure data	NXE	N	NN	NR	ALPHA BETA
	13	14	14	0	.0 .0
	ITS	DTIM	ISTEP	THETA	ITYPE
	10	.00025	2	.5	1
Initial conditions	RAMVEL	STATLD			
	148.8	100.0			
Element and spring data	E	30.E9	15.4E6	30.E9	
		30.E6	(10 times)		
	RHO	6.9E-3	.0	9.67E-4	
		7.32E-4	(10 times)		
	CSA	15.58	(13 times)		
	ELL	120.0	(13 times)		
	RU	.0	(13 times)	2.E5	
	QU	.0	(13 times)	.1	
	JJ	.0	(13 times)	.0125	

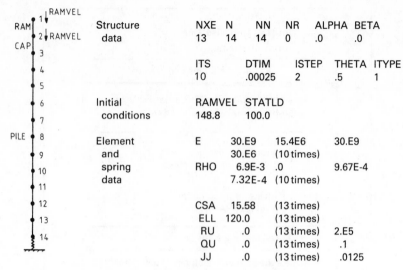

Figure 12.17 Mesh and data for Program 12.3

ISTEP	number of time steps (2)
THETA	θ parameter for implicit time stepping (0.5)
ITYPE	switch—1 for lumped mass, 0 for consistent
RAMVEL	ram velocity on impact (148.8)
STATLD	static load (100.0)

Data constants:

IKM, IEMM	working sizes of arrays KM, EMM
IDOF	number of degrees of freedom per element (2)
NODOF	number of degrees of freedom per node (1)
IW	half-bandwidth (1)

Fixed length arrays:

KM	element stiffness matrix
EMM	element mass matrix
ELD	element nodal displacements
G	element steering vector

Variable length arrays:

E	Young's moduli for elements
RHO	mass density of elements
SX	axial stress in elements
ELL	element lengths
CSA	element cross-sectional areas
RU	ultimate (static) resistance of springs

QU	quake of springs
JJ	damping in springs
F	force in springs
DF	change of force in springs
KV	global stiffness matrix
F1	global stiffness for static solution
MM	global mass matrix
LOADS	system load vector
X0, D1X0, D2X0	displacement, velocity, acceleration (at time 0)
X1, D1X1, D2X1	displacement, velocity, acceleration (at time 1)
NF	node freedom array

PARAMETER statement restrictions are:

IKV ⩾ N∗(IW + I)
ILOADS ⩾ N
INF ⩾ NN

The features of the program have all been described before in other contexts. A static solution is done first as a check, using a load at the top of the pile given by STATLD. Differing initial conditions are often used in pile driving programs. The present one uses an initial velocity applied to the top two nodes of the mesh,

```
 .4068E-03   .4068E-03   .3568E-03   .3567E-03   .3311E-03   .3054E-03 } Static
 .2797E-03   .2540E-03   .2284E-03   .2027E-03   .1770E-03   .1513E-03 } displacement
 .1257E-03   .1000E-03
00025 Time
 .3712E-01   .3711E-01   .4567E-03   .4447E-03   .1819E-04   .7441E-06
 .3044E-07   .1245E-08   .5093E-10   .2083E-11   .8522E-13   .3486E-14
 .1428E-15   .1120E-16   Displacements
 .1481E+03   .1481E+03   .3653E+01   .3557E+01   .1455E+00   .5952E-02
 .2435E-03   .9960E-05   .4074E-06   .1667E-07   .6818E-09   .2789E-10
 .1143E-11   .8957E-13   Velocities
 .0000E+00   .0000E+00   .0000E+00   .0000E+00   .0000E+00   .0000E+00
 .0000E+00   .0000E+00   .0000E+00   .0000E+00   .0000E+00   .0000E+00
 .0000E+00  -.2239E-10   Spring forces
-.2233E+04  -.4703E+04  -.3008E+04  -.1066E+03  -.4361E+01  -.1784E+00
-.7298E-02  -.2985E-03  -.1221E-04  -.4995E-06  -.2043E-07  -.8358E-09
-.3291E-10               Element stresses
00050 Time
 .7388E-01   .7386E-01   .2645E-02   .2620E-02   .1742E-03   .9869E-05
 .5159E-06   .2569E-07   .1239E-08   .5835E-10   .2701E-11   .1233E-12
 .5581E-14   .4774E-15
 .1460E+03   .1460E+03   .1385E+02   .1385E+02   .1103E+01   .6705E-01
 .3640E-02   .1856E-03   .9094E-05   .4334E-06   .2024E-07   .9309E-09
 .4236E-10   .3640E-11
 .0000E+00   .0000E+00   .0000E+00   .0000E+00   .0000E+00   .0000E+00
 .0000E+00   .0000E+00   .0000E+00   .0000E+00   .0000E+00   .0000E+00
 .0000E+00  -.9548E-09
-.4789E+04  -.9140E+04  -.6102E+04  -.6115E+03  -.4109E+02  -.2338E+01
-.1225E+00  -.6113E-02  -.2951E-03  -.1391E-04  -.6444E-06  -.2944E-07
-.1276E-08
```

Figure 12.18 Results from Program 12.3

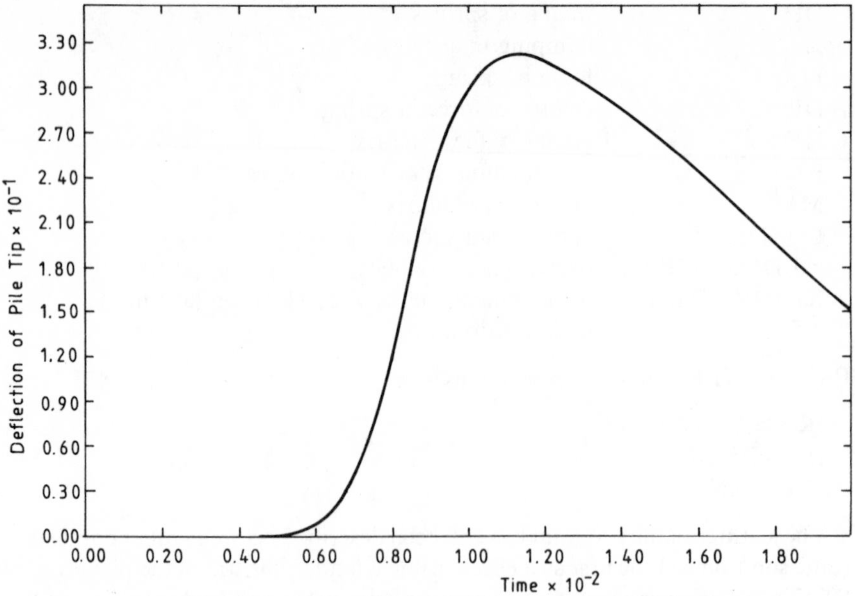

Figure 12.19 Tip displacement versus time from Program 12.3

which represent the hammer. Beneath this is the capblock and then eleven pile elements.

The output is listed as Figure 12.18 and consists of the static displacements followed by the current time and the four arrays containing displacements, velocities, spring forces and stresses in the pile, for two steps in time. A plot of tip displacement with time is shown in Figure 12.19. This reaches a maximum value from which the quake has to be subtracted to give the permanent, plastic 'set' of the tip for that hammer blow.

In the present example the 'set' is $0.322 - 0.1$, that is 0.222. The analysis becomes invalid if continued much beyond this first peak (Smith and Chow, 1982).

12.1 REFERENCES

Chow, Y. K. (1986) Analysis of vertically loaded pile groups. *Int. J. Num. Anal. Meth. Geomechanics*, **10**, 1, 59–72.

Chow, Y. K., and Smith, I. M. (1982) Static/dynamic analysis of an axially loaded pile. Proc. 4th Int. Conf. on Numerical Methods in Geomechanics, Edmonton, pp. 819–24.

Poulos, H. G., and Davis, E. H. (1980) *Pile Foundation Analysis and Design*. John Wiley, New York.

Smith, E. A. L. (1960) Pile driving analysis by the wave equation. *Proc. ASCE, JSMFD*, **86**, SM4, 35–61.

Smith, I. M. (1979) A survey of numerical methods in offshore piling. Proc. lst Int. Conf. Num. Meth. Offshore Piling, Institution of Civil Engineers, London, pp. 1–8.

Smith, I. M., and Chow, Y. K. (1982) Three-dimensional analysis of pile drivability. Proc. 2nd Int. Conf. Num. Meth. Offshore Piling, Austin, pp. 1–19.

APPENDIX 1
Consistent Nodal Loads

Assume that applied stress = 1 unit.

A. PLANAR (TWO-DIMENSIONS), ELEMENT WIDTH = 1 UNIT

1. Three-node triangle

$$F_1 = F_2 = \tfrac{1}{2}$$

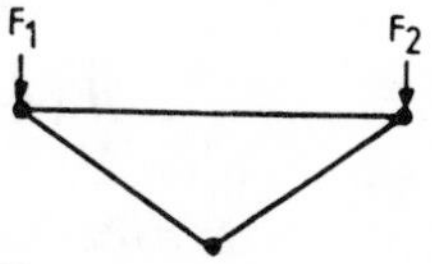

2. Six-node triangle

$$F_1 = F_3 = \tfrac{1}{6}$$
$$F_2 = \tfrac{2}{3}$$

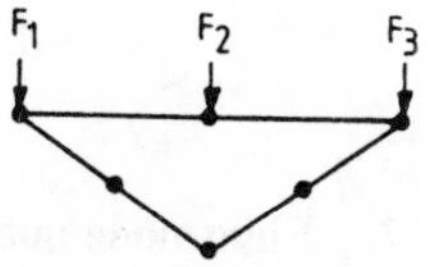

3. Ten-node triangle

$$F_1 = F_4 = \tfrac{1}{8}$$
$$F_2 = F_3 = \tfrac{3}{8}$$

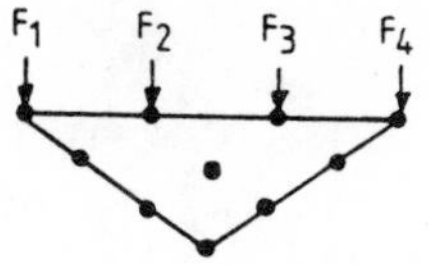

4. Fifteen-node triangle

$$F_1 = F_5 = \tfrac{7}{90}$$
$$F_2 = F_4 = \tfrac{32}{90}$$
$$F_3 = \tfrac{12}{90}$$

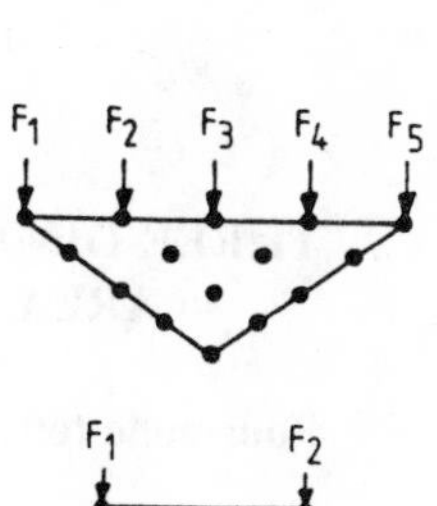

5. Four-node quadrilateral

$$F_1 = F_2 = \tfrac{1}{2}$$

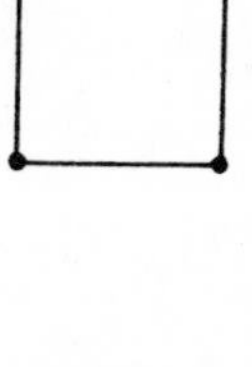

6. Eight-node quadrilateral

$$F_1 = F_3 = \tfrac{1}{6}$$
$$F_2 = \tfrac{2}{3}$$

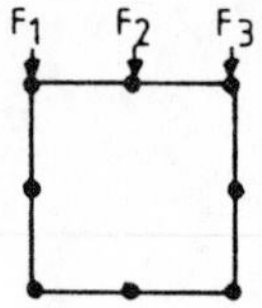

7. Nine-node quadrilateral

$$F_1 = F_3 = \tfrac{1}{6}$$
$$F_2 = \tfrac{2}{3}$$

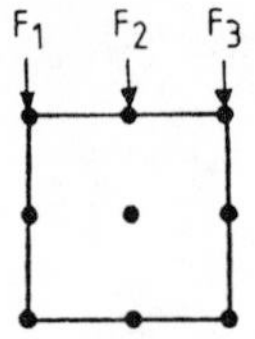

B. AXISYMMETRIC (TWO-DIMENSIONS), LOADING OVER 1 RADIAN

1. Four-node quadrilateral

$$F_1 = \frac{r_1 - r_0}{6}(2r_0 + r_1)$$

$$F_2 = \frac{r_1 - r_0}{6}(r_0 + 2r_1)$$

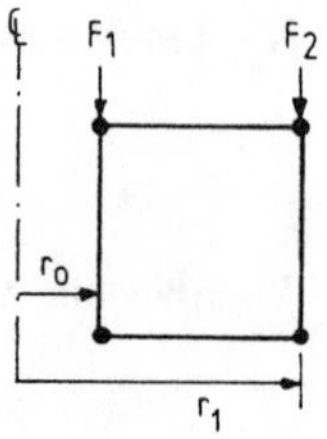

2. Eight-node quadrilateral

$$F_1 = \frac{r_1 - r_0}{6} r_0$$

$$F_2 = \frac{r_1 - r_0}{3}(r_0 + r_1)$$

$$F_3 = \frac{r_1 - r_0}{6} r_1$$

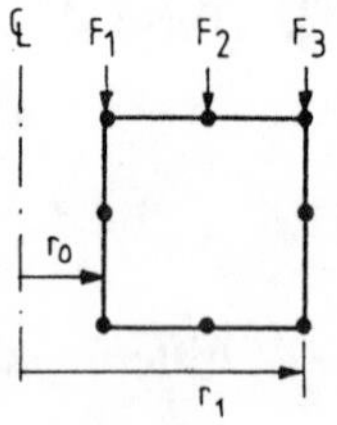

C. THREE-DIMENSIONS, LOADED AREA = 1 UNIT

1. Four-node tetrahedron

$$F_1 = F_2 = F_3 = \tfrac{1}{3}$$

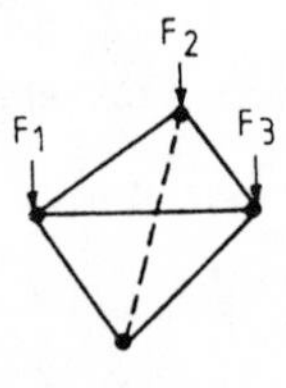

2. Eight-node cube

$$F_1 = F_2 = F_3 = F_4 = \tfrac{1}{4}$$

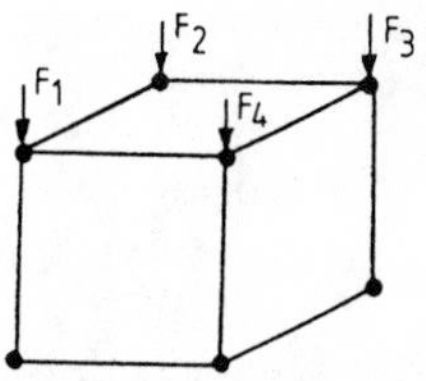

3. Twenty-node cube

$$F_1 = F_3 = F_5 = F_7 = -\tfrac{1}{12}$$
$$F_2 = F_4 = F_6 = F_8 = \tfrac{1}{3}$$

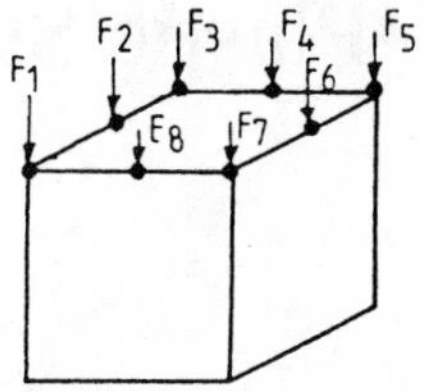

APPENDIX 2

Plastic Stress–Strain Matrices and Plastic Potential Derivatives

A. PLASTIC STRESS–STRAIN MATRICES

1. von Mises

$$\mathbf{D}^{\mathrm{P}} = \frac{2G}{t^2}\begin{bmatrix} s_x^2 & s_x s_y & s_x \tau_{xy} & s_z s_x \\ & s_y^2 & s_y \tau_{xy} & s_y s_z \\ & & \tau_{xy}^2 & s_z \tau_{xy} \\ \text{Symmetrical} & & & s_z^2 \end{bmatrix}$$

where

G = shear modulus

t = second deviatoric stress invariant (equation 6.3)

$s_x = (2\sigma_x - \sigma_y - \sigma_z)/3$, etc.

2. Mohr–Coloumb

If not near a corner, that is $\sin\theta \leqslant 0.49$, then

$$\mathbf{D}^{\mathrm{P}} = \frac{E}{2(1+\nu)(1-2\nu)(1-2\nu+\sin\phi\sin\psi)}\mathbf{A}$$

where

$$\mathbf{A} = \begin{bmatrix} R_1C_1 & R_1C_2 & R_1C_3 & R_1C_4 \\ R_2C_1 & R_2C_2 & R_2C_3 & R_2C_4 \\ R_3C_1 & R_3C_2 & R_3C_3 & R_3C_4 \\ R_4C_1 & R_4C_2 & R_4C_3 & R_4C_4 \end{bmatrix}$$

and

$$C_1 = \sin\phi + k_1(1-2\nu)\sin\alpha$$

$$C_2 = \sin\phi - k_1(1-2\nu)\sin\alpha$$

$$C_3 = k_2(1-2\nu)\cos\alpha$$

$$C_4 = 2\nu \sin\phi$$
$$R_1 = \sin\psi + k_1(1-2\nu)\sin\alpha$$
$$R_2 = \sin\psi - k_1(1-2\nu)\sin\alpha$$
$$R_3 = k_2(1-2\nu)\cos\alpha$$
$$R_4 = 2\nu\sin\psi$$

$$\alpha = \arctan\left|\frac{\sigma_x - \sigma_y}{2\tau_{xy}}\right|$$

$$k_1 = \begin{cases} 1 & \text{if } |\sigma_y| \geqslant |\sigma_x| \\ -1 & \text{if } \sigma_x > \sigma_y \end{cases}$$

$$k_2 = \begin{cases} 1 & \text{if } \tau_{xy} \geqslant 0 \\ -1 & \text{if } \tau_{xy} < 0 \end{cases}$$

If near a corner, that is $\sin\theta > 0.49$, then

$$\mathbf{D}^{\mathrm{P}} = \frac{E}{(1+\nu)(1-2\nu)[K_\phi \sin\psi + C_\phi C_\psi t^2(1-2\nu)]}\mathbf{A}$$

where $\mathbf{A}$ is defined as before with

$$C_1 = K_\phi + C_\phi[(1-\nu)s_x + \nu(s_y + s_z)]$$
$$C_2 = K_\phi + C_\phi[(1-\nu)s_y + \nu(s_z + s_x)]$$
$$C_3 = C_\phi(1-2\nu)\tau_{xy}$$
$$C_4 = K_\phi + C_\phi[(1-\nu)s_z + \nu(s_x + s_y)]$$
$$R_1 = K_\psi + C_\psi[(1-\nu)s_x + \nu(s_y + s_z)]$$
$$R_2 = K_\psi + C_\psi[(1-\nu)s_y + \nu(s_z + s_x)]$$
$$R_3 = C_\psi(1-2\nu)\tau_{xy}$$
$$R_4 = K_\psi + C_\psi[(1-\nu)s_z + \nu(s_x + s_y)]$$

where

$$K_\phi = \frac{\sin\phi}{3}(1+\nu)$$

$$K_\psi = \frac{\sin\psi}{3}(1+\nu)$$

$$C_\phi = \frac{\sqrt{6}}{4t}\left(1 \pm \frac{\sin\phi}{3}\right)$$

$$C_\psi = \frac{\sqrt{6}}{4t}\left(1 \pm \frac{\sin\psi}{3}\right)$$

In the expressions for C_ϕ and C_ψ, the positive sign is valid if $\theta \simeq -30°$ and the negative sign is valid if $\theta \simeq 30°$.

B. PLASTIC POTENTIAL DERIVATIVES

$$\frac{\partial Q}{\partial \boldsymbol{\sigma}} = (\mathrm{DQ1}\,\mathbf{M}^1 + \mathrm{DQ2}\,\mathbf{M}^2 + \mathrm{DQ3}\,\mathbf{M}^3)\boldsymbol{\sigma}$$

where

$$\mathbf{M}^1 = \frac{1}{3(\sigma_x + \sigma_y + \sigma_z)} \begin{bmatrix} 1 & 1 & 1 & 0 & 0 & 0 \\ & 1 & 1 & 0 & 0 & 0 \\ & & 1 & 0 & 0 & 0 \\ & & & 0 & 0 & 0 \\ & & & & 0 & 0 \\ \text{Symmetrical} & & & & & 0 \end{bmatrix}$$

$$\mathbf{M}^2 = \frac{1}{3} \begin{bmatrix} 2 & -1 & -1 & 0 & 0 & 0 \\ & 2 & -1 & 0 & 0 & 0 \\ & & 2 & 0 & 0 & 0 \\ & & & 6 & 0 & 0 \\ & & & & 6 & 0 \\ \text{Symmetrical} & & & & & 6 \end{bmatrix}$$

$$\mathbf{M}^3 = \frac{1}{3} \begin{bmatrix} s_x & s_z & s_z & \tau_{xy} & -2\tau_{yz} & \tau_{zx} \\ & s_y & s_x & \tau_{xy} & \tau_{yz} & -2\tau_{zx} \\ & & s_z & -2\tau_{xy} & \tau_{yz} & \tau_{zx} \\ & & & -3s_z & 3\tau_{zx} & 3\tau_{yz} \\ & & & & -3s_x & 3\tau_{xy} \\ \text{Symmetrical} & & & & & -3s_y \end{bmatrix}$$

$$\boldsymbol{\sigma} = \begin{Bmatrix} \sigma_x \\ \sigma_y \\ \sigma_z \\ \tau_{xy} \\ \tau_{yz} \\ \tau_{zx} \end{Bmatrix}$$

1. von Mises

$$\mathrm{DQ1} = 0$$

$$\mathrm{DQ2} = \sqrt{\frac{3}{2}}\frac{1}{t}$$

$$\mathrm{DQ3} = 0$$

2. Mohr–Coloumb

$$\mathrm{DQ1} = \sin\psi$$

$$DQ2 = \frac{\cos\theta}{\sqrt{2}t}\left[1 + \tan\theta\tan 3\theta + \frac{\sin\psi}{\sqrt{3}}(\tan 3\theta - \tan\theta)\right]$$

$$DQ3 = \frac{\sqrt{3}\sin\theta + \sin\psi\cos\theta}{t^2\cos 3\theta}$$

where

t = second deviatoric stress invariant (equation 6.3)
θ = Lode angle (equation 6.3)
ψ = dilation angle

APPENDIX 3
Geometry Subroutines

As an alternative to rather elaborate mesh generation code, this book uses some twenty small 'geometry' subroutines for simple (usually rectangular) mesh arrangements. They are listed in full below so that users are encouraged to develop similar routines of their own. The routines all use a uniform nomenclature for parameters and have the essential job of delivering as output a real array called COORD, which holds the nodal coordinates of the element and/or an integer array G, the 'steering vector', which holds the nodal freedom numbers, including zeros if a freedom is supressed.

The naming convention is that a geometry subroutine name contains the characters 'GE' and usually the characters 'GEO' or 'GEOM'. The first number encountered in the subroutine name, for example the '3' in the routine GEOM3X, tells how many nodes the element has, and the following character 'X' or 'Y' tells the direction in which the nodes and freedoms are numbered.

Usually, meshes are made up of elements of constant size, but if the character 'V' appears in the name, the subroutine allows elements of varying sizes. Thus GEOV8Y sets up rectangles of eight-noded elements of variable sizes numbered in the *y* direction.

Because plane elements describing non-axisymmetric situations have an extra degree of freedom per node, the characters 'GEN' are used to distinguish them. For three-dimensional elements, the characters '3D' will be found.

Usually there are two freedoms per node in two-dimensional elements, but the suffix '1' or '3' is used if these are one or three freedoms instead. Thus, GEV4X3 creates a rectangular mesh of four-node elements of variable size, numbered in the *x* direction and having three freedoms per node.

The special routines GSTRNG, GEOM, FORMGP, SLOGEO and WELGEO do not fit simply into the above general scheme. The first three refer to simple line structures analysed in Chapters 4 and 12 where generally only a steering vector needs to be created, while the last two create non-rectangular 'slope' and 'well' geometries respectively as required in Chapters 6 and 7.

While the output parameters of all routines are the same, namely COORD and/or G, reference must be made to the calling programs for details of the input parameters. These are usually simple counters, such as IP, IQ, IS, numbers of elements in a given direction, such as NXE, NYE, NDE, constant element dimensions, such as AA, BB, and coordinates of variable mesh lines, such as WIDTH and DEPTH. The node freedom array NF is needed as an input parameter in order that G can be found.

The 21 geometry routines are listed below. In each case a commented section describes the action of the routine.

```
      SUBROUTINE FORMGP(IP,IQ,NYE,G,NF,INF)
C
C      THIS SUBROUTINE FORMS THE 'STEERING' VECTOR FOR A
C      4-NODE RECTANGULAR PLATE BENDING ELEMENT
C
      INTEGER G(*),NF(INF,*),NUM(4)
      I1=(IP-1)*(NYE+1)+IQ
      I2=I1+1
      I3=IP*(NYE+1)+IQ
      I4=I3+1
      DO 1 I=1,4
      G(I)=NF(I1,I)
      G(I+4)=NF(I2,I)
      G(I+8)=NF(I4,I)
      G(I+12)=NF(I3,I)
    1 CONTINUE
      RETURN
      END

      SUBROUTINE GENA8X(IP,IQ,NRE,AA,BB,COORD,ICOORD,G,NF,INF)
C
C      THIS SUBROUTINE FORMS THE COORDINATES AND STEERING VECTOR
C      FOR 8-NODE QUADS COUNTING IN X-DIRECTION (3-FREEDOMS/NODE)
C
      REAL COORD(ICOORD,*)
      INTEGER G(*),NF(INF,*),NUM(8)
      NUM(1)=IQ*(3*NRE+2)+2*IP-1
      NUM(2)=IQ*(3*NRE+2)+IP-NRE-1
      NUM(3)=(IQ-1)*(3*NRE+2)+2*IP-1
      NUM(4)=NUM(3)+1
      NUM(5)=NUM(4)+1
      NUM(6)=NUM(2)+1
      NUM(7)=NUM(1)+2
      NUM(8)=NUM(1)+1
      INC=0
      DO 1 I=1,8
      DO 1 J=1,3
      INC=INC+1
    1 G(INC)=NF(NUM(I),J)
      COORD(1,1)=AA*(IP-1)
      COORD(2,1)=AA*(IP-1)
      COORD(3,1)=AA*(IP-1)
      COORD(5,1)=AA*IP
      COORD(6,1)=AA*IP
      COORD(7,1)=AA*IP
      COORD(4,1)=.5*(COORD(3,1)+COORD(5,1))
      COORD(8,1)=.5*(COORD(7,1)+COORD(1,1))
      COORD(1,2)=-BB*IQ
      COORD(8,2)=-BB*IQ
      COORD(7,2)=-BB*IQ
      COORD(3,2)=-BB*(IQ-1)
```

```
      COORD(4,2)=-BB*(IQ-1)
      COORD(5,2)=-BB*(IQ-1)
      COORD(2,2)=.5*(COORD(1,2)+COORD(3,2))
      COORD(6,2)=.5*(COORD(5,2)+COORD(7,2))
      RETURN
      END

      SUBROUTINE GEOM(IP,NN,NPILE,G)
C
C      THIS SUBROUTINE FORMS THE NODE NUMBERS OF PILE IP
C
      INTEGER G(*)
      DO 1 I=1,NN
    1 G(I)=IP+(I-1)*NPILE
      RETURN
      END

      SUBROUTINE GEOM3X(IP,IQ,NXE,AA,BB,COORD,ICOORD,G,NF,INF)
C
C      THIS SUBROUTINE FORMS THE COORDINATES AND STEERING VECTOR
C      FOR 3-NODE TRIANGLES COUNTING IN X-DIRECTION
C
      REAL COORD(ICOORD,*)
      INTEGER G(*),NF(INF,*),NUM(3)
      IF(MOD(IQ,2).EQ.0)GOTO 1
      NUM(1)=(NXE+1)*(IQ-1)/2+IP
      NUM(2)=(NXE+1)*(IQ+1)/2+IP
      NUM(3)=NUM(1)+1
      COORD(1,1)=(IP-1)*AA
      COORD(1,2)=-(IQ-1)/2*BB
      COORD(2,1)=(IP-1)*AA
      COORD(2,2)=-(IQ+1)/2*BB
      COORD(3,1)=IP*AA
      COORD(3,2)=COORD(1,2)
      GOTO 2
    1 NUM(1)=(NXE+1)*IQ/2+IP+1
      NUM(2)=(NXE+1)*(IQ-2)/2+IP+1
      NUM(3)=NUM(1)-1
      COORD(1,1)=IP*AA
      COORD(1,2)=-IQ/2*BB
      COORD(2,1)=IP*AA
      COORD(2,2)=-(IQ-2)/2*BB
      COORD(3,1)=(IP-1)*AA
      COORD(3,2)=COORD(1,2)
    2 CONTINUE
      INC=0
      DO 3 I=1,3
      DO 3 J=1,2
      INC=INC+1
    3 G(INC)=NF(NUM(I),J)
      RETURN
      END

      SUBROUTINE GEOM6X(IP,IQ,NXE,AA,BB,COORD,ICOORD,G,NF,INF)
C
C      THIS SUBROUTINE FORMS THE COORDINATES AND STEERING VECTOR
C      FOR 6-NODE TRIANGLES COUNTING IN X-DIRECTION
C
      REAL COORD(ICOORD,*)
      INTEGER G(*),NF(INF,*),NUM(6)
```

```
      IF(MOD(IQ,2).EQ.0)GOTO 1
      NUM(1)=(IQ-1)*(2*NXE+1)+2*IP-1
      NUM(2)=(IQ-1)*(2*NXE+1)+2*NXE+2*IP
      NUM(3)=(IQ+1)*(2*NXE+1)+2*IP-1
      NUM(4)=NUM(2)+1
      NUM(5)=NUM(1)+2
      NUM(6)=NUM(1)+1
      COORD(1,1)=(IP-1)*AA
      COORD(1,2)=-(IQ-1)/2*BB
      COORD(3,1)=(IP-1)*AA
      COORD(3,2)=-(IQ+1)/2*BB
      COORD(5,1)=IP*AA
      COORD(5,2)=COORD(1,2)
      GOTO 2
    1 NUM(1)=IQ*(2*NXE+1)+2*IP+1
      NUM(2)=(IQ-2)*(2*NXE+1)+2*NXE+2*IP+2
      NUM(3)=(IQ-2)*(2*NXE+1)+2*IP+1
      NUM(4)=NUM(2)-1
      NUM(5)=NUM(1)-2
      NUM(6)=NUM(1)-1
      COORD(1,1)=IP*AA
      COORD(1,2)=-IQ/2*BB
      COORD(3,1)=IP*AA
      COORD(3,2)=-(IQ-2)/2*BB
      COORD(5,1)=(IP-1)*AA
      COORD(5,2)=COORD(1,2)
    2 DO 3 I=1,2
      COORD(2,I)=.5*(COORD(1,I)+COORD(3,I))
      COORD(4,I)=.5*(COORD(3,I)+COORD(5,I))
    3 COORD(6,I)=.5*(COORD(5,I)+COORD(1,I))
      INC=0
      DO 4 I=1,6
      DO 4 J=1,2
      INC=INC+1
    4 G(INC)=NF(NUM(I),J)
      RETURN
      END

      SUBROUTINE GEOM8X(IP,IQ,NXE,AA,BB,COORD,ICOORD,G,NF,INF)
C
C      THIS SUBROUTINE FORMS THE COORDINATES AND STEERING VECTOR
C      FOR 8-NODE QUADS COUNTING IN X-DIRECTION
C
      REAL COORD(ICOORD,*)
      INTEGER G(*),NF(INF,*),NUM(8)
      NUM(1)=IQ*(3*NXE+2)+2*IP-1
      NUM(2)=IQ*(3*NXE+2)+IP-NXE-1
      NUM(3)=(IQ-1)*(3*NXE+2)+2*IP-1
      NUM(4)=NUM(3)+1
      NUM(5)=NUM(4)+1
      NUM(6)=NUM(2)+1
      NUM(7)=NUM(1)+2
      NUM(8)=NUM(1)+1
      INC=0
      DO 1 I=1,8
      DO 1 J=1,2
      INC=INC+1
    1 G(INC)=NF(NUM(I),J)
      COORD(1,1)=AA*(IP-1)
      COORD(2,1)=AA*(IP-1)
```

```
      COORD(3,1)=AA*(IP-1)
      COORD(5,1)=AA*IP
      COORD(6,1)=AA*IP
      COORD(7,1)=AA*IP
      COORD(4,1)=.5*(COORD(3,1)+COORD(5,1))
      COORD(8,1)=.5*(COORD(7,1)+COORD(1,1))
      COORD(1,2)=-BB*IQ
      COORD(8,2)=-BB*IQ
      COORD(7,2)=-BB*IQ
      COORD(3,2)=-BB*(IQ-1)
      COORD(4,2)=-BB*(IQ-1)
      COORD(5,2)=-BB*(IQ-1)
      COORD(2,2)=.5*(COORD(1,2)+COORD(3,2))
      COORD(6,2)=.5*(COORD(5,2)+COORD(7,2))
      RETURN
      END

      SUBROUTINE GEOM9X(IP,IQ,NXE,AA,BB,COORD,ICOORD,G,NF,INF)
C
C      THIS SUBROUTINE FORMS THE COORDINATES AND STEERING VECTOR
C      FOR 9-NODE QUADS COUNTING IN X-DIRECTION
C
      REAL COORD(ICOORD,*)
      INTEGER G(*),NF(INF,*),NUM(9)
      NUM(1)=IQ*(4*NXE+2)+2*IP-1
      NUM(2)=IQ*(4*NXE+2)+2*IP-NXE-4
      NUM(3)=(IQ-1)*(4*NXE+2)+2*IP-1
      NUM(4)=NUM(3)+1
      NUM(5)=NUM(4)+1
      NUM(6)=NUM(2)+2
      NUM(7)=NUM(1)+2
      NUM(8)=NUM(1)+1
      NUM(9)=NUM(2)+1
      INC=0
      DO 1 I=1,9
      DO 1 J=1,2
      INC=INC+1
    1 G(INC)=NF(NUM(I),J)
      COORD(1,1)=(IP-1)*AA
      COORD(3,1)=(IP-1)*AA
      COORD(5,1)=IP*AA
      COORD(7,1)=IP*AA
      COORD(1,2)=-IQ*BB
      COORD(3,2)=-(IQ-1)*BB
      COORD(5,2)=-(IQ-1)*BB
      COORD(7,2)=-IQ*BB
      COORD(2,1)=.5*(COORD(1,1)+COORD(3,1))
      COORD(2,2)=.5*(COORD(1,2)+COORD(3,2))
      COORD(4,1)=.5*(COORD(3,1)+COORD(5,1))
      COORD(4,2)=.5*(COORD(3,2)+COORD(5,2))
      COORD(6,1)=.5*(COORD(5,1)+COORD(7,1))
      COORD(6,2)=.5*(COORD(5,2)+COORD(7,2))
      COORD(8,1)=.5*(COORD(1,1)+COORD(7,1))
      COORD(8,2)=.5*(COORD(1,2)+COORD(7,2))
      COORD(9,1)=.5*(COORD(2,1)+COORD(6,1))
      COORD(9,2)=.5*(COORD(4,2)+COORD(8,2))
      RETURN
      END
```

```
      SUBROUTINE GEOM4Y(IP,IQ,NYE,AA,BB,COORD,ICOORD,G,NF,INF)
C
C      THIS SUBROUTINE FORMS THE COORDINATES AND STEERING VECTOR
C      FOR 4-NODE QUADS COUNTING IN Y-DIRECTION
C
      REAL COORD(ICOORD,*)
      INTEGER G(*),NF(INF,*),NUM(4)
      NUM(1)=(IP-1)*(NYE+1)+IQ+1
      NUM(2)=NUM(1)-1
      NUM(3)=IP*(NYE+1)+IQ
      NUM(4)=NUM(3)+1
      INC=0
      DO 1 I=1,4
      DO 1 J=1,2
      INC=INC+1
    1 G(INC)=NF(NUM(I),J)
      COORD(1,1)=AA*(IP-1)
      COORD(2,1)=AA*(IP-1)
      COORD(3,1)=AA*IP
      COORD(4,1)=AA*IP
      COORD(1,2)=-BB*IQ
      COORD(2,2)=-BB*(IQ-1)
      COORD(3,2)=-BB*(IQ-1)
      COORD(4,2)=-BB*IQ
      RETURN
      END
      SUBROUTINE GEOM8Y(IP,IQ,NYE,AA,BB,COORD,ICOORD,G,NF,INF)
C
C      THIS SUBROUTINE FORMS THE COORDINATES AND STEERING VECTOR
C      FOR 8-NODE QUADS COUNTING IN THE Y-DIRECTION
C
C
      REAL COORD(ICOORD,*)
      INTEGER NUM(8),G(*),NF(INF,*)
      NUM(1)=(IP-1)*(3*NYE+2)+2*IQ+1
      NUM(2)=NUM(1)-1
      NUM(3)=NUM(2)-1
      NUM(4)=(IP-1)*(3*NYE+2)+2*NYE+IQ+1
      NUM(5)=IP*(3*NYE+2)+2*IQ-1
      NUM(6)=NUM(5)+1
      NUM(7)=NUM(6)+1
      NUM(8)=NUM(4)+1
      INC=0
      DO 1 I=1,8
      DO 1 J=1,2
      INC=INC+1
    1 G(INC)=NF(NUM(I),J)
      COORD(1,1)=(IP-1)*AA
      COORD(2,1)=(IP-1)*AA
      COORD(3,1)=(IP-1)*AA
      COORD(5,1)=IP*AA
      COORD(6,1)=IP*AA
      COORD(7,1)=IP*AA
      COORD(4,1)=(COORD(3,1)+COORD(5,1))*.5
      COORD(8,1)=COORD(4,1)
      COORD(3,2)=-(IQ-1)*BB
      COORD(4,2)=-(IQ-1)*BB
      COORD(5,2)=-(IQ-1)*BB
      COORD(1,2)=-IQ*BB
```

```
      COORD(8,2)=-IQ*BB
      COORD(7,2)=-IQ*BB
      COORD(2,2)=(COORD(1,2)+COORD(3,2))*.5
      COORD(6,2)=COORD(2,2)
      RETURN
      END

      SUBROUTINE GEOV4Y(IP,IQ,NDE,RAD,DEP,COORD,ICOORD,G,NF,INF)
C
C      THIS SUBROUTINE FORMS THE COORDINATES AND STEERING VECTOR
C      FOR EACH ELEMENT (NUMBERING IN THE Y-DIRECTION)
C
      REAL COORD(ICOORD,*),RAD(*),DEP(*)
      INTEGER G(*),NF(INF,*),NUM(4)
      NUM(1)=(IP-1)*(NDE+1)+IQ+1
      NUM(2)=NUM(1)-1
      NUM(3)=IP*(NDE+1)+IQ
      NUM(4)=NUM(3)+1
      INC=0
      DO 1 I=1,4
      DO 1 J=1,2
      INC=INC+1
    1 G(INC)=NF(NUM(I),J)
      COORD(1,1)=RAD(IP)
      COORD(2,1)=RAD(IP)
      COORD(3,1)=RAD(IP+1)
      COORD(4,1)=RAD(IP+1)
      COORD(1,2)=DEP(IQ+1)
      COORD(2,2)=DEP(IQ)
      COORD(3,2)=DEP(IQ)
      COORD(4,2)=DEP(IQ+1)
      RETURN
      END

      SUBROUTINE GEOV8Y(IP,IQ,NYE,WIDTH,DEPTH,COORD,ICOORD,G,NF,INF)
C
C      THIS SUBROUTINE FORMS THE COORDINATES AND STEERING VECTOR
C      FOR 8-NODE QUADRILATERALS NUMBERING IN Y-DIRECTION
C
      REAL COORD(ICOORD,*),WIDTH(*),DEPTH(*)
      INTEGER G(*),NF(INF,*),NUM(8)
      NUM(1)=(IP-1)*(3*NYE+2)+2*IQ+1
      NUM(2)=NUM(1)-1
      NUM(3)=NUM(1)-2
      NUM(4)=(IP-1)*(3*NYE+2)+2*NYE+IQ+1
      NUM(5)=IP*(3*NYE+2)+2*IQ-1
      NUM(6)=NUM(5)+1
      NUM(7)=NUM(5)+2
      NUM(8)=NUM(4)+1
      INC=0
      DO 1 I=1,8
      DO 1 J=1,2
      INC=INC+1
    1 G(INC)=NF(NUM(I),J)
      COORD(1,1)=WIDTH(IP)
      COORD(2,1)=WIDTH(IP)
      COORD(3,1)=WIDTH(IP)
      COORD(5,1)=WIDTH(IP+1)
      COORD(6,1)=WIDTH(IP+1)
      COORD(7,1)=WIDTH(IP+1)
```

```
      COORD(4,1)=.5*(COORD(3,1)+COORD(5,1))
      COORD(8,1)=.5*(COORD(7,1)+COORD(1,1))
      COORD(1,2)=DEPTH(IQ+1)
      COORD(8,2)=DEPTH(IQ+1)
      COORD(7,2)=DEPTH(IQ+1)
      COORD(3,2)=DEPTH(IQ)
      COORD(4,2)=DEPTH(IQ)
      COORD(5,2)=DEPTH(IQ)
      COORD(2,2)=.5*(COORD(1,2)+COORD(3,2))
      COORD(6,2)=.5*(COORD(5,2)+COORD(7,2))
      RETURN
      END

      SUBROUTINE GEO4X1(IP,IQ,NXE,AA,BB,COORD,ICOORD,G,NF,INF)
C
C       THIS SUBROUTINE FORMS THE COORDINATES AND STEERING VECTOR
C       FOR 4-NODE QUADS COUNTING IN X-DIRECTION
C       LAPLACE'S EQUATION   1-FREEDOM PER NODE
C
      REAL COORD(ICOORD,*)
      INTEGER NUM(4),G(*),NF(INF,*)
      NUM(1)=IQ*(NXE+1)+IP
      NUM(2)=(IQ-1)*(NXE+1)+IP
      NUM(3)=NUM(2)+1
      NUM(4)=NUM(1)+1
      DO 1 I=1,4
    1 G(I)=NF(NUM(I),1)
      COORD(1,1)=(IP-1)*AA
      COORD(2,1)=(IP-1)*AA
      COORD(3,1)=IP*AA
      COORD(4,1)=IP*AA
      COORD(1,2)=-IQ*BB
      COORD(2,2)=-(IQ-1)*BB
      COORD(3,2)=-(IQ-1)*BB
      COORD(4,2)=-IQ*BB
      RETURN
      END

      SUBROUTINE GEO83D(IP,IQ,IS,NXE,NZE,AA,BB,CC,COORD,ICOORD,G,NF,INF)
C
C       THIS SUBROUTINE FORMS THE COORDINATES AND STEERING VECTOR
C       FOR 8-NODE BRICK ELEMENTS COUNTING X-Z PLANES IN Y-DIRECTION
C
      REAL COORD(ICOORD,*)
      INTEGER G(*),NF(INF,*),NUM(8)
      NUM(1)=(IQ-1)*(NXE+1)*(NZE+1)+IS*(NXE+1)+IP
      NUM(2)=NUM(1)-NXE-1
      NUM(3)=NUM(2)+1
      NUM(4)=NUM(1)+1
      NUM(5)=NUM(1)+(NXE+1)*(NZE+1)
      NUM(6)=NUM(5)-NXE-1
      NUM(7)=NUM(6)+1
      NUM(8)=NUM(5)+1
      INC=0
      DO 1 I=1,8
      DO 1 J=1,3
      INC=INC+1
    1 G(INC)=NF(NUM(I),J)
      COORD(1,1)=(IP-1)*AA
      COORD(2,1)=(IP-1)*AA
```

```
      COORD(5,1)=(IP-1)*AA
      COORD(6,1)=(IP-1)*AA
      COORD(3,1)=IP*AA
      COORD(4,1)=IP*AA
      COORD(7,1)=IP*AA
      COORD(8,1)=IP*AA
      COORD(1,2)=(IQ-1)*BB
      COORD(2,2)=(IQ-1)*BB
      COORD(3,2)=(IQ-1)*BB
      COORD(4,2)=(IQ-1)*BB
      COORD(5,2)=IQ*BB
      COORD(6,2)=IQ*BB
      COORD(7,2)=IQ*BB
      COORD(8,2)=IQ*BB
      COORD(1,3)=-IS*CC
      COORD(4,3)=-IS*CC
      COORD(5,3)=-IS*CC
      COORD(8,3)=-IS*CC
      COORD(2,3)=-(IS-1)*CC
      COORD(3,3)=-(IS-1)*CC
      COORD(6,3)=-(IS-1)*CC
      COORD(7,3)=-(IS-1)*CC
      RETURN
      END

      SUBROUTINE GEVUPV(IP,IQ,NXE,WIDTH,DEPTH,COORD,ICOORD,
     +                  COORDF,ICORDF,G,NF,INF)
C
C        THIS SUBROUTINE FORMS THE NODAL COORDINATES AND
C        STEERING VECTORFOR A RECTANGULAR MESH OF 4-NODE/8-NODE
C        QUADRILATERAL ELEMENTS NUMBERING IN THE X-DIRECTION
C        (U,P,V NAVIER STOKES FLOW)
C
      REAL COORD(ICOORD,*),COORDF(ICORDF,*),WIDTH(*),DEPTH(*)
      INTEGER NUM(8),G(*),NF(INF,*)
      NUM(1)=IQ*(3*NXE+2)+2*IP-1
      NUM(2)=IQ*(3*NXE+2)+IP-NXE-1
      NUM(3)=(IQ-1)*(3*NXE+2)+2*IP-1
      NUM(4)=NUM(3)+1
      NUM(5)=NUM(4)+1
      NUM(6)=NUM(2)+1
      NUM(7)=NUM(1)+2
      NUM(8)=NUM(1)+1
      INC=0
      DO 1 I=1,8
      INC=INC+1
    1 G(INC)=NF(NUM(I),1)
      DO 2 I=1,7,2
      INC=INC+1
    2 G(INC)=NF(NUM(I),2)
      DO 3 I=1,8
      INC=INC+1
    3 G(INC)=NF(NUM(I),3)
      COORD(1,1)=WIDTH(IP)
      COORD(2,1)=WIDTH(IP)
      COORD(3,1)=WIDTH(IP)
      COORDF(1,1)=WIDTH(IP)
      COORDF(2,1)=WIDTH(IP)
      COORD(5,1)=WIDTH(IP+1)
      COORD(6,1)=WIDTH(IP+1)
```

```
      COORD(7,1)=WIDTH(IP+1)
      COORDF(3,1)=WIDTH(IP+1)
      COORDF(4,1)=WIDTH(IP+1)
      COORD(4,1)=.5*(COORD(3,1)+COORD(5,1))
      COORD(8,1)=.5*(COORD(7,1)+COORD(1,1))
      COORD(1,2)=DEPTH(IQ+1)
      COORD(8,2)=DEPTH(IQ+1)
      COORD(7,2)=DEPTH(IQ+1)
      COORDF(1,2)=DEPTH(IQ+1)
      COORDF(4,2)=DEPTH(IQ+1)
      COORD(3,2)=DEPTH(IQ)
      COORD(4,2)=DEPTH(IQ)
      COORD(5,2)=DEPTH(IQ)
      COORDF(2,2)=DEPTH(IQ)
      COORDF(3,2)=DEPTH(IQ)
      COORD(2,2)=.5*(COORD(1,2)+COORD(3,2))
      COORD(6,2)=.5*(COORD(5,2)+COORD(7,2))
      RETURN
      END

      SUBROUTINE GEVUVP(IP,IQ,NXE,WIDTH,DEPTH,COORD,ICOORD,
     +                  COORDF,ICORDF,G,NF,INF)
C
C      THIS SUBROUTINE FORMS THE NODAL COORDINATES AND STEERING
C      VECTOR FOR A VARIABLE MESH OF 4-NODE/8-NODE
C      QUADRILATERAL ELEMENTS NUMBERING IN THE X-DIRECTION
C      (U,V,P  BIOT CONSOLIDATION)
C
      REAL COORD(ICOORD,*),COORDF(ICORDF,*),WIDTH(*),DEPTH(*)
      INTEGER NUM(8),G(*),NF(INF,*)
      NUM(1)=IQ*(3*NXE+2)+2*IP-1
      NUM(2)=IQ*(3*NXE+2)+IP-NXE-1
      NUM(3)=(IQ-1)*(3*NXE+2)+2*IP-1
      NUM(4)=NUM(3)+1
      NUM(5)=NUM(4)+1
      NUM(6)=NUM(2)+1
      NUM(7)=NUM(1)+2
      NUM(8)=NUM(1)+1
      INC=0
      DO 1 I=1,8
      DO 1 J=1,2
      INC=INC+1
    1 G(INC)=NF(NUM(I),J)
      DO 2 I=1,7,2
      INC=INC+1
    2 G(INC)=NF(NUM(I),3)
      COORD(1,1)=WIDTH(IP)
      COORD(2,1)=WIDTH(IP)
      COORD(3,1)=WIDTH(IP)
      COORDF(1,1)=WIDTH(IP)
      COORDF(2,1)=WIDTH(IP)
      COORD(5,1)=WIDTH(IP+1)
      COORD(6,1)=WIDTH(IP+1)
      COORD(7,1)=WIDTH(IP+1)
      COORDF(3,1)=WIDTH(IP+1)
      COORDF(4,1)=WIDTH(IP+1)
      COORD(4,1)=.5*(COORD(3,1)+COORD(5,1))
      COORD(8,1)=.5*(COORD(7,1)+COORD(1,1))
      COORD(1,2)=DEPTH(IQ+1)
      COORD(8,2)=DEPTH(IQ+1)
```

```
      COORD(7,2)=DEPTH(IQ+1)
      COORDF(1,2)=DEPTH(IQ+1)
      COORDF(4,2)=DEPTH(IQ+1)
      COORD(3,2)=DEPTH(IQ)
      COORD(4,2)=DEPTH(IQ)
      COORD(5,2)=DEPTH(IQ)
      COORDF(2,2)=DEPTH(IQ)
      COORDF(3,2)=DEPTH(IQ)
      COORD(2,2)=.5*(COORD(1,2)+COORD(3,2))
      COORD(6,2)=.5*(COORD(5,2)+COORD(7,2))
      RETURN
      END

      SUBROUTINE GEV15Y(IP,IQ,NYE,WID,DEP,COORD,ICOORD,G,NF,INF)
C
C      THIS SUBROUTINE FORMS THE COORDINATES AND STEERING VECTOR
C      FOR 15-NODE TRIANGLES COUNTING IN Y-DIRECTION
C
      REAL COORD(ICOORD,*),WID(*),DEP(*)
      INTEGER G(*),NF(INF,*),NUM(15)
      IF(MOD(IQ,2).EQ.0)GOTO 1
      FAC1=4*(2*NYE+1)*(IP-1)+2*IQ-1
      NUM(1)=FAC1
      NUM(2)=FAC1+1
      NUM(3)=FAC1+2
      NUM(4)=FAC1+3
      NUM(5)=FAC1+4
      NUM(6)=FAC1+2*NYE+4
      NUM(7)=FAC1+4*NYE+4
      NUM(8)=FAC1+6*NYE+4
      NUM(9)=FAC1+8*NYE+4
      NUM(10)=FAC1+6*NYE+3
      NUM(11)=FAC1+4*NYE+2
      NUM(12)=FAC1+2*NYE+1
      NUM(13)=FAC1+2*NYE+2
      NUM(14)=FAC1+2*NYE+3
      NUM(15)=FAC1+4*NYE+3
      COORD(1,1)=WID(IP)
      COORD(1,2)=DEP((IQ+1)/2)
      COORD(5,1)=WID(IP)
      COORD(5,2)=DEP((IQ+3)/2)
      COORD(9,1)=WID(IP+1)
      COORD(9,2)=DEP((IQ+1)/2)
      GOTO 2
    1 FAC2=4*(2*NYE+1)*(IP-1)+2*IQ+8*NYE+5
      NUM(1)=FAC2
      NUM(2)=FAC2-1
      NUM(3)=FAC2-2
      NUM(4)=FAC2-3
      NUM(5)=FAC2-4
      NUM(6)=FAC2-2*NYE-4
      NUM(7)=FAC2-4*NYE-4
      NUM(8)=FAC2-6*NYE-4
      NUM(9)=FAC2-8*NYE-4
      NUM(10)=FAC2-6*NYE-3
      NUM(11)=FAC2-4*NYE-2
      NUM(12)=FAC2-2*NYE-1
      NUM(13)=FAC2-2*NYE-2
      NUM(14)=FAC2-2*NYE-3
      NUM(15)=FAC2-4*NYE-3
```

```
      COORD(1,1)=WID(IP+1)
      COORD(1,2)=DEP((IQ+2)/2)
      COORD(5,1)=WID(IP+1)
      COORD(5,2)=DEP(IQ/2)
      COORD(9,1)=WID(IP)
      COORD(9,2)=DEP((IQ+2)/2)
    2 DO 3 I=1,2
      COORD(3,I)=.5*(COORD(1,I)+COORD(5,I))
      COORD(7,I)=.5*(COORD(5,I)+COORD(9,I))
      COORD(11,I)=.5*(COORD(9,I)+COORD(1,I))
      COORD(2,I)=.5*(COORD(1,I)+COORD(3,I))
      COORD(4,I)=.5*(COORD(3,I)+COORD(5,I))
      COORD(6,I)=.5*(COORD(5,I)+COORD(7,I))
      COORD(8,I)=.5*(COORD(7,I)+COORD(9,I))
      COORD(10,I)=.5*(COORD(9,I)+COORD(11,I))
      COORD(12,I)=.5*(COORD(11,I)+COORD(1,I))
      COORD(15,I)=.5*(COORD(7,I)+COORD(11,I))
      COORD(14,I)=.5*(COORD(3,I)+COORD(7,I))
      COORD(13,I)=.5*(COORD(2,I)+COORD(15,I))
    3 CONTINUE
      INC=0
      DO 4 I=1,15
      DO 4 J=1,2
      INC=INC+1
    4 G(INC)=NF(NUM(I),J)
      RETURN
      END

      SUBROUTINE GEV4X3(IP,IQ,NXE,WIDTH,DEPTH,COORD,ICOORD,G,NF,INF)
C
C      THIS SUBROUTINE FORMS THE NODAL COORDINATES AND STEERING
C      VECTOR FOR A VARIABLE MESH OF 4-NODE QUADRILATERAL ELEMENTS
C      NUMBERING IN THE X-DIRECTION
C      (U,V,P  BIOT CONSOLIDATION)
C
      REAL COORD(ICOORD,*),WIDTH(*),DEPTH(*)
      INTEGER NUM(4),G(*),NF(INF,*)
      NUM(1)=IQ*(NXE+1)+IP
      NUM(2)=(IQ-1)*(NXE+1)+IP
      NUM(3)=NUM(2)+1
      NUM(4)=NUM(1)+1
      NINC=0
      DO 1 I=1,4
      DO 1 J=1,2
      NINC=NINC+1
    1 G(NINC)=NF(NUM(I),J)
      DO 2 I=1,4
      NINC=NINC+1
    2 G(NINC)=NF(NUM(I),3)
      COORD(1,1)=WIDTH(IP)
      COORD(2,1)=WIDTH(IP)
      COORD(3,1)=WIDTH(IP+1)
      COORD(4,1)=WIDTH(IP+1)
      COORD(2,2)=DEPTH(IQ)
      COORD(3,2)=DEPTH(IQ)
      COORD(4,2)=DEPTH(IQ+1)
      COORD(1,2)=DEPTH(IQ+1)
      RETURN
      END
```

```
      SUBROUTINE GE203D(IP,IQ,IS,NXE,NZE,AA,BB,CC,COORD,ICOORD,G,NF,INF)
C
C      THIS SUBROUTINE FORMS THE STEERING VECTOR AND COORDINATES
C      FOR 20-NODE BRICK ELEMENTS COUNTING X-Z PLANES IN Y-DIRECTION
C
      REAL COORD(ICOORD,*)
      INTEGER G(*),NF(INF,*),NUM(20)
      FAC1=((2*NXE+1)*(NZE+1)+(2*NZE+1)*(NXE+1))*(IQ-1)
      FAC2=((2*NXE+1)*(NZE+1)+(2*NZE+1)*(NXE+1))*IQ
      NUM(1)=FAC1+(3*NXE+2)*IS+2*IP-1
      NUM(2)=FAC1+(3*NXE+2)*IS-NXE+IP-1
      NUM(3)=NUM(1)-3*NXE-2
      NUM(4)=NUM(3)+1
      NUM(5)=NUM(4)+1
      NUM(6)=NUM(2)+1
      NUM(7)=NUM(1)+2
      NUM(8)=NUM(1)+1
      NUM(9)=FAC2-(NXE+1)*(NZE+1)+(NXE+1)*IS+IP
      NUM(10)=NUM(9)-NXE-1
      NUM(11)=NUM(10)+1
      NUM(12)=NUM(9)+1
      NUM(13)=FAC2+(3*NXE+2)*IS+2*IP-1
      NUM(14)=FAC2+(3*NXE+2)*IS-NXE+IP-1
      NUM(15)=NUM(13)-3*NXE-2
      NUM(16)=NUM(15)+1
      NUM(17)=NUM(16)+1
      NUM(18)=NUM(14)+1
      NUM(19)=NUM(13)+2
      NUM(20)=NUM(13)+1
      INC=0
      DO 1 I=1,20
      DO 1 J=1,3
      INC=INC+1
    1 G(INC)=NF(NUM(I),J)
      COORD(1,1)=(IP-1)*AA
      COORD(2,1)=(IP-1)*AA
      COORD(3,1)=(IP-1)*AA
      COORD(9,1)=(IP-1)*AA
      COORD(10,1)=(IP-1)*AA
      COORD(13,1)=(IP-1)*AA
      COORD(14,1)=(IP-1)*AA
      COORD(15,1)=(IP-1)*AA
      COORD(5,1)=IP*AA
      COORD(6,1)=IP*AA
      COORD(7,1)=IP*AA
      COORD(11,1)=IP*AA
      COORD(12,1)=IP*AA
      COORD(17,1)=IP*AA
      COORD(18,1)=IP*AA
      COORD(19,1)=IP*AA
      COORD(4,1)=.5*(COORD(3,1)+COORD(5,1))
      COORD(8,1)=.5*(COORD(1,1)+COORD(7,1))
      COORD(16,1)=.5*(COORD(15,1)+COORD(17,1))
      COORD(20,1)=.5*(COORD(13,1)+COORD(19,1))
      COORD(1,2)=(IQ-1)*BB
      COORD(2,2)=(IQ-1)*BB
      COORD(3,2)=(IQ-1)*BB
      COORD(4,2)=(IQ-1)*BB
      COORD(5,2)=(IQ-1)*BB
      COORD(6,2)=(IQ-1)*BB
```

```
      COORD(7,2)=(IQ-1)*BB
      COORD(8,2)=(IQ-1)*BB
      COORD(13,2)=IQ*BB
      COORD(14,2)=IQ*BB
      COORD(15,2)=IQ*BB
      COORD(16,2)=IQ*BB
      COORD(17,2)=IQ*BB
      COORD(18,2)=IQ*BB
      COORD(19,2)=IQ*BB
      COORD(20,2)=IQ*BB
      COORD(9,2)=.5*(COORD(1,2)+COORD(13,2))
      COORD(10,2)=.5*(COORD(3,2)+COORD(15,2))
      COORD(11,2)=.5*(COORD(5,2)+COORD(17,2))
      COORD(12,2)=.5*(COORD(7,2)+COORD(19,2))
      COORD(1,3)=-IS*CC
      COORD(7,3)=-IS*CC
      COORD(8,3)=-IS*CC
      COORD(9,3)=-IS*CC
      COORD(12,3)=-IS*CC
      COORD(13,3)=-IS*CC
      COORD(19,3)=-IS*CC
      COORD(20,3)=-IS*CC
      COORD(3,3)=-(IS-1)*CC
      COORD(4,3)=-(IS-1)*CC
      COORD(5,3)=-(IS-1)*CC
      COORD(10,3)=-(IS-1)*CC
      COORD(11,3)=-(IS-1)*CC
      COORD(15,3)=-(IS-1)*CC
      COORD(16,3)=-(IS-1)*CC
      COORD(17,3)=-(IS-1)*CC
      COORD(2,3)=.5*(COORD(1,3)+COORD(3,3))
      COORD(6,3)=.5*(COORD(5,3)+COORD(7,3))
      COORD(14,3)=.5*(COORD(13,3)+COORD(15,3))
      COORD(18,3)=.5*(COORD(17,3)+COORD(19,3))
      RETURN
      END

      SUBROUTINE GSTRNG(IP,NODOF,NF,INF,G)
C
C      THIS SUBROUTINE SELECTS THE G-VECTOR FROM THE NF-DATA
C
      INTEGER NF(INF,*),G(*),NODE(2)
      NODE(1)=IP
      NODE(2)=IP+1
      L=0
      DO 1 I=1,2
      DO 1 J=1,NODOF
      L=L+1
    1 G(L)=NF(NODE(I),J)
      RETURN
      END

      SUBROUTINE SLOGEO(IP,IQ,NYE,TOP,BOT,DEPTH,COORD,ICOORD,G,NF,INF)
C
C      THIS SUBROUTINE FORMS THE COORDINATES AND STEERING VECTOR
C      FOR 8-NODE QUADRILATERALS IN A 'SLOPE' GEOMETRY
C      (NUMBERING IN THE Y-DIRECTION)
C
       REAL TOP(*),BOT(*),COORD(ICOORD,*),DEPTH(*)
       INTEGER G(*),NF(INF,*),NUM(8)
```

```
      NUM(1)=(IP-1)*(3*NYE+2)+2*IQ+1
      NUM(2)=NUM(1)-1
      NUM(3)=NUM(1)-2
      NUM(4)=(IP-1)*(3*NYE+2)+2*NYE+IQ+1
      NUM(5)=IP*(3*NYE+2)+2*IQ-1
      NUM(6)=NUM(5)+1
      NUM(7)=NUM(5)+2
      NUM(8)=NUM(4)+1
      INC=0
      DO 1 I=1,8
      DO 1 J=1,2
      INC=INC+1
    1 G(INC)=NF(NUM(I),J)
      FAC1=(BOT(IP)-TOP(IP))/NYE
      FAC2=(BOT(IP+1)-TOP(IP+1))/NYE
      COORD(1,1)=TOP(IP)+IQ*FAC1
      COORD(3,1)=TOP(IP)+(IQ-1)*FAC1
      COORD(5,1)=TOP(IP+1)+(IQ-1)*FAC2
      COORD(7,1)=TOP(IP+1)+IQ*FAC2
      COORD(2,1)=.5*(COORD(1,1)+COORD(3,1))
      COORD(6,1)=.5*(COORD(5,1)+COORD(7,1))
      COORD(4,1)=.5*(COORD(3,1)+COORD(5,1))
      COORD(8,1)=.5*(COORD(7,1)+COORD(1,1))
      COORD(1,2)=DEPTH(IQ+1)
      COORD(8,2)=DEPTH(IQ+1)
      COORD(7,2)=DEPTH(IQ+1)
      COORD(3,2)=DEPTH(IQ)
      COORD(4,2)=DEPTH(IQ)
      COORD(5,2)=DEPTH(IQ)
      COORD(2,2)=.5*(COORD(1,2)+COORD(3,2))
      COORD(6,2)=.5*(COORD(5,2)+COORD(7,2))
      RETURN
      END

      SUBROUTINE WELGEO(IP,IQ,NXE,NYE,WIDTH,SURF,COORD,ICOORD,G,NF,INF)
C
C      THIS SUBROUTINE FORMS THE COORDINATES AND STEERING VECTOR
C      FOR 4-NODE QUADS NUMBERING IN THE X-DIRECTION
C      LAPLACE'S EQUATION,VARIABLE MESH, 1-FREEDOM PER NODE
C
      REAL COORD(ICOORD,*),WIDTH(*),SURF(*)
      INTEGER NUM(4),G(*),NF(INF,*)
      NUM(1)=IQ*(NXE+1)+IP
      NUM(2)=(IQ-1)*(NXE+1)+IP
      NUM(3)=NUM(2)+1
      NUM(4)=NUM(1)+1
      DO 1 I=1,4
    1 G(I)=NF(NUM(I),1)
      BB=SURF(IP)/NYE
      BVAR=SURF(IP+1)/NYE
      COORD(1,1)=WIDTH(IP)
      COORD(2,1)=WIDTH(IP)
      COORD(3,1)=WIDTH(IP+1)
      COORD(4,1)=WIDTH(IP+1)
      COORD(1,2)=(NYE-IQ)*BB
      COORD(2,2)=(NYE-IQ+1)*BB
      COORD(3,2)=(NYE-IQ+1)*BVAR
      COORD(4,2)=(NYE-IQ)*BVAR
      RETURN
      END
```

APPENDIX 4
Alphabetic List of Building Block Subroutines

All routines used in programs, with the exception of the geometry and steering vector routines described in Appendix 3, are listed below. Routines marked with an asterisk are so-called 'black box' routines, mostly concerned with standard linear algebra. Although the coding of these routines is available to users, it is not believed to be of direct interest, and so is not reproduced in this book. However, the remaining routines, not asterisked, are concerned with specific operations in finite element analysis—description of shape functions etc.—and so the code is helpful to users in understanding the programs in the book and in creating new ones. Therefore, these routines are listed in full, in FORTRAN 77, in Appendix 5. In the parameter lists, underlined parameters are those returned by the routine as output. The other parameters are input. Routines in parenthesis are normally used in conjunction with the routines they follow.

Routine name	Parameters	Action of routine
AXIKM	KM, CSA, E, ELL	Forms element stiffness of an axially loaded line, **KM**, from its sectional area CSA, modulus E and length ELL.
BACSUB (BANRED)	BK, LOADS, N, IW	Completes Gaussian forward and backward substitutions on factors **BK**, having N rows and half-bandwidth IW. Right hand side **LOADS** overwritten by solution.
BANDRD* (BISECT)	N, IW, A, IA, D, E, E2	Jacobi tridiagonalisation of symmetric band matrix **A** (dimensions N*(IW + 1)).

		Resulting diagonal is **D** and off-diagonal is **E**. Vector **E2** used for working space.
BANMUL	KB, IKB, LOADS, ANS, N, IN	Multiplies symmetrical band matrix **KB** (dimensions N∗(IW + 1)) by vector **LOADS** with resulting vector **ANS**.
BANRED (BACSUB)	BK, N, IW	Factorises stiffness matrix **BK** with N rows and half-bandwidth IW, stored as a vector. Factors overwrite **BK**.
BANTML	KB, IKB, LOADS, ANS, N, IW	Multiplies unsymmetrical matrix **KB** (dimension N∗(2IW + 1)) by vector **LOADS** with resulting vector **ANS**.
BCMASS	MM, RHO, AREA, IP, COORD, ICOORD	Forms consistent mass matrix **MM** of an inclined two-dimensional beam-column element of sectional area AREA and mass RHO. Counter IP locates nodal coordinates from **COORD**
BEAMKM	KM, EI, ELL	Forms beam element stiffness matrix **KM** for a line element with flexural stiffness EI and length ELL.
BEAMKP	KP, ELL	Forms beam element axial force modification matrix **KP** for a line element of length ELL.
BISECT∗ (BANDRD)	N, ACHEPS, D, E, IFAIL	Finds N eigenvalues of symmetrical tridiagonal matrix. Diagonal in **D**, sub-diagonal in **E**. Machine tolerance ACHEPS. On exit, eigenvalues in **D**. Flag IFAIL set to 1 on entry should be unchanged on exit.

BMCOL2	KM, EA, EI, IP, COORD, ICOORD	Forms element stiffness matrix **KM** of a two-dimensional beam-column element of axial stiffness EA and flexural stiffness EI. Counter IP locates nodal coordinates from **COORD**.
BMCOL3	KM, EA, EIY, EIZ, GJ, IP, COORD, ICOORD	As BMCOL2 but three-dimensional version. Flexural stiffnesses EIY, EIZ and torsional stiffness GJ (see Chapter 4).
BNONAX	BEE, IBEE, DERIV, IDERIV, FUN, COORD, ICOORD, SUM, NOD, IFLAG, LTH	Forms B matrix **BEE** for axisymmetric elements with non-axisymmetric load. Shape functions **FUN**, derivatives **DERIV**, element nodal coordinates **COORD**, nodes NOD, radius SUM. For LTH harmonic, IFLAG is +1 for symmetry, −1 for antisymmetry.
CHECON	LOADS, OLDLDS, N, TOL, ICON	Convergence check for vectors **LOADS** and **OLDLDS** of length N. Flag ICON set to one if TOL achieved.
CHOBAC (CHOLIN)	KB, IKB, LOADS, N, IW	Combination of CHOBK1 and CHOBK2.
CHOBK1* (CHOLIN)	KB, IKB, LOADS, N, IW	Forward substitution using Choleski factors **KB** (dimensions N*(IW + 1)) on right hand side **LOADS**. Result overwrites **LOADS**.
CHOBK2 * (CHOLIN)	KB, IKB, LOADS, N, IW	Backward substitution using Choleski factors **KB** (dimensions N*(IW + 1)) on right hand side **LOADS**. Result overwrites **LOADS**.
CHOLIN	KB, IKB, N, IW	Factorises matrix **KB**

(CHOBAC, CHOBK1, CHOBK2)		(dimensions N∗(IW + 1)) by Choleski method, leaving result in **KB**.
COMBAC ∗ (COMRED)	BK, R, L, IW	Complex version of BACSUB.
COMRED ∗ (COMBAC)	BK, L, IW	Complex version of BANRED.
ECMAT	ECM, IECM, TN, ITN, NT, INT, FUN, NOD, NODOF	Forms element consistent mass matrix **ECM** (dimension (NOD∗NODOF)) from shape functions **FUN**. **TN** and **NT** are working space arrays.
EVECTS ∗ (TRIDIA)	N, ACHEPS, D, E, Z, IZ, IFAIL	Find eigenvalues and eigenvectors of symmetrical tridiagonal matrix of order N. Diagonal is **D** and sub-diagonal is **E**. Machine tolerance ACHEPS. Resulting eigenvalues are in **D** and eigenvectors in **Z**. Flag IFAIL set to 1 on entry should be unchanged on exit.
FDIAGV	BK, KM, IKM, G, KDIAG, IDOF	Forms global matrix (vector) **BK** from constituent diagonal matrices **KM**. Steering vector **G** has length IDOF and **KDIAG** holds lengths of rows for sparse storage.
FKDIAG	KDIAG, G, IDOF	For steering vector **G** of length IDOF, calculates the length of each row of a sparse global matrix and holds in **KDIAG**
FMBEAM	DER2, FUN, SAMP, ISAMP, ELL, I	Forms beam element shape functions **FUN** and their second derivatives **DER2** for an element of length ELL. Gaussian integration data in **SAMP** located by counter

FMBIGK	BIGK, IBIGK, KM, IKM, G, IDOF	Forms uncompacted (square) global stiffness matrix **BIGK** from constituent element matrices **KM**. Steering vector **G** of length IDOF.
FMBRAD	BEE, IBEE, DERIV, IDERIV, FUN, COORD, ICOORD, SUM, NOD	Forms strain-displacement matrix **BEE** for an element with NOD nodes in axisymmetric strain. Element shape functions in **FUN** and derivatives in **DERIV**. Nodal coordinates **COORD** and radius of current Gauss point is SUM.
FMDEPS	DEE, IDEE, E, V	Forms plane strain **DEE** matrix for elastic bodies with Young's modulus E and Poisson's ratio V.
FMDRAD	DEE, IDEE, E, V	Forms axisymmetric strain **DEE** matrix for elastic bodies with Young's modulus E and Poisson's ratio V.
FMDSIG	DEE, IDEE, E, V	Forms plane stress **DEE** matrix for elastic bodies with Young's modulus E and Poisson's ratio V.
FMKDKE	KM, IKM, KP, IKP, C, IC, KE, IKE, KD, IKD, IDOF, NOD, ITOT, THETA	Forms Biot matrices **KD** and **KE**, order ITOT, for coupled problems. **KM** is solid stiffness order IDOF, **KP** the fluid stiffness order NOD and **C** the coupling matrix. Time-stepping control via THETA.
FMLAG9	DER, IDER, FUN, SAMP, ISAMP, I, J	Forms nine-node Lagrangian quadrilateral shape functions **FUN** and derivatives **DER** in local coordinates. Gaussian integration

		data in SAMP, located by counters I, J.
FMLIN3	DER, IDER, FUN, SAMP, ISAMP, I, J, K	Forms eight-node brick shape functions **FUN** and derivatives **DER** in local coordinates. Gaussian integration data in **SAMP** located by counters I, J, K.
FMLUMP	DIAG, EMM, IEMM, G, IDOF	For lumped element matrices in **EMM**, forms global lumped vector **DIAG**. Steering vector **G** of length IDOF.
FMPLAT	FUN, D1X, D1Y, D2X, D2Y, D2XY, SAMP, ISAMP, AA, BB, I, J	Forms shape functions **FUN** and their first and second derivatives for a rectangular plate element dimensions AA*BB. Gaussian integration data in **SAMP** located by counters I, J.
FMQUAD	DER, IDER, FUN, SAMP, ISAMP, I, J	Forms eight-node quadrilateral shape functions **FUN** and derivatives **DER** in local coordinates. Gaussian integration data in **SAMP** located by counters I, J.
FMQUA3	DER, IDER, FUN, SAMP, ISAMP, I, J, K	Forms 20-node brick shape functions **FUN** and derivatives **DER** in local coordinates. Gaussian integration data in **SAMP** located by counters I, J, K.
FMTET4	DER, IDER, FUN, SAMP, ISAMP, I	Forms shape functions **FUN** and their derivatives **DER** for four-noded tetrahedral elements. Gaussian integration data in

		SAMP located by counter I.
FMTRI3	DER, IDER, FUN, SAMP, ISAMP, I	Forms shape functions **FUN** and derivatives **DER** for three-noded triangular elements. Gaussian integration data in **SAMP** located by counter I.
FMTRI6	DER, IDER, FUN, SAMP, ISAMP, I	As FMTRI3 but six-noded triangles.
FMTR15	DER, IDER, FUN, SAMP, ISAMP, I	As FMTRI3 but fifteen-noded triangles.
FORMB	BEE, IBEE, DERIV, IDERIV, NOD	Forms the strain-displacement matrix **BEE** for an element with NOD nodes in plane strain. Element shape function derivatives in **DERIV**.
FORMB3	BEE, IBEE, DERIV, IDERIV, NOD	Forms strain-displacement matrix **BEE** for an element with NOD nodes in three-dimensional strain. Element shape function derivatives in **DERIV**.
FORMD3	DEE, IDEE, E, V	Forms three-dimensional **DEE** matrix for isotropic elastic bodies with Young's modulus E and Poisson's ratio V.
FORMKB	KB, IKB, KM, IKM, G, IW, IDOF	Forms symmetrical global stiffness matrix KB as a rectangle (N∗(IW + 1)). Lower triangle stored. Constituent element matrices **KM** have IDOF freedoms and steering vector is **G**.
FORMKC	BK, KM, IKM, CM, ICM, G, N, IDOF	As FORMKV for complex matrices. Real part of element matrices **KM**, imaginary part **CM**.

FORMKU	KU, IKU, KM, IKM, G, IW, IDOF	As FORMKB but upper triangle stored as rectangle **KU**.
FORMKV	BK, KM, IKM, G, N, IDOF	Forms symmetrical stiffness matrix of order N (as vector **BK**) from constituent element matrices **KM** with IDOF freedoms. Steering vector is **G**. Upper triangle stored.
FORMLN	DER, IDER, FUN, SAMP, ISAMP, I, J	Forms four-node quadrilateral shape functions **FUN** and derivatives **DER** in local coordinates. Gaussian integration date in **SAMP** located by counters I, J.
FORMM	STRESS, M1, M2, M3	Forms **M1**, **M2**, **M3** matrices (see Appendix 2b) from stress components in **STRESS** for plane problems.
FORMM3	STRESS, M1, M2, M3	As FORMM but for three-dimensional problems.
FORMTB	KB, IKB, KM, IKM, G, IW, IDOF	Forms unsymmetrical banded global stiffness **KB** stored as rectangle of bandwidth 2*IW + 1 from element matrices **KM**. Steering vector **G** has length IDOF.
FORMXI	FSOIL, FMAX, RF, RM, RO, XI	Forms part of soil stiffness XI (see Chapter 12). Soil forces FSOIL, FMAX. Hyperbolic parameter RF. Pile and soil radii RO and RM.
FRMUPV	KE, IKE, C11, IC11, C12, IC12, C21, IC21, C23, IC23, C32, IC32, NOD, NODF, ITOT	Forms coupled Navier–Stokes element matrix **KE** from submatrices **CIJ** (see 3.104). Velocity nodes NOD and pressure nodes NODF.

		Total size of **KE** is ITOT.
FSPARV	BK, KM, IKM, G, KDIAG, IDOF	Forms sparse global stiffness matrix (vector) **BK** from constituent **KM** element matrices. Steering vector **G** of length IDOF and number of terms in each row held in **KDIAG**.
GAUSBA (SOLVBA)	PB, IPB, WORK, IWORK, N, IW	Gaussian reduction with partial pivoting of symmetrical band matrix **PB** stored as a rectangle (N∗(2IW + 1)). Working space **WORK** has dimensions (IW + 1)∗N and message printed if singular.
GAUSS	SAMP, ISAMP, NGP	Returns weights and abscissae **SAMP** of sampling points in Gauss–Legendre quadrature for NGP integrating points 1 to 7.
GCOORD	FUN, COORD, ICOORD, NOD, IT, GC	Forms IT Cartesian coordinates of the current point held in **GC**. Shape functions **FUN**, nodal coordinates **COORD**, nodes NOD.
HING2	IP, HOLDR, COORD, IPROP, REACT, ACTION, BMP	Forms end forces and moments **ACTION** for a two-dimensional member when a joint has gone plastic. Current load level **HOLDR**, increment **REACT**, plastic moment BMP. Nodal coordinates in **COORD** located by counter IP.
INVAR	STRESS, SIGM, DSBAR, THETA	For two-dimensional stress components in **STRESS**, finds first invariant

		SIGM, second invariant DSBAR and Lode angle THETA (radians).
INVAR3	STRESS, SIGM DSBAR, THETA	As INVAR but for three-dimensional stress components.
KVDET*	KV, N, IW, DET, KSC	Find determinant DET of global stiffness matrix **KV** stored as a vector of the upper triangle (N*(IW + 1)). KSC monitors negative diagonal values in the factors of **KV**.
LANCZ1* (LANCZ2)	N, EL, ER, ACC, LEIG, LX, LALFA, LP, ITAPE, IFLAG, U, V, EIG, JEIG, NEIG, X, DEL, NU, ALFA, BETA	Lanczos method for eigenvalue problems. See Chapter 10.
LANCZ2* (LANCZ1)	N, LALFA, LP, ITAPE, EIG, JEIG, NEIG, ALFA, BETA, LY, LZ, JFLAG, Y, W, Z	Lanczos method for eigenvalue problems. See Chapter 10.
LBBT*	L, BT, IBT, IW, N	Solves $\mathbf{LA} = \mathbf{B}^{\mathrm{T}}$ where **L** is symmetrical band matrix (lower triangle stored as rectangle N*(IW + 1)). **A** and **B** are N*N and **A** overwrites $\mathbf{B}^{\mathrm{T}}$. Working dimension of **A**, **L**, $\mathbf{B}^{\mathrm{T}}$ is IBT $\geqslant$ N.
LBKBAN*	L, B, K, IK, IW, N	Solves $\mathbf{LB} = \mathbf{K}$ where **L** and **K** are symmetrical band matrices (lower triangles stored as rectangular arrays (N*(IW + 1)). **B** is N*N. Working space via first dimension of **L**, **B** and **K** is IK $\geqslant$ N.
LINMLS	BP, DISPS, LOADS, N, KDIAG	Profile storage version of LINMUL. Length of rows in profile in **KDIAG**.

LINMUL	BK, DISPS, LOADS, N, IW	Symmetrical band matrix **BK** multiplied by vector **DISPS** to give vector **LOADS**. **BK** stored as vector of upper triangle which has N rows and half-bandwidth IW.
LOC2F	LOCAL, GLOBAL, IP, COORD, ICOORD	Transforms global two-dimensional rigid frame end reactions and moments **GLOBAL** to local ones **LOCAL**. Counter IP locates nodal coordinates in **COORD**.
LOC3F	LOCAL, GLOBAL, IP, COORD, ICOORD	Three-dimensional version of LOC2F.
LOC2T	AXIAL, GLOBAL, IP, COORD, ICOORD	Retrieves axial force AXIAL in a two-dimensional pin-jointed element from end reactions in **GLOBAL**. Counter IP locates nodal coordinates in **COORD**.
LOC3T	AXIAL, GLOBAL, IP, COORD, ICOORD	As LOC2T but for three-dimensional pin-jointed line element.
MATADD	A, IA, B, IB, M, N	$\mathbf{A} = \mathbf{A} + \mathbf{B}$ where **A** and **B** are $(M*N)$. Working space restrictions $IA = IB \geqslant M$.
MATCOP	A, IA, B, IB, M, N	**B** is a copy of **A** where both are $M*N$ matrices. Working space restriction $IA = IB \geqslant M$.
MATINV	A, IA, N	Replaces $N*N$ matrix **A** by its inverse. Only use for small N. Working space restriction $IA \geqslant N$.
MATMUL	A, IA, B, IB, C, IC, L, M, N	$\mathbf{C} = \mathbf{A}*\mathbf{B}$ where $\mathbf{A}(L*M)$, $\mathbf{B}(M*N)$ and $\mathbf{C}(L*N)$. Working restrictions $IA = IC \geqslant L$, $IB \geqslant M$.
MATRAN	A, IA, B, IB, M, N	$\mathbf{A} = \mathbf{B}^T$ where **B** is $M*N$. Working restrictions $IB \geqslant M$, $IA \geqslant N$

MOCOPL	PHI, PSI, E, V, STRESS, PL	Forms plastic stress–strain matrix **PL** from the stresses **STRESS**, angle of friction PHI, dilation angle PSI and elastic stiffness E, V for a Mohr–Coulomb material.
MOCOUF	PHI, C, SIGM, DSBAR, THETA, F	Mohr–Coulomb yield function F from invariants SIGM, DSBAR. Lode angle THETA (radians), angle of friction PHI (degrees) and cohesion C.
MOCOUQ	PSI, DSBAR, THETA, DQ1, DQ2, DQ3	Derivatives of Mohr–Coulomb function DQ1, DQ2, DQ3 (Appendix 2b) for angle of dilation PSI, invariant DSBAR and Lode angle THETA.
MSMULT	A, IA, C, M, N	Multiplies M∗N matrix **A** by scalar C, returning result to **A**. Working restriction IA ≥ M.
MVMULT	M, IM, V, K, L, Y	Multiplies K∗L matrix **M** by L vector **V** with result **Y** of length K. Working space IM ≥ K.
NULL	A, IA, M, N	Nulls M∗N array **A**. Working restriction IA ≥ M.
NULL3	A, IA1, IA2, L, M, N	Nulls L∗M∗N array **A**. Working restriction IA1 ≥ L, IA2 ≥ M.
NULVEC	VEC, N	Nulls a vector **VEC** of length N.
NUMIN3	SAMP, ISAMP, WT, NIP	Sampling coordinates **SAMP** and weights **WT** for integration over tetrahedra using NIP integrating points.
NUMINT	S, IS, WT, NIP	Sampling coordinates **S** and weights **WT** for integration over

		triangles using NIP integrating points.
PINJ2	KM, EA, IP, COORD, ICOORD	Forms element stiffness matrix **KM** for a two-dimensional pin-jointed line of axial stiffness EA. Coordinates **COORD** located by counter IP.
PINJ3	KM, EA, IP, COORD, ICOORD	As PINJ2 but three-dimensional pin-jointed element.
PRINTA*	A, IA, M, N	Prints M*N array **A**. Working restriction IA ⩾ M. Tape 6 assumed.
PRINTV*	VEC, N	Prints a vector **VEC** of length N. Tape 6 assumed.
PRNCPL	SIGM, DSBAR, THETA, SIG1, SIG2, SIG3	Forms the principal stresses SIG1, SIG2, SIG3 from first invariant SIGM, second invariant DSBAR and Lode angle THETA.
READNF	NF, INF, NN, NODOF, NR	Creates node freedom array **NF** by reading data for a mesh with NN nodes, NODOF freedoms per node and NR restrained nodes. Tape 5 assumed.
SOLVBA	PB, IPB, COPY, ICOPY, ANS, N, IW	Forward and backward substitution on matrix **PB** factorised by GAUSBA to yield result **ANS**. **PB** has dimensions N*(2*IW + 1) and **COPY** is a copy of **WORK**, dimensions (IW + 1)*N.
SPABAC (SPARIN)	A, B, N, KDIAG	Choleski forward and backward substitution on **A** created by SPARIN to yield solution vector **B**.
SPARIN (SPABAC)	A, N, KDIAG	Choleski factorisation of **A** stored in rows by profile.

		There are N rows whose length is held in **KDIAG**.
STAB2D	KM, EA, EI, IP, COORD, ICOORD, PAX	Forms stiffness matrix **KM** of inclined two-dimensional beam-column with axial stiffness EA, flexural stiffness EI and axial force PAX using 'stability functions'. Nodal coordinates **COORD** located by counter IP.
TREEX3	JAC, IJAC, JAC1, IJAC1, DET	Forms inverse **JAC1** of 3 × 3 matrix **JAC** and its determinant DET.
TRIDIA*	N, ATOL, A, D, E, Z, IZ	Householder reduction of symmetrical N*N array **A** to tridiagonal form. Diagonal in **D**. Sub-diagonal in **E**, rotations in **Z**. Iteration tolerance ATOL.
TWOBY2	JAC, IJAC, JAC1, IJAC1, DET	Forms the inverse **JAC1** of a 2 × 2 matrix **JAC** and its determinant DET.
VECADD	A, B, C, N	Forms vector **C** = **A** + **B** all of length N.
VECCOP	A, B, N	Makes a copy **B** of vector **A** of length N.
VMPL	E, V, STRESS, PL	Forms plastic matrix **PL** for a von Mises material from stress vector **STRESS** and elastic moduli E and V.
VOL2D	BEE, IBEE, VOL, NOD	Forms vector **VOL** for matrix **BEE** enabling volumetric strain in two-dimensional element with NOD nodes to be found.
VVMULT	V1, V2, PROD, IPROD, M, N	Vector product **PROD** = **V1*V2** where **V1** has length M and **V2** has length N. Working restriction IPROD ⩾ M.

APPENDIX 5
Listings of Special Purpose Routines

```
      SUBROUTINE AXIKM(KM,CSA,E,ELL)
C
C      THIS SUBROUTINE FORMS THE STIFFNESS MATRIX FOR AN
C      AXIALLY LOADED LINE ELEMENT
C
      REAL KM(2,2)
      KM(1,1)=CSA*E/ELL
      KM(2,2)=KM(1,1)
      KM(1,2)=-KM(1,1)
      KM(2,1)=KM(1,2)
      RETURN
      END
```

```
      SUBROUTINE BACSUB(BK,LOADS,N,IW)
C
C      THIS SUBROUTINE PERFORMS THE GAUSSIAN BACK-SUBSTITUTION
C
      REAL BK(*),LOADS(*)
      LOADS(1)=LOADS(1)/BK(1)
      DO 1 I=2,N
      SUM=LOADS(I)
      I1=I-1
      NKB=I-IW
      IF(NKB)2,2,3
    2 NKB=1
    3 DO 4 K=NKB,I1
      JN=(I-K)*N+K
      SUM=SUM-BK(JN)*LOADS(K)
    4 CONTINUE
      LOADS(I)=SUM/BK(I)
    1 CONTINUE
      DO 5 JJ=2,N
      I=N-JJ+1
      SUM=0.
      I1=I+1
      NKB=I+IW
      IF(NKB-N)7,7,6
    6 NKB=N
    7 DO 8 K=I1,NKB
      JN=(K-I)*N+I
    8 SUM=SUM+BK(JN)*LOADS(K)
      LOADS(I)=LOADS(I)-SUM/BK(I)
    5 CONTINUE
      RETURN
      END
```

```
      SUBROUTINE BANMUL(KB,IKB,LOADS,ANS,N,IW)
C
C      THIS SUBROUTINE MULTIPLIES A MATRIX BY A VECTOR
C      THE MATRIX IS SYMMETRICAL WITH ITS LOWER TRIANGLE
C      STORED AS A RECTANGLE
C
      REAL KB(IKB,*),LOADS(*),ANS(*)
      DO 1 I=1,N
      X=0.
      J=IW+1
    2 IF(I+J.LE.IW+1)GOTO 3
      X=X+KB(I,J)*LOADS(I+J-IW-1)
    3 J=J-1
      IF(J.NE.0)GOTO 2
      J=IW
    6 IF(I-J.GE.N-IW)GOTO 7
      X=X+KB(I-J+IW+1,J)*LOADS(I-J+IW+1)
    7 J=J-1
      IF(J.NE.0)GOTO 6
      ANS(I)=X
    1 CONTINUE
      RETURN
      END

      SUBROUTINE BANRED(BK,N,IW)
C
C      THIS SUBROUTINE PERFORMS GAUSSIAN REDUCTION OF
C      THE STIFFNESS MATRIX STORED AS A VECTOR BK(N*(IW+1))
C
      REAL BK(*)
      DO 1 I=2,N
      IL1=I-1
      KBL=IL1+IW+1
      IF(KBL-N)3,3,2
    2 KBL=N
    3 DO 1 J=I,KBL
      IJ=(J-I)*N+I
      SUM=BK(IJ)
      NKB=J-IW
      IF(NKB)4,4,5
    4 NKB=1
    5 IF(NKB-IL1)6,6,8
    6 DO 7 M=NKB,IL1
      NI=(I-M)*N+M
      NJ=(J-M)*N+M
    7 SUM=SUM-BK(NI)*BK(NJ)/BK(M)
    8 BK(IJ)=SUM
    1 CONTINUE
      RETURN
      END
```

```
      SUBROUTINE BANTML(KB,IKB,LOADS,ANS,N,IW)
C
C      THIS SUBROUTINE MULTIPLIES AN UNSYMMETRIC BANDED MATRIX
C      'PB' BY THE VECTOR 'LOADS'.
C
      REAL KB(IKB,*),LOADS(*),ANS(*)
      DO 1 I=1,N
      X=0.
      K=IW+2
      L=IW+IW+1
      DO 2 J=1,L
      K=K-1
      IF(I-K+1.GT.N)GOTO 2
      IF(I-K+1.LT.1)GOTO 2
      X=X+KB(I,J)*LOADS(I-K+1)
    2 CONTINUE
      ANS(I)=X
    1 CONTINUE
      RETURN
      END

      SUBROUTINE BCMASS(MM,RHO,AREA,IP,COORD,ICOORD)
C
C      THIS SUBROUTINE FORMS THE CONSISTENT MASS MATRIX OF
C      AN INCLINED 2-D BEAM-COLUMN ELEMENT
C
      REAL MM(6,6),COORD(ICOORD,*)
      X1=COORD(IP,1)
      Y1=COORD(IP,2)
      X2=COORD(IP,3)
      Y2=COORD(IP,4)
      ELL=SQRT((Y2-Y1)**2+(X2-X1)**2)
      C=(X2-X1)/ELL
      S=(Y2-Y1)/ELL
      MM(1,1)=140.*C*C+156.*S*S
      MM(4,4)=MM(1,1)
      MM(2,2)=156.*C*C+140.*S*S
      MM(5,5)=MM(2,2)
      MM(3,3)=4.*ELL*ELL
      MM(6,6)=MM(3,3)
      MM(1,2)=-16.*C*S
      MM(4,5)=MM(1,2)
      MM(1,5)=-MM(1,2)
      MM(2,4)=-MM(1,2)
      MM(1,3)=-22.*ELL*S
      MM(4,6)=-MM(1,3)
      MM(2,3)=22.*ELL*C
      MM(5,6)=-MM(2,3)
      MM(1,4)=70.*C*C+54.*S*S
      MM(1,6)=13.*ELL*S
      MM(3,4)=-MM(1,6)
      MM(2,5)=54.*C*C+70.*S*S
      MM(2,6)=-13.*ELL*C
      MM(3,5)=-MM(2,6)
      MM(3,6)=-3.*ELL*ELL
      FAC=RHO*AREA*ELL/420.
      DO 1 I=1,6
      DO 1 J=I,6
      MM(I,J)=MM(I,J)*FAC
    1 MM(J,I)=MM(I,J)
      RETURN
      END
```

```
      SUBROUTINE BEAMKM(KM,EI,ELL)
C
C      THIS SUBROUTINE FORMS THE STIFFNESS MATRIX OF A
C      HORIZONTAL BEAM ELEMENT(BENDING ONLY)
C
      REAL KM(4,4)
      KM(1,1)=12.*EI/(ELL*ELL*ELL)
      KM(3,3)=KM(1,1)
      KM(1,2)=6.*EI/(ELL*ELL)
      KM(2,1)=KM(1,2)
      KM(1,4)=KM(1,2)
      KM(4,1)=KM(1,4)
      KM(1,3)=-KM(1,1)
      KM(3,1)=KM(1,3)
      KM(3,4)=-KM(1,2)
      KM(4,3)=KM(3,4)
      KM(2,3)=KM(3,4)
      KM(3,2)=KM(2,3)
      KM(2,2)=4.*EI/ELL
      KM(4,4)=KM(2,2)
      KM(2,4)=2.*EI/ELL
      KM(4,2)=KM(2,4)
      RETURN
      END

      SUBROUTINE BEAMKP(KP,ELL)
C
C      THIS SUBROUTINE FORMS THE TERMS OF THE BEAM STIFFNESS
C      MATRIX DUE TO AXIAL LOADING
C
      REAL KP(4,4)
      KP(1,1)=2.0*ELL/15.0
      KP(2,2)=2.0*ELL/15.0
      KP(2,1)=-ELL/30.0
      KP(1,2)=-ELL/30.0
      KP(3,1)=0.1
      KP(1,3)=0.1
      KP(3,2)=0.1
      KP(2,3)=0.1
      KP(4,1)=-0.1
      KP(1,4)=-0.1
      KP(4,2)=-0.1
      KP(2,4)=-0.1
      KP(3,3)=1.2/ELL
      KP(4,4)=1.2/ELL
      KP(4,3)=-KP(3,3)
      KP(3,4)=-KP(3,3)
      RETURN
      END

      SUBROUTINE BMCOL2(KM,EA,EI,IP,COORD,ICOORD)
C
C      THIS SUBROUTINE FORMS THE STIFFNESS MATRIX OF AN
C      INCLINED 2-D BEAM-COLUMN ELEMENT
C
      REAL KM(6,6),COORD(ICOORD,*)
      X1=COORD(IP,1)
      Y1=COORD(IP,2)
      X2=COORD(IP,3)
```

```
      Y2=COORD(IP,4)
      ELL=SQRT((Y2-Y1)**2+(X2-X1)**2)
      C=(X2-X1)/ELL
      S=(Y2-Y1)/ELL
      E1=EA/ELL
      E2=12.*EI/(ELL*ELL*ELL)
      E3=EI/ELL
      E4=6.*EI/(ELL*ELL)
      KM(1,1)=C*C*E1+S*S*E2
      KM(4,4)=KM(1,1)
      KM(1,2)=S*C*(E1-E2)
      KM(2,1)=KM(1,2)
      KM(4,5)=KM(1,2)
      KM(5,4)=KM(4,5)
      KM(1,3)=-S*E4
      KM(3,1)=KM(1,3)
      KM(1,6)=KM(1,3)
      KM(6,1)=KM(1,6)
      KM(3,4)=S*E4
      KM(4,3)=KM(3,4)
      KM(4,6)=KM(3,4)
      KM(6,4)=KM(4,6)
      KM(1,4)=-KM(1,1)
      KM(4,1)=KM(1,4)
      KM(1,5)=S*C*(-E1+E2)
      KM(5,1)=KM(1,5)
      KM(2,4)=KM(1,5)
      KM(4,2)=KM(2,4)
      KM(2,2)=S*S*E1+C*C*E2
      KM(5,5)=KM(2,2)
      KM(2,5)=-KM(2,2)
      KM(5,2)=KM(2,5)
      KM(2,3)=C*E4
      KM(3,2)=KM(2,3)
      KM(2,6)=KM(2,3)
      KM(6,2)=KM(2,6)
      KM(3,3)=4.*E3
      KM(6,6)=KM(3,3)
      KM(3,5)=-C*E4
      KM(5,3)=KM(3,5)
      KM(5,6)=KM(3,5)
      KM(6,5)=KM(5,6)
      KM(3,6)=2.*E3
      KM(6,3)=KM(3,6)
      RETURN
      END

      SUBROUTINE BMCOL3(KM,EA,EIY,EIZ,GJ,IP,COORD,ICOORD)
C
C      THIS SUBROUTINE FORMS THE STIFFNESS MATRIX OF A
C      GENERAL 3-D BEAM-COLUMN ELEMENT
C
      REAL KM(12,12),COORD(ICOORD,*),T(12,12),TT(12,12),RO(3,3),C(12,12)
      PI=4.*ATAN(1.)
      GAMA=COORD(IP,7)*PI/180.
      X1=COORD(IP,1)
      Y1=COORD(IP,2)
      Z1=COORD(IP,3)
      X2=COORD(IP,4)
```

```
      Y2=COORD(IP,5)
      Z2=COORD(IP,6)
      XL=X2-X1
      YL=Y2-Y1
      ZL=Z2-Z1
      ELL=SQRT(XL*XL+YL*YL+ZL*ZL)
      CG=COS(GAMA)
      SG=SIN(GAMA)
      DEN=ELL*SQRT(XL*XL+ZL*ZL)
      DO 1 I=1,12
      DO 1 J=1,12
      KM(I,J)=0.
      T(I,J)=0.
    1 TT(I,J)=0.
      A1=EA/ELL
      A2=12.*EIZ/(ELL*ELL*ELL)
      A3=12.*EIY/(ELL*ELL*ELL)
      A4=6.*EIZ/(ELL*ELL)
      A5=6.*EIY/(ELL*ELL)
      A6=4.*EIZ/ELL
      A7=4.*EIY/ELL
      A8=GJ/ELL
      KM(1,1)=A1
      KM(7,7)=A1
      KM(1,7)=-A1
      KM(7,1)=-A1
      KM(2,2)=A2
      KM(8,8)=A2
      KM(2,8)=-A2
      KM(8,2)=-A2
      KM(3,3)=A3
      KM(9,9)=A3
      KM(3,9)=-A3
      KM(9,3)=-A3
      KM(4,4)=A8
      KM(10,10)=A8
      KM(4,10)=-A8
      KM(10,4)=-A8
      KM(5,5)=A7
      KM(11,11)=A7
      KM(5,11)=.5*A7
      KM(11,5)=.5*A7
      KM(6,6)=A6
      KM(12,12)=A6
      KM(6,12)=.5*A6
      KM(12,6)=.5*A6
      KM(2,6)=A4
      KM(6,2)=A4
      KM(2,12)=A4
      KM(12,2)=A4
      KM(6,8)=-A4
      KM(8,6)=-A4
      KM(8,12)=-A4
      KM(12,8)=-A4
      KM(5,9)=A5
      KM(9,5)=A5
      KM(9,11)=A5
      KM(11,9)=A5
      KM(3,5)=-A5
      KM(5,3)=-A5
```

```
      KM(3,11)=-A5
      KM(11,3)=-A5
      IF(DEN.EQ.0.)GOTO 50
      RO(1,1)=XL/ELL
      RO(1,2)=YL/ELL
      RO(1,3)=ZL/ELL
      RO(2,1)=(-XL*YL*CG-ELL*ZL*SG)/DEN
      RO(2,2)=DEN*CG/(ELL*ELL)
      RO(2,3)=(-YL*ZL*CG+ELL*XL*SG)/DEN
      RO(3,1)=(XL*YL*SG-ELL*ZL*CG)/DEN
      RO(3,2)=-DEN*SG/(ELL*ELL)
      RO(3,3)=(YL*ZL*SG+ELL*XL*CG)/DEN
      GOTO 60
   50 RO(1,1)=0.
      RO(1,3)=0.
      RO(2,2)=0.
      RO(3,2)=0.
      RO(1,2)=1.
      RO(2,1)=-CG
      RO(3,3)=CG
      RO(2,3)=SG
      RO(3,1)=SG
   60 CONTINUE
      DO 2 I=1,3
      DO 2 J=1,3
      X=RO(I,J)
      DO 2 K=0,9,3
      T(I+K,J+K)=X
    2 TT(J+K,I+K)=X
      DO 3 I=1,12
      DO 3 J=1,12
      SUM=0.
      DO 4 K=1,12
    4 SUM=SUM+KM(I,K)*T(K,J)
    3 C(I,J)=SUM
      DO 5 I=1,12
      DO 5 J=1,12
      SUM=0.
      DO 6 K=1,12
    6 SUM=SUM+TT(I,K)*C(K,J)
    5 KM(I,J)=SUM
      RETURN
      END

      SUBROUTINE BNONAX(BEE,IBEE,DERIV,IDERIV,FUN,COORD,ICOORD,
     +                  SUM,NOD,IFLAG,LTH)
C
C      THIS SUBROUTINE FORMS THE STRAIN-DISPLACEMENT MATRIX FOR
C      AXISYMMETRIC SOLIDS SUBJECTED TO NON-AXISYMMETRIC LOADING
C
      REAL BEE(IBEE,*),DERIV(IDERIV,*),FUN(*),COORD(ICOORD,*)
      SUM=0.
      DO 1 K=1,NOD
    1 SUM=SUM+FUN(K)*COORD(K,1)
      DO 2 M=1,NOD
      N=3*M
      K=N-1
      L=K-1
      BEE(1,L)=DERIV(1,M)
```

```
      BEE(2,K)=DERIV(2,M)
      BEE(3,L)=FUN(M)/SUM
      BEE(3,N)=IFLAG*LTH*BEE(3,L)
      BEE(4,L)=DERIV(2,M)
      BEE(4,K)=DERIV(1,M)
      BEE(5,K)=-IFLAG*LTH*FUN(M)/SUM
      BEE(5,N)=DERIV(2,M)
      BEE(6,L)=BEE(5,K)
    2 BEE(6,N)=DERIV(1,M)-FUN(M)/SUM
      RETURN
      END

      SUBROUTINE CHECON(LOADS,OLDLDS,N,TOL,ICON)
C
C      THIS SUBROUTINE SETS ICON TO ZERO IF THE RELATIVE CHANGE
C      IN VECTORS 'LOADS' AND 'OLDLDS' IS GREATER THAN 'TOL'
C
      REAL LOADS(*),OLDLDS(*)
      ICON=1
      BIG=0.
      DO 1 I=1,N
    1 IF(ABS(LOADS(I)).GT.BIG)BIG=ABS(LOADS(I))
      DO 2 I=1,N
      IF(ABS(LOADS(I)-OLDLDS(I))/BIG.GT.TOL)ICON=0
    2 OLDLDS(I)=LOADS(I)
      RETURN
      END

      SUBROUTINE CHOBAC(KB,IKB,LOADS,N,IW)
C
C      THIS SUBROUTINE PERFORMS THE CHOLESKI BACK-SUBSTITUTION
C
      REAL KB(IKB,*),LOADS(*)
      LOADS(1)=LOADS(1)/KB(1,IW+1)
      DO 1 I=2,N
      X=0.0
      K=1
      IF(I.LE.IW+1)K=IW-I+2
      DO 2 J=K,IW
    2 X=X+KB(I,J)*LOADS(I+J-IW-1)
    1 LOADS(I)=(LOADS(I)-X)/KB(I,IW+1)
      LOADS(N)=LOADS(N)/KB(N,IW+1)
      I=N-1
    3 X=0.0
      L=I+IW
      IF(I.GT.N-IW)L=N
      M=I+1
      DO 4 J=M,L
    4 X=X+KB(J,IW+I-J+1)*LOADS(J)
      LOADS(I)=(LOADS(I)-X)/KB(I,IW+1)
      I=I-1
      IF(I)5,5,3
    5 CONTINUE
      RETURN
      END
```

```
      SUBROUTINE CHOLIN(KB,IKB,N,IW)
C
C      THIS SUBROUTINE PERFORMS CHOLESKI REDUCTION OF
C      THE STIFFNESS MATRIX STORED AS AN ARRAY BK(N,IW+1)
C
      REAL KB(IKB,*)
      DO 1 I=1,N
      X=0.
      DO 2 J=1,IW
    2 X=X+KB(I,J)**2
      KB(I,IW+1)=SQRT(KB(I,IW+1)-X)
      DO 3 K=1,IW
      X=0.
      IF(I+K.GT.N)GOTO 3
    6 IF(K.EQ.IW)GOTO 4
    7 L=IW-K
    5 X=X+KB(I+K,L)*KB(I,L+K)
      L=L-1
      IF(L.NE.0)GOTO 5
    4 IA=I+K
      IB=IW-K+1
      KB(IA,IB)=(KB(IA,IB)-X)/KB(I,IW+1)
    3 CONTINUE
    1 CONTINUE
      RETURN
      END

      SUBROUTINE ECMAT(ECM,IECM,TN,ITN,NT,INT,FUN,NOD,NODOF)
C
C      THIS SUBROUTINE FORMS THE CONSISTENT MASS MATRIX
C
      REAL ECM(IECM,*),TN(ITN,*),NT(INT,*),FUN(*)
      IDOF=NOD*NODOF
      DO 1 I=1,IDOF
      DO 1 J=1,NODOF
      NT(I,J)=0.
    1 TN(J,I)=NT(I,J)
      DO 2 I=1,NOD
      DO 2 J=1,NODOF
      NT((I-1)*NODOF+J,J)=FUN(I)
    2 TN(J,(I-1)*NODOF+J)=FUN(I)
      DO 3 I=1,IDOF
      DO 3 J=1,IDOF
      X=0.0
      DO 4 K=1,NODOF
    4 X=X+NT(I,K)*TN(K,J)
      ECM(I,J)=X
    3 CONTINUE
      RETURN
      END

      SUBROUTINE FDIAGV(BK,KM,IKM,G,KDIAG,IDOF)
C
C      THIS SUBROUTINE ASSEMBLES A DIAGONAL ELEMENT MATRIX
C      INTO THE GLOBAL SYSTEM
C
      REAL BK(*),KM(IKM,*)
      INTEGER G(*),KDIAG(*)
      DO 1 I=1,IDOF
```

```
      J=G(I)
      IF(J.EQ.0)GO  TO 1
      K=KDIAG(J)
      BK(K)=BK(K)+KM(I,I)
    1 CONTINUE
      RETURN
      END

      SUBROUTINE FKDIAG(KDIAG,G,IDOF)
C
C      THIS SUBROUTINE FINDS THE MAXIMUM BANDWIDTH
C      FOR EACH FREEDOM
C
      INTEGER KDIAG(*),G(*)
      DO 1 I=1,IDOF
      IWP1=1
      IF(G(I).EQ.0)GOTO 1
      DO 2 J=1,IDOF
      IF(G(J).EQ.0)GOTO 2
      IM=G(I)-G(J)+1
      IF(IM.GT.IWP1)IWP1=IM
    2 CONTINUE
      K=G(I)
      IF(IWP1.GT.KDIAG(K))KDIAG(K)=IWP1
    1 CONTINUE
      RETURN
      END

      SUBROUTINE FMBEAM(DER2,FUN,SAMP,ISAMP,ELL,I)
C
C      THIS SUBROUTINE FORMS THE BEAM SHAPE FUNCTIONS
C      AND THEIR 2ND DERIVATIVES IN LOCAL COORDINATES
C
      REAL DER2(*),FUN(*),SAMP(ISAMP,*)
      XI=SAMP(I,1)
      XI2=XI*XI
      XI3=XI2*XI
      FUN(1)=.25*(XI3-3.*XI+2.)
      FUN(2)=.125*ELL*(XI3-XI2-XI+1.)
      FUN(3)=.25*(-XI3+3.*XI+2.)
      FUN(4)=.125*ELL*(XI3+XI2-XI-1.)
      DER2(1)=1.5*XI
      DER2(2)=.25*ELL*(3.*XI-1.)
      DER2(3)=-1.5*XI
      DER2(4)=.25*ELL*(3.*XI+1.)
      RETURN
      END

      SUBROUTINE FMBIGK(BIGK,IBIGK,KM,IKM,G,IDOF)
C
C       THIS SUBROUTINE ASSEMBLES ELEMENT MATRICES INTO A
C       FULL GLOBAL MATRIX
C
      REAL BIGK(IBIGK,*),KM(IKM,*)
      INTEGER G(*)
      DO 1 I=1,IDOF
      IF(G(I).EQ.0) GO TO 1
    2 DO 3 J=1,IDOF
      IF(G(J).EQ.0) GO TO 3
```

```
    4 BIGK(G(I),G(J))=BIGK(G(I),G(J))+KM(I,J)
    3 CONTINUE
    1 CONTINUE
      RETURN
      END

      SUBROUTINE FMBRAD(BEE,IBEE,DERIV,IDERIV,FUN,COORD,ICOORD,SUM,NOD)
C
C      THIS SUBROUTINE FORMS THE STRAIN/DISPLACEMENT MATRIX
C      FOR AXISYMMETRIC STRAIN
C
      REAL BEE(IBEE,*),DERIV(IDERIV,*),FUN(*),COORD(ICOORD,*)
      SUM=0.
      DO 1 I=1,NOD
    1 SUM=SUM+FUN(I)*COORD(I,1)
      DO 2 M=1,NOD
      K=2*M
      L=K-1
      X=DERIV(1,M)
      BEE(1,L)=X
      BEE(3,K)=X
      Y=DERIV(2,M)
      BEE(2,K)=Y
      BEE(3,L)=Y
      BEE(4,L)=FUN(M)/SUM
    2 CONTINUE
      RETURN
      END

      SUBROUTINE FMDEPS(DEE,IDEE,E,V)
C
C      THIS SUBROUTINE FORMS THE ELASTIC PLANE STRAIN
C      STRESS/STRAIN MATRIX
C
      REAL DEE(IDEE,*)
      V1=1.-V
      C=E/((1.+V)*(1.-2.*V))
      DEE(1,1)=V1*C
      DEE(2,2)=V1*C
      DEE(3,3)=.5*C*(1.-2.*V)
      DEE(1,2)=V*C
      DEE(2,1)=V*C
      DEE(1,3)=0.
      DEE(3,1)=0.
      DEE(2,3)=0.
      DEE(3,2)=0.
      RETURN
      END

      SUBROUTINE FMDRAD(DEE,IDEE,E,V)
C
C      THIS SUBROUTINE FORMS THE ELASTIC AXISYMMETRIC
C      STRESS/STRAIN MATRIX
C
      REAL DEE(IDEE,*)
      V1=1.-V
      C=E/((1.+V)*(1.-2.*V))
      DEE(1,1)=V1*C
      DEE(2,2)=V1*C
```

```
      DEE(3,3)=.5*C*(1.-2.*V)
      DEE(4,4)=V1*C
      DEE(1,2)=V*C
      DEE(2,1)=V*C
      DEE(1,3)=0.
      DEE(3,1)=0.
      DEE(1,4)=V*C
      DEE(4,1)=V*C
      DEE(2,3)=0.
      DEE(3,2)=0.
      DEE(2,4)=V*C
      DEE(4,2)=V*C
      DEE(4,3)=0.
      DEE(3,4)=0.
      RETURN
      END

      SUBROUTINE FMDSIG(DEE,IDEE,E,V)
C
C      THIS SUBROUTINE FORMS THE ELASTIC PLANE STRESS
C      STRESS/STRAIN MATRIX
C
      REAL DEE(IDEE,*)
      C=E/(1.-V*V)
      DEE(1,1)=C
      DEE(2,2)=C
      DEE(3,3)=.5*C*(1.-V)
      DEE(1,2)=V*C
      DEE(2,1)=V*C
      DEE(1,3)=0.
      DEE(3,1)=0.
      DEE(3,2)=0.
      DEE(2,3)=0.
      RETURN
      END

      SUBROUTINE FMKDKE(KM,IKM,KP,IKP,C,IC,KE,IKE,KD,IKD,
     +                  IDOF,NOD,ITOT,THETA)
C
C      THIS SUBROUTINE FORMS THE ELEMENT COUPLED STIFFNESS
C      MATRICES KE AND KD FROM THE ELASTIC STIFFNESS KM,
C      THE FLUID 'STIFFNESS' KP AND COUPLING MATRIX C
C
      REAL KM(IKM,*),KP(IKP,*),C(IC,*),KE(IKE,*),KD(IKD,*)
      DO 11 I=1,IDOF
      DO 12 J=1,IDOF
   12 KE(I,J)=KM(I,J)
      DO 13 K=1,NOD
      KE(I,IDOF+K)=C(I,K)
   13 KE(IDOF+K,I)=C(I,K)
   11 CONTINUE
      DO 14 I=1,NOD
      DO 16 K=1,NOD
   16 KE(IDOF+I,IDOF+K)=KP(I,K)
   14 CONTINUE
      DO 17 I=1,IDOF
      DO 17 J=1,ITOT
      KD(I,J)=(THETA-1.)*KE(I,J)
   17 KE(I,J)=THETA*KE(I,J)
      M=IDOF+1
```

```
      DO 18 I=M,ITOT
      DO 18 J=1,IDOF
      KD(I,J)=KE(I,J)*THETA
   18 KE(I,J)=KD(I,J)
      DO 19 I=M,ITOT
      DO 19 J=M,ITOT
      KD(I,J)=KE(I,J)*(1.-THETA)*THETA
   19 KE(I,J)=-KE(I,J)*THETA*THETA
      RETURN
      END

      SUBROUTINE FMLAG9(DER,IDER,FUN,SAMP,ISAMP,I,J)
C
C      THIS SUBROUTINE FORMS THE SHAPE FUNCTIONS AND
C      THEIR DERIVATIVES FOR 9-NODED QUADRILATERAL ELEMENTS
C
      REAL DER(IDER,*),FUN(*),SAMP(ISAMP,*)
      ETA=SAMP(I,1)
      XI=SAMP(J,1)
      ETAM=ETA-1.
      XIM=XI-1.
      ETAP=ETA+1.
      XIP=XI+1.
      X2P1=2.*XI+1.
      X2M1=2.*XI-1.
      E2P1=2.*ETA+1.
      E2M1=2.*ETA-1.
      FUN(1)=.25*XI*XIM*ETA*ETAM
      FUN(2)=-.5*XI*XIM*ETAP*ETAM
      FUN(3)=.25*XI*XIM*ETA*ETAP
      FUN(4)=-.5*XIP*XIM*ETA*ETAP
      FUN(5)=.25*XI*XIP*ETA*ETAP
      FUN(6)=-.5*XI*XIP*ETAP*ETAM
      FUN(7)=.25*XI*XIP*ETA*ETAM
      FUN(8)=-.5*XIP*XIM*ETA*ETAM
      FUN(9)=XIP*XIM*ETAP*ETAM
      DER(1,1)=.25*X2M1*ETA*ETAM
      DER(1,2)=-.5*X2M1*ETAP*ETAM
      DER(1,3)=.25*X2M1*ETA*ETAP
      DER(1,4)=-XI*ETA*ETAP
      DER(1,5)=.25*X2P1*ETA*ETAP
      DER(1,6)=-.5*X2P1*ETAP*ETAM
      DER(1,7)=.25*X2P1*ETA*ETAM
      DER(1,8)=-XI*ETA*ETAM
      DER(1,9)=2.*XI*ETAP*ETAM
      DER(2,1)=.25*XI*XIM*E2M1
      DER(2,2)=-XI*XIM*ETA
      DER(2,3)=.25*XI*XIM*E2P1
      DER(2,4)=-.5*XIP*XIM*E2P1
      DER(2,5)=.25*XI*XIP*E2P1
      DER(2,6)=-XI*XIP*ETA
      DER(2,7)=.25*XI*XIP*E2M1
      DER(2,8)=-.5*XIP*XIM*E2M1
      DER(2,9)=2.*XIP*XIM*ETA
      RETURN
      END
```

```
      SUBROUTINE FMLIN3(DER,IDER,FUN,SAMP,ISAMP,I,J,K)
C
C      THIS SUBROUTINE FORMS THE SHAPE FUNCTIONS AND THEIR
C      DERIVATIVES FOR 8-NODED BRICK ELEMENTS
C
      REAL DER(IDER,*),FUN(*),SAMP(ISAMP,*)
      ETA=SAMP(I,1)
      XI=SAMP(J,1)
      ZETA=SAMP(K,1)
      ETAM=1.-ETA
      XIM=1.-XI
      ZETAM=1.-ZETA
      ETAP=ETA+1.
      XIP=XI+1.
      ZETAP=ZETA+1.
      FUN(1)=.125*XIM*ETAM*ZETAM
      FUN(2)=.125*XIM*ETAM*ZETAP
      FUN(3)=.125*XIP*ETAM*ZETAP
      FUN(4)=.125*XIP*ETAM*ZETAM
      FUN(5)=.125*XIM*ETAP*ZETAM
      FUN(6)=.125*XIM*ETAP*ZETAP
      FUN(7)=.125*XIP*ETAP*ZETAP
      FUN(8)=.125*XIP*ETAP*ZETAM
      DER(1,1)=-.125*ETAM*ZETAM
      DER(1,2)=-.125*ETAM*ZETAP
      DER(1,3)=.125*ETAM*ZETAP
      DER(1,4)=.125*ETAM*ZETAM
      DER(1,5)=-.125*ETAP*ZETAM
      DER(1,6)=-.125*ETAP*ZETAP
      DER(1,7)=.125*ETAP*ZETAP
      DER(1,8)=.125*ETAP*ZETAM
      DER(2,1)=-.125*XIM*ZETAM
      DER(2,2)=-.125*XIM*ZETAP
      DER(2,3)=-.125*XIP*ZETAP
      DER(2,4)=-.125*XIP*ZETAM
      DER(2,5)=.125*XIM*ZETAM
      DER(2,6)=.125*XIM*ZETAP
      DER(2,7)=.125*XIP*ZETAP
      DER(2,8)=.125*XIP*ZETAM
      DER(3,1)=-.125*XIM*ETAM
      DER(3,2)=.125*XIM*ETAM
      DER(3,3)=.125*XIP*ETAM
      DER(3,4)=-.125*XIP*ETAM
      DER(3,5)=-.125*XIM*ETAP
      DER(3,6)=.125*XIM*ETAP
      DER(3,7)=.125*XIP*ETAP
      DER(3,8)=-.125*XIP*ETAP
      RETURN
      END
```

```
      SUBROUTINE FMLUMP(DIAG,EMM,IEMM,G,IDOF)
C
C      THIS SUBROUTINE FORMS THE GLOBAL MASS MATRIX
C      AS VECTOR DIAG
C
      REAL DIAG(*),EMM(IEMM,*)
      INTEGER G(*)
      DO 1 I=1,IDOF
      IF(G(I).EQ.0)GOTO 1
      DIAG(G(I))=DIAG(G(I))+EMM(I,I)
    1 CONTINUE
      RETURN
      END

      SUBROUTINE FMPLAT(FUN,D1X,D1Y,D2X,D2Y,D2XY,SAMP,ISAMP,AA,BB,I,J)
C
C      THIS SUBROUTINE FORMS THE SHAPE FUNCTIONS AND THEIR 1ST
C      AND 2ND DERIVATIVES FOR RECTANGULAR PLATE BENDING ELEMENTS
C
      REAL FUN(*),D1X(*),D1Y(*),D2X(*),D2Y(*),D2XY(*),SAMP(ISAMP,*)
      X=SAMP(I,1)
      E=SAMP(J,1)
      XP1=X+1.
      XP12=XP1*XP1
      XP13=XP12*XP1
      EP1=E+1.
      EP12=EP1*EP1
      EP13=EP12*EP1
      P1=1.-.75*XP12+.25*XP13
      Q1=1.-.75*EP12+.25*EP13
      P2=.5*AA*XP1*(1.-XP1+.25*XP12)
      Q2=.5*BB*EP1*(1.-EP1+.25*EP12)
      P3=.25*XP12*(3.-XP1)
      Q3=.25*EP12*(3.-EP1)
      P4=.25*AA*XP12*(.5*XP1-1.)
      Q4=.25*BB*EP12*(.5*EP1-1.)
      FUN(1)=P1*Q1
      FUN(2)=P2*Q1
      FUN(3)=P1*Q2
      FUN(4)=P2*Q2
      FUN(5)=P1*Q3
      FUN(6)=P2*Q3
      FUN(7)=P1*Q4
      FUN(8)=P2*Q4
      FUN(9)=P3*Q3
      FUN(10)=P4*Q3
      FUN(11)=P3*Q4
      FUN(12)=P4*Q4
      FUN(13)=P3*Q1
      FUN(14)=P4*Q1
      FUN(15)=P3*Q2
      FUN(16)=P4*Q2
      DP1=1.5*XP1*(.5*XP1-1.)
      DQ1=1.5*EP1*(.5*EP1-1.)
      DP2=AA*(.5-XP1+.375*XP12)
      DQ2=BB*(.5-EP1+.375*EP12)
      DP3=1.5*XP1*(1.-.5*XP1)
      DQ3=1.5*EP1*(1.-.5*EP1)
      DP4=.5*AA*XP1*(.75*XP1-1.)
      DQ4=.5*BB*EP1*(.75*EP1-1.)
```

```
D2P1=1.5*X
D2P2=.25*AA*(3.*X-1.)
D2P3=-D2P1
D2P4=.25*AA*(3.*X+1.)
D2Q1=1.5*E
D2Q2=.25*BB*(3.*E-1.)
D2Q3=-D2Q1
D2Q4=.25*BB*(3.*E+1.)
D1X(1)=DP1*Q1
D1X(2)=DP2*Q1
D1X(3)=DP1*Q2
D1X(4)=DP2*Q2
D1X(5)=DP1*Q3
D1X(6)=DP2*Q3
D1X(7)=DP1*Q4
D1X(8)=DP2*Q4
D1X(9)=DP3*Q3
D1X(10)=DP4*Q3
D1X(11)=DP3*Q4
D1X(12)=DP4*Q4
D1X(13)=DP3*Q1
D1X(14)=DP4*Q1
D1X(15)=DP3*Q2
D1X(16)=DP4*Q2
D1Y(1)=P1*DQ1
D1Y(2)=P2*DQ1
D1Y(3)=P1*DQ2
D1Y(4)=P2*DQ2
D1Y(5)=P1*DQ3
D1Y(6)=P2*DQ3
D1Y(7)=P1*DQ4
D1Y(8)=P2*DQ4
D1Y(9)=P3*DQ3
D1Y(10)=P4*DQ3
D1Y(11)=P3*DQ4
D1Y(12)=P4*DQ4
D1Y(13)=P3*DQ1
D1Y(14)=P4*DQ1
D1Y(15)=P3*DQ2
D1Y(16)=P4*DQ2
D2X(1)=D2P1*Q1
D2X(2)=D2P2*Q1
D2X(3)=D2P1*Q2
D2X(4)=D2P2*Q2
D2X(5)=D2P1*Q3
D2X(6)=D2P2*Q3
D2X(7)=D2P1*Q4
D2X(8)=D2P2*Q4
D2X(9)=D2P3*Q3
D2X(10)=D2P4*Q3
D2X(11)=D2P3*Q4
D2X(12)=D2P4*Q4
D2X(13)=D2P3*Q1
D2X(14)=D2P4*Q1
D2X(15)=D2P3*Q2
D2X(16)=D2P4*Q2
D2Y(1)=P1*D2Q1
D2Y(2)=P2*D2Q1
D2Y(3)=P1*D2Q2
D2Y(4)=P2*D2Q2
```

```
      D2Y(5)=P1*D2Q3
      D2Y(6)=P2*D2Q3
      D2Y(7)=P1*D2Q4
      D2Y(8)=P2*D2Q4
      D2Y(9)=P3*D2Q3
      D2Y(10)=P4*D2Q3
      D2Y(11)=P3*D2Q4
      D2Y(12)=P4*D2Q4
      D2Y(13)=P3*D2Q1
      D2Y(14)=P4*D2Q1
      D2Y(15)=P3*D2Q2
      D2Y(16)=P4*D2Q2
      D2XY(1)=DP1*DQ1
      D2XY(2)=DP2*DQ1
      D2XY(3)=DP1*DQ2
      D2XY(4)=DP2*DQ2
      D2XY(5)=DP1*DQ3
      D2XY(6)=DP2*DQ3
      D2XY(7)=DP1*DQ4
      D2XY(8)=DP2*DQ4
      D2XY(9)=DP3*DQ3
      D2XY(10)=DP4*DQ3
      D2XY(11)=DP3*DQ4
      D2XY(12)=DP4*DQ4
      D2XY(13)=DP3*DQ1
      D2XY(14)=DP4*DQ1
      D2XY(15)=DP3*DQ2
      D2XY(16)=DP4*DQ2
      RETURN
      END

      SUBROUTINE FMQUAD(DER,IDER,FUN,SAMP,ISAMP,I,J)
C
C      THIS SUBROUTINE FORMS THE SHAPE FUNCTIONS AND
C      THEIR DERIVATIVES FOR 8-NODED QUADRILATERAL ELEMENTS
C
      REAL DER(IDER,*),FUN(*),SAMP(ISAMP,*)
      ETA=SAMP(I,1)
      XI=SAMP(J,1)
      ETAM=.25*(1.-ETA)
      ETAP=.25*(1.+ETA)
      XIM=.25*(1.-XI)
      XIP=.25*(1.+XI)
      FUN(1)=4.*ETAM*XIM*(-XI-ETA-1.)
      FUN(2)=32.*ETAM*XIM*ETAP
      FUN(3)=4.*ETAP*XIM*(-XI+ETA-1.)
      FUN(4)=32.*XIM*XIP*ETAP
      FUN(5)=4.*ETAP*XIP*(XI+ETA-1.)
      FUN(6)=32.*ETAP*XIP*ETAM
      FUN(7)=4.*XIP*ETAM*(XI-ETA-1.)
      FUN(8)=32.*XIM*XIP*ETAM
      DER(1,1)=ETAM*(2.*XI+ETA)
      DER(1,2)=-8.*ETAM*ETAP
      DER(1,3)=ETAP*(2.*XI-ETA)
      DER(1,4)=-4.*ETAP*XI
      DER(1,5)=ETAP*(2.*XI+ETA)
      DER(1,6)=8.*ETAP*ETAM
      DER(1,7)=ETAM*(2.*XI-ETA)
      DER(1,8)=-4.*ETAM*XI
      DER(2,1)=XIM*(XI+2.*ETA)
```

```
      DER(2,2)=-4.*XIM*ETA
      DER(2,3)=XIM*(2.*ETA-XI)
      DER(2,4)=8.*XIM*XIP
      DER(2,5)=XIP*(XI+2.*ETA)
      DER(2,6)=-4.*XIP*ETA
      DER(2,7)=XIP*(2.*ETA-XI)
      DER(2,8)=-8.*XIM*XIP
      RETURN
      END

      SUBROUTINE FMQUA3(DER,IDER,FUN,SAMP,ISAMP,I,J,K)
C
C      THIS SUBROUTINE FORMS THE SHAPE FUNCTIONS AND THEIR
C      DERIVATIVES FOR 20-NODED BRICK ELEMENTS
C
      REAL DER(IDER,*),FUN(*),SAMP(ISAMP,*)
      INTEGER XII(20),ETAI(20),ZETAI(20)
      XI=SAMP(I,1)
      ETA=SAMP(J,1)
      ZETA=SAMP(K,1)
      XII(1)=-1
      XII(2)=-1
      XII(3)=-1
      XII(9)=-1
      XII(10)=-1
      XII(13)=-1
      XII(14)=-1
      XII(15)=-1
      XII(4)=0
      XII(8)=0
      XII(16)=0
      XII(20)=0
      XII(5)=1
      XII(6)=1
      XII(7)=1
      XII(11)=1
      XII(12)=1
      XII(17)=1
      XII(18)=1
      XII(19)=1
      DO 1 L=1,8
    1 ETAI(L)=-1
      DO 2 L=9,12
    2 ETAI(L)=0
      DO 3 L=13,20
    3 ETAI(L)=1
      ZETAI(1)=-1
      ZETAI(7)=-1
      ZETAI(8)=-1
      ZETAI(9)=-1
      ZETAI(12)=-1
      ZETAI(13)=-1
      ZETAI(19)=-1
      ZETAI(20)=-1
      ZETAI(2)=0
      ZETAI(6)=0
      ZETAI(14)=0
      ZETAI(18)=0
      ZETAI(3)=1
      ZETAI(4)=1
```

```
      ZETAI(5)=1
      ZETAI(10)=1
      ZETAI(11)=1
      ZETAI(15)=1
      ZETAI(16)=1
      ZETAI(17)=1
      DO 4 L=1,20
      XIO=XI*XII(L)
      ETAO=ETA*ETAI(L)
      ZETAO=ZETA*ZETAI(L)
      IF(L.EQ.4.OR.L.EQ.8.OR.L.EQ.16.OR.L.EQ.20)THEN
      FUN(L)=.25*(1.-XI*XI)*(1.+ETAO)*(1.+ZETAO)
      DER(1,L)=-.5*XI*(1.+ETAO)*(1.+ZETAO)
      DER(2,L)=.25*ETAI(L)*(1.-XI*XI)*(1.+ZETAO)
      DER(3,L)=.25*ZETAI(L)*(1.-XI*XI)*(1.+ETAO)
      ELSE IF(L.GE.9.AND.L.LE.12)THEN
      FUN(L)=.25*(1.+XIO)*(1.-ETA*ETA)*(1.+ZETAO)
      DER(1,L)=.25*XII(L)*(1.-ETA*ETA)*(1.+ZETAO)
      DER(2,L)=-.5*ETA*(1.+XIO)*(1.+ZETAO)
      DER(3,L)=.25*ZETAI(L)*(1.+XIO)*(1.-ETA*ETA)
      ELSE IF(L.EQ.2.OR.L.EQ.6.OR.L.EQ.14.OR.L.EQ.18)THEN
      FUN(L)=.25*(1.+XIO)*(1.+ETAO)*(1.-ZETA*ZETA)
      DER(1,L)=.25*XII(L)*(1.+ETAO)*(1.-ZETA*ZETA)
      DER(2,L)=.25*ETAI(L)*(1.+XIO)*(1.-ZETA*ZETA)
      DER(3,L)=-.5*ZETA*(1.+XIO)*(1.+ETAO)
      ELSE
      FUN(L)=.125*(1.+XIO)*(1.+ETAO)*(1.+ZETAO)*(XIO+ETAO+ZETAO-2.)
      DER(1,L)=.125*XII(L)*(1.+ETAO)*(1.+ZETAO)*(2.*XIO+ETAO+ZETAO-1.)
      DER(2,L)=.125*ETAI(L)*(1.+XIO)*(1.+ZETAO)*(XIO+2.*ETAO+ZETAO-1.)
      DER(3,L)=.125*ZETAI(L)*(1.+XIO)*(1.+ETAO)*(XIO+ETAO+2.*ZETAO-1.)
      END IF
    4 CONTINUE
      RETURN
      END

      SUBROUTINE FMTET4(DER,IDER,FUN,SAMP,ISAMP,I)
C
C       THIS SUBROUTINE FORMS THE SHAPE FUNCTIONS AND
C       THEIR DERIVATIVES FOR 4-NODE TETRAHEDRON ELEMENTS
C
      REAL DER(IDER,*),SAMP(ISAMP,*),FUN(*)
      FUN(1)=SAMP(I,1)
      FUN(2)=SAMP(I,2)
      FUN(3)=SAMP(I,3)
      FUN(4)=1.-FUN(1)-FUN(2)-FUN(3)
      DO 1 M=1,3
      DO 1 N=1,4
    1 DER(M,N)=0.
      DER(1,1)=1.
      DER(2,2)=1.
      DER(3,3)=1.
      DER(1,4)=-1.
      DER(2,4)=-1.
      DER(3,4)=-1.
      RETURN
      END
```

```
      SUBROUTINE FMTRI3(DER,IDER,FUN,SAMP,ISAMP,I)
C
C      THIS SUBROUTINE FORMS THE SHAPE FUNCTIONS AND
C      THEIR DERIVATIVES FOR 3-NODED TRIANGULAR ELEMENTS
C
      REAL DER(IDER,*),SAMP(ISAMP,*),FUN(*)
      FUN(1)=SAMP(I,1)
      FUN(2)=SAMP(I,2)
      FUN(3)=1.-FUN(1)-FUN(2)
      DER(1,1)=1.
      DER(1,2)=0.
      DER(1,3)=-1.
      DER(2,1)=0.
      DER(2,2)=1.
      DER(2,3)=-1.
      RETURN
      END

      SUBROUTINE FMTRI6(DER,IDER,FUN,SAMP,ISAMP,I)
C
C      THIS SUBROUTINE FORMS THE SHAPE FUNCTIONS AND
C      THEIR DERIVATIVES FOR 6-NODED TRIANGULAR ELEMENTS
C
      REAL DER(IDER,*),SAMP(ISAMP,*),FUN(*)
      C1=SAMP(I,1)
      C2=SAMP(I,2)
      C3=1.-C1-C2
      FUN(1)=(2.*C1-1.)*C1
      FUN(2)=4.*C1*C2
      FUN(3)=(2.*C2-1.)*C2
      FUN(4)=4.*C2*C3
      FUN(5)=(2.*C3-1.)*C3
      FUN(6)=4.*C3*C1
      DER(1,1)=4.*C1-1.
      DER(1,2)=4.*C2
      DER(1,3)=0.
      DER(1,4)=-4.*C2
      DER(1,5)=-(4.*C3-1.)
      DER(1,6)=4.*(C3-C1)
      DER(2,1)=0.
      DER(2,2)=4.*C1
      DER(2,3)=4.*C2-1.
      DER(2,4)=4.*(C3-C2)
      DER(2,5)=-(4.*C3-1.)
      DER(2,6)=-4.*C1
      RETURN
      END

      SUBROUTINE FMTR15(DER,IDER,FUN,SAMP,ISAMP,I)
C
C      THIS SUBROUTINE FORMS THE SHAPE FUNCTIONS AND
C      THEIR DERIVATIVES FOR 15-NODED TRIANGULAR ELEMENTS
C
      REAL DER(IDER,*),SAMP(ISAMP,*),FUN(*)
      C1=SAMP(I,1)
      C2=SAMP(I,2)
      C3=1.-C1-C2
      T1=C1-.25
      T2=C1-.5
      T3=C1-.75
```

```
T4=C2-.25
T5=C2-.5
T6=C2-.75
T7=C3-.25
T8=C3-.5
T9=C3-.75
FUN(1)=32./3.*C1*T1*T2*T3
FUN(2)=128./3.*C1*C2*T1*T2
FUN(3)=64.*C1*C2*T1*T4
FUN(4)=128./3.*C1*C2*T4*T5
FUN(5)=32./3.*C2*T4*T5*T6
FUN(6)=128./3.*C2*C3*T4*T5
FUN(7)=64.*C2*C3*T4*T7
FUN(8)=128./3.*C2*C3*T7*T8
FUN(9)=32./3.*C3*T7*T8*T9
FUN(10)=128./3.*C3*C1*T7*T8
FUN(11)=64.*C3*C1*T1*T7
FUN(12)=128./3.*C3*C1*T1*T2
FUN(13)=128.*C1*C2*T1*C3
FUN(14)=128.*C1*C2*C3*T4
FUN(15)=128.*C1*C2*C3*T7
DER(1,1)=32./3.*(T2*T3*(T1+C1)+C1*T1*(T3+T2))
DER(1,2)=128./3.*C2*(T2*(T1+C1)+C1*T1)
DER(1,3)=64.*C2*T4*(T1+C1)
DER(1,4)=128./3.*C2*T4*T5
DER(1,5)=0.
DER(1,6)=-128./3.*C2*T4*T5
DER(1,7)=-64.*C2*T4*(T7+C3)
DER(1,8)=-128./3.*C2*(T8*(T7+C3)+C3*T7)
DER(1,9)=-32./3.*(T8*T9*(T7+C3)+C3*T7*(T8+T9))
DER(1,10)=128./3.*(C3*T7*T8-C1*(T8*(T7+C3)+C3*T7))
DER(1,11)=64.*(C3*T7*(T1+C1)-C1*T1*(T7+C3))
DER(1,12)=128./3.*(C3*(T2*(T1+C1)+C1*T1)-C1*T1*T2)
DER(1,13)=128.*C2*(C3*(T1+C1)-C1*T1)
DER(1,14)=128.*C2*T4*(C3-C1)
DER(1,15)=128.*C2*(C3*T7-C1*(T7+C3))
DER(2,1)=0.0
DER(2,2)=128./3.*C1*T1*T2
DER(2,3)=64.*C1*T1*(T4+C2)
DER(2,4)=128./3.*C1*(T5*(T4+C2)+C2*T4)
DER(2,5)=32./3.*(T5*T6*(T4+C2)+C2*T4*(T6+T5))
DER(2,6)=128./3.*((C3*(T5*(T4+C2)+C2*T4))-C2*T4*T5)
DER(2,7)=64.*(C3*T7*(T4+C2)-C2*T4*(T7+C3))
DER(2,8)=128./3.*(C3*T7*T8-C2*(T8*(T7+C3)+C3*T7))
DER(2,9)=-32./3.*(T8*T9*(T7+C3)+C3*T7*(T8+T9))
DER(2,10)=-128./3.*C1*(T8*(T7+C3)+C3*T7)
DER(2,11)=-64.*C1*T1*(T7+C3)
DER(2,12)=-128./3.*C1*T1*T2
DER(2,13)=128.*C1*T1*(C3-C2)
DER(2,14)=128.*C1*(C3*(T4+C2)-C2*T4)
DER(2,15)=128.*C1*(C3*T7-C2*(C3+T7))
RETURN
END
```

```
      SUBROUTINE FORMB(BEE,IBEE,DERIV,IDERIV,NOD)
C
C      THIS SUBROUTINE FORMS THE STRAIN/DISPLACEMENT MATRIX
C      FOR PLANE STRAIN
C
      REAL BEE(IBEE,*),DERIV(IDERIV,*)
      DO 1 M=1,NOD
      K=2*M
      L=K-1
      X=DERIV(1,M)
      BEE(1,L)=X
      BEE(3,K)=X
      Y=DERIV(2,M)
      BEE(2,K)=Y
      BEE(3,L)=Y
    1 CONTINUE
      RETURN
      END

      SUBROUTINE FORMB3(BEE,IBEE,DERIV,IDERIV,NOD)
C
C      THIS SUBROUTINE FORMS THE 3-D STRAIN-DISPLACEMENT MATRIX
C
      REAL BEE(IBEE,*),DERIV(IDERIV,*)
      DO 1 M=1,NOD
      N=3*M
      K=N-1
      L=K-1
      X=DERIV(1,M)
      BEE(1,L)=X
      BEE(4,K)=X
      BEE(6,N)=X
      Y=DERIV(2,M)
      BEE(2,K)=Y
      BEE(4,L)=Y
      BEE(5,N)=Y
      Z=DERIV(3,M)
      BEE(3,N)=Z
      BEE(5,K)=Z
      BEE(6,L)=Z
    1 CONTINUE
      RETURN
      END

      SUBROUTINE FORMD3(DEE,IDEE,E,V)
C
C      THIS SUBROUTINE FORMS THE 3-D STRAIN
C      STRESS/STRAIN MATRIX
C
      REAL DEE(IDEE,*)
      V1=V/(1.-V)
      VV=(1.-2.*V)/(1.-V)*.5
      DO 1 I=1,6
      DO 1 J=1,6
    1 DEE(I,J)=0.
      DEE(1,1)=1.
      DEE(2,2)=1.
      DEE(3,3)=1.
      DEE(1,2)=V1
      DEE(2,1)=V1
```

```
      DEE(1,3)=V1
      DEE(3,1)=V1
      DEE(2,3)=V1
      DEE(3,2)=V1
      DEE(4,4)=VV
      DEE(5,5)=VV
      DEE(6,6)=VV
      DO 2 I=1,6
      DO 2 J=1,6
    2 DEE(I,J)=DEE(I,J)*E/(2.*(1.+V)*VV)
      RETURN
      END

      SUBROUTINE FORMKB(KB,IKB,KM,IKM,G,IW,IDOF)
C
C      THIS SUBROUTINE FORMS THE GLOBAL STIFFNESS MATRIX
C      STORING THE LOWER TRIANGLE AS AN ARRAY BK(N,IW+1)
C
      REAL KB(IKB,*),KM(IKM,*)
      INTEGER G(*)
      DO 1 I=1,IDOF
      IF(G(I))1,1,2
    2 DO 3 J=1,IDOF
      IF(G(J))3,3,4
    4 ICD=G(J)-G(I)+IW+1
      IF(ICD-IW-1)5,5,3
    5 KB(G(I),ICD)=KB(G(I),ICD)+KM(I,J)
    3 CONTINUE
    1 CONTINUE
      RETURN
      END

      SUBROUTINE FORMKC(BK,KM,IKM,CM,ICM,G,N,IDOF)
C
C      THIS SUBROUTINE ASSEMBLES COMPLEX ELEMENT MATRICES INTO A
C      SYMMETRICAL BAND GLOBAL MATRIX(STORED AS A VECTOR)
C
      COMPLEX BK(*)
      REAL KM(IKM,*),CM(ICM,*)
      INTEGER G(*)
      DO 1 I=1,IDOF
      IF(G(I).EQ.0) GO TO 1
    2 DO 5 J=1,IDOF
      IF(G(J).EQ.0) GO TO 5
    3 ICD=G(J)-G(I)+1
      IF(ICD-1)5,4,4
    4 IVAL=N*(ICD-1)+G(I)
      BK(IVAL)=BK(IVAL)+CMPLX(KM(I,J),CM(I,J))
    5 CONTINUE
    1 CONTINUE
      RETURN
      END

      SUBROUTINE FORMKU(KU,IKU,KM,IKM,G,IW,IDOF)
C
C      THIS SUBROUTINE ASSEMBLES ELEMENT MATRICES INTO SYMMETRICAL
C      GLOBAL MATRIX(STORED AS AN UPPER RECTANGLE)
C
      REAL KU(IKU,*),KM(IKM,*)
      INTEGER G(*)
```

```
      DO 1 I=1,IDOF
      IF(G(I).EQ.0)GO TO 1
      DO 2 J=1,IDOF
      IF(G(J).EQ.0)GO TO 2
      ICD=G(J)-G(I)+1
      IF(ICD.LT.1)GO TO 2
      KU(G(I),ICD)=KU(G(I),ICD)+KM(I,J)
    2 CONTINUE
    1 CONTINUE
      RETURN
      END

      SUBROUTINE FORMKV(BK,KM,IKM,G,N,IDOF)
C
C      THIS SUBROUTINE FORMS THE GLOBAL STIFFNESS MATRIX
C      STORING THE UPPER TRIANGLE AS A VECTOR BK(N*(IW+1))
C
      REAL BK(*),KM(IKM,*)
      INTEGER G(*)
      DO 1 I=1,IDOF
      IF(G(I).EQ.0)GOTO 1
      DO 5 J=1,IDOF
      IF(G(J).EQ.0)GOTO 5
      ICD=G(J)-G(I)+1
      IF(ICD-1)5,4,4
    4 IVAL=N*(ICD-1)+G(I)
      BK(IVAL)=BK(IVAL)+KM(I,J)
    5 CONTINUE
    1 CONTINUE
      RETURN
      END

      SUBROUTINE FORMLN(DER,IDER,FUN,SAMP,ISAMP,I,J)
C
C      THIS SUBROUTINE FORMS THE SHAPE FUNCTIONS AND
C      THEIR DERIVATIVES FOR 4-NODED QUADRILATERAL ELEMENTS
C
      REAL DER(IDER,*),FUN(*),SAMP(ISAMP,*)
      ETA=SAMP(I,1)
      XI=SAMP(J,1)
      ETAM=.25*(1.-ETA)
      ETAP=.25*(1.+ETA)
      XIM=.25*(1.-XI)
      XIP=.25*(1.+XI)
      FUN(1)=4.*XIM*ETAM
      FUN(2)=4.*XIM*ETAP
      FUN(3)=4.*XIP*ETAP
      FUN(4)=4.*XIP*ETAM
      DER(1,1)=-ETAM
      DER(1,2)=-ETAP
      DER(1,3)=ETAP
      DER(1,4)=ETAM
      DER(2,1)=-XIM
      DER(2,2)=XIM
      DER(2,3)=XIP
      DER(2,4)=-XIP
      RETURN
      END
```

```
      SUBROUTINE FORMM(STRESS,M1,M2,M3)
C
C      THIS SUBROUTINE FORMS THE DERIVATIVES OF THE INVARIANTS
C      WITH RESPECT TO THE STRESSES
C
      REAL STRESS(*),M1(4,4),M2(4,4),M3(4,4)
      SX=STRESS(1)
      SY=STRESS(2)
      TXY=STRESS(3)
      SZ=STRESS(4)
      DX=(2.*SX-SY-SZ)/3.
      DY=(2.*SY-SZ-SX)/3.
      DZ=(2.*SZ-SX-SY)/3.
      SIGM=(SX+SY+SZ)/3.
      DO 1 I=1,4
      DO 1 J=1,4
      M1(I,J)=0.
      M2(I,J)=0.
    1 M3(I,J)=0.
      M1(1,1)=1.
      M1(1,2)=1.
      M1(2,1)=1.
      M1(1,4)=1.
      M1(4,1)=1.
      M1(2,2)=1.
      M1(2,4)=1.
      M1(4,2)=1.
      M1(4,4)=1.
      DO 2 I=1,4
      DO 2 J=1,4
    2 M1(I,J)=M1(I,J)/(9.*SIGM)
      M2(1,1)=.6666666666666666
      M2(2,2)=.6666666666666666
      M2(4,4)=.6666666666666666
      M2(3,3)=2.
      M2(2,4)=-.3333333333333333
      M2(4,2)=-.3333333333333333
      M2(1,2)=-.3333333333333333
      M2(2,1)=-.3333333333333333
      M2(1,4)=-.3333333333333333
      M2(4,1)=-.3333333333333333
      M3(1,1)=DX/3.
      M3(2,4)=DX/3.
      M3(4,2)=DX/3.
      M3(2,2)=DY/3.
      M3(1,4)=DY/3.
      M3(4,1)=DY/3.
      M3(4,4)=DZ/3.
      M3(1,2)=DZ/3.
      M3(2,1)=DZ/3.
      M3(3,3)=-DZ
      M3(3,4)=-2.*TXY/3.
      M3(4,3)=-2.*TXY/3.
      M3(1,3)=TXY/3.
      M3(3,1)=TXY/3.
      M3(2,3)=TXY/3.
      M3(3,2)=TXY/3.
      RETURN
      END
```

```
      SUBROUTINE FORMM3(STRESS,M1,M2,M3)
C
C      THIS SUBROUTINE FORMS THE DERIVATIVES OF THE INVARIANTS
C      WITH RESPECT TO THE STRESSES (3-D)
C
      REAL STRESS(*),M1(6,6),M2(6,6),M3(6,6)
      SX=STRESS(1)
      SY=STRESS(2)
      SZ=STRESS(3)
      TXY=STRESS(4)
      TYZ=STRESS(5)
      TZX=STRESS(6)
      SIGM=(SX+SY+SZ)/3.
      DX=SX-SIGM
      DY=SY-SIGM
      DZ=SZ-SIGM
      DO 1 I=1,6
      DO 1 J=I,6
      M1(I,J)=0.
    1 M2(I,J)=0.
      DO 2 I=1,3
      DO 2 J=1,3
    2 M1(I,J)=1./(3.*SIGM)
      DO 3 I=1,3
      M2(I,I)=2.
    3 M2(I+3,I+3)=6.
      M2(1,2)=-1.
      M2(1,3)=-1.
      M2(2,3)=-1.
      M3(1,1)=DX
      M3(1,2)=DZ
      M3(1,3)=DY
      M3(1,4)=TXY
      M3(1,5)=-2.*TYZ
      M3(1,6)=TZX
      M3(2,2)=DY
      M3(2,3)=DX
      M3(2,4)=TXY
      M3(2,5)=TYZ
      M3(2,6)=-2.*TZX
      M3(3,3)=DZ
      M3(3,4)=-2.*TXY
      M3(3,5)=TYZ
      M3(3,6)=TZX
      M3(4,4)=-3.*DZ
      M3(4,5)=3.*TZX
      M3(4,6)=3.*TYZ
      M3(5,5)=-3.*DX
      M3(5,6)=3.*TXY
      M3(6,6)=-3.*DY
      DO 4 I=1,6
      DO 4 J=I,6
      M1(I,J)=M1(I,J)/3.
      M1(J,I)=M1(I,J)
      M2(I,J)=M2(I,J)/3.
      M2(J,I)=M2(I,J)
      M3(I,J)=M3(I,J)/3.
    4 M3(J,I)=M3(I,J)
      RETURN
      END
```

```
      SUBROUTINE FORMTB(KB,IKB,KM,IKM,G,IW,IDOF)
C
C      THIS SUBROUTINE ASSEMBLES THE ELEMENT MATRICES INTO AN
C      UNSYMMETRICAL BANDED MATRIX 'KB'.
C
      REAL KB(IKB,*),KM(IKM,*)
      INTEGER G(*)
      DO 1 I=1,IDOF
      IF(G(I).EQ.0)GOTO 1
      DO 2 J=1,IDOF
      IF(G(J).EQ.0)GOTO 2
      ICD=G(J)-G(I)+IW+1
      KB(G(I),ICD)=KB(G(I),ICD)+KM(I,J)
    2 CONTINUE
    1 CONTINUE
      RETURN
      END

      SUBROUTINE FORMXI(FSOIL,FMAX,RF,RM,RO,XI)
C
C      THIS SUBROUTINE FORMS PART OF THE SPRING STIFFNESS TERM
C      FOR PROGRAM 12.2
C
      PHI=FSOIL*RO*RF/FMAX
      XI=LOG((RM-PHI)/(RO-PHI))+PHI*(RM-RO)/((RM-PHI)*(RO-PHI))
      RETURN
      END

      SUBROUTINE FRMUPV(KE,IKE,C11,IC11,C12,IC12,C21,IC21,
     +C23,IC23,C32,IC32,NOD,NODF,ITOT)
C
C      THIS SUBROUTINE FORMS THE UNSYMMETRICAL STIFFNESS MATRIX
C      FOR THE U-V-P VERSION OF THE NAVIER STOKES EQUATIONS
C
      REAL KE(IKE,*),C11(IC11,*),C21(IC21,*),C23(IC23,*),
     +C32(IC32,*),C12(IC12,*)
      K=NOD+NODF
      DO 1 I=1,NOD
      DO 1 J=1,NOD
    1 KE(I,J)=C11(I,J)
      DO 2 I=1,NOD
      DO 2 J=NOD+1,K
    2 KE(I,J)=C12(I,J-NOD)
      DO 3 I=NOD+1,K
      DO 3 J=1,NOD
    3 KE(I,J)=C21(I-NOD,J)
      DO 4 I=NOD+1,K
      DO 4 J=K+1,ITOT
    4 KE(I,J)=C23(I-NOD,J-K)
      DO 5 I=K+1,ITOT
      DO 5 J=NOD+1,K
    5 KE(I,J)=C32(I-K,J-NOD)
      DO 6 I=K+1,ITOT
      DO 6 J=K+1,ITOT
    6 KE(I,J)=C11(I-K,J-K)
      RETURN
      END
```

```
      SUBROUTINE FSPARV(BK,KM,IKM,G,KDIAG,IDOF)
C
C      THIS SUBROUTINE ASSEMBLES THE ELEMENT STIFFNESS MATRIX INTO
C      THE GLOBAL MATRIX STORED AS A VECTOR ACCOUNTING FOR A
C      VARIABLE BANDWIDTH
C
      INTEGER KDIAG(*),G(*)
      REAL BK(*),KM(IKM,*)
      DO 1 I=1,IDOF
      K=G(I)
      IF(K.EQ.0)GOTO 1
      DO 2 J=1,IDOF
      IF(G(J).EQ.0)GOTO 2
      IW=K-G(J)
      IF(IW.LT.0)GOTO 2
      IVAL=KDIAG(K)-IW
      BK(IVAL)=BK(IVAL)+KM(I,J)
    2 CONTINUE
    1 CONTINUE
      RETURN
      END

      SUBROUTINE GAUSBA(PB,IPB,WORK,IWORK,N,IW)
C
C      THIS SUBROUTINE PERFORMS GAUSSIAN REDUCTION OF AN
C      UNSYMMETRIC BANDED MATRIX 'PB' .  ARRAY 'WORK'
C      USED AS WORKING SPACE.
C
      REAL PB(IPB,*),WORK(IWORK,*)
      IWP1=IW+1
      IQ=2*IWP1-1
      IQP=IWP1
      IWP11=IWP1-1
      DO 1 I=1,IWP11
      DO 1 J=1,IQ
      IF(J.GE.IWP1+I)GOTO 2
      PB(I,J)=PB(I,J+IWP1-I)
      GOTO 1
    2 PB(I,J)=0.
      PB(N-I+1,J)=0.
    1 CONTINUE
      DO 3 K=1,N
      L=K+IWP1-1
      IF(L.GT.N)L=N
      IP=0
      S=1.E-10
      DO 4 I=K,L
      IF(ABS(PB(I,1)).LE.S)GOTO 4
      S=ABS(PB(I,1))
      IP=I
    4 CONTINUE
      IF(IP.EQ.0)GOTO 5
      IF(K.EQ.N)GOTO 11
      WORK(IWP1,K)=IP
      IQP=IQP-1
      J=IWP1+IP-K
      IF(IQP.LT.J)IQP=J
      IF(J.EQ.IWP1)GOTO 6
      DO 7 J=1,IQP
```

```
      S=PB(K,J)
      PB(K,J)=PB(IP,J)
      PB(IP,J)=S
    7 CONTINUE
    6 K1=K+1
      DO 8 I=K1,L
      S=PB(I,1)/PB(K,1)
      DO 9 J=2,IQ
      IF(J.GT.IQP)GOTO 10
      PB(I,J-1)=PB(I,J)-S*PB(K,J)
      GOTO 9
   10 PB(I,J-1)=PB(I,J)
    9 CONTINUE
      PB(I,IQ)=0.
      WORK(I-K,K)=S
    8 CONTINUE
    3 CONTINUE
    5 WRITE(6,'(" SINGULAR")')
   11 RETURN
      END

      SUBROUTINE GAUSS(SAMP,ISAMP,NGP)
C
C      THIS SUBROUTINE PROVIDES THE WEIGHTS AND SAMPLING POINTS
C      FOR GAUSS-LEGENDRE QUADRATURE
C
      REAL SAMP(ISAMP,*)
      GO TO(1,2,3,4,5,6,7),NGP
    1 SAMP(1,1)=0.
      SAMP(1,2)=2.
      GOTO 100
    2 SAMP(1,1)=1./SQRT(3.)
      SAMP(2,1)=-SAMP(1,1)
      SAMP(1,2)=1.
      SAMP(2,2)=1.
      GO TO 100
    3 SAMP(1,1)=.2*SQRT(15.)
      SAMP(2,1)=.0
      SAMP(3,1)=-SAMP(1,1)
      SAMP(1,2)=5./9.
      SAMP(2,2)=8./9.
      SAMP(3,2)=SAMP(1,2)
      GO TO 100
    4 SAMP(1,1)=.861136311594053
      SAMP(2,1)=.339981043584856
      SAMP(3,1)=-SAMP(2,1)
      SAMP(4,1)=-SAMP(1,1)
      SAMP(1,2)=.347854845137454
      SAMP(2,2)=.652145154862546
      SAMP(3,2)=SAMP(2,2)
      SAMP(4,2)=SAMP(1,2)
      GO TO 100
    5 SAMP(1,1)=.906179845938664
      SAMP(2,1)=.538469310105683
      SAMP(3,1)=.0
      SAMP(4,1)=-SAMP(2,1)
      SAMP(5,1)=-SAMP(1,1)
      SAMP(1,2)=.236926885056189
      SAMP(2,2)=.478628670499366
```

```
      SAMP(3,2)=.568888888888889
      SAMP(4,2)=SAMP(2,2)
      SAMP(5,2)=SAMP(1,2)
      GO TO 100
    6 SAMP(1,1)=.932469514203152
      SAMP(2,1)=.661209386466265
      SAMP(3,1)=.238619186083197
      SAMP(4,1)=-SAMP(3,1)
      SAMP(5,1)=-SAMP(2,1)
      SAMP(6,1)=-SAMP(1,1)
      SAMP(1,2)=.171324492379170
      SAMP(2,2)=.360761573048139
      SAMP(3,2)=.467913934572691
      SAMP(4,2)=SAMP(3,2)
      SAMP(5,2)=SAMP(2,2)
      SAMP(6,2)=SAMP(1,2)
      GO TO 100
    7 SAMP(1,1)=.949107912342759
      SAMP(2,1)=.741531185599394
      SAMP(3,1)=.405845151377397
      SAMP(4,1)=.0
      SAMP(5,1)=-SAMP(3,1)
      SAMP(6,1)=-SAMP(2,1)
      SAMP(7,1)=-SAMP(1,1)
      SAMP(1,2)=.129484966168870
      SAMP(2,2)=.279705391489277
      SAMP(3,2)=.381830050505119
      SAMP(4,2)=.417959183673469
      SAMP(5,2)=SAMP(3,2)
      SAMP(6,2)=SAMP(2,2)
      SAMP(7,2)=SAMP(1,2)
  100 CONTINUE
      RETURN
      END

      SUBROUTINE GCOORD(FUN,COORD,ICOORD,NOD,IT,GC)
C
C      THIS SUBROUTINE OBTAINS THE CARTESIAN COORDINATES OF THE
C      GAUSS-POINTS FROM THE SHAPE FUNCTIONS
C
      REAL GC(*),FUN(*),COORD(ICOORD,*)
      DO 1 I=1,IT
      GC(I)=0.
      DO 1 J=1,NOD
    1 GC(I)=GC(I)+COORD(J,I)*FUN(J)
      RETURN
      END

      SUBROUTINE HING2(IP,HOLDR,COORD,IPROP,REACT,ACTION,BMP)
C
C      THIS SUBROUTINE FORMS THE END FORCES AND MOMENTS TO BE
C      APPLIED TO A MEMBER IF A JOINT HAS GONE PLASTIC
C
      REAL HOLDR(IPROP,*),COORD(IPROP,*),REACT(*),ACTION(*)
      X1=COORD(IP,1)
      Y1=COORD(IP,2)
      X2=COORD(IP,3)
      Y2=COORD(IP,4)
      ELL=SQRT((Y2-Y1)**2+(X2-X1)**2)
```

```
      CSCH=(X2-X1)/ELL
      SNCH=(Y2-Y1)/ELL
      BM1=0.
      BM2=0.
      S1=HOLDR(IP,3)+REACT(3)
      S2=HOLDR(IP,6)+REACT(6)
      IF(ABS(S1).LE.BMP)GOTO 2
      IF(S1.GT.0.)BM1=BMP-S1
      IF(S1.LE.0.)BM1=-BMP-S1
    2 CONTINUE
      IF(ABS(S2).LE.BMP)GOTO 1
      IF(S2.GT.0.)BM2=BMP-S2
      IF(S2.LE.0.)BM2=-BMP-S2
    1 CONTINUE
      ACTION(1)=-(BM1+BM2)*SNCH/ELL
      ACTION(2)= (BM1+BM2)*CSCH/ELL
      ACTION(3)=BM1
      ACTION(4)=-ACTION(1)
      ACTION(5)=-ACTION(2)
      ACTION(6)=BM2
      RETURN
      END

      SUBROUTINE INVAR(STRESS,SIGM,DSBAR,THETA)
C
C      THIS SUBROUTINE FORMS THE STRESS INVARIANTS (2-D)
C
      REAL STRESS(*)
      SX=STRESS(1)
      SY=STRESS(2)
      TXY=STRESS(3)
      SZ=STRESS(4)
      SIGM=(SX+SY+SZ)/3.
      DSBAR=SQRT((SX-SY)**2+(SY-SZ)**2+(SZ-SX)**2+6.*TXY**2)/SQRT(2.)
      IF(DSBAR.EQ.0.)THEN
      THETA=0.
      ELSE
      DX=(2.*SX-SY-SZ)/3.
      DY=(2.*SY-SZ-SX)/3.
      DZ=(2.*SZ-SX-SY)/3.
      XJ3=DX*DY*DZ-DZ*TXY**2
      SINE=-13.5*XJ3/DSBAR**3
      IF(SINE.GT.1.)SINE=1.
      IF(SINE.LT.-1.)SINE=-1.
      THETA=ASIN(SINE)/3.
      END IF
      RETURN
      END

      SUBROUTINE INVAR3(STRESS,SIGM,DSBAR,THETA)
C
C      THIS SUBROUTINE FORMS THE STRESS INVARIANTS FROM THE
C      STRESS COMPONENTS IN 3-D
C
      REAL STRESS(*)
      SQ3=SQRT(3.)
      S1=STRESS(1)
      S2=STRESS(2)
      S3=STRESS(3)
```

```
      S4=STRESS(4)
      S5=STRESS(5)
      S6=STRESS(6)
      SIGM=(S1+S2+S3)/3.
      D2=((S1-S2)**2+(S2-S3)**2+(S3-S1)**2)/6.+S4*S4+S5*S5+S6*S6
      DS1=S1-SIGM
      DS2=S2-SIGM
      DS3=S3-SIGM
      D3=DS1*DS2*DS3-DS1*S5*S5-DS2*S6*S6-DS3*S4*S4+2.*S4*S5*S6
      DSBAR=SQ3*SQRT(D2)
      IF(DSBAR.EQ.0.)THEN
      THETA=0.
      ELSE
      SINE=-3.*SQ3*D3/(2.*SQRT(D2)**3)
      IF(SINE.GT.1.)SINE=1.
      IF(SINE.LT.-1.)SINE=-1.
      THETA=ASIN(SINE)/3.
      END IF
      RETURN
      END

      SUBROUTINE LINMLS(BP,DISPS,LOADS,N,KDIAG)
C
C      THIS SUBROUTINE FORMS THE PRODUCT OF A MATRIX AND A VECTOR
C      WHERE THE MATRIX IS STORED IN A SKYLINE VECTOR
C
      REAL BP(*),LOADS(*),DISPS(*)
      INTEGER KDIAG(*)
      DO 2 I=1,N
      X=0.
      LUP=KDIAG(I)
      IF(I.EQ.1)LOW=LUP
      IF(I.NE.1)LOW=KDIAG(I-1)+1
      DO 3 J=LOW,LUP
    3 X=X+BP(J)*DISPS(I+J-LUP)
      LOADS(I)=X
      IF(I.EQ.1)GOTO 2
      LUP=LUP-1
      DO 4 J=LOW,LUP
      K=I+J-LUP-1
    4 LOADS(K)=LOADS(K)+BP(J)*DISPS(I)
    2 CONTINUE
      RETURN
      END

      SUBROUTINE LINMUL(BK,DISPS,LOADS,N,IW)
C
C      THIS SUBROUTINE MULTIPLIES A MATRIX BY A VECTOR
C      THE MATRIX IS SYMMETRICAL WITH ITS UPPER TRIANGLE
C      STORED AS A VECTOR
C
      REAL BK(*),DISPS(*),LOADS(*)
      DO 1 I=1,N
      X=0.
      DO 2 J=1,IW+1
      IF(I+J.LE.N+1)X=X+BK(N*(J-1)+I)*DISPS(I+J-1)
    2 CONTINUE
      DO 3 J=2,IW+1
      IF(I-J+1.GE.1)X=X+BK((N-1)*(J-1)+I)*DISPS(I-J+1)
```

```
    3 CONTINUE
      LOADS(I)=X
    1 CONTINUE
      RETURN
      END

      SUBROUTINE LOC2F(LOCAL,GLOBAL,IP,COORD,ICOORD)
C
C      THIS SUBROUTINE TRANSFORMS THE END REACTIONS AND MOMENTS
C      INTO THE ELEMENT'S LOCAL COORDINATE SYSTEM (2-D)
C
      REAL COORD(ICOORD,*),LOCAL(*),GLOBAL(*)
      X1=COORD(IP,1)
      Y1=COORD(IP,2)
      X2=COORD(IP,3)
      Y2=COORD(IP,4)
      ELL=SQRT((X2-X1)**2+(Y2-Y1)**2)
      C=(X2-X1)/ELL
      S=(Y2-Y1)/ELL
      LOCAL(1)=C*GLOBAL(1)+S*GLOBAL(2)
      LOCAL(2)=C*GLOBAL(2)-S*GLOBAL(1)
      LOCAL(3)=GLOBAL(3)
      LOCAL(4)=C*GLOBAL(4)+S*GLOBAL(5)
      LOCAL(5)=C*GLOBAL(5)-S*GLOBAL(4)
      LOCAL(6)=GLOBAL(6)
      RETURN
      END

      SUBROUTINE LOC3F(LOCAL,GLOBAL,IP,COORD,ICOORD)
C
C      THIS SUBROUTINE TRANSFORMS THE END REACTION AND MOMENTS
C      INTO THE ELEMENT'S LOCAL COORDINATE SYSTEM (3-D)
C
      REAL LOCAL(*),GLOBAL(*),COORD(ICOORD,*),RO(3,3),T(12,12)
      DO 1 I=1,12
      DO 1 J=1,12
    1 T(I,J)=0.
      X1=COORD(IP,1)
      Y1=COORD(IP,2)
      Z1=COORD(IP,3)
      X2=COORD(IP,4)
      Y2=COORD(IP,5)
      Z2=COORD(IP,6)
      PI=4.*ATAN(1.)
      GAMA=COORD(IP,7)*PI/180.
      CG=COS(GAMA)
      SG=SIN(GAMA)
      XL=X2-X1
      YL=Y2-Y1
      ZL=Z2-Z1
      ELL=SQRT(XL*XL+YL*YL+ZL*ZL)
      DEN=ELL*SQRT(XL*XL+ZL*ZL)
      IF(DEN.EQ.0.)GOTO 50
      RO(1,1)=XL/ELL
      RO(1,2)=YL/ELL
      RO(1,3)=ZL/ELL
      RO(2,1)=(-XL*YL*CG-ELL*ZL*SG)/DEN
      RO(2,2)=DEN*CG/(ELL*ELL)
      RO(2,3)=(-YL*ZL*CG+ELL*XL*SG)/DEN
```

```
      RO(3,1)=(XL*YL*SG-ELL*ZL*CG)/DEN
      RO(3,2)=-DEN*SG/(ELL*ELL)
      RO(3,3)=(YL*ZL*SG+ELL*XL*CG)/DEN
      GOTO 60
   50 RO(1,1)=0.
      RO(1,3)=0.
      RO(2,2)=0.
      RO(3,2)=0.
      RO(1,2)=1.
      RO(2,1)=-CG
      RO(3,3)=CG
      RO(2,3)=SG
      RO(3,1)=SG
   60 CONTINUE
      DO 2 I=1,3
      DO 2 J=1,3
      X=RO(I,J)
      DO 2 K=0,9,3
    2 T(I+K,J+K)=X
      DO 3 I=1,12
      SUM=0.
      DO 4 J=1,12
    4 SUM=SUM+T(I,J)*GLOBAL(J)
    3 LOCAL(I)=SUM
      RETURN
      END

      SUBROUTINE LOC2T(AXIAL,GLOBAL,IP,COORD,ICOORD)
C
C      THIS SUBROUTINE RETRIEVES THE AXIAL FORCE IN A
C      2-D PIN-JOINTED ELEMENT FROM END REACTIONS (COMP -VE)
C
      REAL GLOBAL(*),COORD(ICOORD,*)
      X1=COORD(IP,1)
      Y1=COORD(IP,2)
      X2=COORD(IP,3)
      Y2=COORD(IP,4)
      ELL=SQRT((X2-X1)**2+(Y2-Y1)**2)
      C=(X2-X1)/ELL
      S=(Y2-Y1)/ELL
      AXIAL=C*GLOBAL(3)+S*GLOBAL(4)
      RETURN
      END

      SUBROUTINE LOC3T(AXIAL,GLOBAL,IP,COORD,ICOORD)
C
C      THIS SUBROUTINE RETRIEVES THE AXIAL FORCE IN A
C      3-D PIN-JOINTED ELEMENT FROM END REACTIONS (COMP -VE)
C
      REAL GLOBAL(*),COORD(ICOORD,*)
      X1=COORD(IP,1)
      Y1=COORD(IP,2)
      Z1=COORD(IP,3)
      X2=COORD(IP,4)
      Y2=COORD(IP,5)
      Z2=COORD(IP,6)
      XL=X2-X1
      YL=Y2-Y1
      ZL=Z2-Z1
```

```
      ELL=SQRT(XL*XL+YL*YL+ZL*ZL)
      XL=XL/ELL
      YL=YL/ELL
      ZL=ZL/ELL
      AXIAL=XL*GLOBAL(4)+YL*GLOBAL(5)+ZL*GLOBAL(6)
      RETURN
      END

      SUBROUTINE MATADD(A,IA,B,IB,M,N)
C
C      THIS SUBROUTINE ADDS TWO EQUAL SIZED ARRAYS
C
      REAL A(IA,*),B(IB,*)
      DO 1 I=1,M
      DO 1 J=1,N
    1 A(I,J)=A(I,J)+B(I,J)
      RETURN
      END

      SUBROUTINE MATCOP(A,IA,B,IB,M,N)
C
C      THIS SUBROUTINE COPIES ARRAY B INTO ARRAY A
C
      REAL A(IA,*),B(IB,*)
      DO 1 I=1,M
      DO 1 J=1,N
    1 B(I,J)=A(I,J)
      RETURN
      END

      SUBROUTINE MATINV(A,IA,N)
C
C      THIS SUBROUTINE FORMS THE INVERSE OF A MATRIX
C      USING GAUSS-JORDAN TRANSFORMATION
C
      REAL A(IA,*)
      DO 1 K=1,N
      CON=A(K,K)
      A(K,K)=1.
      DO 2 J=1,N
    2 A(K,J)=A(K,J)/CON
      DO 1 I=1,N
      IF(I.EQ.K)GOTO 1
      CON=A(I,K)
      A(I,K)=0.
      DO 3 J=1,N
    3 A(I,J)=A(I,J)-A(K,J)*CON
    1 CONTINUE
      RETURN
      END

      SUBROUTINE MATMUL(A,IA,B,IB,C,IC,L,M,N)
C
C      THIS SUBROUTINE FORMS THE PRODUCT OF TWO MATRICES
C
      REAL A(IA,*),B(IB,*),C(IC,*)
      DO 1 I=1,L
      DO 1 J=1,N
      X=0.0
      DO 2 K=1,M
```

```
    2 X=X+A(I,K)*B(K,J)
      C(I,J)=X
    1 CONTINUE
      RETURN
      END

      SUBROUTINE MATRAN(A,IA,B,IB,M,N)
C
C      THIS SUBROUTINE FORMS THE TRANSPOSE OF A MATRIX
C
      REAL A(IA,*),B(IB,*)
      DO 1 I=1,M
      DO 1 J=1,N
    1 A(J,I)=B(I,J)
      RETURN
      END

      SUBROUTINE MOCOPL(PHI,PSI,E,V,STRESS,PL)
C
C      THIS SUBROUTINE FORMS THE PLASTIC STRESS/STRAIN MATRIX
C      FOR A MOHR-COULOMB MATERIAL  (PHI,PSI IN DEGREES)
C
      REAL STRESS(4),ROW(4),COL(4),PL(4,4)
      SX=STRESS(1)
      SY=STRESS(2)
      TXY=STRESS(3)
      SZ=STRESS(4)
      PI=4.*ATAN(1.)
      PHIR=PHI*PI/180.
      PSIR=PSI*PI/180.
      SNPH=SIN(PHIR)
      SNPS=SIN(PSIR)
      SQ3=SQRT(3.)
      CC=1.-2.*V
      DX=(2.*SX-SY-SZ)/3.
      DY=(2.*SY-SZ-SX)/3.
      DZ=(2.*SZ-SX-SY)/3.
      D2=SQRT(-DX*DY-DY*DZ-DZ*DX+TXY*TXY)
      D3=DX*DY*DZ-DZ*TXY*TXY
      TH=-3.*SQ3*D3/(2.*D2**3)
      IF(TH.GT.1.)TH=1.
      IF(TH.LT.-1.)TH=-1.
      TH=ASIN(TH)/3.
      SNTH=SIN(TH)
      IF(ABS(SNTH).GT..49)THEN
      SIG=-1.
      IF(SNTH.LT.0.)SIG=1.
      RPH=SNPH*(1.+V)/3.
      RPS=SNPS*(1.+V)/3.
      CPS=.25*SQ3/D2*(1.+SIG*SNPS/3.)
      CPH=.25*SQ3/D2*(1.+SIG*SNPH/3.)
      COL(1)=RPH+CPH*((1.-V)*DX+V*(DY+DZ))
      COL(2)=RPH+CPH*((1.-V)*DY+V*(DZ+DX))
      COL(3)=CPH*CC*TXY
      COL(4)=RPH+CPH*((1.-V)*DZ+V*(DX+DY))
      ROW(1)=RPS+CPS*((1.-V)*DX+V*(DY+DZ))
      ROW(2)=RPS+CPS*((1.-V)*DY+V*(DZ+DX))
      ROW(3)=CPS*CC*TXY
      ROW(4)=RPS+CPS*((1.-V)*DZ+V*(DX+DY))
```

```
      EE=E/((1.+V)*CC*(RPH*SNPS+2.*CPH*CPS*D2*D2*CC))
      ELSE
      ALP=ATAN(ABS((SX-SY)/(2.*TXY)))
      CA=COS(ALP)
      SA=SIN(ALP)
      DD=CC*SA
      S1=1.
      S2=1.
      IF((SX-SY).LT..0)S1=-1.
      IF(TXY.LT..0)S2=-1.
      COL(1)=SNPH+S1*DD
      COL(2)=SNPH-S1*DD
      COL(3)=S2*CC*CA
      COL(4)=2.*V*SNPH
      ROW(1)=SNPS+S1*DD
      ROW(2)=SNPS-S1*DD
      ROW(3)=S2*CC*CA
      ROW(4)=2.*V*SNPS
      EE=E/(2.*(1.+V)*CC*(SNPH*SNPS+CC))
      END IF
      DO 1 I=1,4
      DO 1 J=1,4
    1 PL(I,J)=EE*ROW(I)*COL(J)
      RETURN
      END

      SUBROUTINE MOCOUF(PHI,C,SIGM,DSBAR,THETA,F)
C
C      THIS SUBROUTINE CALCULATES THE VALUE OF THE YIELD FUNCTION
C      FOR A MOHR-COULOMB MATERIAL (PHI IN DEGREES)
C
      PHIR=PHI*4.*ATAN(1.)/180.
      SNPH=SIN(PHIR)
      CSPH=COS(PHIR)
      CSTH=COS(THETA)
      SNTH=SIN(THETA)
      F=SNPH*SIGM+DSBAR*(CSTH/SQRT(3.)-SNTH*SNPH/3.)-C*CSPH
      RETURN
      END

      SUBROUTINE MOCOUQ(PSI,DSBAR,THETA,DQ1,DQ2,DQ3)
C
C      THIS SUBROUTINE FORMS THE DERIVATIVES OF A MOHR-COULOMB
C      POTENTIAL FUNCTION WITH RESPECT TO THE THREE INVARIANTS
C      PSI IN DEGREES
C
      PSIR=PSI*4.*ATAN(1.)/180.
      SNTH=SIN(THETA)
      SNPS=SIN(PSIR)
      SQ3=SQRT(3.)
      DQ1=SNPS
      IF(ABS(SNTH).GT..49)THEN
      C1=1.
      IF(SNTH.LT.0.)C1=-1.
      DQ2=(SQ3*.5-C1*SNPS*.5/SQ3)*SQ3*.5/DSBAR
      DQ3=0.
      ELSE
      CSTH=COS(THETA)
      CS3TH=COS(3.*THETA)
```

```
      TN3TH=TAN(3.*THETA)
      TNTH=SNTH/CSTH
      DQ2=SQ3*CSTH/DSBAR*((1.+TNTH*TN3TH)+SNPS*(TN3TH-TNTH)/SQ3)*.5
      DQ3=1.5*(SQ3*SNTH+SNPS*CSTH)/(CS3TH*DSBAR*DSBAR)
      END IF
      RETURN
      END

      SUBROUTINE MSMULT(A,IA,C,M,N)
C
C      THIS SUBROUTINE MULTIPLIES A MATRIX BY A SCALAR
C
      REAL A(IA,*)
      DO 1 I=1,M
      DO 1 J=1,N
    1 A(I,J)=A(I,J)*C
      RETURN
      END

      SUBROUTINE MVMULT(M,IM,V,K,L,Y)
C
C      THIS SUBROUTINE MULTIPLIES A MATRIX BY A VECTOR
C
      REAL M(IM,*),V(*),Y(*)
      DO 1 I=1,K
      X=0.
      DO 2 J=1,L
    2 X=X+M(I,J)*V(J)
      Y(I)=X
    1 CONTINUE
      RETURN
      END

      SUBROUTINE NULL(A,IA,M,N)
C
C      THIS SUBROUTINE NULLS A 2-D ARRAY
C
      REAL A(IA,*)
      DO 1 I=1,M
      DO 1 J=1,N
    1 A(I,J)=0.0
      RETURN
      END

      SUBROUTINE NULL3(A,IA1,IA2,L,M,N)
C
C      THIS SUBROUTINE NULLS A 3-D MATRIX
C
      REAL A(IA1,IA2,*)
      DO 1 I=1,L
      DO 1 J=1,M
      DO 1 K=1,N
    1 A(I,J,K)=0.
      RETURN
      END
```

```
      SUBROUTINE NULVEC(VEC,N)
C
C      THIS SUBROUTINE NULLS A COLUMN VECTOR
C
      REAL VEC(*)
      DO 1 I=1,N
    1 VEC(I)=0.
      RETURN
      END

      SUBROUTINE NUMIN3(SAMP,ISAMP,WT,NGP)
C
C      THIS SUBROUTINE FORMS THE SAMPLING POINTS AND
C      WEIGHTS FOR INTEGRATION OVER A TETRAHEDRON
C
      REAL SAMP(ISAMP,*),WT(*)
      IF(NGP.EQ.1)GOTO 10
      IF(NGP.EQ.4)GOTO 40
      IF(NGP.EQ.5)GOTO 50
   10 SAMP(1,1)=.25
      SAMP(1,2)=.25
      SAMP(1,3)=.25
      WT(1)=1.
      GOTO 99
   40 SAMP(1,1)=.58541020
      SAMP(1,2)=.13819660
      SAMP(1,3)=SAMP(1,2)
      SAMP(2,2)=SAMP(1,1)
      SAMP(2,3)=SAMP(1,2)
      SAMP(2,1)=SAMP(1,2)
      SAMP(3,3)=SAMP(1,1)
      SAMP(3,1)=SAMP(1,2)
      SAMP(3,2)=SAMP(1,2)
      SAMP(4,1)=SAMP(1,2)
      SAMP(4,2)=SAMP(1,2)
      SAMP(4,3)=SAMP(1,2)
      WT(1)=.25
      WT(2)=.25
      WT(3)=.25
      WT(4)=.25
      GOTO 99
   50 SAMP(1,1)=.25
      SAMP(1,2)=.25
      SAMP(1,3)=.25
      SAMP(2,1)=.5
      SAMP(2,2)=1./6.
      SAMP(2,3)=SAMP(2,2)
      SAMP(3,2)=.5
      SAMP(3,3)=1./6.
      SAMP(3,1)=SAMP(3,3)
      SAMP(4,3)=.5
      SAMP(4,1)=1./6.
      SAMP(4,2)=SAMP(4,1)
      SAMP(5,1)=1./6.
      SAMP(5,2)=SAMP(5,1)
      SAMP(5,3)=SAMP(5,1)
      WT(1)=-.8
      WT(2)=9./20.
      WT(3)=WT(2)
```

```
      WT(4)=WT(2)
      WT(5)=WT(2)
   99 CONTINUE
      RETURN
      END

      SUBROUTINE NUMINT(S,IS,WT,NGP)
C
C      THIS SUBROUTINE FORMS THE SAMPLING POINTS AND
C      WEIGHTS FOR INTEGRATION OVER A TRIANGULAR AREA
C
      REAL S(IS,*),WT(*)
      GOTO(1,1,3,4,4,6,7,7,7,7,7,12,12,12,12,16),NGP
    1 S(1,1)=1./3.
      S(1,2)=1./3.
      WT(1)=1.
      GOTO 99
    3 S(1,1)=.5
      S(1,2)=.5
      S(2,1)=.5
      S(2,2)=0.
      S(3,1)=0.
      S(3,2)=.5
      WT(1)=1./3.
      WT(2)=WT(1)
      WT(3)=WT(1)
      GOTO 99
    4 S(1,1)=1./3.
      S(1,2)=1./3.
      S(2,1)=.6
      S(2,2)=.2
      S(3,1)=.2
      S(3,2)=.6
      S(4,1)=.2
      S(4,2)=.2
      WT(1)=-9./16.
      WT(2)=25./48.
      WT(3)=WT(2)
      WT(4)=WT(2)
      GOTO 99
    6 S(1,1)=.816847572980459
      S(1,2)=.091576213509771
      S(2,1)=S(1,2)
      S(2,2)=S(1,1)
      S(3,1)=S(1,2)
      S(3,2)=S(1,2)
      S(4,1)=.108103018168070
      S(4,2)=.445948490915965
      S(5,1)=S(4,2)
      S(5,2)=S(4,1)
      S(6,1)=S(4,2)
      S(6,2)=S(4,2)
      WT(1)=.109951743655322
      WT(2)=WT(1)
      WT(3)=WT(1)
      WT(4)=.223381589678011
      WT(5)=WT(4)
      WT(6)=WT(4)
      GOTO 99
```

```
 7 S(1,1)=1./3.
   S(1,2)=1./3.
   S(2,1)=.797426985353087
   S(2,2)=.101286507323456
   S(3,1)=S(2,2)
   S(3,2)=S(2,1)
   S(4,1)=S(2,2)
   S(4,2)=S(2,2)
   S(5,1)=.470142064105115
   S(5,2)=.059715871789770
   S(6,1)=S(5,2)
   S(6,2)=S(5,1)
   S(7,1)=S(5,1)
   S(7,2)=S(5,1)
   WT(1)=.225
   WT(2)=.125939180544827
   WT(3)=WT(2)
   WT(4)=WT(2)
   WT(5)=.132394152788506
   WT(6)=WT(5)
   WT(7)=WT(5)
   GOTO 99
12 S(1,1)=.873821971016996
   S(1,2)=.063089014491502
   S(2,1)=S(1,2)
   S(2,2)=S(1,1)
   S(3,1)=S(1,2)
   S(3,2)=S(1,2)
   S(4,1)=.501426509658179
   S(4,2)=.249286745170910
   S(5,1)=S(4,2)
   S(5,2)=S(4,1)
   S(6,1)=S(4,2)
   S(6,2)=S(4,2)
   S(7,1)=.636502499121399
   S(7,2)=.310352451033785
   S(8,1)=S(7,1)
   S(8,2)=.053145049844816
   S(9,1)=S(7,2)
   S(9,2)=S(7,1)
   S(10,1)=S(7,2)
   S(10,2)=S(8,2)
   S(11,1)=S(8,2)
   S(11,2)=S(7,1)
   S(12,1)=S(8,2)
   S(12,2)=S(7,2)
   WT(1)=.050844906370207
   WT(2)=WT(1)
   WT(3)=WT(1)
   WT(4)=.116786275726379
   WT(5)=WT(4)
   WT(6)=WT(4)
   WT(7)=.082851075618374
   WT(8)=WT(7)
   WT(9)=WT(7)
   WT(10)=WT(7)
   WT(11)=WT(7)
   WT(12)=WT(7)
   GOTO 99
```

```
   16 S(1,1)=1./3.
      S(1,2)=1./3.
      S(2,1)=.658861384496478
      S(2,2)=.170569307751761
      S(3,1)=S(2,2)
      S(3,2)=S(2,1)
      S(4,1)=S(2,2)
      S(4,2)=S(2,2)
      S(5,1)=.898905543365938
      S(5,2)=.050547228317031
      S(6,1)=S(5,2)
      S(6,2)=S(5,1)
      S(7,1)=S(5,2)
      S(7,2)=S(5,2)
      S(8,1)=.081414823414554
      S(8,2)=.459292588292723
      S(9,1)=S(8,2)
      S(9,2)=S(8,1)
      S(10,1)=S(8,2)
      S(10,2)=S(8,2)
      S(11,1)=.008394777409958
      S(11,2)=.263112829634638
      S(12,1)=S(11,1)
      S(12,2)=.728492392955404
      S(13,1)=S(11,2)
      S(13,2)=S(11,1)
      S(14,1)=S(11,2)
      S(14,2)=S(12,2)
      S(15,1)=S(12,2)
      S(15,2)=S(11,1)
      S(16,1)=S(12,2)
      S(16,2)=S(11,2)
      WT(1)=.144315607677787
      WT(2)=.103217370534718
      WT(3)=WT(2)
      WT(4)=WT(2)
      WT(5)=.032458497623198
      WT(6)=WT(5)
      WT(7)=WT(5)
      WT(8)=.095091634267284
      WT(9)=WT(8)
      WT(10)=WT(8)
      WT(11)=.027230314174435
      WT(12)=WT(11)
      WT(13)=WT(11)
      WT(14)=WT(11)
      WT(15)=WT(11)
      WT(16)=WT(11)
   99 CONTINUE
      RETURN
      END

      SUBROUTINE PINJ2(KM,EA,IP,COORD,ICOORD)
C
C      THIS SUBROUTINE FORMS THE STIFFNESS MATRIX FOR AN
C      INCLINED 2-D PIN-JOINTED ELEMENT
C
      REAL KM(4,4),COORD(ICOORD,*)
      X1=COORD(IP,1)
```

```
      Y1=COORD(IP,2)
      X2=COORD(IP,3)
      Y2=COORD(IP,4)
      ELL=SQRT((Y2-Y1)**2+(X2-X1)**2)
      CS=(X2-X1)/ELL
      SN=(Y2-Y1)/ELL
      A=CS*CS
      B=SN*SN
      C=CS*SN
      KM(1,1)=A
      KM(3,3)=A
      KM(1,3)=-A
      KM(3,1)=-A
      KM(2,2)=B
      KM(4,4)=B
      KM(2,4)=-B
      KM(4,2)=-B
      KM(1,2)=C
      KM(2,1)=C
      KM(3,4)=C
      KM(4,3)=C
      KM(1,4)=-C
      KM(4,1)=-C
      KM(2,3)=-C
      KM(3,2)=-C
      DO 1 I=1,4
      DO 1 J=1,4
    1 KM(I,J)=KM(I,J)*EA/ELL
      RETURN
      END

      SUBROUTINE PINJ3(KM,EA,IP,COORD,ICOORD)
C
C      THIS SUBROUTINE FORMS THE STIFFNESS MATRIX FOR A
C      GENERAL 3-D PIN-JOINTED ELEMENT
C
      REAL KM(6,6),COORD(ICOORD,*)
      X1=COORD(IP,1)
      Y1=COORD(IP,2)
      Z1=COORD(IP,3)
      X2=COORD(IP,4)
      Y2=COORD(IP,5)
      Z2=COORD(IP,6)
      XL=X2-X1
      YL=Y2-Y1
      ZL=Z2-Z1
      ELL=SQRT(XL*XL+YL*YL+ZL*ZL)
      XL=XL/ELL
      YL=YL/ELL
      ZL=ZL/ELL
      A=XL*XL
      B=YL*YL
      C=ZL*ZL
      D=XL*YL
      E=YL*ZL
      F=ZL*XL
      KM(1,1)=A
      KM(4,4)=A
      KM(2,2)=B
```

```
      KM(5,5)=B
      KM(3,3)=C
      KM(6,6)=C
      KM(1,2)=D
      KM(2,1)=D
      KM(4,5)=D
      KM(5,4)=D
      KM(2,3)=E
      KM(3,2)=E
      KM(5,6)=E
      KM(6,5)=E
      KM(1,3)=F
      KM(3,1)=F
      KM(4,6)=F
      KM(6,4)=F
      KM(1,4)=-A
      KM(4,1)=-A
      KM(2,5)=-B
      KM(5,2)=-B
      KM(3,6)=-C
      KM(6,3)=-C
      KM(1,5)=-D
      KM(5,1)=-D
      KM(2,4)=-D
      KM(4,2)=-D
      KM(2,6)=-E
      KM(6,2)=-E
      KM(3,5)=-E
      KM(5,3)=-E
      KM(1,6)=-F
      KM(6,1)=-F
      KM(3,4)=-F
      KM(4,3)=-F
      DO 1 I=1,6
      DO 1 J=1,6
    1 KM(I,J)=KM(I,J)*EA/ELL
      RETURN
      END

      SUBROUTINE PRNCPL(SIGM,DSBAR,THETA,SIG1,SIG2,SIG3)
C
C      THIS SUBROUTINE FORMS THE PRINCIPAL STRESSES FROM
C      THE THREE INVARIANTS
C
      PI=4.*ATAN(1.)
      SIG1=SIGM+2.*DSBAR/3.*SIN(THETA-2.*PI/3.)
      SIG2=SIGM+2.*DSBAR/3.*SIN(THETA)
      SIG3=3.*SIGM-SIG1-SIG2
      RETURN
      END

      SUBROUTINE READNF(NF,INF,NN,NODOF,NR)
C
C      THIS SUBROUTINE READS THE NODAL FREEDOM DATA
C
      INTEGER NF(INF,*)
      DO 1 I=1,NN
      DO 1 J=1,NODOF
```

```
    1 NF(I,J)=1
      IF(NR.GT.0)READ(5,*)(K,(NF(K,J),J=1,NODOF),I=1,NR)
      N=0
      DO 2 I=1,NN
      DO 2 J=1,NODOF
      IF(NF(I,J).NE.0)THEN
      N=N+1
      NF(I,J)=N
      ENDIF
    2 CONTINUE
      RETURN
      END

      SUBROUTINE SOLVBA(PB,IPB,COPY,ICOPY,ANS,N,IW)
C
C      THIS SUBROUTINE PERFORMS THE GAUSSIAN BACK-SUBSTITUTION
C      ON THE REDUCED MATRIX 'PB'.
C
      REAL PB(IPB,*),COPY(ICOPY,*),ANS(*)
      IWP1=IW+1
      IQ=2*IWP1-1
      N1=N-1
      DO 1 IV=1,N1
      I=INT(COPY(IWP1,IV)+.5)
      IF(I.EQ.IV)GOTO 2
      S=ANS(IV)
      ANS(IV)=ANS(I)
      ANS(I)=S
    2 L=IV+IWP1-1
      IF(L.GT.N)L=N
      IV1=IV+1
      DO 3 I=IV1,L
    3 ANS(I)=ANS(I)-COPY(I-IV,IV)*ANS(IV)
    1 CONTINUE
      ANS(N)=ANS(N)/PB(N,1)
      IV=N-1
    6 S=ANS(IV)
      L=IQ
      IF(IV+L-1.GT.N)L=N-IV+1
      DO 4 I=2,L
      S=S-PB(IV,I)*ANS(IV+I-1)
    4 ANS(IV)=S/PB(IV,1)
      IV=IV-1
      IF(IV.NE.0)GOTO 6
    5 CONTINUE
      RETURN
      END

      SUBROUTINE SPABAC(A,B,N,KDIAG)
C
C      THIS SUBROUTINE PERFORMS THE CHOLESKI BACK-SUBSTITUTION
C      ON THE VARIABLE BANDWIDTH STIFFNESS MATRIX
C
      REAL A(*),B(*)
      INTEGER KDIAG(*)
      B(1)=B(1)/A(1)
      DO 1 I=2,N
      KI=KDIAG(I)-I
      L=KDIAG(I-1)-KI+1
```

```
      X=B(I)
      IF(L.EQ.I)GOTO 1
      M=I-1
      DO 2 J=L,M
    2 X=X-A(KI+J)*B(J)
    1 B(I)=X/A(KI+I)
      DO 3 IT=2,N
      I=N+2-IT
      KI=KDIAG(I)-I
      X=B(I)/A(KI+I)
      B(I)=X
      L=KDIAG(I-1)-KI+1
      IF(L.EQ.I)GOTO 3
      M=I-1
      DO 4 K=L,M
    4 B(K)=B(K)-X*A(KI+K)
    3 CONTINUE
      B(1)=B(1)/A(1)
      RETURN
      END

      SUBROUTINE SPARIN(A,N,KDIAG)
C
C      THIS SUBROUTINE PERFORMS CHOLESKI REDUCTION OF THE
C      VARIABLE-BANDWIDTH STIFFNESS MATRIX STORED AS A VECTOR
C
      REAL A(*)
      INTEGER KDIAG(*)
      A(1)=SQRT(A(1))
      DO 1 I=2,N
      KI=KDIAG(I)-I
      L=KDIAG(I-1)-KI+1
      DO 2 J=L,I
      X=A(KI+J)
      KJ=KDIAG(J)-J
      IF(J.EQ.1)GOTO 2
      LBAR=KDIAG(J-1)-KJ+1
      LBAR=MAX0(L,LBAR)
      IF(LBAR.EQ.J)GOTO 2
      M=J-1
      DO 3 K=LBAR,M
    3 X=X-A(KI+K)*A(KJ+K)
    2 A(KI+J)=X/A(KJ+J)
    1 A(KI+I)=SQRT(X)
      RETURN
      END

      SUBROUTINE STAB2D(KM,EA,EI,IP,COORD,ICOORD,PAX)
C
C      THIS SUBROUTINE FORMS THE STIFFNESS MATRIX OF AN
C      INCLINED 2-D BEAM-COLUMN ELEMENT TAKING ACCOUNT
C      OF THE EFFECTS OF AXIAL FORCES
C
      REAL KM(6,6),COORD(ICOORD,*)
      X1=COORD(IP,1)
      Y1=COORD(IP,2)
      X2=COORD(IP,3)
      Y2=COORD(IP,4)
      ELL=SQRT((X2-X1)**2+(Y2-Y1)**2)
```

```
C=(X2-X1)/ELL
S=(Y2-Y1)/ELL
ALP=.5*ELL*SQRT(ABS(PAX)/EI)
IF(PAX.GT.(5.E-5*EI/ELL**2))THEN
SBAR=ALP*(1.-2.*ALP/TANH(2.*ALP))/(TANH(ALP)-ALP)
CBAR=(2.*ALP-SINH(2.*ALP))/(SINH(2.*ALP)-2.*ALP*COSH(2.*ALP))
ELSE IF(PAX.LT.(-5.E-5*EI/ELL**2))THEN
SBAR=ALP*(1.-2.*ALP/TAN(2.*ALP))/(TAN(ALP)-ALP)
CBAR=(2.*ALP-SIN(2.*ALP))/(SIN(2.*ALP)-2.*ALP*COS(2.*ALP))
ELSE
SBAR=4.
CBAR=.5
END IF
BET1=2.*SBAR*(1.+CBAR)+PAX*ELL**2/EI
BET2=SBAR*(1.+CBAR)
E1=EA/ELL
E2=BET1*EI/(ELL*ELL*ELL)
E3=EI/ELL
E4=BET2*EI/(ELL*ELL)
KM(1,1)=C*C*E1+S*S*E2
KM(4,4)=KM(1,1)
KM(1,2)=S*C*(E1-E2)
KM(2,1)=KM(1,2)
KM(4,5)=KM(1,2)
KM(5,4)=KM(4,5)
KM(1,3)=-S*E4
KM(3,1)=KM(1,3)
KM(1,6)=KM(1,3)
KM(6,1)=KM(1,6)
KM(3,4)=S*E4
KM(4,3)=KM(3,4)
KM(4,6)=KM(3,4)
KM(6,4)=KM(4,6)
KM(1,4)=-KM(1,1)
KM(4,1)=KM(1,4)
KM(1,5)=S*C*(-E1+E2)
KM(5,1)=KM(1,5)
KM(2,4)=KM(1,5)
KM(4,2)=KM(2,4)
KM(2,2)=S*S*E1+C*C*E2
KM(5,5)=KM(2,2)
KM(2,5)=-KM(2,2)
KM(5,2)=KM(2,5)
KM(2,3)=C*E4
KM(3,2)=KM(2,3)
KM(2,6)=KM(2,3)
KM(6,2)=KM(2,6)
KM(3,3)=SBAR*E3
KM(6,6)=KM(3,3)
KM(3,5)=-C*E4
KM(5,3)=KM(3,5)
KM(5,6)=KM(3,5)
KM(6,5)=KM(5,6)
KM(3,6)=SBAR*CBAR*E3
KM(6,3)=KM(3,6)
RETURN
END
```

```
      SUBROUTINE TREEX3(JAC,IJAC,JAC1,IJAC1,DET)
C
C      THIS SUBROUTINE FORMS THE INVERSE OF A 3 BY 3 MATRIX
C
      REAL JAC(IJAC,*),JAC1(IJAC1,*)
      DET=JAC(1,1)*(JAC(2,2)*JAC(3,3)-JAC(3,2)*JAC(2,3))
      DET=DET-JAC(1,2)*(JAC(2,1)*JAC(3,3)-JAC(3,1)*JAC(2,3))
      DET=DET+JAC(1,3)*(JAC(2,1)*JAC(3,2)-JAC(3,1)*JAC(2,2))
      JAC1(1,1)=JAC(2,2)*JAC(3,3)-JAC(3,2)*JAC(2,3)
      JAC1(2,1)=-JAC(2,1)*JAC(3,3)+JAC(3,1)*JAC(2,3)
      JAC1(3,1)=JAC(2,1)*JAC(3,2)-JAC(3,1)*JAC(2,2)
      JAC1(1,2)=-JAC(1,2)*JAC(3,3)+JAC(3,2)*JAC(1,3)
      JAC1(2,2)=JAC(1,1)*JAC(3,3)-JAC(3,1)*JAC(1,3)
      JAC1(3,2)=-JAC(1,1)*JAC(3,2)+JAC(3,1)*JAC(1,2)
      JAC1(1,3)=JAC(1,2)*JAC(2,3)-JAC(2,2)*JAC(1,3)
      JAC1(2,3)=-JAC(1,1)*JAC(2,3)+JAC(2,1)*JAC(1,3)
      JAC1(3,3)=JAC(1,1)*JAC(2,2)-JAC(2,1)*JAC(1,2)
      DO 1 K=1,3
      DO 1 L=1,3
      JAC1(K,L)=JAC1(K,L)/DET
    1 CONTINUE
      RETURN
      END

      SUBROUTINE TWOBY2(JAC,IJAC,JAC1,IJAC1,DET)
C
C      THIS SUBROUTINE FORMS THE INVERSE OF A 2 BY 2 MATRIX
C
      REAL JAC(IJAC,*),JAC1(IJAC1,*)
      DET=JAC(1,1)*JAC(2,2)-JAC(1,2)*JAC(2,1)
      JAC1(1,1)=JAC(2,2)
      JAC1(1,2)=-JAC(1,2)
      JAC1(2,1)=-JAC(2,1)
      JAC1(2,2)=JAC(1,1)
      DO 1 K=1,2
      DO 1 L=1,2
    1 JAC1(K,L)=JAC1(K,L)/DET
      RETURN
      END

      SUBROUTINE VECADD(A,B,C,N)
C
C      THIS SUBROUTINE ADDS VECTORS  A+B=C
C
      REAL A(*),B(*),C(*)
      DO 1 I=1,N
    1 C(I)=A(I)+B(I)
      RETURN
      END

      SUBROUTINE VECCOP(A,B,N)
C
C      THIS SUBROUTINE COPIES VECTOR A INTO VECTOR B
C
      REAL A(*),B(*)
      DO 1 I=1,N
    1 B(I)=A(I)
      RETURN
      END
```

```
      SUBROUTINE VMPL(E,V,STRESS,PL)
C
C      THIS SUBROUTINE FORMS THE PLASTIC MATRIX FOR A
C      VON-MISES MATERIAL
C
      REAL STRESS(*),TERM(4),PL(4,4)
      SX=STRESS(1)
      SY=STRESS(2)
      TXY=STRESS(3)
      SZ=STRESS(4)
      DSBAR=SQRT((SX-SY)**2+(SY-SZ)**2+(SZ-SX)**2+6.*TXY**2)/SQRT(2.)
      EE=1.5*E/((1.+V)*DSBAR*DSBAR)
      TERM(1)=(2.*SX-SY-SZ)/3.
      TERM(2)=(2.*SY-SZ-SX)/3.
      TERM(3)=TXY
      TERM(4)=(2.*SZ-SX-SY)/3.
      DO 1 I=1,4
      DO 1 J=I,4
      PL(I,J)=TERM(I)*TERM(J)*EE
    1 PL(J,I)=PL(I,J)
      RETURN
      END

      SUBROUTINE VOL2D(BEE,IBEE,VOL,NOD)
C
C      THIS SUBROUTINE FORMS A VECTOR CONTAINING THE
C      DERIVATIVES OF THE SHAPE FUNCTIONS (PLANE 2-D)
C
      REAL BEE(IBEE,*),VOL(*)
      DO 1 M=1,NOD
      K=2*M
      L=K-1
      VOL(L)=BEE(1,L)
      VOL(K)=BEE(2,K)
    1 CONTINUE
      RETURN
      END

      SUBROUTINE VVMULT(V1,V2,PROD,IPROD,M,N)
C
C      THIS SUBROUTINE FORMS A VECTOR PRODUCT
C
      REAL V1(*),V2(*),PROD(IPROD,*)
      DO 1 I=1,M
      DO 1 J=1,N
    1 PROD(I,J)=V1(I)*V2(J)
      RETURN
      END
```

Author Index

Italic page numbers indicate references in full.

Subject Index